W9-ADT-919

Asymptotics of Operator and Pseudo-Differential Equations

CONTEMPORARY SOVIET MATHEMATICS

Series Editor: Revaz Gamkrelidze, *Steklov Institute, Moscow, USSR*

ASYMPTOTICS OF OPERATOR AND PSEUDO-DIFFERENTIAL EQUATIONS
V. P. Maslov and V. E. Nazaikinskii

COHOMOLOGY OF INFINITE-DIMENSIONAL LIE ALGEBRAS
D. B. Fuks

DIFFERENTIAL GEOMETRY AND TOPOLOGY
A. T. Fomenko

LINEAR DIFFERENTIAL EQUATIONS OF PRINCIPAL TYPE
Yu. V. Egorov

OPTIMAL CONTROL
V. M. Alekseev, V. M. Tikhomirov, and S. V. Fomin

THEORY OF SOLITONS: The Inverse Scattering Method
S. Novikov, S. V. Manakov, L. P. Pitaevskii, and V. E. Zakharov

TOPICS IN MODERN MATHEMATICS: Petrovskii Seminar No. 5
Edited by O. A. Oleinik

Asymptotics of Operator and Pseudo-Differential Equations

V. P. Maslov and
V. E. Nazaikinskii
Moscow Institute of Electronic Engineering
Moscow, USSR

CONSULTANTS BUREAU • NEW YORK AND LONDON

Library of Congress Cataloging in Publication Data

Maslov, V. P.
[Asimptoticheskie metody resheniia psevdodifferentsial 'nykh uravnenii. English]
Asymptotics of operator and pseudo-differential equations / V. P. Maslov and V. E. Nazaikinskii.
p. cm.—(Contemporary Soviet mathematics)
Translation of: Asimptoticheskie metody resheniia psevdodifferentsial'nykh uravnenii.
Bibliography: p.
Includes index.
ISBN 0-306-11014-8
1. Operator equations—Asymptotic theory. 2. Differential equations—Asymptotic theory. I. Nazaikinskii, V. E. II. Title. III. Series.
QA329.M3613 1988
515.7'24—dc19 88-3984
CIP

This translation is published under an agreement with the Copyright Agency of the USSR (VAAP).

© 1988 Consultants Bureau, New York
A Division of Plenum Publishing Corporation
233 Spring Street, New York, N.Y. 10013

All rights reserved

No part of this book may be reproduced, stored in a retrieval system, or transmitted in any form or by any means, electronic, mechanical, photocopying, microfilming, recording, or otherwise, without written permission from the Publisher

Printed in the United States of America

Contents

CHAPTER IV
QUASI-INVERSION THEOREM FOR FUNCTIONS OF A TUPLE OF NON-COMMUTING OPERATORS

I
Introduction

1. EXAMPLES AND GENERAL STATEMENT OF ASYMPTOTIC PROBLEMS FOR LINEAR EQUATIONS. WHAT DO WE CALL CHARACTERISTICS? RELATIONSHIPS BETWEEN CHARACTERISTICS AND ASYMPTOTICS

A. The Klein-Gordon Equation

Consider the equation

$$\Box u - \frac{m^2c^4}{h^2} u \equiv -\frac{\partial^2 u}{\partial t^2} + \sum_{j=1}^{3} \frac{\partial^2 u}{\partial x_j^2} - \frac{m^2c^4}{h^2} u = 0 \tag{1}$$

(the Klein-Gordon equation for a scalar classical field with the nonzero mass m*), and pose the question: what is the characteristic equation that naturally corresponds to the equation (1)? One may obtain, at least, two different answers to this question.

From the physical viewpoint, one should write the Hamilton-Jacobi equation describing the free motion of a relativistic particle with the rest mass m, namely,

$$\left(\frac{\partial S}{\partial t}\right)^2 - \sum_{j=1}^{3} \left(\frac{\partial S}{\partial x_j}\right)^2 - m^2c^4 = 0. \tag{2}$$

On the other hand, from the viewpoint of mathematicians specializing in the theory of hyperbolic equations, the characteristic equation for (1) would read

$$\left(\frac{\partial \Phi}{\partial t}\right)^2 - \sum_{j=1}^{3} \left(\frac{\partial \Phi}{\partial x_j}\right)^2 = 0. \tag{3}$$

The question is: which of equations (2) and (3) is the right characteristic equation for the equation (1) in the standard sense? We give the following answer: neither of (2) and (3) is, or, more carefully, both of them are right characteristic equations for (1). The nature of discrepancy between

---

* c is the velocity of light, h is Planck's constant divided by 2π. In order not to restrict ourselves to equations with constant coefficients, we assume, although there may be no evident physical interpretation, that c may depend on x,t: $c = c(x,t)$.

(2) and (3) is easy to see. It is due to the fact that two different asymptotic problems for equation (1) are considered.

Treat first the problem of the so-called quasi-classical approximation to the solution of (1). Quasi-classical approximation (originating from quantum mechanics) is none other than the asymptotic expansion of the solution in the case when h may be regarded as a small parameter, i.e., it is the asymptotics of the solution for $h \to 0$.

One usually seeks the quasi-classical asymptotics in the form

$$u(x,t,h) = e^{(i/h)S(x,t)}[\phi_0(x,t) + h\phi_1(x,t) + \ldots] \tag{4}$$

(or as linear combination of such functions), where $S(x,t), \phi_j(x,t)$, $j = 0, 1, \ldots$ are smooth functions; $S(x,t)$ is real-valued. We see that $S(x,t)/h$ is a rapidly varying phase, and $\phi(x,t,h) = \phi_0(x,t) + h\phi_1(x,t) + \ldots$ is a slowly varying amplitude of the "wave" $u(x,t,h)$. Thus, $u(x,t,h)$ is rapidly oscillating as $h \to 0$ (the wavelength is of order $|\text{grad } S|/h$).

The characteristic equation is the equation which determines the evolution of the phase $S(x,t)$. To obtain it, substitute the function $u = \exp(iS/h)\phi$ into the equation (1). This substitution yields

$$\begin{aligned} \Box u - \frac{m^2c^4}{h^2} u &= h^{-2}e^{(i/h)S(x,t)}\Big[\Big(\frac{\partial S}{\partial t} - ih\frac{\partial}{\partial t}\Big)^2 - \sum_{j=1}^{3}\Big(\frac{\partial S}{\partial x_j} - ih\frac{\partial}{\partial x_j}\Big)^2 - \\ &- m^2c^4\Big]\phi = h^{-2}e^{(i/h)S(x,t)}\Big\{\Big[\Big(\frac{\partial S}{\partial t}\Big)^2 - \sum_{j=1}^{3}\Big(\frac{\partial S}{\partial x_j}\Big)^2 - m^2c^4\Big]\phi - \\ &- ih\Big[\frac{\partial^2 S}{\partial t^2} + \frac{\partial S}{\partial t}\frac{\partial}{\partial t} - \sum_{j=1}^{3}\Big(\frac{\partial^2 S}{\partial x_j^2} + \frac{\partial S}{\partial x_j}\frac{\partial}{\partial x_j}\Big)\Big]\phi - h^2\Big[\frac{\partial^2}{\partial t^2} - \sum_{j=1}^{3}\frac{\partial^2}{\partial x_j^2}\Big]\phi\Big\} = 0. \end{aligned} \tag{5}$$

Equating to zero the coefficient of h^{-2} in (5) readily gives the Hamilton-Jacobi equation (2).* The equation (2) is a nonlinear partial differential equation of the first order. The method to solve such equations is well known; we recall it in item C and now we turn the reader's attention to other problems for equation (1).

Namely, consider the problem on propagation of singularities for the equation (1): given the initial data with prescribed discontinuities, we intend to find the solution to within smooth functions (asymptotics with respect to smoothness) and thus to determine the evolution of initial data discontinuity.

We assume that the initial data are smooth except for the surface $\{\Phi_0(x) = 0\}$ and have a jump discontinuity on this surface (we assume that $\Phi_0 \in C^\infty(\mathbb{R}^3)$ and $\text{grad } \Phi_0 \neq 0$).

Any such function, up to a smooth function, may be written in the form

$$u_0(x) = \theta(\Phi_0(x))\phi(x), \tag{6}$$

where $\phi(x)$ is smooth, $\theta(z)$ is the Heaviside θ-function,

---

* The solution of (2) being found, the equations for $\phi_1, \phi_2, \phi_3, \ldots$ may be obtained by equating to zero the coefficients of $h^{-1}, h^0, h, \ldots$ We do not discuss this matter in the Introduction.

$$\theta(z) = \begin{cases} 0, & z \leqslant 0, \\ 1, & z > 0. \end{cases} \tag{7}$$

We seek the solution as the linear combination of functions of the form

$$u(x,t) = \theta(\Phi(x,t))\phi(x,t) \equiv \theta(\Phi(x,t))\phi_o(x,t) + \\ + \Phi(x,t)\theta(\Phi(x,t))\phi_1(x,t) + (\Phi(x,t))^2\theta(\Phi(x,t))\phi_2(x,t) + \dots . \tag{8}$$

It is useful to note that each term in the series (8) is generally more smooth than the previous one; thus, the first term $\theta(\Phi)\phi_o$ has a jump discontinuity at the surface $\{\Phi(x,t) = 0\}$, the second term $\Phi\theta(\Phi)\phi_1$ is continuous but its derivatives of the first order have jump discontinuities, etc. Substitute the function (8) into the equation (1). We have

$$\frac{\partial^2}{\partial t^2}[\theta(\Phi(x,t))\phi(x,t)] = \theta(\Phi(x,t))\frac{\partial^2\phi(x,t)}{\partial t^2} + \\ + 2\delta(\Phi(x,t))\frac{\partial\Phi}{\partial t}(x,t)\frac{\partial\phi}{\partial t}(x,t) + \delta(\Phi(x,t))\frac{\partial^2\Phi}{\partial t^2}(x,t)\times \\ \times\phi(x,t) + \delta'(\Phi(x,t))(\frac{\partial\Phi}{\partial t}(x,t))^2\phi(x,t), \tag{9}$$

where $\delta(z)$ is the Dirac δ-function, $\delta'(z)$ is its derivative. Calculating the x-derivatives in a similar way we obtain

$$\Box u - \frac{m^2c^4}{h^2}u = \delta'(\Phi)[-(\frac{\partial\Phi}{\partial t})^2 + \sum_{j=1}^{3}(\frac{\partial\phi}{\partial x_j})^2]\phi + \\ + \delta(\Phi)[-2\frac{\partial\Phi}{\partial t}\frac{\partial\phi}{\partial t} + 2\sum_{j=1}^{3}\frac{\partial\Phi}{\partial x_j}\frac{\partial\phi}{\partial x_j} - \frac{\partial^2\Phi}{\partial t^2}\phi + \sum_{j=1}^{3}\frac{\partial^2\Phi}{\partial x_j^2}\phi] + \\ + \theta(\Phi)[-\frac{\partial^2\phi}{\partial t^2} + \sum_{j=1}^{3}\frac{\partial^2\phi}{\partial x^2} - \frac{m^2c^4}{h^2}\phi]. \tag{10}$$

Equating to zero the factor at the main singularity $\delta'(\Phi)$ in (10), we obtain for $\Phi(x,t)$ the characteristic equation (3), determining the evolution of the discontinuity surface for solutions of equation (1), with initial data $\Phi(x,0) = \Phi_o(x)$.

Assume now that we intend to construct the asymptotics with respect to smoothness for arbitrary initial data and show that this problem all the same leads to characteristic equation (3).

Note first of all that the Cauchy problem with arbitrary initial data for equation (1) reduces to the following one:

$$\Box G - \frac{m^2c^4}{h^2}G = 0, \\ G|_{t=0} = 0, \left.\frac{\partial G}{\partial t}\right|_{t=0} = \delta(x-y) \equiv \delta(x_1-y_1)\delta(x_2-y_2)\delta(x_3-y_3); \tag{11}$$

the function $G(x,y,t)$ is called the fundamental solution of the Cauchy problem for (1). Thus, it suffices to construct the asymptotics with respect to smoothness of the fundamental solution, i.e., the so-called parametrix. To do this, we make use of the Fourier transform and represent $\delta(x-y)$ as the superposition of plane waves:

$$\delta(x-y) = \frac{1}{(2\pi)^3}\int e^{ip(x-y)}d^3p. \tag{12}$$

Thus, we should solve the Klein-Gordon equation with the initial condition $e^{ip(x-y)}$ and then integrate over p; the properties of the Fourier transform make it clear that $|p| = \sqrt{p_1^2 + p_2^2 + p_3^2}$ should be regarded as a large parameter, with respect to which the asymptotics should be constructed. Then set

$$\omega_j = p_j/|p|, \ j = 1,2,3; \ \sum_{j=1}^{3} \omega_j^2 = 1. \tag{13}$$

We seek the solution in the form of a linear combination of functions

$$G(x,y,p,t) = e^{i|p|\Phi(x,y,\omega,t)}(\phi_o(x,y,\omega,t) + |p|^{-1}\phi_1(x,y,\omega,t) + \ldots). \tag{14}$$

Substituting (14) into (1), we obtain

$$(\square - \frac{m^2c^4}{h^2})G(x,y,p,t) = |p|^2 e^{i|p|\Phi}[(\frac{\partial\Phi}{\partial t})^2 - \sum_{j=1}^{3}(\frac{\partial\Phi}{\partial x_j})^2] + \tag{15}$$

$$+ \text{ terms of lower order with respect to } |p|.$$

Formula (15) leads immediately to (3) with initial data $\Phi|_{t=0} = \Sigma_{j=1}^{3}\omega_j \times (x_j - y_j)$. Thus, there exist two types of characteristics for equations with a small parameter at senior derivatives which describe, respectively, asymptotic expansions with respect to powers of the parameter (rapidly oscillating asymptotics) and with respect to smoothness (parametrices).

In fact there is a close connection between them, as one sees when comparing (4) and (14), and the Fourier transform helps to establish this connection. It is noteworthy that despite such evident correspondence between these two kinds of asymptotics they were studied independently for a long period of time.

It is quite natural to pose the problem of constructing a mixed asymptotic expansion for this class of equations, namely, the expansions which give solutions modulo functions which are smooth and simultaneously decaying as the parameter tends to zero. It turns out that for the Klein-Gordon equation the characteristics corresponding to this latter problem are defined by the family of Hamilton-Jacobi equations, depending on the parameter $\omega \in [0,1]$:

$$(\frac{\partial S}{\partial t})^2 - \sum_{j=1}^{3}(\frac{\partial S}{\partial x_j})^2 - \omega^2 m^2 c^4 = 0. \tag{16}$$

For $\omega = 0$ these characteristics coincide with those for the problem of constructing asymptotics with respect to smoothness and for $\omega = 1$ they coincide with those for rapidly oscillating asymptotics. To demonstrate this, consider the solution of the Klein-Gordon equation (1) as a function $u = u(x,t,\lambda)$ of the variables x,t and $\lambda = 1/h$, and perform the Fourier transformation with respect to λ, i.e., set

$$v(x,t,x_4) = (2\pi)^{-1/2}\int_{-\infty}^{\infty} e^{-i\lambda x_4}u(x,t,\lambda)d\lambda. \tag{17}$$

We see that, u(x,t) being the solution of (1), $v(x,t,x_4)$ satisfies the hyperbolic equation

$$-\frac{\partial^2 v}{\partial t^2} + \sum_{j=1}^{3}\frac{\partial^2 v}{\partial x_j^2} + m^2c^4\frac{\partial^2 v}{\partial x_4^2}, \tag{18}$$

and that the mixed asymptotic expansion of the solution $u(x,t,\lambda)$ of the Klein-Gordon equation is equivalent to the expansion with respect to smoothness of solution $v(x,t,x_4)$ of the equation (18).

Similar to the above, we seek the parametrix for (18) as an integral over $p = (p_1, p_2, p_3, p_4)$ of a linear combination of the functions of the form

$$e^{i|p|\Phi(x,x_4,y,y_4,\omega,t)}(\phi_0(x,x_4,y,y_4,\omega,t) + + |p|^{-1}\phi_1(x,x_4,y,y_4,\omega,t) + \dots),$$

where $\omega = (\omega_1, \omega_2, \omega_3, \omega_4)$, and obtain the equation for $\Phi(x,x_4,y,y_4,\omega,t)$

$$\left(\frac{\partial\Phi}{\partial t}\right)^2 - \sum_{j=1}^{3}\left(\frac{\partial\Phi}{\partial x_j}\right)^2 - m^2c^4\left(\frac{\partial\Phi}{\partial x_4}\right)^2 = 0, \tag{19}$$

with initial conditions

$$\Phi|_{t=0} = \sum_{j=1}^{4}\omega_j(x_j - y_j). \tag{20}$$

Set

$$\Phi(x,x_4,y,y_4,\omega,t) = S(x,y,\omega,t) + \omega_4(x_4 - y_4), \tag{21}$$

where S does not depend on x_4, y_4 (such substitution is possible since coefficients of the equation (19) do not depend on x_4). The equation for S reads:

$$\left(\frac{\partial S}{\partial t}\right)^2 - \sum_{j=1}^{3}\left(\frac{\partial S}{\partial x_j}\right)^2 - m^2c^4\omega_4^2 = 0, \tag{22}$$

i.e., we obtain (16) (the condition $\omega_4^2 \in [0,1]$ follows from the fact that $\Sigma_{j=1}^4\omega_j^2 = 1$).

The mixed asymptotics and corresponding characteristics play an essential role in applications. For example, the mixed asymptotics is well adapted to the mathematical description of effects of Cherenkov type in which the tail of rapid oscillations (light beaming) is observed behind the particle (described by the δ-function) moving at the velocity exceeding the light velocity in the medium. The domain of rapid oscillations is defined by a corresponding family of characteristics [52].

B. General Statement of the Problem of Finding Asymptotics with Respect to Smoothness and Parameter

Now we may give the more formal and general definition for asymptotics of the type described above. Let H_s denote the Sobolev space of index s:

$$\|u\|_{H_s} = \{\int_{\mathbb{R}^n}\bar{u}(x)(1-\Delta)^s u(x)dx\}^{1/2}, \tag{1}$$

where Δ is the Laplacian in $\mathbb{R}^n$.

The problem of construction of asymptotics with respect to smoothness or to the parameter may be considered as follows: given a (differential) operator L,

$$L : H_s \to H_{s-m} \text{ for } s \in \mathbb{R}, \tag{2}$$

we seek "almost inverse" operator R_N such that

$$LR_N = 1 + Q_N, \tag{3}$$

where Q_N is either a smoothing operator

$$\|Q_N\|_{H_s \to H_{s+N}} < \infty, \; s \in \mathbb{R} \tag{4}$$

(this corresponds to the case of asymptotics with respect to smoothness), or a "small" operator

$$\| Q_N \|_{H_s \to H_s} \leqslant \text{const}\cdot h^N, \quad s \in \mathbb{R}, \tag{5}$$

in the case of asymptotics with respect to a small parameter h.

For mixed asymptotics we require that the operator Q_N be simultaneously a "small" operator and a smoothing one:

$$\| Q_N \|_{H_s \to H_{s+N}} \leqslant \text{const}\cdot h^N, \tag{6}$$

in the latter case, once the asymptotics is constructed, the original equation

$$Lu = v \tag{7}$$

is reduced to the equation

$$(1 + Q_N)\phi = v, \tag{8}$$

or

$$\phi(x) = v(x) - \int Q_N(x,y)\phi(y)dy, \tag{9}$$

where $Q_N(x,y)$ is the kernel of the operator Q_N. Equation (9) is an integral equation of the second kind with the kernel, which is not only smooth but also small, as the parameter tends to zero. This fact enables existence theorems to be proved for the equation (7) and to construct estimates of the solution, which are uniform with respect to the parameter.

The mixed asymptotics can be most naturally interpreted as the construction of the asymptotics in the scale,* generated by a set of commuting operators $A_o = \lambda$ (multiplication by the large parameter $\lambda = 1/h$) and $A_j = -i(\partial/\partial x_j)$, $j = 1,\ldots,n$. More precisely, define on $C_o^\infty(\mathbb{R}^{n+1})$ the norm

$$\| \phi \|_s^2 = \int \bar{\phi}(x,\lambda)(1 - \Delta + \lambda^2)^s \phi(x,\lambda)dxd\lambda, \tag{10}$$

and denote by X_s the closure of $C_o^\infty(\mathbb{R}^{n+1})$ with respect to this norm (in particular, $X_o = L_2(\mathbb{R}^{n+1})$).

Then the mixed asymptotics may be interpreted as the construction of an "almost inverse" operator R_N in the scale $\{X_s\}$, such that (3) holds and

$$\| Q_N \|_{X_s \to X_{s+N}} < \infty, \quad s \in \mathbb{R}. \tag{11}$$

Of course, the Fourier transform with respect to λ transfers this scale into the usual Sobolev scale.

C. A Little More on Characteristic Equations

In item A we have obtained some characteristic equations for the Klein-Gordon equation, corresponding to different kinds of asymptotics. Here we outline briefly the well-known method of solving characteristic equations (detailed considerations and proofs are contained in Chapter 3).

The characteristic equation is a nonlinear first-order partial differential equation. We assume it may be resolved for the t-derivative of the unknown function and written in the form

---

* By "scale" we mean here nothing more than a family of embedded linear Banach spaces (see Chapter 2 for further information on scales).

$$\frac{\partial S}{\partial t} + H(x, \frac{\partial S}{\partial t}, t) = 0. \qquad (1)$$

The function $H(x,p,t)$ is called a Hamiltonian function. In problems of item A it has the form

$$H(x,p,t) = \pm\sqrt{(\sum_{j=1}^{3} p_j^2 + m^2c^4(x,t)\omega^2)} \qquad (2)$$

for various values of parameter $\omega \in [0,1]$.

Let the initial condition

$$S(x,0) = S_o(x) \qquad (3)$$

for equation (1) be given. The method of solution goes as follows. Consider the Hamiltonian system

$$\dot{x} = \frac{\partial H}{\partial p}(x,p,t),$$
$$\dot{p} = \frac{\partial H}{\partial x}(x,p,t). \qquad (4)$$

Equation (4) is a system of 2n ordinary differential equations of the first order. Its solutions are usually called the bicharacteristics of the Hamiltonian function $H(x,p,t)$ (and of the asymptotic problem considered). Prescribe the initial conditions

$$x(0) = x_o, \; p(0) = \frac{\partial S_o}{\partial x}(x_o), \qquad (5)$$

and denote the corresponding bicharacteristic by $(x(x_o,t),p(x_o,t))$.

The solution $S(x,t)$ may be obtained in the following way. Set

$$W(x_o,t) = S_o(x_o) + \int_o^t \left(\sum_{j=1}^{n} p_j \frac{\partial H}{\partial p_j} - H\right)\Bigg|_{\substack{x=x(x_o,\tau)\\ p=p(x_o,\tau)}} d\tau. \qquad (6)$$

Solve the equation

$$x = x(x_o,t) \qquad (7)$$

for x_o: $x_o = x_o(x,t)$ (by the implicit function theorem it is always possible for sufficiently small t, since the Jacobian $\det[\partial x(x_o,t)/\partial x_o]$ is equal to unity for $t = 0$).

The function

$$S(x,t) = W(x_o(x,t),t)$$

satisfies the equation (1) with initial data (3).

The characteristics appear to be connected with rich and complicated geometric structures in the phase space in which the bicharacteristics are defined. The questions of the type "what is the form of the solution to the asymptotic problem if (7) cannot be solved for x_o ($\det(\partial x/\partial x_o)$ vanishes) and therefore $S(x,t)$ does not exist" lead eventually to quantization conditions in which the topology of Lagrangian manifolds (i.e., surfaces, formed by bicharacteristics) plays an essential role. We cannot detail this matter further in the Introduction and refer the reader to Chapter 3.

D. System of Equations of Crystal Lattice Oscillations

Analogous mixed asymptotics and corresponding characteristics may be constructed for difference and difference-differential equations (see [52]). Consider here a simple example of a differential-difference equation, namely, a system of equations describing the atom oscillations in a one-dimensional crystal lattice with the step h on a circle of the length $2\pi r_N = Nh$ (here N is the number of atoms). This system has the form

$$\frac{d^2u_n}{dt^2} = \frac{c^2}{h^2}(u_{n+1} - 2u_n + u_{n-1}); \quad n = 1,\dots,N; \tag{1}$$

in (1) we set $u_o = u_N$, $u_1 = u_{N+1}$; u_n denotes the displacement of the n-th atom (with respect to its equilibrium position), and the constant c is the velocity of sound in the crystal.

We assume that the radius r_N of the circle remains finite, while $N \sim 1/h$ is a very large number, and look for an asymptotic solution under such assumptions.

Consider a smooth $2\pi r_N$-periodic function $u(x,t)$ taking the values u_j in the lattice points. Equation (1) may then be rewritten in the form

$$\frac{\partial^2 u_n}{\partial t^2} = \frac{c^2}{h^2}[u(x+h,t) - 2u(x,t) + u(x-h,t)] \tag{2}$$

or, using the identity

$$e^{i[-ih(\partial/\partial x)]}u(x,t) = u(x+h,t), \tag{3}$$

which may be verified either by a Taylor expansion, or by means of the Fourier transform, in the form

$$h^2\frac{\partial^2 u}{\partial t^2} + 4c^2\sin^2\left(-\frac{ih}{2}\frac{\partial}{\partial x}\right)u = 0. \tag{4}$$

The family of characteristic equations corresponding to the posed problem even in this simple case is not standard and follows from the first author's general concept [50]. It has the form

$$\left(\frac{\partial S}{\partial t}\right)^2 - 4\frac{c^2}{\omega^2}\sin^2\left(\frac{\omega}{2}\frac{\partial S}{\partial x}\right) = 0. \tag{5}$$

The parameter ω varies in the interval $[0, 2\pi r_N]$ (it is connected with the fact that the lattice characters $(2\pi k/N)r_N$ vary on this interval). In the three-dimensional case the parameters in the characteristic equation vary in the Brillouin zone provided the lattice corresponds to an Abelian discrete group. If the lattice corresponds to a non-Abelian group, the parameters in the characteristic equations vary in some extraordinary domains.

E. An Example of a Difference Scheme

Closely related with those presented in the previous item are the results for a simple finite-difference equation, namely, the difference scheme approximating the wave equation

$$\frac{\partial^2 u(x,t)}{\partial t^2} = c^2\frac{\partial^2 u(x,t)}{\partial x^2}. \tag{1}$$

Using the grid with the step h both in x and t axes and denoting

$$u_n^m \overset{\text{def}}{=} u(nh, mh), \tag{2}$$

we come to the following difference scheme

$$u_n^{m+1} - 2u_n^m + u_n^{m-1}/h^2 = c^2(u_{n+1}^m - 2u_n^m + u_{n-1}^m/h\ ). \tag{3}$$

The family of characteristic equations for the scheme (3) has the form

$$\frac{\partial S}{\partial t} \pm \frac{2}{\omega}\arcsin[c^2 \sin^2](\frac{\omega}{2}\frac{\partial S}{\partial x})] = 0, \tag{4}$$

where the parameter ω varies on the interval $[0,2\pi]$. These characteristics define the spread zone for the so-called unity error, i.e., for the solution which at the initial moment equals unity at some grid point and equals zero at all other points. Such characteristics for different schemes were first introduced in [51].

F. The Mixed Asymptotics with Respect to Smoothness and Decreasing at Infinity

In item B the statement of a mixed asymptotics problem was outlined for the case of commuting operators $A_o = \lambda$, $A_j = -i(\partial/\partial x_j), j = 1,\ldots,n$ with respect to powers of which the asymptotics is constructed. A somewhat more complicated situation arises when we make an attempt to construct asymptotics with respect to a tuple of non-commuting operators.

As an example, consider here differential equations with coefficients growing at infinity, i.e., the equations containing both the powers of the differentiation operator $A_1 = -i(\partial/\partial x)$ and of the operator $A_2 = x$ (multiplication by x).

We consider the following example:

$$Lu = -\frac{\partial^2 u}{\partial t^2} + \frac{\partial^2 u}{\partial x^2} - x^2(1 + b(X))u, \quad b \in C_0^\infty(\mathbb{R}^1). \tag{1}$$

It is natural to pose the following problem for the equation (1): an "almost inverse" operator R_N should be constructed, such that $LR_N = 1 + Q_N$, where the remainder Q_N has a kernel $Q_N(x, y)$ not only smooth, but also vanishing as $x \to \infty$. This requirement may be more precisely formulated as follows: consider the scale $\{X_s\}$, where X_s is the closure of $C_0^\infty(\mathbb{R})$ with respect to the norm

$$\|u\|_s^2 = \int_{-\infty}^{\infty} \bar{u}(x)(1 + x^2 - \frac{\partial^2}{\partial x^2})^s u(x)dx. \tag{2}$$

In the scale $\{X_s\}$ the remainder Q_N satisfies the estimates (11) of item B, thus the "almost inverse" operator R_N gives, for non-smooth and non-vanishing at infinity initial data and right-hand sides, the solution to within functions that are smooth enough and decaying as $|x| \to \infty$,

The characteristics for the problem (1) are defined by a tuple of non-commuting operators A_1 and A_2. The following Hamilton-Jacobi equation occurs as the characteristic equation (cf. [52]):

$$(\frac{\partial S}{\partial t})^2 = (\frac{\partial S}{\partial x})^2 + \omega(1 + b(x)), \tag{3}$$

where ω is a parameter varying in the interval $[0,1]$.

It is no wonder that both terms in the right-hand side of (1) contribute to the Hamilton-Jacobi equation. Indeed, assume for a moment that $b(x) \equiv 0$ in (1). After the Fourier transform with respect to x both the equation (1) and the system of norms (2) remain unaffected. Thus, the first and the second terms on the right-hand side of (2) are equipollent; the asymmetry in (3) is due to the fact that Fourier transform does affect the model solutions such as (4) or (8) of item A; this symmetry disappears if we consider the general form of solution given by the canonical operator — [50,52] and Chapter 3 of the present book.

Most noteworthy is the fact that the characteristic equation (3) looks as if $\partial/\partial x$ and x (but not $\partial/\partial x$ and $b(x)$!) in (1) were commutative. In fact the impact of non-commutativity displays itself in lower-order terms only, and this is why the asymptotics may be constructed.

For general equations with growing coefficients, analogous characteristics may be constructed and global asymptotics of the solution with respect to smoothness and to the growth at infinity is obtained (i.e., the almost inverse operator in the scale $\{X_s\}$ defined by the norms (2) is constructed (see [52,54]). Examples and general techniques of constructing asymptotics in this case are briefly considered also in Section 1 of Chapter 4 of the present book.

G. Example of Constructing Asymptotics for Equations with Degenerate Characteristics

If we consider the equation $Lu = f$ with singularities, or an equation with degenerate standard characteristics, it is sometimes possible to find the appropriate operators $A_1,\ldots,A_n$ such that the non-standard characteristics of the equation with respect to these operators are non-singular. In this case it would be possible to construct an almost inverse operator in the scale generated by operators $A_1,\ldots,A_n$. This scale is defined by the sequence of norms

$$\|u\|_s = \|(1 + A_1^2 + \ldots + A_n^2)^{s/2}u\| \tag{1}$$

(we restrict ourselves to the case when A_j-s are self-adjoint operators in some Hilbert space).

The almost inverse operator R_N of the operator L satisfies the equation

$$LR_N = 1 + Q_N, \quad \|Q_N\|_{s\to s+N} < \infty, \tag{2}$$

and can be expressed explicitly in terms of functions of non-commuting model operators $A_1,\ldots,A_n$ provided that they form a representation of a nilpotent Lie algebra or nonlinear Poisson algebra ([52,53,36,35] and Chapters 2 and 4 of this book).

Even in the case of geometrically trivial characteristics, namely, for some degenerate elliptic equations, this idea has led to profound and unexpected results (see [17,70,68,26,69] and other papers where the almost inverse operator in this case is constructed in the scale, generated by the tuple of vector fields $A_1,\ldots,A_n$ which form a nilpotent Lie algebra).

Here we show by the simplest example how the ideas discussed above enable us to solve the problem of oscillating solutions for a hyperbolic equation with degeneracy in characteristics.

Consider the following problem:

$$\frac{\partial^2 u(x,t,h)}{\partial t^2} - c^2(x)\Delta u(x,t,h) = f(x,t,h), \tag{3}$$

$$u|_{t=0} = u'_t|_{t=0} = 0, \tag{4}$$

where $c(x) \in C^\infty(\mathbb{R}^n)$, $c(x) \geq \delta > 0$ and state the question of finding asymptotics for the solution of (3) and (4) as $h \to 0$ for the right-hand side $f(x,t,h)$ of the form

$$f(x,t,h) = e^{iS_0(x)/h}\phi_0(x), \quad S_0(x) = x^2 \equiv \sum_{j=1}^n x_j^2, \tag{5}$$

where $\phi_0 \in C_0^\infty(\mathbb{R}^n)$, $\phi_0(0) \neq 0$.

By means of Duhamel's principle the solution of the problem (3) and (4) may be expressed by the formula

$$u(x,t,h) = \int_0^t v(x,t-\tau,h)d\tau, \tag{6}$$

where $v(x,t,h)$ satisfies the Cauchy problem

$$\frac{\partial^2 v}{\partial t^2} - c^2(x)\Delta v = 0, \tag{7}$$

$$v|_{t=0} = 0, \ v_t|_{t=0} = e^{iS_o(x)/h}. \tag{8}$$

The standard scheme of constructing the asymptotic solution to the latter problem is as follows (cf. item A): the asymptotic solution is represented in the form of a linear combination of functions of the form $e^{iS(x,t)/h}\phi(x,t,h)$. Evidently, the result of action of the wave operator on such a trial function may be written as follows:

$$\begin{aligned}&[\frac{\partial^2}{\partial t^2} - c^2(x)\Delta]e^{iS(x,t)/h}\phi(x,t,h) = -h^2\{[(\frac{\partial S}{\partial t})^2 - c^2(x)(\frac{\partial S}{\partial x})^2] \times \\ &\times \phi - ih[2\frac{\partial S}{\partial t}\frac{\partial \phi}{\partial t} - 2c(x)\frac{\partial S}{\partial x}\frac{\partial \phi}{\partial x} + (\frac{\partial^2 S}{\partial t^2} - c^2\Delta S)\phi] - h^2[\frac{\partial^2\phi}{\partial t^2} - c^2(x)\Delta\phi]\}.\end{aligned} \tag{9}$$

The characteristic equation takes the form

$$(\frac{\partial S}{\partial t})^2 - c^2(x)(\frac{\partial S}{\partial x})^2 = 0, \tag{10}$$

with the initial data

$$S|_{t=0} = S_o(x). \tag{11}$$

We have $(\partial S_o/\partial x)|_{x=0} = 0$; therefore the solutions of the characteristic equation (10) with initial data (11) are not smooth functions at this point. Thus the standard scheme of constructing rapidly oscillating asymptotics fails in the considered case.

It is evident, however, that the characteristics of the problem (7) - - (8) corresponding to the asymptotics with respect to smoothness have no degeneracy. Indeed, the wave operator is homogeneous with respect to operators $-i(\partial/\partial x_j)$, hence the bicharacteristics start from the sphere $|p| = 1$ so that the degeneracy point $|p| = 0$ falls out (cf. items A and C).

Therefore we are able to construct a parametrix for the problem (7) and finally, solving the integral equation of the second kind, represent the solution of the problem (3) - (4) in the form

$$u(x,t,h) = (R_N f)(x,t,h) + (r_N f)(x,t,h). \tag{12}$$

In (12) the operator R_N may be calculated explicitly and all that we know about r_N is that r_N has a sufficiently smooth kernel, so that r_N is a smoothing operator ($R_N : H_s \to H_{s+1}$ and $r_N : H_s \to H_{s+N}$ are continuous operators). In our example we have in the space H_N

$$u(x,t,h) - (R_N f)(x,t,h) = O(h^{n/2}), \tag{13}$$

that is, the leading term of the solution as $h \to 0$ is given by $R_N f$. Indeed, the difference (13) equals $r_N f$ and is a convolution of the exponent $\exp(iS_o(x)/h)\phi_o(x)$ with the (sufficiently smooth) kernel of the operator r_N, and the desired estimate follows from the stationary phase method.

In order to obtain the next terms of the expansion of $r_N f$ (and therefore the solution $u(x,t,h)$) with respect to powers of h, one should substitute $R_N f$ into the equation. This procedure yields a wave equation with zero Cauchy data and smooth right-hand side. The latter problem differs from the initial one in that the wave operator needs to be inverted on functions, smooth (non-oscillating) with respect to the small parameter. This task can be easily performed by a computer.

However, it should be mentioned that the accuracy of approximation of the solution by the leading term $R_N f$ depends on the type of initial conditions. For example, if $S_0(x) = |x|^4$, then the accuracy is $O(h^{n/4})$. Thus, the estimates of the leading term of asymptotics are connected with individual conditions.

The leading term has a rather complicated form, namely, it may be asymptotically represented as an integral with integrand having a singularity and simultaneously oscillating with respect to the parameter h.

For example, in the three-dimensional case and for $c \equiv 1$, it has the form

$$R_N f = \frac{1}{4\pi} \int_{|x-\xi| \leqslant t} \frac{e^{i|\xi|^2/h} \phi_0(\xi)}{|x - \xi|} d\xi. \tag{14}$$

In the case of variable coefficients and arbitrary dimension the leading term has the same properties.

The outlined scheme is applicable as well in the case of general operator equations which have degenerate characteristics with respect to one tuple of operators and non-degenerate characteristics with respect to another tuple. In particular cases this scheme was magnificently realized in the papers [18,11,58,24,25,21,5].

H. General Statement of the Problem of Constructing Asymptotics

The examples considered above evidently lead us to the general definition of asymptotics and consequently to the general statement of the asymptotic problems. From the abstract operator point of view, all the variety of asymptotics considered above can be treated by a single approach as the asymptotics in some Banach scale $\{H_s\}$.

Specifically, by an asymptotic solution of equation

$$Lu = v \in H_s, \tag{1}$$

we shall mean a sequence of elements $u_N \in H_{s_1}$ for some s_1, $N = 1,2,\dots$, such that

$$Lu_N - v \in H_{s+N} \tag{2}$$

for any N. The operator R_N, which takes v into u_N will be called an "almost inverse" operator; thus, for an almost inverse operator

$$LR_N = 1 + Q_N; \quad Q_N : H_s \to H_{s+N} \tag{3}$$

is continuous for any s, and the problem of constructing asymptotics is the problem of finding the almost inverse operator in the scale. Usually the Banach scale $\{H_s\}$ is generated by some operator A or, more generally, by a set of operators $A_1,\dots,A_n$ defined in the space H_0 in such a way that

$$u \in H_s \text{ if and only if } u \in H_{s-1} \text{ and } A_j u \in H_{s-1}, j = 1,\dots,n. \tag{4}$$

Sometimes, for example, if A_j are the self-adjoint operators in a Hilbert space H_0, the case of n operators may be reduced to the case of a single operator by setting

$$A = (1 + A_1^2 + \ldots + A_n^2)^{1/2}. \tag{5}$$

The space H_s then is none other than the domain $D(A^s)$ of the operator A^s (cf. [50,52]).

Two main conclusions follow from the above considerations of this section. First of all, characteristics play the most essential role in the process of constructing the asymptotics for given problems. They define the geometry of the problem and lead eventually to explicit expressions for the terms of asymptotic expansion of the solution (or at least for the leading term). Moreover, the role of characteristics is not restricted only to asymptotic expansions and related topics. The characteristics say to specialists something else and contribute essentially to their intuitive understanding of the problem, e.g., enabling them to say what kind of problems (Cauchy problem, boundary-value problem, etc.) it is natural to impose for given equations. A number of simple concrete examples considered above show that it is necessary to give the general definition of characteristics. This necessity is also emphasized by close connection of physical and mathematical problems concerning the notion of characteristics.

In the second place, the characteristics themselves depend on what asymptotic problem is considered for the given equation, and this in turn is determined by the choice of tuple of operators, with respect to which the asymptotic expansion is constructed. In other words, in order to define characteristics we need a tuple of operators (the model operators) with respect to which the asymptotics is constructed. Further, the initial equation should be expressed in terms of these model operators. This procedure itself demanded considerable preliminary investigation, namely, construction of the calculus of non-commuting operators [52]. Then the definition of characteristics is given in terms of the symplectic structure of a phase manifold corresponding to the commutation relations for the model operators (see [36]).

It may happen that "standard" (i.e., commonly used) characteristics have degeneracies, while "non-standard" characteristics, obtained by appropriate choice of the operator tuple, are well behaved. For individual initial data the asymptotics in the new sense may produce the old-fashioned one.

Thus, the general theory of asymptotics deals with two principal ranges of ideas:

a) The geometric and analytic concepts connected with the notion of characteristics.
b) Functions of commuting and non-commuting operators in functional spaces and related topics.

They are preliminarily considered in Sections 2 and 3 of the present chapter respectively.

2. QUANTIZATION PROCEDURE AND QUANTIZATION CONDITIONS FOR GENERAL SYMPLECTIC MANIFOLDS

The following difficulty arises both for asymptotic problems in terms of a parameter or smoothness and for mixed asymptotics in terms of a tuple of operators. The phase space for these problems is not necessarily a cotangent bundle. Thus, the construction of asymptotics demands the application of general phase manifolds and, in particular, the construction of the appropriate calculus of pseudo-differential operators. Here we consider this problem for h-pseudo-differential operators (and, respectively, asymptotics with respect to a small parameter).

If the configuration space U of the mechanical system is compact, then the phase space of the corresponding quantum system possesses a discrete structure.

Let, for example, $U = S^1$ be a circumference of unit radius. The corresponding phase space is not a cylinder $T^*U = S^1 \times \mathbb{R}^1$ but rather its submanifold, diffeomorphic to $S^1 \times \mathbb{Z}$ and consisting of circumferences, lying on this cylinder with the distance h (the Planck constant) between them. Indeed, the ψ-function $\psi(x)$ defined for $x \in S^1$ may be expanded in the Fourier series

$$\psi(x) = \sum_{n=-\infty}^{\infty} \psi_n e^{inx}, \tag{1}$$

rather than the Fourier integral and therefore the possible values of momenta, namely, points of the spectrum of the operator $\hat{p} = -ih(\partial/\partial x)$, form a discrete set of the form

$$p = kh, \quad k = 0, \pm 1, \pm 2, \ldots . \tag{2}$$

Note that the values (2) are the very values satisfying the well-known Bohr-Sommerfeld quantization condition

$$\frac{1}{2\pi h} \int_\gamma p dq \in \mathbb{Z} \tag{3}$$

(here γ is any closed 1-cycle in the phase space).

Similarly, if some classical phase space X is compact with respect to momenta, the coordinates become discrete after quantization and vice versa. One may assert that if in some field theory analogous to the Einstein theory, the classical phase space is curvilinear and compact with respect to momenta, then its quantization will necessarily lead to discretization of coordinates (the fundamental length). Non-trivial examples arise, for instance, when the phase space is non-compact both in coordinates and in momenta.

Consider again the system of equations for crystal lattice oscillations (see Section 1). The phase space even in this simplest case is not a cotangent bundle. Indeed, the configuration space for this problem is a two-parametric family of circumferences depending on a continuous parameter $h > 0$ and discrete parameter $N \in \mathbb{Z}_+$ (the length of the circumference equal to $2\pi r_N = Nh$). The coordinate x being discrete-valued, the momenta space is also a circumference, the length of which equals 2π. The phase space is therefore a two-dimensional torus T^2, for which the equality holds

$$\frac{1}{2\pi h} \int_{T^2} dp \wedge dx = N \quad \text{(the number of atoms)}. \tag{4}$$

It turns out that the condition that the left-hand side of (4) be an integer is necessary and sufficient for the existence of the correspondence which takes any symbol f, i.e., any smooth function on the torus, into the operator $\hat{f}$, acting on functions defined on the circumference $0 \leqslant x < 2\pi r_N$, so that the following properties are valid: (a) commutation formula

$$[\hat{f}, \hat{g}] \equiv \hat{f}\hat{g} - \hat{g}\hat{f} = - ih\widehat{\{f,g\}} + O(h^2) \tag{5}$$

for any symbols f and g ($\{f,g\}$ is a Poisson bracket on the torus, corresponding to the form $dp \wedge dx$),

$$\{f,g\} = \left(\frac{\partial f}{\partial x}\frac{\partial g}{\partial p} - \frac{\partial g}{\partial x}\frac{\partial f}{\partial p}\right), \tag{6}$$

and (b) formula of operator action on the rapidly oscillating exponential:

$$e^{-(i/h)S(x)}\hat{f}(e^{(i/h)S(x)}\phi) = f(x, \frac{\partial S}{\partial x})\phi(x) -$$

$$- ih[\frac{\partial f}{\partial p}(x, \frac{\partial S}{\partial x})\frac{\partial \phi}{\partial x} + \frac{1}{2}\phi\frac{\partial}{\partial x}(\frac{\partial f}{\partial p}(x, \frac{\partial S}{\partial x})) - \tag{7}$$

$$- \frac{1}{2}\phi\frac{\partial^2 f}{\partial x \partial p}(x, \frac{\partial S}{\partial x})] + O(h^2).$$

Here $\phi(x)$ and $S(x)$ are arbitrary smooth functions on the circumference, $S(x)$ is a real-valued function, and the estimate $O(h^2)$ is in the norm $L^2(-\pi r_N, \pi r_N)$. The operators $\hat{f}$ will be called, in analogy with the case of Euclidean space, the h-pseudo-differential operators (with symbols defined on the torus).

The construction of calculus of h-pseudo-differential operators may be performed on an arbitrary phase manifold (this will be made in Chapter 3:2). It leads to the analogue of condition (4), and this "two-dimensional quantization condition" is what we intend to discuss below.

Let X be an arbitrary symplectic manifold, i.e., a manifold on which a closed non-degenerate 2-form ω^2 is defined. As in the case of the torus, consider, instead of a single manifold, the family of diffeomorphic manifolds, depending on the parameter μ, varying in the compact set. It is more convenient to assume that the manifold X is fixed, and the symplectic form ω^2 depends on the parameter μ, $\omega^2 = \omega^{(\mu)}$, or equivalently, the Poisson bracket $\{\cdot,\cdot\}^{(\mu)}$ depends on the parameter μ. Recall that the Poisson bracket may be defined in the following way: by Darboux theorem [2] the form ω^2 has, in the appropriate coordinate system, the form $\omega^2 = dp \wedge dx$, and then the Poisson bracket in this coordinate system is given by (6).

The condition in question reads: for $\mu = \mu(h)$, for any two-dimensional cycle Σ on X the integral of the symplectic form divided by $2\pi h$ should coincide modulo integer numbers with half of the value of the second Stiefel-Whitney class, i.e.,

$$\frac{1}{2\pi h}[\omega^{(\mu(h))}] \equiv \frac{1}{2}W_2 (\mathrm{mod}\, \mathbb{Z}). \tag{8}$$

The condition (8) being satisfied, it is possible to establish the correspondence $f \to \hat{f}$, where f is a smooth function on X, and $\hat{f}$ is an operator in the space of sections of some sheaf, so that (5) is valid with $\{\cdot,\cdot\} = \{\cdot,\cdot\}^{\mu(h)}$ and (7) holds locally (see [36,37,38] and detailed reproduction in [55], and the Spanish translation of [52]).

Call the equation (8) the quantization condition for coordinate-momenta. In analogy with the Bohr-Sommerfeld quantization condition

$$\frac{1}{2\pi h}\oint p\,dq - \mu/2 \in \mathbb{Z}, \tag{9}$$

the class $\frac{1}{2}W_2$ in (8) will be called the <u>vacuum correction</u>. Recall that the vacuum correction in the Bohr-Sommerfeld quantization condition, which at first seemed to contradict the evident expectations based on the foundations of quantum mechanics, further obtained its interpretation as the energy of vacuum and led to explanation of such delicate phenomena as Lamb's displacement. The physical sense of the vacuum correction term in quantization condition (8) is not yet clear, but one may expect that it is also connected with some delicate intrinsic properties of the field theory models with a non-plane phase space.

It is remarkable that condition (8) (with zero vacuum correction $W_2 = 0$) already arose when constructing a rather narrow class of pseudo-differential operators with symbols, locally linear in momenta, namely, differential operators of the first order within the framework of the geometric quantization [46,73,74,75].

For the nonzero vacuum correction the condition (8) within the context of the first order operators was obtained for Kaehler manifolds [10] and in the case of general real manifolds [27]; in both cases the existence of complex polarization was required; this requirement is absent in our approach.

When constructing the calculus of pseudo-differential operators on the orbits of a compact Lie group, the condition (8) numerates the irreducible representations of the group. Thus, the coordinate-momenta quantization coincides with the Weyl rule of integrity for the major weights of irreducible representations [7,42]. The vacuum correction in this case equals zero.

It should be noted that for general phase manifolds the coordinate-momenta quantization condition is a sufficient condition for the existence of the canonical operator on real and complex characteristics. This fact enables one to apply on general phase manifolds the theory of global asymptotic solutions for h-pseudo-differential equations which is constructed in detail for the case of phase space $\mathbb{R}^{2n}$ in [50].

The global calculus of h-pseudo-differential operators can be defined also on symplectic V-manifolds [37].

We should note that when constructing the calculus of common pseudo-differential operators on homogeneous symplectic manifolds (see [8,23]), the quantization conditions (8) disappear.*

In Chapter 3:2 we give the construction of the calculus of h-pseudo-differential operators on a symplectic manifold X, which, in particular, yields the quantization conditions (8). We wish to mention here that this construction elucidates the relations between the vacuum correction $\frac{1}{2} W_2$ in (8) and the index class [50,3] in the one-dimensional quantization condition: the relation between these two classes is just the same as between the classes of $dp \wedge dq$ in (8) and pdq in the one-dimensional case. More precisely, the situation is as follows. The one-dimensional classes correspond to the Lagrangian manifolds and the two-dimensional ones correspond to symplectic manifolds. The graphs of canonical transformations are considered, which define the transition diffeomorphisms from one canonical coordinate system to another system. If the intersection of three coordinate charts is considered, we may construct a closed one-parametric family of canonical transformations which passes successively through diffeomorphisms corresponding to coordinate changes. The graph of a canonical diffeomorphism may be regarded both as a Lagrangian manifold as well as a symplectic manifold; the correspondence between two-dimensional classes and one-dimensional classes is established then via the familiar Stokes formula:

$$\int_{\partial\Omega} \omega = \int_{\Omega} d\omega. \tag{10}$$

This relation might probably give a new outlook on the integer two-dimensional class occurring in the quantization condition for coordinate-momenta.

* However, these conditions take place if the phase space is homogeneous only with respect to some of the variables. These conditions guarantee the existence of regularized canonical operators [52].

3. FEYNMAN APPROACH TO OPERATOR CALCULUS: ITS PROPERTIES AND ADVANTAGES

To this end, the aim of this book has become clear: we intend to study various kinds of asymptotics for linear equations within the limits of the unified approach bound up with characteristics in the general operator context.

Here we make a preliminary discussion of the main tool that we use to solve this problem.

A. Feynman Ordering of Operators

In quantum mechanics one deals with operators and commutation relations between them from the very beginning - these are essential basic notions of the theory. In general problems of constructing asymptotics we also come to the necessity of dealing with systems of non-commuting operators and functions of them. The ordered operator calculus proved to be a very convenient technique to treat the functions of non-commuting operators.

It bases on Feynman's observation [15] that once we supply the non-commuting operators with numbers which indicate the order in which they act, we may treat them as if they were commuting ones. Later this idea was thoroughly developed in [52].

Illustrate this by a simple example. We write indicating numbers over the operators, the operators with smaller numbers act before the operators with greater ones; coinciding numbers over operators, which do not commute, are not allowed. Thus, for example, if

$$f(x,y,z) = \Sigma\, c_{k\ell m} x^k y^\ell z^m,$$

then a substitution of the operators $\overset{1}{A},\overset{2}{B},\overset{3}{C}$ into f yields the operator

$$f(\overset{1}{A},\overset{2}{B},\overset{3}{C}) = \Sigma\, c_{k\ell m} C^m B^\ell A^k.$$

Consider the problem [15]: expand the product of exponents $e^B e^A$ into the sum of polynomials homogeneous in B and A. In our notations we have

$$e^B e^A = e^{\overset{1}{A} + \overset{2}{B}} = \sum_{n=0}^{\infty} \frac{(\overset{1}{A} + \overset{2}{B})^n}{n!} \; ;$$

we see that the Feynman notations allowed combinatorial computations (though simple in this example) to be avoided and to obtain a solution readily. The same is the situation in the complicated problems. There is a list of simple rules of calculations and transformations for functions of Feynman-ordered operators which simplify greatly the computations and arguments in numerous problems concerned with non-commuting operators. These rules will be derived and proved completely in Chapter 2; all these rules are evident for polynomials and used in examples below without rigorous proof.

We now show how the method works for differential equations. In the theory of linear ordinary differential equations the Heaviside method is well known. For the equation

$$y'' - y = \sin \frac{x}{h} \,, \tag{1}$$

this method goes as follows. Denote the differentiation operator by D, $D = d/dx$; the equation reads now:

$$(D^2 - 1)y = \sin \frac{x}{h} \,.$$

Then the solution is

$$y = \frac{1}{D^2 - 1} (\sin \frac{x}{h}) = \frac{1}{D^2 - 1} \frac{1}{2i} (e^{ix/h} - e^{-ix/h}) =$$

$$= \frac{1}{2i} e^{ix/h} \frac{1}{(D + i/h)^2 - 1} 1 - e^{-ix/h} \frac{1}{(D - i/h)^2 - 1} 1$$

(here we used the "permutation with exponent" rule

$$f(D)e^{\lambda x} = e^{\lambda x} f(D + \lambda),$$

which is easily verified for polynomials $f(z)$). The latter expression up to solutions of the homogeneous equation $y'' - y = 0$ reads

$$y = \frac{1}{2i} \{e^{ix/h} \frac{1}{-1 - h^{-2}} - e^{-ix/h} \frac{1}{-1 - h^{-2}}\} =$$

$$= -\frac{1}{2i(1 + h^{-2})} (e^{ix/h} - e^{-ix/h}) = -\frac{h^2 \sin(x/h)}{1 + h^2} .$$

It is easy to see that we have obtained the precise solution of equation (1).

Consider now the equation with variable coefficients

$$y'' - x^2 y = \sin \frac{x}{h} , \tag{2}$$

or, in above notations,

$$(D^2 - x^2)y = \sin \frac{x}{h} .$$

The Heaviside method fails now; however, we may write the numbers over operators and try the solution of the form

$$y = \frac{1}{\overset{1}{D}{}^2 - \overset{2}{x}{}^2} (\sin \frac{x}{h}).$$

Using the "permutation with exponent" rule

$$f(\overset{2}{x},\overset{1}{D})e^{i\lambda x} = e^{i\lambda x} f(\overset{2}{x},\overset{1}{D} + \lambda), \tag{3}$$

and the identity

$$f(\overset{2}{x},\overset{1}{D})1 = f(x,0)$$

since 1 is an eigenfunction of D with the eigenvalue 0, we obtain

$$y = \frac{1}{\overset{1}{D}{}^2 - \overset{2}{x}{}^2} \frac{1}{2i} (e^{ix/h} - e^{-ix/h}) = \frac{1}{2i} \{e^{ix/h} \frac{1}{(\overset{1}{D} + i/h)^2 - \overset{2}{x}{}^2} 1 -$$

$$- e^{-ix/h} \frac{1}{(\overset{1}{D} - i/h)^2 - \overset{2}{x}{}^2} 1\} = -\frac{1}{2i} \{\frac{e^{ix/h}}{x^2 + h^{-2}} - \frac{e^{-ix/h}}{x^2 + h^{-2}}\} =$$

$$= -\frac{h^2 \sin(x/h)}{1 + h^2 x^2} .$$

Substitution of this into (2) yields

$$y'' - x^2y = -h^2\{-h^{-2}\sin\frac{x}{h}\frac{1}{1+h^2x^2} - h^{-1}\cos\frac{x}{h}\frac{2h^2x}{(1+h^2x^2)^2} +$$

$$+ \sin\frac{x}{h}\frac{2h^2x\cdot 2h^2x(1+h^2x^2) - 2h^2(1+h^2x^2)^2}{(1+h^2x^2)^4}\} + \frac{h^2x^2\sin(x/h)}{1+h^2x^2} =$$

$$= \sin\frac{x}{h} + \frac{2h^3x\cos(x/h)}{(1+h^2x^2)^2} + \frac{2h^4x^2\sin(x/h)}{(1+h^2x^2)^3} - \frac{2h^2\sin(x/h)}{(1+h^2x^2)^2} =$$

$$= \sin\frac{x}{h} + \frac{1}{x^2+h^{-2}}\{\frac{2xh}{1+(xh)^2}\cos\frac{x}{h} + \frac{2(xh)^2}{(1+(xh)^2)^2}\sin\frac{x}{h} - \frac{2\sin(x/h)}{1+(xh)^2}\}.$$

The quantity in the curly brackets is bounded uniformly with respect to $x \in \mathbb{R}$, $h \in [0,1]$. Thus the function (3) is not a precise solution, but an "almost solution," which satisfies the equation up to the "disturbance" which tends to zero as $h \to 0$ and decays as x tends to infinity, as $1/(x^2 + h^{-2})$. The ordered operators method enables the construction of all the subsequent terms of such mixed asymptotic expansion with respect to parameter $h \to 0$ and growth at infinity, but this technique is beyond the scope of the Introduction.

The method of ordered operators has a widespread area of application, from obtaining of the so-called regular representations (see Chapter 2) and defining the characteristics in asymptotic problems (cf. B and C of this section where two examples are considered) to deduction of certain identities in the theory of nonlinear equations such as the KdV equation, etc.

Of course, the ordered operators method is only a technique and often the results obtained may be proved without it; but it is a very convenient technique, which provides great computational economy and lucidity of discussion in problems that this book is concerned with. In these problems, to give up the use of the operator method would be just the same as if one were dealing with the limits of increment ratios instead of derivatives throughout the theory of differential equations.

The operator technique is extremely fruitful in various concrete examples. In general situations, it is also much more convenient to formulate and to prove the results in these terms.

Feynman himself was dealing with continuous families of ordered operators rather than finite ones (T-products, the corresponding continual integrals in quantum field theory, etc.). This matter is not considered in the book; however, it is quite natural that consideration of operator calculus for a finite number of ordered operators, to which Chapter 2 of this book is devoted, may serve as a useful preliminary step in thus becoming acquainted with these questions and as an introduction to Feynman operator calculus. It is interesting to note that the relations between operator calculus for a finite number of operators and Feynman's investigations are not confined to that mentioned above. For the equations of quantum mechanics and quantum field theory the method of ordered operators gives asymptotic approximations to precise solutions (formally) expressed via Feynman continual integration technique.

The definition (2) of the function of ordered operators is purely illustrative; it is valid only for polynomials or convergent power series.

The definition in terms of Fourier transform

$$f(\overset{1}{A_1},\ldots,\overset{n}{A_n}) = (2\pi)^{-n/2}\int \tilde{f}(t_1,\ldots,t_n)e^{iA_nt_n}\ldots e^{iA_1t_1}dt_1\ldots dt_n \tag{4}$$

where $e^{iA_jt_j}$ is a group of linear operators generated by the operator A_j, is most appropriate for our aims, permitting the consideration of non-analytic symbols f and unbounded operators. It is thoroughly studied in Chapter 2.

Consider now the Feynman approach in a very brief comparison with some other existing methods of constructing the operator calculus.

We begin with the simple observation that, in fact, in the theory of differential (and later pseudo-differential) operators the ordered form of notation was always used. Indeed, if

$$\hat{L} = \sum_{|\alpha| \leqslant m} a_\alpha(x)\left(-i \frac{\partial}{\partial x}\right)^\alpha \tag{5}$$

is a differential operator, then its symbol is usually considered to be

$$L(x,\xi) = \sum_{|\alpha| \leqslant m} a_\alpha(x)\xi^\alpha, \tag{6}$$

and this means that in our notations

$$\hat{L} = L(\overset{2}{x}, -i \overset{1}{\frac{\partial}{\partial x}}). \tag{7}$$

The same form of notations is often used in quantum mechanics, when some given operator $\hat{P}$ is expressed in terms of creation and annihilation operators a^* and a. The "normal form" reads

$$\hat{P} = \sum_{m,n} p_{mn}(a^*)^n a^m, \tag{8}$$

i.e.,

$$\hat{P} = P(\overset{1}{a}, \overset{2}{a^*}), \tag{9}$$

where $P(z_1,z_2) = \sum_{m,n} p_{mn} z_1^m z_2^n$ is a Wick symbol of the operator $\hat{P}$ [6].

As for pseudo-differential operators, they are often written in the form (see [30]):

$$\hat{P}u(x) = \frac{1}{(2\pi)^{n/2}} \int_{\mathbb{R}^n} p(x,\xi)e^{ix\xi}\tilde{u}(\xi)d\xi, \tag{10}$$

where $p(x,\xi)$ is the symbol of $\hat{P}$, i.e., $\hat{P} = p(\overset{2}{x}, \overset{1}{-i(\partial/\partial x)})$ (see Chapter 2 for detailed explanations); sometimes the opposite order $\overset{1}{x}$, $\overset{2}{-i\partial/\partial x}$ is used (cf. [45]).

However, there exist different approaches to construction of operator calculus. We do not mention here those which are based on the theory of analytic functions and the Cauchy integral formula since the class of analytic functions is too narrow for asymptotic problems to be considered and also these approaches are mainly oriented for functions of bounded operators.

Among the methods which could be used also for our aims, the Weyl approach to functional calculus should be mentioned (see [1]). The characteristic feature of this approach is that if

$$f(x_1,\ldots,x_n) = g(\alpha_1 x_1 + \ldots + \alpha_n x_n), \tag{11}$$

then

$$f_W(A_1,\ldots,A_n) = g(\alpha_1 A_1 + \ldots + \alpha_n A_n), \tag{12}$$

where in the right-hand side of (12) we have the function of a single operator $\alpha_1 A_1 + \dots + \alpha_n A_n$.

Compare the Weyl and Feynman approach on an example of polynomial functions of two operators A and B:

Function	Feynman-ordered operator	Weyl operator
$f(x,y)$	$f(\overset{1}{A},\overset{2}{B})$	$f_W(A,B)$
$(x+y)^2$	$(\overset{1}{A}+\overset{2}{B})^2 = B^2 + 2BA + A^2$	$(A+B)^2_W = (A+B)^2 = A^2 + AB + BA + B^2$
xy	$\overset{1}{A}\overset{2}{B} = BA$	$(AB)_W = \frac{1}{4}[(A+B)^2 - (A-B)^2] = \frac{1}{2}(AB+BA)$
x^2y	$\overset{1}{A^2}\overset{2}{B} = BA^2$	$(A^2B)_W = \frac{1}{3}(A^2B + ABA + BA^2)$
x^2y^2	$\overset{1}{A^2}\overset{2}{B^2} = B^2A^2$	$(A^2B^2)_W = \frac{1}{6}(A^2B^2 + B^2A^2 + AB^2A + BA^2B + BABA + ABAB)$

(in the last columns we used the property (11) - (12) for computations, representing the polynomial in question in the form of the sum of powers of linear functions; for example,

$$x^2y = \frac{1}{6}[(x+y)^3 - (x-y)^3 - 2y^3]), \tag{13}$$

Assume that the operators A and B satisfy the commutation relation

$$[A,B] \equiv AB - BA = 1, \tag{14}$$

and compare Feynman and Weyl symbols for some polynomial operators. We present also the Jordan symbol defined by

$$f_J(A,B) = \frac{1}{2}[f(\overset{1}{A},\overset{2}{B}) + f(\overset{2}{A},\overset{1}{B})]. \tag{15}$$

Operator	Feynman symbol		Weyl symbol		Jordan symbol
$\hat{P}$	$f(x,y)$		$g(x,y)$		$h(x,y)$
	$(\hat{P} = f(\overset{1}{A},\overset{2}{B})$	$=$	$g_W(A,B)$	$=$	$h_J(A,B))$
AB	$xy + 1$		$xy + \frac{1}{2}$		$xy + \frac{1}{2}$
BA^2	x^2y		$x^2y - x$		$x^2y - x$
B^2A^2	x^2y^2		$x^2y^2 - 2xy + \frac{1}{2}$		$x^2y^2 - 2xy$

The comparison yields the conclusion that the Weyl calculus, being more "symmetrical," is more complicated from the computational viewpoint.

The symmetry of Weyl calculus is a useful property: If all the operators $A_1,\dots,A_n$ are self-adjoint and $f(x_1,\dots,x_n)$ is real-valued, then $f_W(A_1,\dots,A_n)$ is also a self-adjoint operator. The same property is valid for Jordan calculus.

However, we have chosen the Feynman ordering of operators for the following reasons:

a) All the computational formulae are essentially simpler for the case of Feynman ordering.
b) In the case of unbounded operators the Weyl calculus imposes much more rigid functional-analytic requirements on the operators $A_1,\ldots,A_n$ which do not seem to be natural for applications except for the case when all the operators $A_1,\ldots,A_n$ are self-adjoint.

In the subsequent two items we present two examples in which the introduction of operator notation simplifies greatly all the exposition.

B. Example 1. Commutation Formula for the Schrödinger Operator and Rapidly Oscillating Exponential

In the problem of quasi-classical approximation to the solution of quantum-mechanical equations the main role belongs to the characteristic equation, which was derived in Section 1:A for the particular case of the Klein-Gordon equation. Here we consider the Schrödinger equation:

$$\hat{H}u \equiv H(\overset{2}{x}, -ih\overset{1}{\frac{\partial}{\partial x}})u(x,h) = 0 \tag{1}$$

with the arbitrary Schrödinger operator $\hat{H}$. We seek $u(x,h)$ in the form

$$u(x,h) = e^{(i/h)S(x)}\phi(x,h), \tag{2}$$

where $\phi(x,h)$ may be expanded in a series with respect to powers of h, and the problem is to obtain the result of application of the operator $\hat{H}$ to the function (2). We do not give rigorous proofs here, but only demonstrate that the operator approach gives the result readily after simple calculations. We intend to obtain the operator $\mathcal{P}$ such that

$$\hat{H}e^{(i/h)S(x)}\phi(x,h) = e^{(i/h)S(x)}\mathcal{P}\phi(x,h); \tag{3}$$

in other words

$$\mathcal{P} = e^{-(i/h)S(x)}\hat{H}e^{(i/h)S(x)}. \tag{4}$$

Proposition 1. For an invertible operator $\mathcal{U}$,

$$\mathcal{U}^{-1} \circ f(\overset{1}{A_1},\ldots,\overset{n}{A_n}) \circ \mathcal{U} = f(\overset{1}{\overline{\mathcal{U}^{-1}A_1\mathcal{U}}},\ldots,\overset{n}{\overline{\mathcal{U}^{-1}A_n\mathcal{U}}}). \tag{5}$$

Sketch of the proof. We have

$$\begin{aligned}\mathcal{U}^{-1} \circ f(\overset{1}{A_1},\ldots,\overset{n}{A_n}) \circ \mathcal{U} &= (2\pi)^{-n/2} \int \tilde{f}(t_1,\ldots,t_n)\mathcal{U}^{-1}e^{iA_nt_n} \ldots \times \\ \times e^{iA_1t_1}\mathcal{U}dt_1\ldots dt_n &= (2\pi)^{-n/2} \int \tilde{f}(t_1,\ldots,t_n)\mathcal{U}^{-1}e^{iA_nt_n}\mathcal{U} \times \\ &\times \ldots \mathcal{U}^{-1}e^{iA_1t_1}\mathcal{U}dt_1\ldots dt_n,\end{aligned} \tag{6}$$

so it suffices to demonstrate that

$$\mathcal{U}^{-1}e^{iAt}\mathcal{U} = e^{i\mathcal{U}^{-1}A\mathcal{U}t}. \tag{7}$$

This is true, since

$$\mathcal{U}^{-1}e^{iAt}\mathcal{U}\big|_{t=0} = 1, \ \frac{d}{dt}(\mathcal{U}^{-1}e^{iAt}\mathcal{U}) = i\mathcal{U}^{-1}Ae^{iAt}\mathcal{U} = -i\mathcal{U}^{-1}A\mathcal{U} \cdot \mathcal{U}^{-1}e^{iAt}\mathcal{U}, \tag{8}$$

thus the validity of Proposition 1 is demonstrated. Since

$$e^{(-i/h)S(x)} x e^{(i/h)S(x)} = x,$$
$$e^{(-i/h)S(x)}(-ih \frac{\partial}{\partial x}) e^{(i/h)S(x)} = -ih \frac{\partial}{\partial x} + \frac{\partial S(x)}{\partial x}, \tag{9}$$

we obtain

$$H(\overset{2}{x}, \overset{1}{-ih \frac{\partial}{\partial x}})(e^{(i/h)S(x)} \phi(x,h)) = e^{(i/h)S(x)} H(\overset{2}{x}, \overline{\overline{\frac{\partial S}{\partial x} - ih \frac{\partial}{\partial x}}}^{1}) \phi(x,h); \tag{10}$$

this formula is the generalization of the permutation formula (3) of item A. Thus the equation for $\phi(x,h)$ reads

$$H(\overset{2}{x}, \overline{\overline{\frac{\partial S}{\partial x} - ih \frac{\partial}{\partial x}}}^{1}) \phi(x,h) = 0; \tag{11}$$

in particular, in the principal term we obtain the characteristic equation

$$H(x, \frac{\partial S}{\partial x}) = 0. \tag{12}$$

It is also easy to obtain the subsequent terms of expansion of

$$H(\overset{2}{x}, \overline{\overline{\frac{\partial S}{\partial x} - ih \frac{\partial}{\partial x}}}^{1})$$

with respect to powers of h (see below). Note that the formula of commutation of Hamiltonian and exponential (10), which is readily obtained in ordered operators notions, was derived in a very complicated way (see [50], for Fourier integral operators see [29]) by means of the stationary phase method.

Next, we obtain the second term of asymptotic expansion of

$$H(\overset{2}{x}, \overline{\overline{\frac{\partial S}{\partial x} - ih \frac{\partial}{\partial x}}}^{1})$$

with respect to powers of h. By the way, we introduce some simple rules for treating a function of non-commuting operators.

Proposition 2. The formula is valid

$$f(A + B) = f(A) + \overset{2}{B} \frac{\delta f}{\delta x} (\overset{1}{A}, \overline{\overline{A + B}}^{3}), \tag{13}$$

where

$$\frac{\delta f}{\delta x} (x,y) = \frac{f(x) - f(y)}{x - y}. \tag{14}$$

Proof. We use the following rules of the ordered operator calculus (they are proved in Chapter 2):

a) The operator $f(\overset{s_1}{A_1}, \ldots, \overset{s_n}{A_n})$ does not change, if we replace the numbers $s_1, \ldots, s_n$ by other ones so that the mutual position of s_j, s_k on the real axis remains unaffected for any pair (j,k) such that $[A_j, A_k] \neq 0$.

b) ("shifting indices")

$$f(\ldots, \overset{m}{A}, \ldots, \overset{m}{A}, \ldots) = g(\ldots, \overset{m}{A}, \ldots), \tag{15}$$

where

$$g(\ldots, x, \ldots) = f(\ldots, x, \ldots, y, \ldots)|_{y=x}. \tag{16}$$

c) If $s_1,\dots,s_m \in [a,b]$, $r_1,\dots,r_s \notin [a,b]$, where $[a,b] \subset \mathbb{R}$ is some given segment, then

$$f(\overset{s_1}{A_1},\dots,\overset{s_m}{A_m})\,g(\overset{r_1}{A_{m+1}},\dots,\overset{r_s}{A_{m+s}}) = \overset{\lambda}{C}\,g(\overset{r_1}{A_{m+1}},\dots,\overset{r_s}{A_{m+s}}), \tag{17}$$

where $\lambda \in [a,b]$ is arbitrary,

$$C = f(\overset{s_1}{A_1},\dots,\overset{s_m}{A_m}). \tag{18}$$

All these rules are evident (the reader can easily verify them for polynomial symbols).

We also use a convenient notation. If we need to use some function $f(\overset{s_1}{A_1},\dots,\overset{s_m}{A_m})$ as a single operator in some more complicated operator expression, we put it into the so-called "autonomous brackets" [52]:

$$[\![\, f(\overset{s_1}{A_1},\dots,\overset{s_n}{A_n}) \,]\!] \;;$$

then the numbers $s_1,\dots,s_m$ are "valid" only inside the brackets and do not "interact" with numbers outside the brackets; the number, under which the whole operator acts, is written over the left bracket. In this notation, the equality (17) may be written in the form

$$f(\overset{s_1}{A_1},\dots,\overset{s_m}{A_m})\,g(\overset{r_1}{A_{m+1}},\dots,\overset{r_s}{A_{m+s}}) = \overset{\lambda}{[\![}\, f(\overset{s_1}{A_1},\dots,\overset{s_m}{A_m}) \,]\!]\, g(\overset{r_1}{A_{m+1}},\dots,\overset{r_s}{A_{m+s}}). \tag{19}$$

Now we prove (13). We have

$$\begin{aligned} f(A+B) &= f(A) + f(A+B) - f(A) = f(A) + f(\overset{3}{\overline{A+B}}) - f(\overset{1}{A}) = \\ &\overset{/b/}{=} f(A) + (\overset{3}{\overline{A+B}} - \overset{1}{A})\,\frac{\delta f}{\delta x}(\overset{1}{A},\overset{3}{\overline{A+B}}) \overset{/a/}{=} f(A) + (\overset{3}{\overline{A+B}} - \overset{1}{A})\,\frac{\delta r}{\delta x}(\overset{0}{A},\overset{4}{\overline{A+B}}) = \\ &\overset{/c/}{=} f(A) + \overset{2}{[\![}\, \overset{3}{\overline{A+B}} - \overset{1}{A} \,]\!]\,\frac{\delta f}{\delta x}(\overset{0}{A},\overset{4}{\overline{A+B}}) = f(A) + \overset{2}{B}\,\frac{\delta f}{\delta x}(\overset{0}{A},\overset{4}{\overline{A+B}}) = \\ &\overset{/a/}{=} f(A) + \overset{2}{B}\,\frac{\delta f}{\delta x}(\overset{1}{A},\overset{3}{\overline{A+B}}), \quad \text{Q.E.D.} \end{aligned} \tag{20}$$

Obviously the proposition remains valid if f depends on additional operator arguments. Now we have

$$\begin{aligned} H(\overset{2}{x}, \overset{1}{\overline{\frac{\partial S}{\partial x} - ih\frac{\partial}{\partial x}}}) &= H(\overset{2}{x}, \overset{1}{\frac{\partial S}{\partial x}}) + (-ih\overset{2}{\frac{\partial}{\partial x}})\,\frac{\delta H}{\delta p}(\overset{4}{x}; \overset{1}{\frac{\partial S}{\partial x}}, \overset{3}{\overline{\frac{\partial S}{\partial x} - ih\frac{\partial}{\partial x}}}) = \\ &= H(\overset{2}{x}, \overset{1}{\frac{\partial S}{\partial x}}) - ih\,\overset{2}{\frac{\partial}{\partial x}}\,\frac{\delta H}{\delta p}(\overset{4}{x}, \overset{1}{\frac{\partial S}{\partial x}}, \overset{3}{\frac{\partial S}{\partial x}}) - \\ &\quad -h^2\,\overset{4}{\frac{\partial}{\partial x}}\,\overset{2}{\frac{\partial}{\partial x}}\,\frac{\delta^2 H}{\delta p^2}(\overset{6}{x}; \overset{1}{\frac{\partial S}{\partial x}}, \overset{3}{\frac{\partial S}{\partial x}}, \overset{5}{\overline{\frac{\partial S}{\partial x} - ih\frac{\partial}{\partial x}}}). \end{aligned} \tag{21}$$

Thus the second term of asymptotic expansion of

$$H(\overset{2}{x}, \overset{1}{\overline{\frac{\partial s}{\partial x} - ih\frac{\partial}{\partial x}}})$$

with respect to powers of h has the form: $-ih\,\overset{2}{\frac{\partial}{\partial x}}\,\frac{\delta H}{\delta p}(\overset{4}{x}; \overset{1}{\frac{\partial S}{\partial x}}, \overset{3}{\frac{\partial S}{\partial x}})$.

Using the rules a) - c), we may commute $\overset{2}{\frac{\partial}{\partial x}}$ and $\overset{1}{\frac{\partial S}{\partial x}}$, obtaining

$$- ih \overset{2}{\frac{\partial}{\partial x}} \frac{\delta H}{\delta p} (\overset{4}{x}; \overset{1}{\frac{\partial S}{\partial x}}, \overset{3}{\frac{\partial S}{\partial x}}) = - ih \frac{\partial H}{\partial p} (x, \frac{\partial s}{\partial x}) \frac{\partial}{\partial x} - \frac{ih}{2} \frac{\partial^2 H}{\partial p^2} (x, \frac{\partial s}{\partial x}) \frac{\partial^2 S}{\partial x^2} \tag{22}$$

(we leave the calculation as an exercise to the reader). Thus we have obtained the commutation formula

$$H(\overset{2}{x}, -i \overset{1}{\frac{\partial}{\partial x}}) e^{(i/h)S(x)} \phi(x,h) =$$

$$= e^{(i/h)S(x)} \{H(x, \frac{\partial S}{\partial x}) \phi(x,h) - ih[\frac{\partial H}{\partial p}(x, \frac{\partial S}{\partial x}) \frac{\partial \phi(x,h)}{\partial x} + \tag{23}$$

$$+ \frac{1}{2} \frac{\partial^2 H}{\partial p^2} (x, \frac{\partial S}{\partial x}) \frac{\partial^2 S}{\partial x^2} \cdot \phi(x,h)] + O(h^2)\},$$

from which the characteristic equation and equation for $\phi_o(x) = \phi(x,0)$ easily follow.

C. Example 2. Characteristic Equation for an Equation with Growing Coefficients

In the previous item we used the operator method to derive the characteristic equation in the familiar case, namely, for quasi-classical asymptotics of the Schrödinger equation. Here we show that in an analogous way the characteristic equation for the non-standard asymptotics (the mixed asymptotics with respect to smoothness and growth at infinity) may be obtained.

We consider the equation (cf. Section 1:F):

$$\frac{\partial^2 u}{\partial t^2} - \frac{\partial^2 u}{\partial x^2} + (1 + b(x))x^2 u = 0, \; u|_{t=0} = 0, \; u_t|_{t=0} = u_o(x), \tag{1}$$

where $b(x) \in C_o^\infty(\mathbb{R}^1)$. Introduce the operators

$$A_1 = - i \frac{\partial}{\partial x}, \; A_2 = x, \; B = x. \tag{2}$$

We seek the asymptotic solution of (1) in the form

$$u(x,t) = G(t, \overset{1}{A_1}, \overset{2}{A_2}, \overset{2}{B}) u_o(x), \tag{3}$$

where $G(t,x_1,x_2,\alpha)$ is a linear combination of functions of the form

$$f(t,x_1,x_2,\alpha) = e^{i\Lambda(x)S(\omega_1,\omega_2,\alpha,t)}[\phi_o(\omega_1,\omega_2,\alpha,t) + \\ + \Lambda(x)^{-1}\phi_1(\omega_1,\omega_2,\alpha,t) + \dots], \tag{4}$$

where $\Lambda(x) = \{x_1^2 + x_2^2\}^{1/2}$, $\omega_1 = x_1/\Lambda(x)$, $\omega_2 = x_2/\Lambda(x)$. Thus, the symbol $G(t,x_1,x_2,\alpha)$ of the resolving operator for the Cauchy problem (1) is sought in the form of an asymptotic expansion with respect to powers of $\Lambda(x_1,x_2) = \{x_1^2 + x_2^2\}^{1/2}$. The fact that such expansion gives the mixed smoothness-growth-at-infinity asymptotics follows from the estimates proved in Chapter 2.

The equation takes the form (where $c(x) = 1 + b(x)$):

$$[\![\frac{\partial^2}{\partial t^2} + A_1^2 + c(B)A_2^2]\!] [\![f(t, \overset{1}{A_1}, \overset{2}{A_2}, \overset{2}{B})]\!] \simeq 0, \tag{5}$$

where $\approx$ denotes the equality modulo operators with kernel sufficiently smooth and decaying at infinity.

Proposition 1.

$$[\![\frac{\partial^2}{\partial t^2} + A_1^2 + c(B)A_2^2]\!] [\![f(t,\overset{1}{A_1},\overset{2}{A_2},\overset{2}{B})]\!] = f_1(t,\overset{1}{A_1},\overset{2}{A_2},\overset{2}{B}), \tag{6}$$

where

$$f_1(t,x_1,x_2,\alpha) = \frac{\partial^2 f}{\partial t^2} + (x_1 - i\frac{\partial}{\partial x_2} - i\frac{\partial}{\partial \alpha})^2 f + c(\alpha)x_2^2 f. \tag{7}$$

Proof. Using rules a) - b) from item B, we obtain

$$\begin{aligned}
A_1 [\![F(\overset{1}{A_1},\overset{2}{A_2},\overset{2}{B})]\!] &= \overset{3}{A_1} F(\overset{1}{A_1},\overset{2}{A_2},\overset{2}{B}) = \\
&= \overset{3}{A_1} F(\overset{1}{A_1},\overset{2}{A_2},\overset{4}{B}) + \overset{3}{A_1}(\overset{2}{B} - \overset{4}{B}) \frac{\delta F}{\delta \alpha}(\overset{1}{A_1};\overset{2}{A_2};\overset{2}{B},\overset{4}{B}) = \\
&= \overset{3}{A_1} F(\overset{1}{A_1},\overset{2}{A_2},\overset{4}{B}) + [\![\overset{3}{A_1}(\overset{3}{B} - \overset{4}{B})]\!] \frac{\delta F}{\delta \alpha}(\overset{1}{A_1};\overset{2}{A_2};\overset{2}{B},\overset{4}{B}) = \\
&= \overset{3}{A_1} F(\overset{1}{A_1},\overset{2}{A_2},\overset{4}{B}) - i\frac{\partial F}{\partial \alpha}(\overset{1}{A_1},\overset{2}{A_2},\overset{2}{B}) = \ldots = \\
&= \overset{1}{A_1} F(\overset{1}{A_1},\overset{2}{A_2},\overset{2}{B}) - i\frac{\partial F}{\partial \alpha}(\overset{1}{A_1},\overset{2}{A_2},\overset{2}{B}) - i\frac{\partial F}{\partial x_2}(\overset{1}{A_1},\overset{2}{A_2},\overset{2}{B}).
\end{aligned} \tag{8}$$

Thus the multiplication by A_1 from the left is equivalent to action by the operator $(x_1 - i\frac{\partial}{\partial x_2} - i\frac{\partial}{\partial \alpha})$ on the symbol. Since A_2 and B commute, we have

$$\begin{aligned}
A_2 [\![F(\overset{1}{A_1},\overset{2}{A_2},\overset{2}{B})]\!] &= \overset{2}{A_2} F(\overset{1}{A_1},\overset{2}{A_2},\overset{2}{B}); \\
c(B) [\![F(\overset{1}{A_1},\overset{2}{A_2},\overset{2}{B})]\!] &= c(\overset{2}{B}) F(\overset{1}{A_1},\overset{2}{A_2},\overset{2}{B}).
\end{aligned} \tag{9}$$

Therefore, we come to (7). Calculate now the function (7) if f is given by (4). We have

$$\begin{aligned}
&[\frac{\partial^2}{\partial t^2} + (x_1 - i\frac{\partial}{\partial x_2} - i\frac{\partial}{\partial \alpha})^2 + c(\alpha)x_2^2] e^{i\Lambda(x)S(\omega_1,\omega_2,\alpha,t)} \times \\
&\times [\phi_o(\omega_1,\omega_2,\alpha,t) + \ldots] = e^{i\Lambda(x)S(\omega_1,\omega_2,\alpha,t)} \times \\
&\times [-(\Lambda\frac{\partial S}{\partial t} - i\frac{\partial}{\partial t})^2 + (x_1 + \frac{\partial}{\partial x_2}(\Lambda S) + \Lambda\frac{\partial S}{\partial \alpha} - i\frac{\partial}{\partial x_2} - i\frac{\partial}{\partial \alpha})^2 + \\
&+ c(\alpha)x_2^2][\phi_o(\omega_1,\omega_2,\alpha,t) + \ldots].
\end{aligned} \tag{10}$$

We extract from (10) the leading term (with respect to powers of Λ), which determines the characteristic equation. It has the form

$$-\Lambda^2(\frac{\partial S}{\partial t})^2 + (x_1 + \Lambda\frac{\partial S}{\partial \alpha})^2 + c(\alpha)x_2^2 = \Lambda^2[-(\frac{\partial S}{\partial t})^2 + (\omega_1 + \frac{\partial S}{\partial \alpha})^2 + c(\alpha)\omega_2^2]. \tag{11}$$

Thus after the change $S = \Phi - \alpha\omega_1$ the characteristic equation takes the form (recall that $c(\alpha) = 1 + b(\alpha)$):

$$(\frac{\partial \Phi}{\partial t})^2 - (\frac{\partial \Phi}{\partial \alpha})^2 - \omega_2^2(1 + b(\alpha)) = 0, \quad \omega_2^2 \in [0,1]. \tag{12}$$

The detailed consideration of mixed asymptotics for equations with growing coefficients is given in Chapter 4, Section 1.

4. THE OUTLINE OF THE BOOK. PRELIMINARY KNOWLEDGE NECESSARY TO READ THE BOOK. THE SECTION DEPENDENCE SCHEME

The main part of the book consists of three chapters: Chapter 2 through to and including Chapter 4.

Chapter 2, entitled "Functional calculus" is devoted to thorough development of functional-analytic baseground for ordered operator calculus and construction of asymptotic solutions. No special preliminary knowledge on these subjects is required; the reader should only be acquainted with basic notions of functional analysis such as Banach and Hilbert spaces, linear operators, etc. The exposition is detailed and as elementary as possible. We begin with the theory of semigroups and functions of a single operator. These questions are considered in Section 2, including functions of self-adjoint operators and functions of operators depending on parameters. Section 3 is concerned with functions of several operators. Various symbol spaces are introduced and investigated, the functions of several operators are considered in Banach spaces and in Banach scales as well (the later notion is introduced and discussed in detail). The main results are in Section 3:D where the mapping

$$\mu : f(x_1,\ldots,x_n) \to f(\overset{1}{A_1},\ldots,\overset{n}{A_n}) \tag{1}$$

is introduced and the main rules of operator calculus are proved then for operators in Banach scales.

In Section 4 regular presentations of tuples of non-commuting operators are introduced and investigated. By definition, the (left) regular representation of the system of operators $A_1,\ldots,A_n$ is the system of operators $L_1,\ldots,L_n$, acting on functions in such a way that

$$(L_j f)(\overset{1}{A_1},\ldots,\overset{n}{A_n}) = A_j \circ f(\overset{1}{A_1},\ldots,\overset{n}{A_n}). \tag{2}$$

The regular representation enables the composition law in the set of functions of $\overset{1}{A_1},\ldots,\overset{n}{A_n}$ to be established and thus to reduce general asymptotic problems to pseudo-differential ones. The functional-calculus conditions of existence of regular representations are studied, the case of Lie algebras is considered as an example; some boundedness theorems are proved which result from the existence of regular representations.

In Chapter 3 the analytic apparatus necessary for construction of almost inverse operators is treated. Two cases are considered: the case of small parameter asymptotics (which stands somewhat separately) and the correspondent technique in the homogeneous case when the action of the group $\mathbb{R}_+$ of positive number is given.

Only elementary knowledge of mathematical analysis is necessary to understand the material of this Chapter. All the notions used which are out of this framework are introduced and explained in the text.

The analytic apparatus developed includes the canonical sheaves on symplectic manifolds (both with $\mathbb{R}_+$-action and without it), the calculus of pseudo-differential operators, and the canonical operator in general symplectic manifolds.

In Chapter 4 the results of previous chapters are applied to the general problem of construction of almost inverse operators. The operator approach to this problem is illustrated in Section 1 by an example of constructing asymptotic solution to equations with growing coefficients. In Section 2 the notion of Poisson algebra is introduced and discussed (for the case of the small parameter theory). In Section 3 the same notion is used in the $\mathbb{R}_+$-homogeneous case and the central object of the theory, namely, the μ-structure on the Poisson algebra, is defined in its local version. In Section 4 the global version of the introduced notions and the symplectic manifold of the Poisson algebra are defined and representations of the canonical sheaf on this latter manifold are used to prove the main theorem, which establishes the relationships between the geometric properties of the symbol $f(x_1,\dots,x_n)$ of the operator $f(\overset{1}{A}_1,\dots,\overset{n}{A}_n)$ and the existence of the quasi-inverse operator (and its explicit expression). The proof of this theorem ends the book.

The exposition is self-closed; the only exception is the theorem on solution of pseudo-differential equation (Chapter 3:4), the proof of which is based on the complex germ techniques [52] not considered in the book. But we present the material in such a way that if one merely takes this theorem as given, then none of the other statements will require the information not contained in the book.

It should be noted that geometric and analytic concepts of Chapters 3 and 4 (symplectic manifold of the Poisson algebra, canonical sheaves and the canonical operator on general symplectic manifolds, etc.) were not developed on the blank space. We try here to review the history of the question briefly.

The sheaves on symplectic manifolds analogous to those used by us are well known in literature. They primarily occur in geometric quantization and in the method of orbits in the theory of representations of Lie groups (see [41,43,73,4,74], etc.). In all of these papers the existence of polarization was required or, more generally, the condition that the Chern class $\varepsilon \in H^2(M,\mathbb{Z})$ be even was required, i.e., instead of the quantization condition

$$\frac{1}{2\pi}[\omega] - \frac{\varepsilon}{2} \in H^2(M,\mathbb{Z}), \tag{3}$$

where $[\omega]$ is the class of the symplectic form, it was separately required that

$$\frac{1}{2\pi h}[\omega] \in H^2(M,\mathbb{Z}); \quad \varepsilon \equiv 0 \pmod 2). \tag{4}$$

The condition (3) was first written for odd ε on Koehler manifolds with the complex polarization in [10]. In [27] the same procedure was held for arbitrary real manifolds (with complex polarization). It was discovered that in general situations the second Stiefel-Whitney class W_2 plays the role of ε in (3).

The analogue of condition (3) was obtained independently in [36] for the procedure of construction of pseudo-differential operators calculus mod $O(h)$ on general symplectic manifolds. The condition reads

$$\frac{1}{2\pi h}[\omega] - \frac{\nu}{4} \in H^2(M,\mathbb{Z}_4), \tag{5}$$

where the class $\nu \in H^2(M,\mathbb{Z}_4)$ is calculated via indices of paths on Lagrangian manifolds [50,3]. No polarization was required. The class in (5) is in fact even [55]; however, it may be odd for manifolds with conic singularities [37].

The canonical operator on general symplectic manifolds was constructed primarily in [38] (where also the pseudo-differential operators calculus was constructed modulo $O(h^2)$). The main novelty is that the classes [pdq] and one-dimensional index are not defined separately. The application of this developed apparatus to asymptotic problems begins from the papers [35] and [33].

The notion of the symplectic manifold of the Poisson algebra together with the notion of the asymptotic group algebra was introduced in the paper [36]. Also the formulae for symbols of regular representation operators was announced in this paper (the complete proof is contained in [34]).

Thus all the main topologic and geometric notions used in Chapters 3 and 4 were recently developed in the cited papers for the case of small parameter asymptotics.

However, the algebraic structure of considered objects being unchanged, the technique of proofs and especially of estimating the remainder terms is completely different (and much more complicated) for the case of homogeneous manifolds and asymptotics with respect to general type of operators. These estimates are the main results of Chapters 2 and 3, since they lead to proof of the quasi-invertibility theorem. Interesting independent results in the homogeneous case may be found in the papers [8,23].

Reproduced here is a scheme of dependence of chapters and sections (dotted lines mean "weak" dependence).

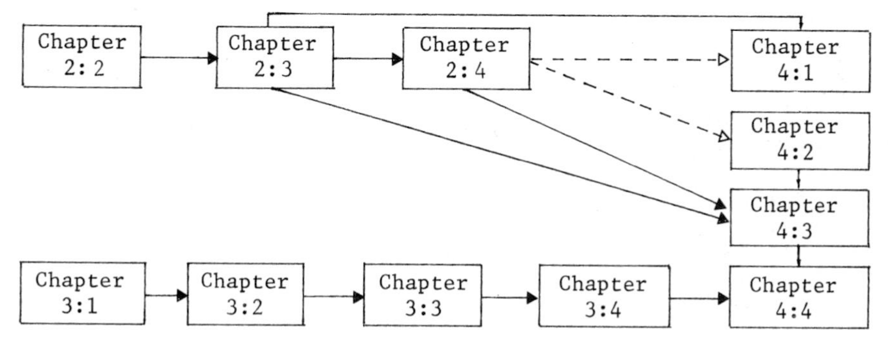

Scheme 1. Dependence of the sections.

II
Functional calculus

1. INTRODUCTION

In this chapter we develop the apparatus which is used in this book for obtaining asymptotics in functional spaces. This apparatus is the he functional calculus for several non-commuting operators acting in a Banach space, or, more generally, in a Banach scale. The aim of the introduction is to give in short the exposition of the main results presented in the chapter and to clarify the main ideas. The bibliographical notes are gathered at the end of the Introduction. The reader, who is not interested in precise formulations and detailed proofs, may restrict himself to reading only the Introduction of this chapter and then go to the subsequent chapters.

We assume that the reader has become acquainted with the basic notions of functional analysis such as normed spaces, closed operators, resolvents, spectral expansion of a self-adjoint operator, etc. No information is assumed to be known about semigroups in functional spaces. We assume that the foundations of theory of Lie groups are known to the reader. Sometimes we use the results of distribution theory.

A. Functions of Several Operators. Feynman and Weyl Orderings

Let $A_1,\ldots,A_n$ be linear operators in some functional space H. Let also $p(y_1,\ldots,y_n)$ be a polynomial

$$p(y_1,\ldots,y_n) = \sum_{|\alpha|\leqslant M} a_\alpha y_1^{\alpha_1} \ldots y_n^{\alpha_n}.$$

Generally speaking, the result of substitution of $A_1,\ldots,A_n$ instead of $y_1,\ldots,y_n$ into the polynomial p is not uniquely defined if the operators $A_1,\ldots,A_n$ do not commute. The ordering of operators is the method by which to avoid such indeterminacy.

(a) Let $\pi_1,\ldots,\pi_n$ be a set of real numbers, such that $\pi_i \neq \pi_j$ if A_i does not commute with A_j. These numbers may be rearranged in the non-decreasing order: $\pi_{j_1} \leqslant \pi_{j_2} \leqslant \ldots \leqslant \pi_{j_n}$. We set

$$p(\overset{\pi}{A_1},\ldots,\overset{\pi}{A_n}) = \sum_{|\alpha|\leqslant M} a_\alpha A_{j_n}^{\alpha_{j_n}} \times \ldots \times A_{j_1}^{\alpha_{j_1}}. \tag{1}$$

Since $[A_i,A_j] \equiv A_iA_j - A_jA_i = 0$ for $\pi_i = \pi_j$, the definition does not depend on the choice of the rearrangement. The ordering (1) was first introduced by R. Feynman. The numbers $\pi_1,\ldots,\pi_n$ indicate the order of action of the operators $A_1,\ldots,A_n$; an operator with the lower index acts first. The mapping

$$\mu_A \equiv \mu_{\overset{\pi_1}{A_1},\ldots,\overset{\pi_n}{A_n}} : p \to p(\overset{\pi_1}{A_1},\ldots,\overset{\pi_n}{A_n})$$

defined by (1) is called the (Feynman) ordered quantization corresponding to the ordered set of operators $A = (\overset{\pi_1}{A_1},\ldots,\overset{\pi_n}{A_n})$, and the polynomial p is called the symbol of the operator (1). Feynman ordering possesses simple properties:

(i) the quantization depends only on the order of numbers π_1, $\ldots$, π_n on the real axis; in particular, if $p(y_1, \ldots, y_n) = q(y_{j_1}, \ldots, y_{j_n})$, then $p(\overset{\pi_1}{A_1},\ldots,\overset{\pi_n}{A_n}) = q(\overset{1}{A_{j_1}},\ldots,\overset{n}{A_{j_n}})$.

(ii) If $p(y_1, \ldots, y_n) = p_1(y_{j_1}, \ldots, y_{j_s}, y_{j_{r+1}}, y_{j_{r+2}}, \ldots, y_{j_n})p_2\cdot(y_{j_{s+1}}, \ldots, y_{j_r})$, $1 \le s < r \le n$, then

$$p(\overset{\pi_1}{A_1},\ldots,\overset{\pi_n}{A_n}) = p_1(\overset{1}{A_{j_1}},\ldots,\overset{s}{A_{j_s}},\overset{r+1}{A_{j_{r+1}}},\overset{r+2}{A_{j_{r+2}}},\ldots,\overset{n}{A_{j_n}}) \times$$
$$\times \overset{\pi}{[\![} p_2(\overset{s+1}{A_{j_{s+1}}},\ldots,\overset{r}{A_{j_r}})]\!], \tag{2}$$

where $\pi \in [s + 1, r]$ is arbitrary.

We used a convenient notation in (2): The expression contained in the "autonomous brackets" $[\![\ ]\!]$ is considered as a single operator, and the place on which it acts is indicated by the number over the left bracket. Thus, there is no correspondence between the numbers over operators inside and outside the brackets, and the notation $\overset{\pi}{[\![} p_2(\overset{s+1}{A_{j_{s+1}}},\ldots,\overset{r}{A_{j_r}})]\!]$ is equivalent to $\overset{\pi}{c}$, where $c = p_2(\overset{s+1}{A_{j_{s+1}}},\ldots,\overset{r}{A_{j_r}})$.

(iii) If $A_{j_k} = A_{j_{k+1}}$, then

$$p(\overset{\pi_1}{A_1},\ldots,\overset{\pi_n}{A_n}) = q(\overset{1}{A_{j_1}},\ldots,\overset{k}{A_{j_k}},\overset{k+2}{A_{j_{k+2}}},\ldots,\overset{n}{A_{j_n}}), \tag{3}$$

where

$$q(y_{j_1},\ldots,y_{j_k},y_{j_{k+2}},\ldots,y_{j_n}) = p(y_1,\ldots,y_n)|_{y_{j_{k+1}}=y_{j_k}}.$$

The properties (i) - (iii) yield that the mapping $p \to p(A)$ is the homomorphism of the algebra of polynomials of one variable into the algebra of operators in $\underline{H}$ and that if $p(y_1 \ldots y_n) = p_1(y_1) \times \ldots \times p_n(y_n)$, then

$$p(\overset{1}{A_1},\ldots,\overset{n}{A_n}) = p_n(A_n) \times p_{n-1}(A_{n-1}) \times \ldots p_1(A_1). \tag{4}$$

The described properties are basic for Feynman ordering, and many useful algebraic identities could be derived starting from these properties.

(b) There is another way of ordering the operators, which is called the Weyl ordering. We set into correspondence to polynomial $p(y_1,\ldots,y_n)$ the operator

$$p_W(A_1,\ldots,A_n) = a_o I + \sum_{0<|\alpha|\leqslant M} a_\alpha \frac{\alpha_1!\ldots\alpha_n!}{|\alpha|!} (\sum_\sigma A_{\sigma(1)} \cdots A_{\sigma(|\alpha|)}), \tag{5}$$

where the sum in round brackets is taken over all the mappings $\sigma : \{1,\ldots,|\alpha|\} \to \{1,\ldots,n\}$ taking each value $j \in \{1,\ldots,n\}$ exactly α_j times. Weyl ordering possesses the properties:

(i) for a single operator A (n = 1) we have

$$p_W(A) = p(A).$$

(ii) (Affine covariance). If $M : \mathbb{C}^n \to \mathbb{C}^n$, $Mz = Bz + b$ is an affine transformation, then for any polynomial $p(y_1, \ldots, y_n)$

$$(M^*p)_W(A_1,\ldots,A_n) = p_W((MA)_1,\ldots,(MA)_n), \tag{6}$$

where $(M^*p)(y) = p(My)$, $(MA)_i = \sum_j B_{ij}A_j + b_i$. In particular, if $p(y_1, \ldots,y_n) = q(y_1 + \ldots + y_n)$, then

$$p_W(A_1,\ldots,A_n) = q(A_1 + \ldots + A_n).$$

The polynomials do not form the class of symbols large enough to be applicable in the theory of asymptotics. There are various possibilities of broadening this class of symbols for which the quantization is defined; for example, holomorphic functional calculus based on Taylor expansions or on Cauchy-type integrals, etc. It turns out that the most convenient for the needs of asymptotic theory is the approach to construction of functional calculus based on exploitation of Fourier transformation. This approach enables us to consider the non-analytic symbols and to obtain substantial results for a wide class of operators.

Assume that operators $A_1,\ldots,A_n$ generate strongly continuous one-parametric groups in H, i.e., Cauchy problems

$$-i\,\frac{\partial x(t)}{\partial t} = A_j x(t),\ x(0) = x_o \in D_{A_j} \tag{7}$$

have the unique solutions $x(t) = \mathcal{U}_j(t)x_o$, where $\mathcal{U}_j(t)$ is a strongly continuous family of continuous operators in H defined for $-\infty \leqslant t \leqslant \infty$ (in this chapter we give a comprehensive analysis of properties of such operators which will be called generators).

Given a function $f(y_1,\ldots,y_n)$, we define the operator $\mu_A(f)$ by virtue of the formula

$$f(\overset{1}{A_1},\ldots,\overset{n}{A_n}) = (2\pi)^{-n/2} \int_{\mathbb{R}^n} \tilde{f}(t)\mathcal{U}_n(t_n)\times\ldots\times\mathcal{U}_1(t_1)dt; \tag{8}$$

the integral in (8) being understood in the sense of, say, strong or weak convergence in H. As for Weyl quantization, it may be defined by

$$f_W(A_1,\ldots,A_n) = (2\pi)^{-n/2} \int_{\mathbb{R}^n} \tilde{f}(t)\mathcal{U}(t)dt, \tag{9}$$

where for each $\omega \in S^{n-1}$, $\mathcal{U}(\omega\tau)$, $\tau \in \mathbb{R}$, is a strongly continuous group generated by $\omega\cdot A \equiv \sum_{j=1}^{n} \omega_j A_j$. Of course, we should require in this case that $\omega\cdot A$ be generators for all $\omega \in S^{n-1}$. The question is whether these definitions make sense and whether the mentioned properties of the quantizations remain valid for such definition.

B. Functional Spaces and Symbol Classes

In applications the functional space H in which the operators A_j act turns out to be a Banach space, or, more generally, a total space of some Banach scale, i.e., the union of the family $\{H_\delta\}_{\delta\in\Delta}$ of Banach spaces, enumerated by elements of the bidirected* set Δ and connected by continuous embeddings $H_\delta \subset H_{\delta'}$ for $\delta \geqslant \delta'$. The important example of the scale appears in the problem of constructing asymptotics with respect to powers of given unbounded operators $B_1,\dots,B_n$ acting in a Banach space X. For $x \in X$ we set $\|x\|_0 = \|x\|$, where $\|\cdot\|$ is the norm function in X and we set

$$\|x\|_{k+1} = \|x\|_k + \sum_{j=1}^{n} \|B_j x\|_k, \quad k = 0,1,2,\dots . \tag{10}$$

Under certain assumptions (see Section 4:C) the subspace $X_k \subset X$ of such $x \in X$, $\|x\|_k$ is finite, is a dense subspace which is a Banach space with respect to the norm $\|\cdot\|_k$ and these spaces form a Banach scale. If X is a Hilbert space, the spaces X_k may be defined for negative k as dual spaces to X_{-k}. The construction of the scale $\{X_k\}$ given above admits generalizations which are also described in Section 4. We also discuss there the conditions under which the operator A in X gives rise to a generator in the scale $\{X_k\}$.

To define $f(\overset{1}{A}_1,\dots,\overset{n}{A}_n)$ by means of (8) we make the assumption that the groups $U_j(t) \equiv \exp(iA_j t)$ generated by A_j have tempered growth as $t \to \infty$ (in the case when H is a Banach space, it means simply that

$$\|\exp(iA_j t)\| \leqslant c(1+|t|)^p, \quad t \in \mathbb{R} \tag{11}$$

for some $c > 0$, $p \geqslant 0$. If H is a Banach scale we have the collection of estimates of the type (11).) If the operators A_j satisfy the described property, formula (8) defines $f(\overset{1}{A}_1,\dots,\overset{n}{A}_n)$ for any f whose Fourier transform $\tilde{f}$ is a continuous functional on the correspondent space of functions slowly growing at infinity (details in Sections 3 and 4). In particular, the expression $f(\overset{1}{A}_1,\dots,\overset{n}{A}_n)$ makes sense for $f \in S^\infty(\mathbb{R}^n)$ (the latter inclusion means that for some $m \in \mathbb{R}$ and any multi-index α the function $(1+|y|)^{-m} \cdot \partial^\alpha f/\partial y^\alpha$ is bounded in $\mathbb{R}^n$. The properties (i) - (iii) of the Feynman ordering remain valid in the case considered (however, the precise formulations become more complicated). Under similar assumptions the formula (9) defines correctly the Weyl-ordered operator $f_w(A_1,\dots,A_n)$. The novelty is the requirement that $U(t)$ be strongly continuous in $t \in \mathbb{R}^n$ (this property holds automatically for $U_n(t) \times \dots \times U_1(t)$, i.e., that the group $\exp(\tau\omega A)$ converge to the group $\exp(\tau\omega' A)$ as $\omega \to \omega'$. The conditions under which this convergence takes place are investigated in Section 2:G.

The important particular case of the general constructions is the calculus of pseudo-differential operators which may be regarded as functions of operators x_j of multiplication by independent variables in $\mathbb{R}^n$ and also operators of differentiation: $-i(\partial/\partial x_j)$. As will be shown below, most of the problems concerned with solution of equations, including functions of several operators, may be reduced to pseudo-differential equations for symbols (or systems of such equations). Thus we give in Section 4:E the construction and the analysis of pseudo-differential operators almost independent of the other material of the chapter. To finish with, we emphasize that the problem of estimates is the crucial one in functional calculus, and purely algebraic consideration is not sufficient for constructing the asymptotics.

---

* The set Δ is called bidirected (or directed in both sides) if it is partially ordered and any pair (δ_1,δ_2) of its elements has a majorant and a minorant.

C. Regular Representations

Fix the ordered tuple $A = (\overset{1}{A}_1,\dots,\overset{n}{A}_n)$ of generators in a functional space H. In general, the product $[\![f(\overset{1}{A}_1,\dots,\overset{n}{A}_n)]\!]\cdot[\![g(\overset{1}{A}_1,\dots,\overset{n}{A}_n)]\!]$ of two Feynman-quantized operators cannot be written in the form

$$[\![f(\overset{1}{A}_1,\dots,\overset{n}{A}_n)]\!]\cdot[\![g(\overset{1}{A}_1,\dots,\overset{n}{A}_n)]\!] = k(\overset{1}{A}_1,\dots,\overset{n}{A}_n) \qquad (12)$$

for some symbol $k(y_1,\dots,y_n)$. But if the tuple A satisfies some additional conditions, the representation (12) exists and

$$k(y_1,\dots,y_n) = f(y_1,\dots,y_n) * g(y_1,\dots,y_n), \qquad (13)$$

where $*$ is a bilinear operation in the symbol space. If (13) holds, we may reduce the equation

$$f(\overset{1}{A}_1,\dots,\overset{n}{A}_n)u = v \in H \qquad (14)$$

to the equation in the space of symbols. Namely, we seek u in the form

$$u = g(\overset{1}{A}_1,\dots,\overset{n}{A}_n)v,$$

and we obtain the equation for g:

$$f(y_1,\dots,y_n) * g(y_1,\dots,y_n) = 1. \qquad (15)$$

If we solve the equation (15) not exactly, but modulo some "good" class of symbols, we obtain the asymptotics of the initial equation (for example, in classical PDE theory "good" symbols decay at infinity and correspond to smoothing operators).

The structure of operation $f * g$ turns out to be rather simple. Set

$$(L_j g)(y_1,\dots,y_n) = y_j * g,$$
$$(R_j g)(y_1,\dots,y_n) = g * y_j, \quad j = 1,\dots,n. \qquad (16)$$

Then

$$(f * g)(y_1,\dots,y_n) = [f(\overset{1}{L}_1,\dots,\overset{n}{L}_n)g](y_1,\dots,y_n) =$$
$$= [g(\overset{n}{R}_1,\dots,\overset{1}{R}_n)f](y_1,\dots,y_n). \qquad (17)$$

(Note that (17) obviously follows from (16) if we consider only polynomial symbols.)

The n-tuple $L = (L_1,\dots,L_n)$ is called the left regular representation of the tuple $A = (A_1,\dots,A_n)$, and the n-tuple $R = (R_1,\dots,R_n)$ is called the right regular representation of A. We give here a simple example of calculation of regular representation. Let $A_1 = -i(\partial/\partial x)$, $A_2 = x$ be operators acting in the space of functions of one variable x. Using the properties (i) - (iii), we may calculate the left regular representation for the tuple $(\overset{1}{A}_1,\overset{2}{A}_2)$ as follows. Take an arbitrary symbol g; we have

$$A_2[\![g(\overset{1}{A}_1,\overset{2}{A}_2)]\!] = \overset{3}{A}_2 g(\overset{1}{A}_1,\overset{2}{A}_2) = \overset{2}{A}_2 g(\overset{1}{A}_1,\overset{2}{A}_2),$$

so $L_2 = y_2$ (the multiplication operator). Next,

$$\overset{}{A_1}[\![g(\overset{1}{A_1},\overset{2}{A_2})]\!] = \overset{3}{A_1} g(\overset{1}{A_1},\overset{2}{A_2}) = \overset{3}{A_1} g(\overset{1}{A_1},\overset{4}{A_2}) +$$

$$+ \overset{3}{A_1}(\overset{2}{A_2} - \overset{4}{A_2})[g(\overset{1}{A_1},\overset{2}{A_2}) - g(\overset{1}{A_1},\overset{4}{A_2}))/(\overset{2}{A_2} - \overset{4}{A_2})]$$

(note that the symbol $[g(y_1,y_2) - g[(y_1,y_2')/y_2 - y_2']$ has no singularities). We may write then

$$A_1[\![g(\overset{1}{A_1},\overset{2}{A_2})]\!] = \overset{3}{A_1}(\overset{2}{A_2} - \overset{4}{A_2})[(g(\overset{0}{A_1},\overset{1}{A_2}) - g(\overset{0}{A_1},\overset{5}{A_2}))/(\overset{2}{A_2} - \overset{5}{A_2})] +$$

$$+ \overset{3}{A_1} g(\overset{1}{A_1},\overset{4}{A_2}) = \overset{1}{A_1} g(\overset{1}{A_1},\overset{2}{A_2}) + \overset{3}{[\![} \overset{3}{A_1}(\overset{2}{A_2} - \overset{4}{A_2})]\!] \times$$

$$\times [(g(\overset{1}{A_1},\overset{2}{A_2}) - g(\overset{1}{A_1},\overset{4}{A_2}))/(\overset{2}{A_2} - \overset{4}{A_2})] = \overset{1}{A_1} g(\overset{1}{A_1},\overset{2}{A_2}) -$$

$$-i[(g(\overset{1}{A_1},\overset{2}{A_2}) - g(\overset{2}{A_1},\overset{4}{A_2})/\overset{2}{A_2} - \overset{4}{A_2})] = \overset{1}{A_1} g(\overset{1}{A_1},\overset{2}{A_2}) - i(\partial g/\partial y_2)(\overset{1}{A_1},\overset{2}{A_2}).$$

Thus $L_1 = y_1 - i(\partial/\partial y_2)$. Similarly $R_1 = y_1$, $R_2 = y_2 - i(\partial/\partial y_1)$.

The existence of regular representation in this example is connected with the fact that there is a commutation relation

$$[A_1,A_2] = - iI$$

(I denotes the identity operator). The general situation is quite similar. We require that

(a) $A_1,\dots,A_n$ satisfy the "rich enough" system of commutation relations;
(b) some agreement conditions of the functional-analytic nature are then satisfied (see Section 3:B for description of these conditions and also for example of what pathologies may appear when these conditions are not satisfied).

If (a) and (b) are satisfied, the regular representation exists. In this chapter we give the proof of existence of regular representation in the case when the operators $A_1,\dots,A_n$ generate a nilpotent Lie algebra Γ. By the way, this case gives an explanation of the terminology: under the Fourier transform the operators L_j are mapped onto the generators of the regular representation of the Lie group G associated with Γ (recall that this representation acts in the space of functions on G by virtue of left shifts). The case of more complicated commutation relations and the related geometric constructions are considered in Chapter 4.

D. Bibliographical Notes

The main textbooks on functional analysis which we recommend to the reader studying this chapter are [13, 28, 79]. The presentation of the semigroup theory in this chapter is based to a great extent on these sources and also on original papers [77,78]. The material of Section 2:E,H and Section 3:A,B,D develops the ideas of functional calculus constructions stated in [50,52,54]. Example 1 of Section 4:B is taken from the paper [62], and Theorem 2 of Section 4:B is an improved version of the Krein-Shikvatov theorem [47]. The material of Section 4:E is based on the paper [39].

2. FUNCTIONS OF A SINGLE OPERATOR

A. Semigroups and Generators

Let $\mathcal{X}$ be a Banach space.

Definition 1. A semigroup (of bounded linear operators) in $\mathcal{X}$ is a family $\{\mathcal{U}(t)\}$ of operators $\mathcal{U}(t) : \mathcal{X} \to \mathcal{X}$, defined for $t \in [0, +\infty)$, such that (i) for each t, $\mathcal{U}(t)$ is bounded; (ii) $\mathcal{U}(t)\mathcal{U}(\tau) = \mathcal{U}(t + \tau)$ for all t, τ; (iii) $\mathcal{U}(0) = I$ (the identity operator in $\mathcal{X}$); and (iv) $\mathcal{U}(t)$ is strongly continuous in t (this means that for each $x \in \mathcal{X}$, $\mathcal{U}(t)x$ is a continuous $\mathcal{X}$-valued function of the variable $t \in [0, +\infty)$).

Remark. More precisely, $\mathcal{U}(t)$ defined above is a strongly continuous semigroup with the C_0-condition. It was omitted in our definition because semigroups without C_0-condition are of no interest to us. We deal further only with semigroups, which are in fact groups, i.e., which are defined for all $t \in \mathbb{R}$ and satisfy (i) - (iv). Nevertheless, most propositions are formulated in this item for semigroups.

Proposition 1. Let $\{\mathcal{U}(t)\}$ be a semigroup in $\mathcal{X}$. Then

$$\|\mathcal{U}(t)\| \leqslant Me^{\omega t} \tag{1}$$

for some M, ω.

Proof. (a) $\|\mathcal{U}(t)\|$ is bounded for $t \in [0,1]$. Indeed, otherwise we should have $\|\mathcal{U}(t_n)\| \to \infty$ for some sequence $\{t_n\}$, $t_n \in [0,1]$, $\lim_{n\to\infty} t_n = t_0 \in [0,1]$. By the resonance theorem [79], $\|\mathcal{U}(t_n)x\|$ is unbounded for some $x \in \mathcal{X}$, contradicting the strong continuity.

(b) Now $\mathcal{U}(t) = \mathcal{U}(t - [t])\mathcal{U}([t]) = \mathcal{U}(t)(\mathcal{U}(1))^{[t]}$. Therefore $\|\mathcal{U}(t)\| \leqslant$ $\leqslant \|\mathcal{U}(t - [t])\| \, \|\mathcal{U}(1)\|^{[t]} \leqslant \|\mathcal{U}(t - [t])\| \, \|\mathcal{U}(1)\|^{t}$ and it suffices to set $M = \sup_{\tau \in [0,1]} \|\mathcal{U}(\tau)\|$, $\omega = \ln \|\mathcal{U}(1)\|$.

Definition 2. The type of semigroup $\mathcal{U}(t)$ is the greatest lower bound of the set of ω, satisfying (1) for some M. $\mathcal{U}(t)$ satisfies the norm condition (for given ω), if (1) holds with $M = 1$.

Now we define the generator for the given semigroup $\mathcal{U}(t)$. Let $D_A \subset \mathcal{X}$ be the set of all $x \in \mathcal{X}$, such that the strong limit $\lim_{t\to 0}(it)^{-1}(\mathcal{U}(t)x - x)$ exists. Clearly D_A is a linear subset of $\mathcal{X}$. Consider an operator A in $\mathcal{X}$ with the domain D_A, defined by the formula

$$A = \lim_{t\to +0} \frac{1}{it} (\mathcal{U}(t)x - x), \; x \in D_A. \tag{2}$$

Definition 3. A is called the generator of the semigroup or simply the generator.

Theorem 1. (a) The subset D_A is dense in $\mathcal{X}$ and A is a closed operator.

(b) $D_A = \mathcal{X}$ and A is a bounded operator iff the semigroup $\mathcal{U}(t)$ is uniformly* continuous in t.

---

*This means that $\mathcal{U}(t) - \mathcal{U}(\tau)\| \to 0$ as $t \to \tau$; in other words, $\mathcal{U}(t)x \to \mathcal{U}(\tau)x$ uniformly in x, $\|x\| \leqslant 1$. The reader should avoid confusion with uniformity in t.

(c) $\mathcal{U}(t)D_A \subset D_A$ for all t, and A commutes with $\mathcal{U}(t)$: $A\mathcal{U}(t)x = \mathcal{U}(t)Ax$, $x \in D_A$. For $\lambda \in \rho(A)$, $(\lambda - A)^{-1}$ also commutes with $\mathcal{U}(t)$.

(d) For each $x_o \in D_A$ the $\mathcal{X}$-valued function $x(t) = \mathcal{U}(t)x_o$ is differentiable. It may be defined as the unique solution of the Cauchy problem

$$i \frac{\partial x(t)}{\partial t} + Ax(t) = 0, \quad x(0) = x_o. \tag{3}$$

(e) If the norm $\|\mathcal{U}(t)\|$ satisfies (1), each $\lambda \in \mathbb{C}$ such that $-\mathrm{Im}\lambda > \omega$ belongs to the resolvent set $\rho(A)$. The resolvent $R_\lambda(A)$ is given by the formula

$$R_\lambda(A)x \overset{\mathrm{def}}{=} (\lambda I - A)^{-1}x = i\int_o^\infty e^{-i\lambda t}\mathcal{U}(t)x\,dt, \quad x \in \mathcal{X}, \ -\mathrm{Im}\lambda > \omega, \tag{4}$$

and satisfies the following estimates

$$\|(R_\lambda(A))^m\| \leqslant M(-\mathrm{Im}\lambda - \omega)^{-m}, \quad -\mathrm{Im}\lambda > \omega, \ m = 1,2,\ldots; \tag{5}$$

M,ω being the same as in (1).

(f) Conversely, let A be a closed densely defined operator in $\mathcal{X}$, such that the resolvent $R_\lambda(A)$ exists for $-\mathrm{Im}\lambda > \omega$ and satisfies the estimates (5).* Then A is a generator of the semigroup $\mathcal{U}(t)$, satisfying (1). $\mathcal{U}(t)$ is given by the formulae

$$\mathcal{U}(t)x = \lim_{\mu\to\infty} (-i\mu R_{-i\mu}(A))^{[\mu t]}x, \quad x \in \mathcal{X}; \tag{6}$$

$$\mathcal{U}(t)x = \lim_{\mu\to\infty} e^{itA_\mu}x, \quad x \in \mathcal{X}; \tag{7}$$

where

$$A_\mu = -i\mu A R_{-i\mu}(A) \tag{8}$$

is a bounded operator in $\mathcal{X}$. Convergence in (6) and (7) is locally uniform with respect to t.

Proof. Let D_{A^n} denote the domain of the operator A^n (it is defined inductively: $x \in D_{A^n}$ if $x \in D_A$ and $Ax \in D_{A^{n-1}}$). First, we prove that the intersection $D = \bigcap_{n=1}^\infty D_{A^n}$ (and hence D_A) is dense in $\mathcal{X}$. We assert that the element

$$x_\phi = \int_o^\infty \phi(t)\mathcal{U}(t)x\,dt \quad \text{(Bochner integral)}$$

lies in D for arbitrary $x \in \mathcal{X}$, $\phi \in C_o^\infty(\mathbb{R}_+ \setminus \{0\})$. Indeed,

$$\frac{1}{it}(\mathcal{U}(t)x_\phi - x_\phi) = \frac{1}{it}\int_o^\infty \phi(t')[\mathcal{U}(t+t')x - \mathcal{U}(t')x]dt' =$$

$$= \int_o^\infty \frac{\phi(t'-t) - \phi(t')}{it}\,\mathcal{U}(t')x\,dt',$$

so the limit (2) exists and $Ax_\phi = x_{i\phi'}$. By induction, $x_\phi \in D$ and $(A^n)x_\phi = x_{(i^n\phi^{(n)})}$. The set $\{x_\phi\}_{x \in \mathcal{X}, \phi \in C_o^\infty}$ is dense in $\mathcal{X}$ (if the sequence ϕ_n converges to the Dirac δ-function, x_{ϕ_n} converges to x), thus D is also dense. Now let $x_o \in D_A$, $x(t) = \mathcal{U}(t)x_o$. Then

* It is sufficient if this requirement holds for pure imaginary λ, $\mathrm{Im}\lambda > \omega$.

$$\frac{1}{i\tau}(\mathcal{U}(\tau)x(t)-x(t)) = \frac{1}{i\tau}(\mathcal{U}(\tau+t)x_o-\mathcal{U}(t)x_o) =$$

$$= \mathcal{U}(t)\frac{1}{i\tau}(\mathcal{U}(\tau)x_o-x_o).$$

Since $\mathcal{U}(t)$ is bounded, $x(t) \in D_A$ and $Ax(t) = \mathcal{U}(t)Ax_o$, thus (c) is proved. (The commutability of $\mathcal{U}(t)$ and $(\lambda - A)^{-1}$ is an easy consequence of the fact that $A\mathcal{U}(t)x = \mathcal{U}(t)Ax$, $x \in D_A$.) Moreover, we see that $Ax(t) = \lim\limits_{\tau\to+0}\frac{1}{i\tau}\times [x(t+\tau)-x(t)]$ and, since $Ax(t)$ is continuous in t, $x(t)$ is differentiable in t, then (3) holds.

To verify (e) we show at first that the right-hand side of (4) is well defined. Indeed, the estimate

$$\|e^{-i\lambda t}\mathcal{U}(t)x\| \leqslant Me^{(\mathrm{Im}\lambda+\omega)t}$$

implies the convergence of the integral for $-\mathrm{Im}\lambda > \omega$. Further

$$iA\int_o^\infty e^{-i\lambda t}\mathcal{U}(t)x\,dt = \lim_{\tau\to+0}\frac{1}{\tau}(\mathcal{U}(\tau)-I)\int_o^\infty e^{-i\lambda t}\mathcal{U}(t)x\,dt =$$

$$= \lim_{\tau\to+0}\frac{1}{\tau}\int_o^\infty e^{-i\lambda t}[\mathcal{U}(\tau+t)-\mathcal{U}(t)]x\,dt = \lim_{\tau\to 0}\frac{1}{\tau}[\int_\tau^\infty e^{-i\lambda(t-\tau)}\mathcal{U}(t)x\,dt -$$

$$-\int_o^\infty e^{-i\lambda t}\mathcal{U}(t)x\,dt] = \lim_{\tau\to 0}\frac{e^{i\lambda\tau}-1}{i\tau}\,i\int_o^\infty e^{-i\lambda t}\mathcal{U}(t)x\,dt -$$

$$-\lim_{\tau\to 0}\frac{1}{\tau}\int_o^\tau e^{-i\lambda(t-\tau)}\mathcal{U}(t)x\,dt.$$

Both limits exist obviously when $-\mathrm{Im}\lambda > \omega$. Then we obtain

$$A(i\int_o^\infty e^{-i\lambda t}\mathcal{U}(t)x\,dt) = \lambda\cdot i\int_o^\infty e^{-i\lambda t}\mathcal{U}(t)x\,dt - x$$

for an arbitrary $x \in X$. The operator $x \to i\int_o^\infty e^{-i\lambda t}\mathcal{U}(t)x\,dt$ is bounded for $-\mathrm{Im}\lambda > \omega$ (its norm does not exceed therefore $M\int_o^\infty e^{(\omega+\mathrm{Im}\lambda)t}dt$). Hence for $x \in D_A$

$$i\int_o^\infty e^{-i\lambda t}\mathcal{U}(t)Ax\,dt = \lim_{\tau\to 0}\frac{1}{\tau}\int_o^\infty e^{-i\lambda t}[\mathcal{U}(\tau+t)-\mathcal{U}(t)]x\,dt =$$

$$= i\int_o^\infty e^{-i\lambda t}\mathcal{U}(t)\lambda x\,dt - x,$$

just as above. Thus (4) is proved. $R_\lambda(A)$ is bounded and defined everywhere, therefore closed. So $A = \lambda - (R_\lambda(A))^{-1}$ is closed as well and the proof of (a) is complete. To prove (5) we observe that

$$\frac{d^n}{d\lambda^n}R_\lambda(A)x = (-1)^n n!(R_\lambda(A))^{n+1}x, \quad \lambda \in \rho(A),\ x \in X \tag{9}$$

(indeed, the resolvent of a closed operator A is a holomorphic function on $\rho(A)$. It may be represented by the Neumann series

$$R_\lambda(A) = \sum_{n=0}^\infty (\lambda_o-\lambda)^n(R_{\lambda_o}(A))^{n+1}, \quad \lambda_o \in \rho(A), \tag{10}$$

convergent in the circle $|\lambda-\lambda_o| < \|R_{\lambda_o}(A)\|^{-1}$; see [40]). Thus

$$(R_\lambda(A))^m x = i\,\frac{(-1)^{m-1}}{(m-1)!}\,\frac{d^{m-1}}{d\lambda^{m-1}}\int_0^\infty e^{-i\lambda t}\,\mathcal{U}(t)x\,dt = \qquad (11)$$
$$= \frac{i^m}{(m-1)!}\int_0^\infty e^{-i\lambda t}\,t^{m-1}\,\mathcal{U}(t)x\,dt.$$

(The differentiation of the integrand is correct, since convergence of the resulting integral is locally uniform with respect to λ in the domain under consideration.) Since

$$\int_0^\infty e^{(\omega + \mathrm{Im}\lambda)t}\,t^{m-1}dt = \frac{(m-1)!}{(-\mathrm{Im}\lambda - \omega)^m}, \quad \omega + \mathrm{Im}\lambda < 0,$$

we obtain (5), and (e) is proved. Now we move on to the proof of uniqueness in (d). Let $x_o = 0$ and $x(t)$ satisfy (3). We assert that $x(t)$ vanishes identically. To prove it, consider an auxiliary X-valued function

$$y(t,\tau) = \mathcal{U}(t-\tau)x(\tau) \text{ defined for } 0 \leqslant \tau \leqslant t.$$

We have $y(t, 0) = 0$ and $y(t, t) = x(t)$. It turns out that $dy/d\tau$ exists and vanishes identically. This immediately yields $x(t) \equiv y(t,t) = y(t,0) = 0$. Let $\tau \in [0,t]$ be fixed, $\varepsilon \in [-\tau, t-\tau]$ (so that $\tau + \varepsilon \in [0,t]$). By (3) we have $x(\tau) \in D_A$ and

$$x(\tau+\varepsilon) = x(\tau) + i\varepsilon A x(\tau) + \varepsilon r(\varepsilon), \qquad (12)$$

where $\| r(\varepsilon)\| \to 0$ as $\varepsilon \to 0$. Thus we obtain

$$y(t,\tau+\varepsilon) - y(t,\tau) = [\mathcal{U}(t-\tau-\varepsilon) - \mathcal{U}(t-\tau)]x(t) +$$
$$+ \mathcal{U}(t-\tau-\varepsilon)[x(\tau+\varepsilon) - x(\tau)] = -i\varepsilon A\mathcal{U}(t-\tau)x(\tau) +$$
$$+ \varepsilon r_1(\varepsilon) + \mathcal{U}(t-\tau-\varepsilon)[i\varepsilon A x(\tau) + \varepsilon r(\varepsilon)],$$

where $\| r_1(\varepsilon)\| \to 0$ as $\varepsilon \to 0$ (the expansion of the first term analogous to (12) is valid since $x(\tau) \in D_A$, $\mathcal{U}(t-\tau-\varepsilon)x(\tau)$ being therefore differentiable in ε, as proved above). Since (c) has already been proved, we may rewrite the formula obtained:

$$y(t,\tau+\varepsilon) - y(t,\tau) = \varepsilon[i(\mathcal{U}(t-\tau-\varepsilon) - \mathcal{U}(t-\tau))Ax(\tau) +$$
$$+ r_1(\varepsilon) + \mathcal{U}(t-\tau-\varepsilon)r(\varepsilon)] \equiv \varepsilon R(\varepsilon).$$

Since $\mathcal{U}(t)$ is strongly continuous and bounded locally uniformly in t, then $\| R(\varepsilon)\| \to 0$ as $\varepsilon \to 0$. Hence $y(t,\tau)$ is differentiable in τ and $dy/d\tau \equiv 0$, as we desired.

If A is a bounded operator, the semigroup $\mathcal{U}(t)$ has the form $\mathcal{U}(t) = \exp(itA)$, where $\exp(B)$ is defined as the sum of the convergent series:

$$\exp(B) \equiv e^B = \sum_{k=0}^{\infty} \frac{1}{k!} B^k. \qquad (13)$$

Hence we obtain, using the inequality $|t^k - \tau^k| \leqslant |t-\tau|\cdot(|t| + |\tau|)^{k-1}$:

$$\| \exp(itA) - \exp(i\tau A)\| \leqslant \Big(\sum_{k=1}^{\infty} \frac{1}{k!}(|t| + |\tau|)^{k+1}\| A\|^k\Big)\cdot|t-\tau|.$$

Since the sum in the parentheses is finite, we conclude that $\mathcal{U}(t)$ is uniformly continuous. Vice versa, let $\mathcal{U}(t)$ be uniformly continuous. We assert that

$$\lim_{\mu\to\infty} \| -i\mu R_{-i\mu}(A) - I\| = 0. \qquad (14)$$

If it were already proved, we may conclude that the operator $-i\mu R_{-i\mu}(A)$ has a bounded inverse for sufficiently large μ and therefore $A = -i\mu - (R_{-i\mu}(A))^{-1}$ is bounded. The proof of (14) is as follows. By (4)

$$-i\mu R_{-i\mu}(A)x = \int_0^\infty \mu e^{-\mu t}\mathcal{U}(t)x dt.$$

Choose an arbitrary $\varepsilon > 0$. Since $\mathcal{U}(t)$ is uniformly continuous, there exists $\delta > 0$ such that $\|\mathcal{U}(t) - I\| \leqslant \varepsilon/2$, when $0 \leqslant t \leqslant \delta$. Since

$$\int_0^\infty \mu e^{-\mu t} dt = 1,$$

$$-i\mu R_{-i\mu}(A)x - x = \int_0^\infty \mu e^{-\mu t}(\mathcal{U}(t) - I)x dt =$$

$$= \int_0^\delta \mu e^{-\mu t}(\mathcal{U}(t) - I)x dt + \int_\delta^\infty \mu e^{-\mu t}(\mathcal{U}(t) - I)x dt.$$

The estimate of the first integral is:

$$\|\int_0^\delta \mu e^{-\mu t}(\mathcal{U}(t) - I)x dt\| \leqslant \frac{\varepsilon}{2}\|x\| \int_0^\delta \mu e^{-\mu t} dt \leqslant \frac{\varepsilon}{2}\|x\| \int \mu e^{-\mu t} dt = \frac{\varepsilon}{2}\|x\| .$$

Now we estimate the second integral:

$$\|\int_\delta^\infty \mu e^{-\mu t}(\mathcal{U}(t) - I)x dt\| \leqslant \mu e^{-\mu\delta}\|x\| \int_0^\infty e^{-\mu t}(1 + Me^{\omega(t+\delta)})dt \leqslant \frac{\varepsilon}{2}\|x\|$$

for sufficiently large μ. Thus (14) is proved. Now only item (f) of the theorem remains unproved. Let the assumptions of this item be satisfied. Set $T_\mu = -i\mu R_{-i\mu}(A)$. We notice at first that $\lim_{\mu\to\infty} T_\mu x = x$ for every $x \in X$. Indeed, if $x \in D_A$,

$$T_\mu x - x = \frac{i}{\mu} T_\mu A x \to 0$$

as $\mu \to \infty$, since (5) yields that $\|T_\mu\|$ remains bounded while $\mu \to \infty$. Now D_A is dense in X, and it follows from above that $T_\mu x \to x$ everywhere in X. Now set

$$\mathcal{U}_\mu(t) = e^{itA_\mu} = e^{itAT_\mu}.$$

(Note that $T_\mu X \subseteq D_A$, so that AT_μ is everywhere defined. Further $AT_\mu = i\mu(I - T_\mu)$ is bounded.) We estimate now the norm of $\mathcal{U}_\mu(t)$:

$$\|\mathcal{U}_\mu(t)\| = \|e^{-t\mu(I-T_\mu)}\| = e^{-t\mu}\|e^{t\mu T_\mu}\| \leqslant e^{-t\mu} \sum_{k=0}^\infty \frac{1}{k!}(t\mu)^k\|T_\mu^k\| .$$

By (5) for large μ

$$\|T_\mu^k\| = \mu^k\|(R_{-i\mu}(A))^k\| \leqslant M\cdot(\frac{\mu}{\mu-\omega})^k, \tag{15}$$

hence

$$\|\mathcal{U}_\mu(t)\| \leqslant Me^{-t\mu}\sum_{k=0}^\infty \frac{1}{k!}(\frac{t\mu^2}{\mu-\omega})^k = Me^{t(\frac{\mu^2}{\mu-\omega} - \mu)} = Me^{t(\frac{\mu\omega}{\mu-\omega})};$$

in particular, $\mathcal{U}_\mu(t)$ is bounded uniformly as $\mu \to \infty$ and t lies in a fixed compact set $K \subset [0,\infty)$. Let $x \in D_A$. We have

$$\|\mathcal{U}_\mu(t)x - \mathcal{U}_\nu(t)x\| = \|\int_0^t \frac{d}{ds}(\mathcal{U}_\nu(t-s)\mathcal{U}_\mu(s)x)ds\| =$$

$$= \|\int_0^t \mathcal{U}_\nu(t-s)\mathcal{U}_\mu(s)(AT_\mu - AT_\nu)x ds\| =$$

$$= \|\int_0^t \mathcal{U}_\nu(t-s)\mathcal{U}_\mu(s)(T_\mu - T_\nu)Ax ds\| \leqslant C\|T_\mu y - T_\nu y\| ,$$

where $y = Ax$; the constant C depends on t but does not depend on μ and ν for large μ,ν. Since $T_\mu y \overset{\mu\to\infty}{\to} y$, the right-hand side of the above inequality tends to zero as $\mu,\nu \to \infty$ and we obtain that there exists the limit

$$\mathcal{U}(t)x \overset{\text{def}}{=} \lim_{\mu\to\infty} \mathcal{U}_\mu(t)x, \tag{16}$$

the convergence in (16) being locally uniform with respect to the variable t. The uniform boundedness of the family $\{\mathcal{U}_\mu(t)\}$ yields that the limit (16) exists for all $x \in X$, $\mathcal{U}(t)$ is strongly continuous,

$$\|\mathcal{U}(t)\| \leqslant \underline{\lim}_{\mu\to\infty} \|\mathcal{U}_\mu(t)\| \leqslant \lim_{\mu\to\infty} Me^{(t\mu\omega/\mu-\omega)} = Me^{\omega t}.$$

Next we have to show that $\mathcal{U}(t)$ is a semigroup with the generator A. We have, for an arbitrary $x \in X$,

$$\|\mathcal{U}(t+\tau)x - \mathcal{U}(t)\mathcal{U}(\tau)x\| \leqslant \|\mathcal{U}(t+\tau)x - \mathcal{U}_\mu(t+\tau)x\| +$$

$$+ \|\mathcal{U}(t)\mathcal{U}(\tau)x - \mathcal{U}_\mu(t)\mathcal{U}(\tau)x\| + \|\mathcal{U}_\mu(t)\mathcal{U}(\tau)x - \mathcal{U}_\mu(t)\mathcal{U}_\mu(\tau)x\|,$$

since the semigroup property is valid for $\mathcal{U}_\mu(t)$. The left-hand side of this inequality does not depend on μ, while its right-hand side tends to zero as $\mu \to \infty$. It follows that $\mathcal{U}(t+\tau)x = \mathcal{U}(t)\mathcal{U}(\tau)x$, i.e., $\mathcal{U}(t)$ is a semigroup. Let $x \in D_A$, then

$$\frac{d}{dt}\mathcal{U}_\mu(t)x = iA_\mu\mathcal{U}_\mu(t)x = i\mathcal{U}_\mu(t)T_\mu Ax$$

converges to $i\mathcal{U}(t)Ax$ locally uniformly with respect to t. Thus we may then conclude that

$$\frac{d}{dt}\mathcal{U}(t)x = i\mathcal{U}(t)Ax \text{ and, particularly, } \frac{d}{dt}\mathcal{U}(t)x|_{t=0} = Ax,\ x \in D_A.$$

Hence the generator of $\mathcal{U}(t)$, say $\tilde{A}$, is a closed extension of A. For μ large enough $\tilde{A} + i\mu I$ is the one-to-one mapping of $D_{\tilde{A}}$ onto X. By assumption A has the same property. Hence $D_{\tilde{A}} = D_A$ and $\tilde{A} = A$. Now we are in position to prove (6). For the sake of simplicity we rewrite (6) in the form

$$\mathcal{U}(t)x = \lim_{\mu\to\infty}(T_\mu)^{[\mu t]}x = \lim_{\mu\to\infty}(I - \frac{i}{\mu}A)^{-[\mu t]}x. \tag{17}$$

We recall that for real p, [p] denotes the integer part of p, i.e., the greatest of the integer not exceeding p. By (15),

$$\|(T_\mu)^{[\mu t]}\| \leqslant M(\frac{\mu}{\mu-\omega})^{[\mu t]} \leqslant M(\frac{\mu}{\mu-\omega})^{\mu t} = M(1 - \frac{\omega}{\mu})^{\mu t} \to Me^{\omega t}, \tag{18}$$

as $\mu \to \infty$; in particular $\|(T_\mu)^{[\mu t]}\|$ is uniformly bounded ($\mu \to \infty$, t belongs to a fixed compact subset of $[0,\infty)$). So it is sufficient to prove (17) for x belonging to some dense subset of X, say to D_{A^2}. Let $x \in D_{A^2}$. We have then

$$\|\mathcal{U}(t)x - (I - \frac{i}{\mu}A)^{-[\mu t]}x\| \leqslant \|\mathcal{U}(t)x - \mathcal{U}(t')x\| +$$

$$+ \|\mathcal{U}(t')x - (I - \frac{i}{\mu}A)^{-\mu t'}x\| = \|\mathcal{U}(t)x - \mathcal{U}(t')x\| + \tag{19}$$

$$+ \|\mathcal{U}(t')x - (I - \frac{it'}{n}A)^{-n}x\|,$$

where $t' = [\mu t]/\mu$, $n = \mu t' = [\mu t]$. Now we estimate both terms on the right-hand side of (19). We have

$$\| U(t)x - U(t')x \| = \| \int_{t'}^{t} \frac{d}{d\tau} U(\tau)x d\tau \| \leq \int_{t'}^{t} \| U(\tau)Ax \| \, d\tau \leq$$

$$\leq M\| Ax\| \int_{t'}^{t} e^{\omega\tau} d\tau \leq M\| Ax\| \frac{e^{\omega t}}{\mu}, \text{ since } t' \leq t, \; t - t' \leq 1/\mu.$$

Further,

$$\| U(t')x - (I - \frac{it'}{n} A)^{-n} x\| \leq \int_{o}^{t'} \| \frac{d}{ds} [(I - \frac{i(t'-s)}{n} A)^{-n} U(s)x] \| ds.$$

We obtain from (3) and (9) that

$$\frac{d}{ds} [(I - \frac{i(t'-s)}{n} A)^{-n} U(s)x] = i(I - i \frac{(t'-s)}{n} A)^{-n} \times$$

$$\times U(s)(Ax - (I - \frac{i(t'-s)}{n} A)^{-1} Ax) =$$

$$= (I - i \frac{(t'-s)}{n} A)^{-n} U(s) \frac{t'-s}{n} (I - i \frac{(t'-s)}{n} A)^{t-1} A^2 x =$$

$$= \frac{t'-s}{n} (I - i \frac{(t'-s)}{n} A)^{-n} U(s) A^2 x, \quad x \in D_{A^2}.$$

(The accurate proof uses expansions of the type (12), and we leave it to the reader.) It follows immediately that

$$\| U(t')x - (I - \frac{it'}{n} A)^{-n} x\| \leq C \cdot \frac{t'^2}{n} = C_1/\mu,$$

where C_1 is bounded as $\mu \to \infty$ and t belongs to a compact set. Thus both terms in (19) tend to zero locally uniformly with respect to t as $\mu \to \infty$. So we obtain (6), and the proof of the theorem is complete.

Notation. From now on we denote the semigroup $U(t)$ with the generator A by $\exp(iAt)$ or e^{iAt} whether A is bounded or not. If the reader is attentive, this will not lead to any confusion.

We formulated Theorem 1 for semigroups. The simple reformulation of some of its items gives us the following:

Theorem 2. Let A be a closed densely defined operator in a Banach space X. Then A is the generator of a strongly continuous group $U(t)$ iff there exist M and ω, such that $\lambda \in \rho(A)$ for $|\mathrm{Im}\lambda| > \omega$ and the following estimates are valid:

$$\| (\lambda I - A)^{-m} \| \leq M(|\mathrm{Im}\lambda| - \omega)^{-m}, \quad |\mathrm{Im}\lambda| > \omega, \; m = 1,2,\dots \; . \tag{20}$$

These estimates yield $\| e^{iAt} \| \leq Me^{\omega|t|}$, and vice versa.

Proof. If $U(t)$ is a strongly continuous group in X, then the formula $U_{\pm}(t) \overset{\text{def}}{=} U(\pm t)$, $t \geq 0$ defines a pair of strongly continuous semigroups. The generators of $U_+(t)$ and $U_-(t)$ are A and -A, respectively; hence we obtain (20). Conversely, let A satisfy (20). It follows that both A and -A are the generators, and we denote by $U_+(t)$ and $U_-(t)$ the corresponding semigroups. The identity $U_+(t)U_-(t) = I$ holds; the proof of it is quite similar to that of uniqueness in the item (d) of Theorem 1. Thus we can define e^{iAt} by the formula

$$e^{iAt} = \begin{cases} U_+(t), & t \geq 0 \\ U_-(-t), & t \leq 0, \end{cases} \tag{21}$$

and all the desired properties are clearly satisfied. Theorem 2 is therefore proved.

Given a semigroup $\mathcal{U}(t)$ it is often difficult or undesirable to describe precisely the domain D_A of its generator A. The following theorem gives us a convenient tool to avoid such difficulties (the effectiveness of it will be seen in sections devoted to functions of several operators).

Definition 4. Let A be a closed linear operator in a Banach space X (or, more generally, an operator between the Banach spaces X and Y) with the dense domain $D_A \subseteq X$. The linear subset $D \subset D_A$ is called a core of A if A coincides with the closure of its restriction on D.

Theorem 3. Let $\mathcal{U}(t)$ be a semigroup in the Banach space X with the generator A. Suppose that $D \subset D_A$ is a dense linear subset of X, such that $\mathcal{U}(t)D \subset D$ for all $t \in [0,\infty)$. Then D is a core of A.

Proof. We establish first that $(A + i\mu)D$ is dense in X for sufficiently large μ, namely for $\mu > \omega$, where ω is the type of the semigroup $\mathcal{U}(t)$. Since D is dense, it suffices to prove that for an arbitrary $x \in D$ there exists a sequence $x_n \in D$, $n = 1,2,\ldots$, such that $\lim_{n\to\infty} (Ax_n + i\mu x_n) = x$. Let $\mu > \omega$ be fixed. Then the integral

$$x = (-A - i\mu)R_{-i\mu}(A)x = -i\int_0^\infty e^{-\mu t}(A + i\mu)\mathcal{U}(t)x\,dt \tag{22}$$

converges and has the continuous integrand. It follows that there exists a collection of numbers $c_{ns} \in \mathbb{C}$, $t_{ns} \in [0,\infty)$, $n = 1,2,\ldots$, $s = 1,\ldots,S_n$ such that

$$x = \lim_{n\to\infty} \left(\sum_{s=1}^{S_n} c_{ns}(A + i\mu)\mathcal{U}(t_{ns})x \right) = \lim_{n\to\infty} (A + i\mu)x_n, \tag{23}$$

where $x_n = \sum_{s=1}^{S_n} c_{ns}\mathcal{U}(t_{ns})x \in D$ by our assumptions. Thus $(A + i\mu)D \equiv \tilde{D}$ is dense. $R_{-i\mu}(A)$ being bounded, $\tilde{D}$ is a core of $R_{-i\mu}(A)$. Now the desired result is an immediate consequence of the following:

Lemma 1. If B is a closed operator in the Banach space X and B^{-1} exists, $D \subset X$ is a core of B iff BD is a core of B^{-1}. Indeed, the closure of the operator is equivalent to the closure of its graph in $X \times X$, and the graph of B is transformed into the graph of B^{-1} after the interchange of the factors. We obtain that D is a core of $A + i\mu$, hence of A. Thus the theorem is proved. In the sequel we sometimes make use of the following evident result:

Lemma 2. If B is a closed operator in the Banach space X and a dense linear subset $D \subset D_B$ is invariant under $R_\lambda(B)$ for some $\lambda \in \rho(B)$, then D is a core of B. The proof is obvious.

Next we formulate a rather useful criterion, which enables us to conclude that the closure of a given operator is the generator of a strongly continuous semigroup.

Theorem 4. Let D be a dense linear subset of the Banach space X, $A : D \to D$ be a given linear operator. Suppose that for each $x_0 \in D$ there exists a unique function $x : [0,\infty) \to D$, $t \to x(t)$ such that*

$$x(0) = x_0, \quad i\frac{\partial x(t)}{\partial t} + Ax(t) = 0, \quad t \geqslant 0, \tag{24}$$

* Derivation in (24) is in the sense of topology of X.

$$\|x(t)\| \leqslant C(t)\|x_o\|, \tag{25}$$

where $C(t)$ does not depend on x_o and remains bounded when t belongs to a compact set. Then A is closurable, and its closure is the generator of a strongly continuous semigroup $\mathcal{U}(t)$, and besides the solution of (24) has the form $x(t) = \mathcal{U}(t)x_o$.

Proof. Let $\mathcal{U}(t)$ be the unique bounded operator, coinciding with the operator $x_o \to x(t)$ on D. By (25) $\|\mathcal{U}(t)\| \leqslant C(t)$, and since $\mathcal{U}(t)$ is strongly continuous on D and bounded locally uniformly, $\mathcal{U}(t)$ is strongly continuous on $\mathcal{X}$. Since the solution of (24) is unique, we easily obtain that $\mathcal{U}(t)\mathcal{U}(\tau) = \mathcal{U}(t + \tau)$. Thus $\mathcal{U}(t)$ is a strongly continuous semigroup and A is the restriction of its generator on the subset D. By Theorem 3, the closure of A is the generator itself, and the proof is complete.

Our last theorem is a collection of almost obvious properties of the powers of a generator.

Theorem 5. Let A be a generator of the semigroup $\mathcal{U}(t)$ in the Banach space $\mathcal{X}$. The following statements are valid:

(a) $\mathcal{U}(t)x$ is ℓ-times differentiable iff $x \in D_{A^\ell}$. In this case the ℓ-th derivative is continuous and may be given by the formula

$$\frac{d^\ell}{dt^\ell}(\mathcal{U}(t)x) = i^\ell \mathcal{U}(t)A^\ell x, \; x \in D_{A^\ell}. \tag{26}$$

(b) Let $p(y) = \sum_{j=0}^{\ell} p_j y^\ell$, $p_\ell \neq 0$, be a polynomial of degree ℓ. Then the operator

$$p(A) = \sum_{j=0}^{\ell} p_j A^j, \; D_{p(A)} = D_{A^\ell} \tag{27}$$

is closed.

(c) Let $D \subset D_{A^s}$ be a subset of $\mathcal{X}$, satisfying the conditions of Theorem 3. Then D is a core of $p(A)$ for each polynomial p of degree $\leqslant s$.

Proof. (a) For $k = 1$ this statement follows from Theorem 1, (d) and the definition of the generator. The induction on k gives the general result.

(b) This is a general property of closed operators with the non-empty resolvent set. Considering $A + \lambda$ instead of A, if necessary, we may assume that $0 \in \rho(A)$, i.e., A^{-1} exists and is bounded. We proceed by induction on ℓ (for $\ell = 1$ the proposition is valid, see Theorem 1, (a)). Let $x_n \in D_{A^\ell}$, $n = 1,2,\ldots$, $x_n \underset{n\to\infty}{\to} x$, $p(A)x_n \underset{n\to\infty}{\to} z$. We show that $x \in D_{A^\ell}$. Indeed, set

$$p_1(y) = \sum_{j=1}^{\ell} p_j y^j = p(y) - p_o, \; q(y) = p_1(y)/y = \sum_{j=1}^{\ell} p_j y^{j-1}. \tag{28}$$

Then $p_1(A)x_n \underset{n\to\infty}{\to} z - p_o x \equiv z_1$ and $A^{-1}p_1(A)x_n = q(A)x_n \underset{n\to\infty}{\to} A^{-1}z_1$. Next $q(A)$ is closed by the induction assumption, so $x \in D_{q(A)} = D_{A^{\ell-1}}$ and $q(A)x = A^{-1}z_1 \in D_A$. It follows that $x \in D_{A^\ell}$ and $p(A)x = Aq(A)x + p_o x = z_1 + p_o x = z$.

(c) Let $\mathrm{Im}\lambda < -\omega$, where ω is the type of the semigroup $\mathcal{U}(t)$. Since every polynomial $p(y)$ of degree $\leqslant s$ may be represented in the form $p(y) = \sum_{j=0}^{s} c_j (y - \lambda)^j$, it suffices to prove that D is a core of $(A - \lambda)^j$ for

$j \le s$. Just as in the proof of Theorem 3, we establish that $(A - \lambda)^j D$ is dense in X. Namely, we have instead of (22)

$$x = (\lambda - A)^j (R_\lambda(A))^j x = \frac{i^j}{(j-1)!} \int_0^\infty e^{-i\lambda t} t^{j-1} U(t)(\lambda - A)^j x dt \tag{29}$$

for $x \in D$. Further, the proof is identical to that of Theorem 3.

B. Examples of Semigroups

The abstract notion of semigroups, introduced above, is supported with some examples in this item. In particular, we give the proof of Stone's theorem, having in mind that functions of self-adjoint operators are of special interest to us. The theorem mentioned asserts that the notion of a self-adjoint operator is identical with that of unitary strongly continuous group generator. We begin with the following:

Example 1. Let $X = L^2(\mathbb{R}^1)$ be a space of measurable square-summable complex-valued functions on the real line with the usual norm. Define the group $U(t)$ in X by

$$U(t)f(x) = f(x+t), \quad t \in \mathbb{R}, \ f \in L^2(\mathbb{R}^1). \tag{1}$$

The operators $U(t)$ are unitary in X and clearly satisfy the semigroup property. $U(t)$ is strongly continuous in t on $C_0^\infty(\mathbb{R}^1)$ and therefore on X, due to the uniform boundedness. The generator of the semigroup (1) is the operator $A = -i(\partial/\partial x)$ with the domain D_A, consisting of all absolutely continuous functions $f \in L^2(\mathbb{R}^1)$ such that $\partial f/\partial x \in L^2(\mathbb{R}^1)$. The operator S so defined is self-adjoint. Indeed, $C_0^\infty(\mathbb{R}^1)$ is invariant under the action of the semigroup (1), and

$$\left(-i \frac{d}{dt} [U(t)f(x)]\right)\Big|_{t=0} = -i \frac{\partial f(x)}{\partial x}, \quad f \in C_0^\infty(\mathbb{R}^1), \tag{2}$$

so that the restriction of A on $C_0^\infty(\mathbb{R}^1)$ is $-i(\partial/\partial x)$. It follows from Theorem 3 of the preceding item, that A is the closure in $L^2(\mathbb{R}^1)$ of $-i(\partial/\partial x)$, defined on $C_0^\infty(\mathbb{R}^1)$. As for the proof of the given description of the domain D_A of this closure, we refer the reader to standard textbooks on functional analysis. The fact that A is self-adjoint is a consequence of Stone's theorem to be proved later in this item.

Example 2. (Dissipative operators) We suppose that X is a Hilbert space.

Definition 1. A contraction semigroup is a strongly continuous semigroup satisfying the condition $\|U(t)\| \leqslant 1$ for all $t > 0$. A dissipative operator in X is a closed densely defined operator A such that $\mathrm{Re}(Ax,x) \leqslant 0$ for arbitrary $x \in D_A$.

Theorem 1. A is a generator of the contraction semigroup iff iA is a dissipative operator and $R_{(I-iA)} = X$ (here R_B denotes the range of B, i.e., the set of all $y \in X$ such that $y = Bx$ for some $x \in D_B$).

Proof. Let A be a generator of the contraction semigroup $U(t)$. The theorem yields that A is closed and the point $\lambda = -i$ belongs to $\rho(A)$, thus $R_{(I-iA)} = X$. If $x \in D_A$, we have

$$\mathrm{Re}(iAx,x) = \lim_{t\to+0} \frac{1}{t} \mathrm{Re}(U(t)x - x, x) = \lim_{t\to+0} \frac{1}{t} \{\mathrm{Re}(U(t)x,x) - (x,x)\} \leqslant$$

$$\leqslant 0, \text{ since } \mathrm{Re}(U(t)x,x) \leqslant |(U(t)x,x)| \leqslant \|x\|^2 = (x,x).$$

Conversely, let iA be a dissipative operator with $R_{(I-iA)} = X$. If Reλ > > 0, we have

$$\mathrm{Re}\lambda\cdot\|x\|^2 \leqslant \mathrm{Re}[\lambda\|x\|^2 - (iAx,x)] \leqslant \|\lambda x - iAx\|\cdot\|x\|, \quad x\in D_A$$

or

$$\|x\| \leqslant \|(\lambda I - iA)x\|\cdot\frac{1}{\mathrm{Re}\lambda}, \quad x\in D_A, \tag{3}$$

so that if for some λ_0 with Reλ_0 > 0, $R_{(\lambda_0 I-iA)} = X$, then $\lambda_0 \in \rho(iA)$ and $\|R_{\lambda_0}(iA)\| \leqslant 1/\mathrm{Re}\lambda_0$. Moreover, this estimate for the norm yields (taking (10) of item A into consideration) that $\rho(A)$ contains the open disk with the radius Reλ_0 and the center λ_0. We start with $\lambda_0 = 1$ and consequently apply the above arguments to points lying in disks constructed before. One can easily see that we are able to construct the set of such disks covering the right half-plane of the complex variable λ. Thus the right half-plane is contained in $\rho(iA)$ and $\|R_\lambda(iA)\| \leqslant (\mathrm{Re}\lambda)^{-1}$ for Reλ > 0. Applying Theorem 1 of item A, we obtain the desired result. The theorem is proved. The analogous theorem for Banach spaces may be found in more special literature (see bibliographic remarks in Section 1).

Example 3. (Self-adjoint operators) We assume again that X is a Hilbert space.

Theorem 2. (Stone) Let $U(t)$ be a strongly continuous group of unitary operators in X. Then its generator A is a self-adjoint operator in X. Conversely, each self-adjoint operator in X is a generator of a strongly continuous group of unitary operators in X.

Proof. We prove at first that A is symmetric. Let $x,y\in D_A$. Then we have

$$(U(t)x,y) = (x,U^*(t)y) = (x,U(-t)y),$$

hence

$$(Ax,y) = (-i\frac{d}{dt}U(t)x,y)|_{t=0} = -i\frac{d}{dt}(U(t)x,y)|_{t=0} =$$

$$= -i\frac{d}{dt}(x,U(-t)y)|_{t=0} = (x,i\frac{d}{dt}U(-t)y)|_{t=0} = (x,Ay),$$

as it was claimed. Further, Theorem 1 of item A implies that A is closed and all non-real λ belong to $\rho(A)$. Thus A has defect indices (0,0) and is therefore self-adjoint [40]. Conversely, if A is self-adjoint, all non-real λ belong to $\rho(A)$ and

$$|((A-\lambda)x,x)|^2 = |(A-\mathrm{Re}\lambda)x,x)|^2 + (\mathrm{Im}\lambda)^2\|x\|^4 \geqslant (\mathrm{Im}\lambda)^2\|x\|^4,$$

$$x\in D_A.$$

It follows that $\|Ax-\lambda x\| \geqslant \mathrm{Im}\lambda\|x\|$ or $\|R_\lambda(A)\| \leqslant (\mathrm{Im}\lambda)^{-1}$.

Applying Theorem 1 of item A, we obtain that A is a generator of the group $U(t)$, satisfying $\|U(t)\| \leqslant 1$. Since $U(t)U(-t) \equiv 1$, $U(t)$ appear to be unitary operators, as desired. Theorem 2 is proved.

Note. One may define the self-adjoint operator in the Banach space x as the generator of a strongly continuous group satisfying $\|U(t)\| \equiv 1$. Stone's theorem shows that in Hilbert spaces this definition coincides with the usual one.

C. Convergence of Semigroups

In this item we establish some general results concerning the relation between convergence of semigroups and convergence of their generators. These results play an important role in our development of operator calculus, especially of functions of several operators. We apply them, in particular, to the investigation of continuity properties of dependence of functions of generators on generators themselves. We have already proven one of such results in item A. Namely, it was proved that, A being the generator, $\exp\{itA\} = s-\lim_{\mu\to\infty} \exp\{itA_\mu\}$, $t \geq 0$, where $A_\mu = -i\mu A R_{-i\mu}(A)$ is a family of bounded operators. Much more complicated situations are discussed here. We consider general family of generators, continuous in some special sense and prove the continuity of the corresponding family of semigroups. It appears later that the type of continuity introduced is quite natural in problems connected with functions of non-commutating operators calculus. Here and further, we formulate our statements for the case of sequences of semigroups (resp. generators). One must have in mind that all these statements are valid (proofs being just the same) for families depending on parameter, lying in a topological space (instead of sequences). We avoid such considerations, our aim being to clarify the exposition. $\mathcal{X}$ denotes a given Banach space throughout the item, all the operators in question acting in $\mathcal{X}$.

Now we come on to precise definitions and statements. Following [40], we introduce a convenient notation:

Notation. Let A be a closed operator in a Banach space $\mathcal{X}$. We write $A \in \mathcal{G}(M,\omega)$, if A is a generator of a strongly continuous semigroup which satisfies the estimate

$$\| e^{iAt} \| \leqslant Me^{\omega t} \text{ for all } t \geqslant 0. \tag{1}$$

We shall also make use of the following:

Definition 1. Let $\mathcal{D}$ be a subset of $\mathbb{R}^n(\mathbb{C}^n)$. Given a sequence $\{B_n(\alpha)\}_{\alpha\in\mathcal{D}}$, $n = 1, 2, \ldots$ of families of bounded linear operators in $\mathcal{X}$, we say that it strongly converges to the family $\{B(\alpha)\}_{\alpha\in\mathcal{D}}$, locally uniformly with respect to α, if for each $x \in \mathcal{X}$, $B_n(\alpha)x \to B(\alpha)x$ and this convergence is uniform when α lies in an arbitrary fixed compact subset of $\mathcal{D}$.

$$\text{Notation: } B_n(\alpha) \overset{s}{\rightrightarrows} B(\alpha), \quad \alpha \in \mathcal{D}.$$

Further, the families $B_n(\alpha)$ will be either semigroups or resolvent families. To begin with, we establish simple sufficient and necessary conditions for locally uniform strong convergence of semigroups. Next we study them in detail and derive sufficient conditions which are much easier to check (we make no investigation of their necessity). At the end of the item these results are applied to strongly continuous groups, in particular to groups of unitary operators in a Hilbert space.

Theorem 1. Let $A_n \in \mathcal{G}(M,\omega)$, $n = 1,2,\ldots$. In order that

$$\exp(itA_n) \overset{s}{\rightrightarrows} \mathcal{U}(t), \quad t \geqslant 0 \tag{2}$$

as $n \to \infty$ for some family $\mathcal{U}(t)$ of operators in $\mathcal{X}$, it is necessary that

$$R_\lambda(A_n) \overset{s}{\rightrightarrows} R_\lambda(A), \quad \operatorname{Im}\lambda < -\omega \tag{3}$$

for some densely defined operator A in $\mathcal{X}$, and sufficient that*

$$R_{\lambda_o}(A_n) \overset{s}{\to} R_{\lambda_o}(A) \tag{4}$$

for some λ_o, $\operatorname{Im}\lambda_o < -\omega$. In this case $A \in G(M,\omega)$ (in particular, A is then closed) and $\mathcal{U}(t) = \exp\{itA\}$.

Proof. We may assume that $\omega = 0$, considering $A_n + i\omega$ instead of A.

(a) Necessity. If (2) holds, $\mathcal{U}(t)$ is a strongly continuous semigroup, satisfying $\|\mathcal{U}(t)\| \leqslant M$. Indeed, continuity and the estimate for the norm immediately follow from (2); the semigroup property is proved in complete analogy with that in Theorem 1, (f) of item A. Thus $\mathcal{U}(t) = \exp\{itA\}$, $A \in G(M,0)$. To prove (3), we note that for arbitrary $x \in \mathcal{X}$,

$$\begin{aligned} R_\lambda(A_n)x - R_\lambda(A)x &= i\int_0^\infty e^{-i\lambda t}[e^{itA_n}x - \mathcal{U}(t)x]dt = \\ &= i(\int_0^C + \int_C^\infty)e^{-i\lambda t}[e^{itA_n}x - \mathcal{U}(t)x]dt \equiv I_1 + I_2, \quad \operatorname{Im}\lambda < 0,\ C > 0 \end{aligned} \tag{5}$$

(see (4), item A). If λ lies in a compact subset of the lower half-plane, $\operatorname{Im}\lambda \leqslant -\delta < 0$. It follows that, given $\varepsilon > 0$, we can find such C that $\|I_2\| < \varepsilon/2$ for all n (the norm of the integrand does not exceed $2M\|x\|\exp(-\delta t)$). $\exp(itA_n)x$ converges to $\mathcal{U}(t)x$ uniformly for $t \in [0,C]$, so $\|I_1\| < \varepsilon/2$ for n large enough, and (3) is therefore proved.

(b) Sufficiency. Let (4) be valid. Then we claim that (3) is valid also. To prove this, we need the following two lemmas which will be used later as well:

Lemma 1. Let A_n, $n = 1,2,\ldots$ be a sequence of closed operators in $\mathcal{X}$. Define $\Delta_b \in \mathbb{C}$ as a set of all $\lambda \in \mathbb{C}$ such that $\lambda \in \rho(A_n)$ for $n \geqslant n_o = n_o(\lambda)$ and $\|R_\lambda(A_n)\| \leqslant M(\lambda) < \infty$, $n \geqslant n_o(\lambda)$. Define also $\Delta_s \subset \mathbb{C}$ as a set of all $\lambda \in \mathbb{C}$ such that $\lambda \in \rho(A_n)$ for $n \geqslant n_o(\lambda)$ and $R_\lambda = s - \lim_{n\to\infty} R_\lambda(A_n)$ exists. Then:

(a) Δ_b is an open subset of $\mathbb{C}$;
(b) $\Delta_s \subset \Delta_b$ and is relatively opened and closed in Δ_b (i.e., Δ_s is a union of connected components of Δ_b);
(c) $R_\lambda(A_n) \overset{s}{\to} R_\lambda$, $\lambda \in \Delta_s$;
(d) R_λ satisfies the resolvent equation;

$$R_\lambda - R_\mu = (\lambda - \mu)R_\lambda R_\mu, \quad \lambda,\mu \in \Delta_s. \tag{6}$$

Lemma 2. Let R_λ, $\lambda \in \Delta \subset \mathbb{C}$ be a family of bounded operators in $\mathcal{X}$, satisfying (6). Then $N = \operatorname{Ker} R_\lambda$ and $R = \operatorname{Im} R_\lambda$ do not depend on λ. R_λ coincides with the resolvent family of some closed operator A iff $N = \{0\}$. In this case $D_A = R$ and

$$A = \lambda I - (R_\lambda)^{-1} \text{ for each } \lambda \in \Delta. \tag{7}$$

Under the conditions of Lemma 1, if $R_\lambda = R_\lambda(A)$, we have $\Delta_s = \Delta_b \cap \rho(A)$.

Proof of Lemma 1. (a) Let $\lambda_o \in \Delta_b$. Using the Neumann series ((10), item A), we easily obtain that $\|R_\lambda(A_n)\| \leqslant M(\lambda_o)(1 - M(\lambda_o)|\lambda - \lambda_o|)^{-1}$ for $|\lambda - \lambda_o| < M(\lambda_o)^{-1}$, $n \geqslant n_o(\lambda)$, i.e., Δ_b contains the disk $|\lambda - \lambda_o| < M(\lambda_o)^{-1}$.

(b) The inclusion $\Delta_s \subset \Delta_b$ is an immediate consequence of the resonance theorem [79]. To prove that Δ_s is open we note that if $\lambda_o \in \Delta_s$, then

---

* $B_n \overset{s}{\to} B$ denotes the usual strong convergence.

there exists $s-\lim_{n\to\infty}(R_{\lambda_o}(A_n))^k = (R_{\lambda_o})^k$, $k = 1,2,\ldots$. The Neumann series for $R_\lambda(A_n)$ is dominated by the convergent series $\Sigma_{k=0}^{\infty} M(\lambda)^{k+1}|\lambda - \lambda_o|^k$, when $|\lambda - \lambda_o| < M(\lambda_o)^{-1}$. Thus $s - \lim_{n\to\infty} R_\lambda(A_n)$ exists for these λ. Note that we have also proved (c). To prove that Δ_s is closed, consider $\lambda \in \Delta_b$ such that $\lambda = \lim_{m\to\infty} \lambda_m$, $\lambda_m \in \Delta_s$. Choose m such that $|\lambda_m - \lambda| < 1/2M(\lambda)$. It follows from above that one may assume $M(\lambda_m) \leqslant M(\lambda)(1 - M(\lambda)|\lambda - \lambda_m|)^{-1} < 2M(\lambda)$. Therefore $|\lambda - \lambda_m| < 1/M(\lambda_m)$ and the above considerations imply that $\lambda \in \Delta_s$. Thus (b) is proved. (d) is valid, since R_λ is a strong limit of operator families, satisfying (6).

Proof of Lemma 2. First we note that (6) implies commutability of the family R_λ. Further

$$R_\lambda u = (I + (\lambda - \mu)R_\lambda)R_\mu u, \text{ so that } \operatorname{Ker} R_\mu \subset \operatorname{Ker} R_\lambda,$$

and

$$R_\mu u = R_\lambda(I + (\mu - \lambda)R_\mu)u, \text{ so that } \operatorname{Im}R_\mu \subset \operatorname{Im}R_\lambda.$$

Interchanging λ and μ, we obtain that $\operatorname{Ker} R_\lambda = \operatorname{Ker} R_\mu$ and $\operatorname{Im}R_\lambda = \operatorname{Im}R_\mu$. If $R_\lambda = (\lambda - A)^{-1}$, then clearly $\operatorname{Ker} R_\lambda = \{0\}$, since $(\lambda - A)R_\lambda = I$. Conversely let $N = \{0\}$. Define the operator A_λ with the domain R by the formula (7). Then clearly A_λ is closed (since R_λ is), $(\lambda I - A_\lambda)R_\lambda = I$, $R_\lambda(\lambda I - A_\lambda) = I|_R$ (the restriction on R of identity operator). It remains to prove that A_λ does not depend on λ. Let $x \in R$, then $x = R_\lambda v$ for some $v \in X$. We have

$$A_\lambda x = A_\lambda R_\lambda v = (\lambda I - (R_\lambda)^{-1})R_\lambda v = \lambda R_\lambda v - v;$$

$$A_\mu x = A_\mu R_\lambda v = (\mu I - (R_\mu)^{-1})R_\lambda v = \mu R_\lambda v - (R_\mu)^{-1}R_\mu v -$$
$$- (R_\mu)^{-1}(\lambda - \mu)R_\lambda R_\mu v = \mu R_\lambda v - v - (\lambda - \mu)R_\lambda v = \lambda R_\lambda v - v = A_\lambda x.$$

Only the last assertion of the lemma remains unproved. If $\lambda,\mu \in \Delta_b \cap \rho(A)$, we have for n large enough (such that resolvents are defined)

$$R_\mu(A_n) - R_\mu(A) = [I + (\mu - \lambda)R_\mu(A_n)] \times (R_\lambda(A_n) - R_\lambda(A))[I + (\mu - \lambda)R_\mu(A)] \quad (8)$$

(this equation is an easy consequence of resolvent identities). If we choose $\lambda \in \Delta_s$ (which is non-empty by assumption), we obtain that $R_\mu(A_n) \overset{s}{\to} R_\mu(A)$ (since $R_\lambda(A_n) \overset{s}{\to} R_\lambda(A)$, and $R_\mu(A_n)$ is uniformly bounded). Thus $\Delta_s = \Delta_b \cap \rho(A)$.

We return to the proof of the theorem. Since $A_n \in G(M,\omega)$, estimates (5) of Theorem 1, item A imply that Δ_b contains the lower half-plane. (4) means that $\lambda_o \in \Delta_s$. Applying Lemma 1, we see that Δ_s also contains the lower half-plane. Now we can apply Lemma 2 and thus we obtain (3). It follows from (3) that $R_\lambda(A)$ satisfies estimates (5) of item A. Hence $A \in G(M,0)$. We set $U(t) = \exp(iAt)$ and prove that (2) is valid. Establish first that

$$R_\lambda(A_n)[e^{itA} - e^{itA_n}]R_\lambda(A) =$$
$$= i\int_o^t e^{i(t-s)A_n}(R_\lambda(A) - R_\lambda(A_n))e^{isA}ds, \quad \operatorname{Im}\lambda < 0 \quad (9)$$

(integral on the right in the sense of strong convergence). Indeed,

$$\frac{d}{ds}[R_\lambda(A_n)e^{i(t-s)A_n}e^{isA}R_\lambda(A)] =$$

$$= -ie^{i(t-s)A_n}A_nR_\lambda(A_n)e^{isA}R_\lambda(A) + ie^{i(t-s)A_n}R_\lambda(A_n)e^{isA}AR_\lambda(A) =$$

$$= ie^{i(t-s)A_n}[(I-\lambda R_\lambda(A_n))R_\lambda(A) - R_\lambda(A_n)(I-\lambda R_\lambda(A))]e^{isA} = \tag{10}$$

$$= ie^{i(t-s)A_n}(R_\lambda(A)-R_\lambda(A_n))e^{isA}.$$

Integrating with respect to s from 0 to t yields (9). We now put λ, $\mathrm{Im}\lambda <$ < 0, fixed. (9) implies

$$\begin{aligned} &\| R_\lambda(A_n)[e^{itA} - e^{itA_n}]R_\lambda(A)x\| \leq \\ &\leq M\int_0^t \| (R_\lambda(A) - R_\lambda(A_n))e^{isA}x\| \, ds. \end{aligned} \tag{11}$$

The integrand in (11) is uniformly bounded as $n \to \infty$, and tends to zero pointwise; by the Dominated Convergence theorem the integral tends to zero as $n \to \infty$. Thus for $n \to \infty$

$$R_\lambda(A_n)[e^{itA} - e^{itA_n}]u \to 0 \tag{12}$$

for $u \in \mathrm{Im}(R_\lambda(A)) = D_A$ and hence for all $u \in X$. Our argument shows that this convergence is locally uniform with respect to t. Further,

$$\begin{aligned} &R_\lambda(A_n)[e^{itA} - e^{itA_n}]u - [e^{itA} - e^{itA_n}]R_\lambda(A)u = \\ &= [R_\lambda(A_n) - R_\lambda(A)]e^{itA}u + e^{itA_n}[R_\lambda(A) - R_\lambda(A_n)]u \to 0 \end{aligned} \tag{13}$$

as $n \to \infty$ locally uniformly with respect to t (the uniformity is evident for the second addend; for the first addend it is a consequence of the fact that $e^{itA}u$ is a continuous function of t and hence for t varying in a bounded subset of $[0,\infty)$ the values of e^{itA} lie in a compact subset of X). It follows from (12) and (13) that $\exp(itA_n)v \to \exp(itA)v$ locally uniformly with respect to t for v lying in $\mathrm{Im}(R_\lambda(A)) = D_A$, hence for all $v \in X$. Thus the theorem is proved.

Our next theorems establish some conditions under which (4) is valid and therefore the semigroups $\exp(itA_n)$ converge. At first we concentrate our attention on the behavior of resolvents.

Theorem 2. Let $A_n \in G(M,\omega)$, $n = 1,2,\ldots$. Suppose also that there exists a limit

$$R_\lambda = s - \lim_{n\to\infty} R_\lambda(A_n), \tag{14}$$

for some $\lambda, \mathrm{Im}\lambda < -\omega$. Then $R_\lambda = R_\lambda(A)$ for some $A \in G(M,\omega)$ and $\exp(itA_n) \overset{s}{\to}$ $\overset{s}{\to} \exp(itA)$, $t \geq 0$, provided that at least one of the following conditions is satisfied:

(a) $\mathrm{Im}R_\lambda$ is dense in X.
(b) $-i\mu R_{-i\mu}(A_n) \overset{s}{\to} I$ uniformly with respect to n, when $\mu \to \infty$.

Proof. In view of Theorem 1, we need only to prove that $R_\lambda = R_\lambda(A)$ for some densely defined operator A. We prove first that (b) implies (a). Applying Lemma 1, just as in the proof of Theorem 1, we obtain that the limit (14) exists for all λ such that $\mathrm{Im}\lambda < -\omega$ and satisfies (6). (b) means that for arbitrary $x \in X$

$$-i\mu R_{-i\mu}(A_n)x \to \chi, \quad \mu \to \infty \tag{15}$$

uniformly with respect to n. Hence we may pass to the limit as $n \to \infty$ in (15), obtaining

$$\underset{\mu\to\infty}{\ell im} - i\mu R_{-i\mu} x = x. \tag{16}$$

Since $R = \mathrm{Im}R_\lambda$ does not depend on λ (Lemma 2), (16) yields $x \in \bar{R}$ ($\bar{R}$ denotes the closure of R). Thus we obtain (a). It remains to show that (a) implies $N = \mathrm{Ker}\, R_\lambda = \{0\}$. Just another application of Lemma 2 completes the proof after this. We may assume that $\omega = 0$, as in the proof of Theorem 1. Then $\| R_\lambda \| \leqslant M/|\mathrm{Im}\lambda|$ for λ lying in the lower half-plane. Thus the family of operators $-i\mu R_{-i\mu}$, $\mu \in (0,\infty)$, is uniformly bounded. Now let $x \in N$, then $-i\mu R_{-i\mu}x = 0$. We shall prove, however, that

$$\underset{\mu\to\infty}{\ell im} - i\mu R_{-i\mu} y = y \text{ for all } y \in X. \tag{17}$$

In particular, $x = \underset{\mu\to\infty}{\ell im}\, 0 = 0$ and we obtain the statement of the theorem. Since R is dense in X, it is enough to prove (17) for $y \in R$. From (6) we obtain

$$\begin{aligned} -i\mu R_{-i\mu} R_\lambda &= \lambda R_\lambda R_{-i\mu} + R_\lambda - R_{-i\mu} = \\ &= \frac{i\lambda}{\mu}(-i\mu R_{-i\mu})R_\lambda + R_\lambda - \frac{i}{\mu}(-i\mu R_{-i\mu}). \end{aligned} \tag{18}$$

Fix some λ, $\mathrm{Im}\lambda < 0$. If $y \in R$, $y \in R_\lambda z$ for some $z \in X$, so that

$$-i\mu R_{-i\mu} y = \frac{i\lambda}{\mu}(-i\mu R_{-i\mu})y + y - \frac{i}{\mu}(-i\mu R_{-i\mu})z. \tag{19}$$

Since $-i\mu R_{-i\mu}$ is a uniformly bounded family, we may pass to the limit as $\mu \to \infty$ in (19), obtaining (17). Theorem 2 is proved.

The following theorem is often rather convenient because no information about resolvents is required; the condition is formulated in terms of strong convergence of generators.

Theorem 3. Let $A_n \in G(M,\omega)$, $n = 1,2,\ldots$. Suppose that for elements of some dense linear subset $D \subset X$ there exist the limits

$$\underset{n\to\infty}{\ell im} A_n x \overset{\mathrm{def}}{=} Ax, \; x \in D. \tag{20}$$

Suppose also that the range $R_{(\lambda-A)}$ is dense in X for some λ, $\mathrm{Im}\lambda < -\omega$. Then A has a closure $\bar{A} \in G(M,\omega)$ and $\exp(iA_n t) \overset{s}{\to} \exp(iAt)$.

Proof. We claim that

$$R_\lambda(A_n) \overset{s}{\to} R_\lambda \text{ as } n \to \infty \tag{21}$$

where R_λ has a dense range and satisfies the conditions,

$$(\lambda - A)R_\lambda \cdot R_{(\lambda-A)} = I|_{R_{(\lambda-A)}}; \; R_\lambda(\lambda - A) = I|_D. \tag{22}$$

Then taking into account Theorem 2, we obtain the required statement. Indeed, the only statement which needs proof is that A has a closure and this closure coincides with the generator of $s - \ell im(\exp(itA_n))$, i.e., with $\lambda I - R_\lambda^{-1}$. But this is evident since (i) (22) implies that $(\lambda I - R_\lambda^{-1})|_D = A$; and (ii) $R_{(\lambda-A)}$ is dense, therefore $D = R_\lambda(R_{(\lambda-A)})$ is a core of $\lambda I - R_\lambda^{-1}$.

Moving on to the proof of (21) - (22), first we prove that $A - \lambda$ is invertible. Since $A_n \in G(M,\omega)$, $\|(A_n - \lambda)x\| \geqslant M^{-1}(|\mathrm{Im}\lambda| - \omega)\|x\|$, $x \in D_{A_n}$. Passing to the limit for $x \in D$ gives the inequality $\|(A - \lambda)x\| \geqslant M^{-1}\cdot(|\mathrm{Im}\lambda| - \omega)\|x\|$, $x \in D$, so we may define an operator $(\lambda - A)^{-1}$ with the domain $R_{(\lambda-A)}$ and the range D, satisfying the estimate $\|(\lambda - A)^{-1}\| \leqslant M/(|\mathrm{Im}\lambda| - \omega)$. Denote its closure by R_λ. R_λ is a bounded everywhere defined operator with dense range (the latter contains D). It remains to prove (21). Let $x \in R_{(\lambda-A)}$. Then $R_\lambda x \in D$ and

$$R_\lambda(A_n)x - R_\lambda x = R_\lambda(A_n)(A_n - A)R_\lambda x \to 0 \text{ as } n \to \infty,$$

since the sequence $\|R_\lambda(A_n)\|$ is bounded. Since $\bar{R}_{(\lambda-A)} = X$, $R_\lambda(A_n)x \to R_\lambda x$ for arbitrary $x \in X$ and the theorem is proved.

The theorems proved may be reformulated for the case of strongly continuous groups, in particular for self-adjoint generators. Some of the corresponding results are given below, others are left to the reader.

Corollary 1. Let $\exp(itA_n)$, $n = 1,2,\ldots$ be a sequence of strongly continuous groups, satisfying the estimate $\|\exp(itA_n)\| \leqslant M\exp(\omega|t|)$. Suppose that the generators A_n strongly converge to an operator A on a dense linear subset D of X and besides $\bar{R}_{(\lambda_1-A)} = \bar{R}_{(\lambda_2-A)} = X$ for some λ_1, λ_2, $\mathrm{Im}\lambda_1 > \omega$, $\mathrm{Im}\lambda_2 < -\omega$. Then A has the closure $\bar{A}$, which is a generator of a strongly continuous group $\exp(itA)$ and $\exp(itA_n) \overset{s}{\to} \exp(itA)$, $t \in \mathbb{R}$ as $n \to \infty$.

Corollary 2. Let A_n be a sequence of self-adjoint operators in the Hilbert space X, and let $A_n x \underset{n\to\infty}{\to} Ax$ for $x \in D_A$, where A is an essentially self-adjoint operator in X. Then $\exp(itA_n) \overset{s}{\to} \exp(it\bar{A})$, $t \in \mathbb{R}$. Both corollaries immediately follow from Theorem 3.

D. Semigroups of Polynomial Growth

The notion of semigroup type, introduced in item A is too rough from the viewpoint of our aim - construction of ordered operator calculus. The inconveniency of this notion is in particular related to the fact that the semigroup of type ω_0 may not satisfy $\|U(t)\| \leqslant Me^{\omega t}$ with $\omega = \omega_0$, since the constant M may grow as $\omega \downarrow \omega_0$.

Example 1. Let X be a Sobolev space $X = W_2^k(\mathbb{R}^1)$, k being an integer (recall that X may be identified with the completion of $C_0^\infty(\mathbb{R}^1)$ with respect to the norm*

$$\|\phi\|_k = \{\int_{-\infty}^{\infty} \bar{\phi}(x)(1 - \frac{\partial^2}{\partial x^2})^k \phi(x)dx\}^{1/2},\ \phi \in C_0^\infty(\mathbb{R}^1); \tag{1}$$

the norm in W_2^k may be also described as

$$\|\phi\|_k = \sup_{\substack{\psi \in C_0^\infty \\ \psi \neq 0}} \frac{|(\phi,\psi)|}{\|\psi\|_{-k}},\ \phi \in C_0^\infty(\mathbb{R}^1), \tag{2}$$

where

$$(\phi,\psi) = \int_{-\infty}^{\infty} \bar{\psi}(x)\phi(x)dx \tag{3}$$

---

* Note that $1 - \frac{\partial^2}{\partial x^2}$ is an invertible operator in $L^2(\mathbb{R}^1)$, so that (1) makes sense for negative k as well.

is the usual L^2 scalar product. Obviously $W_2^0(\mathbb{R}^1) = L^2(\mathbb{R}^1)$.) Consider the operator A in X with the domain $D = C_0^\infty(\mathbb{R}^1)$, defined by

$$[A\phi](x) = x\phi(x), \quad \phi \in C_0^\infty(\mathbb{R}^1). \tag{4}$$

We claim that A has a closure, which will be denoted again by A, and this closure is a generator of a strongly continuous group $\exp(itA)$, $t \in (-\infty,\infty)$, whose action on $\phi \in C_0^\infty(\mathbb{R}^1)$ is given by the formula

$$[\exp(itA)\phi](x) = e^{itx}\phi(x), \quad t \in \mathbb{R}. \tag{5}$$

This group satisfies an estimate

$$\|\exp(itA)\|_k \leqslant C_{|k|}\cdot(1+|t|)^{|k|}. \tag{6}$$

Indeed, denote the right-hand side of (5) by $\mathcal{U}(t)\phi(x)$. First we prove the estimate (6). We have for non-negative k

$$\|\mathcal{U}(t)\phi\|_k \leqslant C \sum_{s=0}^{k} \left\| \frac{\partial^s}{\partial x^s} \mathcal{U}(t)\phi\right\|_0 \leqslant \tilde{C} \sum_{s=0}^{k} \sum_{\ell=0}^{s} |t|^{\ell} \|\phi\|_{s-\ell} \leqslant$$

$$\leqslant \tilde{\tilde{C}}(1+|t|)^k \|\phi\|_k.$$

For negative k, as (2) shows, W_2^k may be identified with $(W_2^{-k})^*$ under pairing (3), and we obtain

$$\|\mathcal{U}(t)\|_k = \|\mathcal{U}(t)^*\|_{-k} = \|\mathcal{U}(-t)\|_{-k} \leqslant C_{|k|}(1+|t|)^{|k|}. \tag{7}$$

Thus we have proved (6), and it is obvious that the exponent $|k|$ in (6) cannot be diminished. The semigroup property and continuity follow immediately from (5), and so does the identity

$$i\frac{\partial}{\partial t}\mathcal{U}(t)\phi + A\mathcal{U}(t)\phi = 0, \quad \phi \in C_0^\infty. \tag{8}$$

It remains to apply Theorem 3 of item A.

Example 2. Let $X = \mathbb{C}^2$, A be a (2×2)-matrix

$$A = \begin{pmatrix} \lambda & 1 \\ 0 & \lambda \end{pmatrix}. \tag{9}$$

Then

$$\exp(itA) = \begin{pmatrix} e^{i\lambda t} & ite^{i\lambda t} \\ 0 & e^{i\lambda t} \end{pmatrix}, \quad t \geqslant 0, \tag{10}$$

is a semigroup of the type $\omega = -\mathrm{Im}\lambda$, but

$$\|\exp(itA)\| \leqslant C(1+t)e^{i\omega t}. \tag{11}$$

It is easily seen that the factor $(1 + t)$ appears in (11) because A has a non-trivial Jordan block of order 2. In the general case there is a deep intrinsic connection between these facts which we have no place to regard (see [52]). In operator calculus we are interested only in strongly continuous groups of type 0 (such as in Example 1), so we give the following:

Definition 1. A closed operator A in a Banach space X is called a generator of degree k, if it is a generator of a strongly continuous group $\mathcal{U}(t) = \exp(itA)$, satisfying the estimate

$$\|\exp(itA)\| \leqslant C(1+|t|)^k, \quad t \in \mathbb{R}. \tag{12}$$

The family $\{A_\mu\}$ of generators of degree k is called uniform if the corresponding groups satisfy (12) with the constant C independent of μ.

The following theorem gives necessary and sufficient conditions for the given operator A to be a generator of degree k:

Theorem 1. In order that a given closed densely defined operator A be a generator of degree k, satisfying (12), it is necessary that $\lambda \in \rho(A)$ for $\mathrm{Im}\lambda \neq 0$ and

$$\| R_\lambda(A)^m \| \leqslant C \sum_{j=0}^{k} \frac{k!(m+j-1)!}{j!(k-j)!(m-1)!} \cdot |\mathrm{Im}\lambda|^{-m-k}, \quad \mathrm{Im}\lambda \neq 0, \ m = 1,2,\ldots, \tag{13}$$

and sufficient that $\lambda \in \rho(A)$ and (13) is valid for $\mathrm{Im}\lambda$ large enough (the constant C is the same as in (12)).

Proof. (a) Necessity. A and -A are clearly the generators of semigroups of type 0, thus Theorem 1 of item A implies that $\lambda \in \rho(A)$ for $\mathrm{Im}\lambda \neq 0$. We prove (13) for $\mathrm{Im}\lambda < 0$, the proof for $\mathrm{Im}\lambda > 0$ being completely analogous. Using the equality (11) of item A and the estimate (12), we obtain

$$\begin{aligned} \| R_\lambda(A)^m \| &\leqslant \frac{C}{(m-1)!} \int_0^\infty e^{-t|\mathrm{Im}\lambda|} t^{m-1} (1+t)^k dt = \\ &= \frac{C}{(m-1)!} \sum_{j=0}^{k} \frac{k!}{j!(k-j)!} \int_0^\infty e^{-t|\mathrm{Im}\lambda|} t^{m+j-1} dt. \end{aligned} \tag{14}$$

Since $\int_0^\infty e^{-ta} t^s dt = s!/a^{s+1}$ $(a > 0)$, we obtain (13).

(b) Sufficiency. Suppose that (13) holds for $|\mathrm{Im}\lambda| > R$. Then $\| R_\lambda(A)^m \| \leqslant M/(|\mathrm{Im}\lambda| - R)^m$, $|\mathrm{Im}\lambda| > R$ for some M. Indeed, as we have just seen, the right-hand side of (13) equals the right-hand side of (14), so

$$\begin{aligned} \| R_\lambda(A)^m \| &\leqslant \frac{C}{(m-1)!} \int_0^\infty e^{-t|\mathrm{Im}\lambda|} t^{m-1} (1+t)^k dt \leqslant \\ &\leqslant \frac{C}{(m-1)!} \int_0^\infty e^{-t(|\mathrm{Im}\lambda| - R)} t^{m-1} [e^{-tR}(1+t)^k] dt \leqslant \frac{M}{(|\mathrm{Im}\lambda| - R)^m}, \end{aligned} \tag{15}$$

where

$$M = C \cdot \sup_{t \in [0,\infty)} [e^{-tR}(1+t)^k] < \infty.$$

Thus by Theorem 2 of item A, A is a generator of a strongly continuous group $\exp(iAt)$. Now we are in position to prove the estimate (12). Let $t > 0$ (the case $t < 0$ is considered analogously). By Theorem 1 (f) of item A

$$\exp(iAt) = s - \lim_{\mu\to\infty} (-i\mu R_{-i\mu}(A))^{[\mu t]} = s - \lim_{n\to\infty} (-i \tfrac{n}{t} R_{-i\frac{n}{t}}(A))^n;$$

thus

$$\begin{aligned} \| \exp(iAt) \| &\leqslant \lim_{n\to\infty} \{ (\tfrac{n}{t})^n \frac{C}{(n-1)!} \int_0^\infty e^{-(\tau/t)n} \tau^{n-1} (1+\tau)^k d\tau \} = \\ &= C \lim_{n\to\infty} [\frac{n^n}{(n-1)!} \int_0^\infty e^{-\tau n} \tau^{n-1} (1+t\tau)^k d\tau] = \\ &= C \lim_{n\to\infty} [\frac{n^n}{(n-1)!} \sum_{j=0}^{k} \frac{k!}{j!(k-j)!} t^j \int_0^\infty e^{-\tau n} \tau^{n+j-1} d\tau] = \end{aligned} \tag{16}$$

$$= C \lim_{n\to\infty} \left[\sum_{j=0}^{k} \frac{k!}{j!(k-j)!} t^j \left(\frac{(n+j-1)!}{(n-1)!n^j}\right)\right] = C(1+t)^k,$$

since the expression in parentheses tends to unity as $n \to \infty$. Thus the theorem is proved.

The groups of polynomial growth being particular type of strongly continuous groups, all the results concerning the latter are valid for the former (one should only remember that dealing with groups it is necessary to study resolvent behavior in both half-planes). In theorems of item C we must substitute "A_n is a uniform sequence of generators of degree k" instead of "$A_n \in G(M,\omega)$." The attentive reader has undoubtedly noticed that, after such substitution, it does not follow directly from these theorems that the limit group satisfies the estimate (12). However, it is the immediate consequence of the inequality

$$\| s - \lim_{n\to\infty} U_n \| \leqslant \lim_{n\to\infty} \| U_n \|.$$

Here we finish our excursion into the semigroup theory and move on to the calculus itself.

E. Functions of a Generator in Banach Space

Let X be a Banach space, A be a generator of degree k in X. We describe here the class of symbols f (y) - functions of a real variable y, for which an operator f(A) in X is well defined. After this we prove that the mapping $f \to f(A)$ has quite natural algebraic properties.

$\mathcal{B}$ being a Banach space, we denote by $C_k^\ell(\mathbb{R},\mathcal{B})$, ($\ell$,k $\geqslant$ 0-integers) the space of all ℓ-smooth (i.e., having ℓ continuous derivatives) mappings ϕ: $\mathbb{R} \to \mathcal{B}$ with the finite norm

$$\|\phi\|_{C_k^\ell} = \sup_{t\in\mathbb{R}} (1+|t|)^{-k} \sum_{j=0}^{\ell} \left\|\frac{\partial^j\phi}{\partial t^j}(t)\right\|. \tag{1}$$

If $\ell = k = 0$, this space will be denoted by $C(\mathbb{R},\mathcal{B})$, while if $\mathcal{B} = \mathbb{C}$, it will be denoted by $C_k^\ell(\mathbb{R})$. Clearly $C_k^\ell(\mathbb{R},\mathcal{B})$ supplied with the norm (1) is a Banach space. Let $C_k^{\ell*}(\mathbb{R})$ denote the space of continuous linear forms on $C_k^\ell(\mathbb{R})$ with the usual norm. The definition of the support of the functional $T \in C_k^{\ell*}(\mathbb{R})$ is not obvious, since one might prove, using the Hahn-Banach theorem, that $C_k^{\ell*}(\mathbb{R})$ contains functionals which vanish on all finite elements of $C_k^{\ell*}(\mathbb{R})$. Nevertheless, we define the notion of a finite functional. $T \in C_k^{\ell*}(\mathbb{R})$ is finite if it takes zero value on elements $\phi \in C_k^\ell(\mathbb{R})$, vanishing for $|t| < R$, $R = R(T)$. Now we define $C_k^{\ell+}(\mathbb{R})$ as the minimal closed subspace of $C_k^{\ell*}(\mathbb{R})$, containing all finite functionals.

Then let $\psi \in C_0^\infty(\mathbb{R})$, $\psi(t) = 1$ for $|t| \leqslant 1$, $\psi_n(t) = \psi(t/n)$.

<u>Lemma 1</u>. A functional T on $C_k^\ell(\mathbb{R})$ belongs to $C_k^{\ell+}(\mathbb{R})$ iff the sequence $T_n = \psi_n T$ converges to T as $n \to \infty$ in the norm of $C_k^{\ell*}(\mathbb{R})$.

<u>Proof</u>. Since T_n are finite functionals, one of the statements of the lemma is evident. To prove the other, suppose that $T \in C_k^{\ell+}(\mathbb{R})$. This means there exists a sequence $T_{(m)}$ of finite functionals convergent to T. The sequence of operators of multiplication by $(1 - \psi_n(t))$ is uniformly bounded in

$C_k^\ell(\mathbb{R})$; denote its bound by M. Given an $\varepsilon > 0$, choose m such that $\|T - T_{(m)}\| < \varepsilon/M$. Then for n large enough $(1 - \psi_n)T_{(m)} = 0$, hence

$$\|T_n - T\| = \|(1-\psi_n)T\| = \|(1-\psi_n)(T - T_{(m)})\| < M\cdot\varepsilon/M = \varepsilon.$$

The lemma is proved. Our next proposition is, in a sense, the main tool in construction of the mapping $f \to f(A)$. Let $\mathcal{B}$ be a Banach space.

Theorem 1. Let $\phi \in C_k^\ell(\mathbb{R},\mathcal{B})$, $T \in C_k^{\ell+}(\mathbb{R})$. Then there exists an element $b \in \mathcal{B}$ such that*

$$T(\langle h,\phi\rangle) = \langle h,b\rangle \tag{2}$$

for each $h \in \mathcal{B}^*$ and

$$\|b\| \leqslant \|T\|_{C_k^{\ell*}}\|\phi\|_{C_k^\ell}. \tag{3}$$

Proof. Note at first that the left-hand side of (2) is defined correctly. Indeed, $\langle h,\phi\rangle(t) \equiv \langle h, \phi(t)\rangle$ clearly is an element of $C_k^\ell(\mathbb{R})$, so that T is applicable. Further, the left-hand side of (2) is linear in R and satisfies the estimate

$$|T(\langle h,\phi\rangle)| \leqslant \|T\|_{C_k^{\ell*}}\|\phi\|_{C_k^\ell}\|h\|. \tag{4}$$

It follows that there exists an (unique) element $b \in \mathcal{B}^{**}$ such that (2) and (3) are valid. It remains to show that in fact b belongs to $\mathcal{B}$ (which is considered a subspace in $\mathcal{B}^{**}$ by means of the standard isometric embedding $\mathcal{B} \to \mathcal{B}^{**}$). Since $\mathcal{B}$ is closed in $\mathcal{B}^{**}$, it suffices to prove the latter proposition for finite T.

Lemma 2. The functional $T \in C_k^{\ell*}(\mathbb{R})$ may be represented in the form:

$$T(f) = \sum_{j=0}^{\ell} T_{(j)}\left(\frac{\partial^j f}{\partial t^j}\right), \quad f \in C_k^\ell(\mathbb{R}), \tag{5}$$

where $T_{(j)} \in C_k^{0*}(\mathbb{R})$. If T is finite, $T_{(j)}$, $j = 0,\ldots,\ell$, also may be chosen finite.

Proof. The latter statement is obvious, since we can multiply (5) by the cut-off function $\psi \in C_0^\infty$ such that $\psi T = T$. To prove that there exists a representation of the form (5), we consider an isometric injection

$$\begin{gathered} C_k^\ell(\mathbb{R}) \to \underbrace{C_k^0(\mathbb{R}) \oplus \ldots \oplus C_k^0(\mathbb{R})}_{(\ell+1)\text{ summands}} \\ f(t) \to \left(f(t), \frac{\partial f}{\partial t}(t), \ldots, \frac{\partial^\ell t}{\partial t^\ell}(t)\right) \end{gathered} \tag{6}$$

and apply the Hahn-Banach theorem.

Lemma 2 gives a reduction to the case when T is a finite functional on $C_k^0(\mathbb{R})$. Let $\Omega \subset \mathbb{R}$ be a support of T (support is well-defined for finite functionals). It is well-known that T may be represented in the form

$$T(f) = \int_\Omega f d\mu, \tag{7}$$

where $\mu = \mu_T$ is a Rhadon measure on Ω. Hence

* Brackets $\langle\,,\rangle$ denote the pairing between the elements of $\mathcal{B}^*$ and $\mathcal{B}$.

$$b = \int_\Omega \phi d\mu \tag{8}$$

is a well defined element of $\mathcal{B}$ (see [71], Chapter 3), and clearly b satisfies (2). The theorem is proved.

Notation. In the situation of the theorem we shall write

$$b \equiv T(\phi) \equiv \int_{-\infty}^{\infty} T(t)\phi(t)dt. \tag{9}$$

Remark. It is clear from the proof of the theorem that if $\mathcal{B}$ is a reflective Banach space (i.e., $\mathcal{B}^{**} = \mathcal{B}$), the statement is valid for arbitrary $T \in C_k^{\ell*}(\mathbb{R})$.

As we are going to define f(A) by the formula $f(A) = (1/\sqrt{2\pi})\int \tilde{f}(t) \times \exp(itA)dt$, we study next the inverse Fourier transform of the elements of $C_k^{\ell+}(\mathbb{R})$.

Note that there is a continuous embedding $S(\mathbb{R}) \subset C_k^{\ell}(\mathbb{R})$, where $S(\mathbb{R})$ is the space of smooth functions, rapidly decaying at infinity with all the derivatives. Thus each element $T \in C_k^{\ell*}(\mathbb{R})$ defines an element of $S'(\mathbb{R})$, i.e., a so-called tempered distribution.

Lemma 3. If $T \in C_k^{\ell+}(\mathbb{R})$ and the corresponding tempered distribution equals zero, then $T = 0$.

Proof. By Lemma 1, $T = \lim_{n\to\infty} T_n$, where $T_n = \psi_n T$. We are going to prove that $T_n = 0$. Indeed, let $f \in C_k^{\ell}(\mathbb{R})$. There exists a sequence $f_m \in S(\mathbb{R})$, m = 1,2,..., such that $f_m \underset{m\to\infty}{\to} \psi_n f$ in $C_k^{\ell}(\mathbb{R})$. Thus $0 = T(f_m) \underset{m\to\infty}{\to} T \times (\psi_n f) = T_n(f)$. So $T_n(f) = 0$, as we claimed. Consequently, $T = 0$. The tempered distribution, corresponding to $T \in C_k^{\ell+}(\mathbb{R})$, will also be denoted by T. The Fourier transform of tempered distribution is a well-known construction (see, for example, [71]) and we give the following:

Definition 1. The inverse Fourier transform of the element $T \in C_k^{\ell+}(\mathbb{R})$ is the (inverse) Fourier transform of the corresponding tempered distribution.

Lemma 4. Let $T \in C_k^{\ell+}(\mathbb{R})$. Then its inverse Fourier transform is a function given by*

$$F^{-1}(T)(y) = \frac{1}{\sqrt{(2\pi)}} T(e^{ity}). \tag{10}$$

Further, $F^{-1}(T) \in C_\ell^k(\mathbb{R})$ and

$$\| F^{-1}(T) \|_{C_\ell^k} \leqslant M \| T \|_{C_k^{\ell*}}, \tag{11}$$

where the constant M does not depend on T.

Proof. Let at first T be a finite functional. Then (10) is valid ([79], Theorem 4, Section 3 of Chapter 6). Moreover,

---

* On the right-hand side of (10), e^{ity} is the function of the variable t, depending on the parameter y. Obviously $e^{ity} \in C_k^{\ell}(\mathbb{R})$.

$$|F^{-1}(T)(y)| \leqslant C \, \|T\|_{C_k^{\ell*}} (1 + |y|)^{\ell} \tag{12}$$

(this inequality immediately follows from (10)). Now let $T \in C_k^{\ell+}(\mathbb{R})$, T_n be the sequence of finite functionals, converging to T. Then $\|T_n\|$ is uniformly bounded, and $F^{-1}(T_n) \to (1/\sqrt{2\pi})T(e^{ity})$ pointwise. Pairing with arbitrary elements of $S(\mathbb{R})$ and applying the Dominated Convergence theorem, we obtain (10). To prove (11), we need the following:

Lemma 5. The mapping

$$f(t) \to (1 + \frac{\partial}{\partial t})^{\ell} (i + t)^{-k} f(z) \tag{13}$$

is an isomorphism of $C_k^{\ell}(\mathbb{R})$ onto $C(\mathbb{R})$. This statement remains valid if we change some of the pluses in parentheses into minuses.

Proof. It is enough to show that the mappings

$$C_k^{\ell}(\mathbb{R}) \to C_k^{\ell-1}(\mathbb{R}),\ f(t) \to (1 + \frac{\partial}{\partial t}) f(t),\ k = 0,1,2,\ldots,\ \ell = 1,2,\ldots, \tag{14}$$

$$C_k^{\ell}(\mathbb{R}) \to C_{k-1}^{\ell}(\mathbb{R}),\ f(t) \to (i + t)^{-1} f(z),\ k = 1,2,\ldots,\ \ell = 0,1,2,\ldots, \tag{15}$$

are isomorphisms. We begin with (14). Direct computation gives

$$\|(1 + \frac{\partial}{\partial t}) f(t)\|_{C_k^{\ell-1}} \leqslant 2 \|f(t)\|_{C_k^{\ell}}. \tag{16}$$

Further, consider the operator

$$[Ih](t) = \int_{-\infty}^{t} e^{\tau - t} h(\tau) d\tau,\ h \in C_k^{\ell-1}(\mathbb{R}). \tag{17}$$

We claim that I maps $C_k^{\ell-1}(\mathbb{R})$ onto $C_k^{\ell}(\mathbb{R})$ and is the two-sided inverse for $(1 + \frac{\partial}{\partial t})$. Indeed, one can easily verify that $(1 + \frac{\partial}{\partial t}) \cdot I = I \cdot (1 + \frac{\partial}{\partial t}) =$ = identity operator. To prove that $Ih \in C_k^{\ell}(\mathbb{R})$, if $h \in C_k^{\ell-1}(\mathbb{R})$, we note that

$$\frac{\partial^j}{\partial t^j} [Ih](t) = \frac{\partial^{j-1}}{\partial t^{j-1}} (h(t) - [Ih](t)). \tag{18}$$

We obtain directly from (17) that

$$\begin{aligned} |[Ih](t)| &\leqslant \|h\|_{C_k^{\ell-1}} \cdot \int_{-\infty}^{t} e^{\tau - t} (1 + |\tau|)^k d\tau \leqslant \\ &\leqslant \|h\|_{C_k^{\ell-1}} (1 + |t|)^k \cdot \int_{-\infty}^{0} e^{\lambda} (1 + |\lambda|)^k d\lambda = \text{const}(1 + |t|)^k. \end{aligned} \tag{19}$$

Using (18), we estimate recursively all the derivatives up to the order ℓ and obtain that $Ih \in C_k^{\ell}(\mathbb{R})$ and $\|Ih\|_{C_k^{\ell}} \leqslant C \|h\|_{C_k^{\ell-1}}$. The investigation of (15) is rather simple and therefore we leave it to the reader. We then return to the proof of Lemma 4. For given $T \in C_k^{\ell+}(\mathbb{R})$ define the functional $\hat{T} \in C^*(\mathbb{R})$ by the equality

$$\hat{T}(f) = T((i + t)^k I^{\ell} f),\ f \in C(\mathbb{R}). \tag{20}$$

We claim that in fact $\hat{T} \in C^+(\mathbb{R})$. To prove this, consider the sequence $\hat{T}_n = \psi_n \hat{T}$, ψ_n being the same as in Lemma 1. We have

$$\hat{T}_n(f) = (\psi_n\hat{T})(f) = \hat{T}(\psi_n f) = T((i+t)^k I^\ell \psi_n f) =$$
$$= (\psi_n T)((i+t)^k I^\ell f) + \sum_{j=0}^{\ell-1} T((i+t)^k I^j [I,\psi_n] I^{\ell-1-j} f) \tag{21}$$

(here $[I,\psi_n] = I\psi_n - \psi_n I$ denotes the commutator of I and the operator of multiplication by ψ_n). The first term on the right of (21) converges to $\hat{T}(f)$, and it suffices to show that the norm of $[I,\psi_n]$ as an operator from $C_o^j(\mathbb{R})$ to $C_o^{j+1}(\mathbb{R})$ tends to zero as $n \to \infty$ for $j = 0,1,\ldots,\ \ell = 1$. We have

$$[I,\psi_n]f(t) = \int_{-\infty}^{o} e^\lambda [\psi_n(\lambda+t) - \psi_n(t)] f(\lambda+t)d\lambda =$$
$$= \frac{1}{n}\int_{-\infty}^{o} \lambda e^\lambda \chi_n(t,\lambda) f(\lambda+t)d\lambda, \tag{22}$$

where $\chi_n(t,\lambda)$ is bounded with all derivatives uniformly with respect to n. Recursive estimates, analogous to that of $[If](t)$, show us that the norm in question tends to zero as $1/n$. We obtain from (20), using the well-known formulae for derivatives of Fourier transform, that

$$F^{-1}(\hat{T})(y) = i^{k+\ell}(1 - \frac{\partial}{\partial y})^k (i+y)^{-\ell} F^{-1}(T)(y), \tag{23}$$

and in view of Lemma 5 it suffices to demonstrate that $F^{-1}(\hat{T}) \in C(\mathbb{R})$ and $\| F^{-1}(\hat{T})\|_C \leqslant M\|\hat{T}\|_{C^*}$. Since $\hat{T} \in C^+(\mathbb{R})$, its inverse Fourier transform is given by (10), so the latter inequality is the consequence of (12). (Recall that $\ell = 0$.) Further $F^{-1}(\hat{T})(y) = \lim_{n\to\infty} F^{-1}(\psi_n\hat{T})(y)$ and this convergence is uniform, since $|F^{-1}(\hat{T})(y) - F^{-1}(\psi_n\hat{T})(y)| \leqslant M\|\hat{T} - \psi_n\hat{T}\|$. $F^{-1}(\psi_n\hat{T})$ is a continuous function, since $F^{-1}(\psi_n\hat{T}) = (1/\sqrt{2\pi})\hat{T}(\psi_n(t)\exp(ity))$ and $\psi_n(t)\exp(ity)$ is a continuous family of elements of $C(\mathbb{R})$. Thus $F^{-1}(\hat{T})(y)$ is continuous, and Lemma 4 is proved.

Definition 2. We denote by $B_k^\ell(\mathbb{R})$ the space of functions of the form $F^{-1}(\phi)$, $\phi \in C_k^{\ell+}(\mathbb{R})$. We supply $B_k^\ell(\mathbb{R})$ with the norm

$$\|f\|_{B_k^\ell} \overset{\text{def}}{=} \|F(f)\|_{C_k^{\ell*}}. \tag{24}$$

Clearly $B_k^\ell(\mathbb{R})$, supplied with the norm (24) is a Banach space. It is also clear that $B_k^\ell \subset B_k^{\ell'}$ for $\ell' \geqslant \ell$, and the embedding is continuous. One may ascertain from the proof of Lemmas 4 and 5 that multiplication by $(i+y)^\ell$ is an isomorphism between B_k^0 and B_k^ℓ, while $(1+\frac{\partial}{\partial y})^k$ realizes an isomorphism between B_k^ℓ and B_o^ℓ. Now we introduce the union

$$B_k^\infty(\mathbb{R}) = \bigcup_{\ell=0}^{\infty} B_k^\ell(\mathbb{R}), \tag{25}$$

which will be the symbol space in functional calculus for a single generator of degree k. Some properties of the symbol space $B_k^\infty(\mathbb{R})$ are described in the following:

Theorem 2. (a) In order that $f(y) \in B_k^\infty(\mathbb{R})$ it is necessary for f(y) to have k continuous derivatives, satisfying for some N the estimates

$$\left|\frac{\partial^j f}{\partial y^j}(y)\right| \leqslant C(1+|y|)^N, \tag{26}$$

$j = 0,1,\dots,k$, and it is sufficient that it has $k + 1$ continuous derivatives satisfying (26), $j = 0,1,\dots,k + 1$.

(b) $\mathcal{B}_k^{\infty}(\mathbb{R})$ is an algebra with respect to pointwise multiplication of the functions. More precisely, if $f_1 \in \mathcal{B}_k^{\ell_1}(\mathbb{R})$, $f_2 \in \mathcal{B}_k^{\ell_2}(\mathbb{R})$, then $f_1,f_2 \in \mathcal{B}_k^{\ell_1+\ell_2}(\mathbb{R})$, and

$$\| f_1 f_2 \|_{\mathcal{B}_k^{\ell_1+\ell_2}} \leqslant \sqrt{2\pi}\, \| f_1 \|_{\mathcal{B}_k^{\ell_1}} \| f_2 \|_{\mathcal{B}_k^{\ell_2}}. \tag{27}$$

In particular, $\mathcal{B}_k^0(\mathbb{R})$ is a Banach algebra.

(c) The function $f(y) = y^s e^{i\tau y}$ belongs to $\mathcal{B}_k^{\ell}(\mathbb{R})$ for $\tau \in \mathbb{R}$, $s = 0,1,2,\dots,\ell \geqslant s$.

(d) Let $\chi \in C^{\infty}(\mathbb{R})$ be a function such that $\chi(y) = \operatorname{sign} y$ for $|y| \geqslant R$. Then χ does not belong to $\mathcal{B}_k^0(\mathbb{R})$. We need the following lemma to prove Theorem 2:

Lemma 6. Denote by $H_{\beta}^{\alpha}(\mathbb{R})$ the completion of $C_0^{\infty}(\mathbb{R})$ with respect to the norm

$$\| \phi \|_{H_{\beta}^{\alpha}} = \| (1 + y^2)^{-\beta/2} (1 - \frac{\partial^2}{\partial y^2})^{\alpha/2} \phi \|_{L^2} \tag{28}$$

(here $\| f \|_{L_2} = \int_{-\infty}^{\infty} |f(y)|^2 dy$ is the usual L^2-norm; clearly $H_0^{\alpha}(\mathbb{R}) = W_2^0(\mathbb{R})$, see example 1 of item D); then for each $\varepsilon > 0$ there are the continuous embeddings

$$H_{\ell}^{k+\varepsilon+1/2}(\mathbb{R}) \subset \mathcal{B}_k^{\ell}(\mathbb{R}) \subset C_{\ell}^{k}(\mathbb{R}) \subset H_{\ell+\varepsilon+1/2}^{k}(\mathbb{R}). \tag{29}$$

Proof. The embedding $\mathcal{B}_k^{\ell}(\mathbb{R}) \subset C_{\ell}^{k}(\mathbb{R})$ was already proved in Lemma 4. For $\alpha \geqslant 0$ the elements of H_{β}^{α} are measurable functions (we do not reproduce a rather standard proof of this fact). So it makes sense to speak about embeddings on the right and on the left of (29). We prove first the right one. Let $f \in C_{\ell}^{k}(\mathbb{R})$. The norm in $H_{\ell+\varepsilon+1/2}^{k}$ is equivalent to the norm

$$\begin{aligned} \|\!| \phi |\!\|_{H_{\ell+\varepsilon+1/2}^{k}} &= \| (1 - \frac{\partial^2}{\partial y^2})^{k/2} (1 + y^2)^{-(\ell+\varepsilon/2)-1/4} \phi \|_{L_2} = \\ &= \{ \sum_{j=0}^{k} \frac{k!}{j!(k-j)!} \cdot \int_{-\infty}^{\infty} | \frac{\partial^j}{\partial y^j} (1 + y^2)^{-(\ell+\varepsilon/2)-1/4} \phi |^2 dy \}^{1/2}. \end{aligned} \tag{30}$$

Since $f \in C_{\ell}^{k}(\mathbb{R})$,

$$| \frac{\partial^j}{\partial y^j} (1 + y^2)^{-(\ell+\varepsilon/2)-1/4} f |^2 \leqslant C \| f \|^2_{C_{\ell}^{k}} (1 + y^2)^{-1/2-\varepsilon}. \tag{31}$$

The right-hand side of (31) being summable in $\mathbb{R}$, $f \in H_{\ell+\varepsilon+1/2}^{k}$ and

$$\| f \|_{H_{\ell+\varepsilon+1/2}^{k}} \leqslant \text{const}\, \| f \|_{C_{\ell}^{k}}.$$

Now let $f \in H_{\ell}^{k+\varepsilon+1/2}(\mathbb{R})$, $f_n \in C_0^{\infty}$, $f_n \underset{n\to\infty}{\to} f$ in $H_{\ell}^{k+\varepsilon+1/2}(\mathbb{R})$. Then the Fourier transforms

$$\tilde{f}_n(t) \equiv F(f_n)(t) = \frac{1}{\sqrt{(2\pi)}} \int e^{-iyt} f_n(y)dy \tag{32}$$

lie in $S(\mathbb{R})$, hence in $C_k^{\ell+}(\mathbb{R})$. It suffices to show that the sequence $\tilde{f}_n(t)$ is fundamental in $C_k^{\ell*}(\mathbb{R})$. Denote $F_{mn}(y) = f_m(y) - f_n(y)$, $\tilde{F}_{mn}(t) = \tilde{f}_m(t) - \tilde{f}_n(t)$. Let $h \in C_k^{\ell}(\mathbb{R})$. Then we obtain by Lemma 5

$$\int_{-\infty}^{\infty} h(t)\tilde{F}_{mn}(t)dt = \int_{-\infty}^{\infty} \frac{h(t)}{(1+t^2)^{k/2+\varepsilon/2+1/4}} [(1+t^2)^{k/2+\varepsilon/2+1/4}\tilde{F}_{mn}(t)]dt =$$

$$= \int_{-\infty}^{\infty} \frac{h(t)}{(1+t^2)^{k/2+\varepsilon/2+1/4}} \{(1+\frac{\partial}{\partial t})^{\ell} I^{\ell}(1+t^2)^{k/2+\varepsilon/2+1/4}\tilde{F}_{mn}(t)\}dt =$$

$$= \int_{-\infty}^{\infty} \{(1-\frac{\partial}{\partial t})^{\ell} \frac{h(t)}{(1+t^2)^{k/2+\varepsilon/2+1/4}}\}\{I^{\ell}(1+t^2)^{k/2+\varepsilon/2+1/4}\tilde{F}_{mn}(t)\}dt \leqslant$$

$$\leqslant \text{const}\|h\|_{C_k^{\ell}} \left\|\frac{1}{(1+t^2)^{1/4+\varepsilon/2}}\right\|_{L^2} \|(I^{\ell}(1+t^2)^{k/2+\varepsilon/2+1/4}\tilde{F}_{mn}(t)\|_{L^2} =$$

$$= \text{const}\,\|h\|_{C_k^{\ell}} \|(i+y)^{-\ell}(1-\frac{\partial^2}{\partial y^2})^{k/2+\varepsilon/2+1/4}F_{mn}(y)\|_{L^2} \leqslant$$

$$\leqslant \text{const}\,\|h\|_{C_k^{\ell}} \|F_{mn}\|_{H_{\ell}^{k+\varepsilon+1/2}}. \tag{33}$$

Since $\|F_{mn}\|_{H_{\ell}^{k+\varepsilon+1/2}} \to 0$ as $m,n \to \infty$, we obtain that $\tilde{f}_n(t)$ is a fundamental sequence in $\tilde{C}_k^{\ell*}(\mathbb{R})$. Thus the lemma is proved.

<u>Proof of Theorem 2</u>. (a) If $f \in B_k^{\infty}(\mathbb{R})$, then $f \in B_k^N(\mathbb{R})$ for some N and, by (29), $f \in C_N^k(\mathbb{R})$. Conversely, let $f \in C_N^{k+1}(\mathbb{R})$. Twice using (29), we write the chain of inclusions (we set $\varepsilon = 1/2$):

$$f \in C_N^{k+1}(\mathbb{R}) \subset H_{N+1}^{k+1}(\mathbb{R}) \subset B_k^{N+1}(\mathbb{R}) \subset B_k^{\infty}(\mathbb{R}), \tag{34}$$

thus (a) is proved. To prove (b) we introduce the following:

<u>Definition 3</u>. Let $T_j \in C_k^{\ell_j+}(\mathbb{R})$, $j = 1,2$. The convolution T of T_1 and T_2 is an element of $C_k^{(\ell_1+\ell_2)+}(\mathbb{R})$, defined as

$$T(\phi) \equiv (T_1 * T_2)(\phi) = T_{1t}[T_{2\tau}(\phi(t+\tau))], \tag{35}$$

where the subscript t (resp. τ) denotes that the corresponding functional acts on functions of a variable t (resp. τ). We claim that the definition is correct and the following inequality holds

$$\|T_1 * T_2\|_{C_k^{\ell_1+\ell_2}} \leqslant \|T_1\|_{C_k^{\ell_1}} \|T_2\|_{C_k^{\ell_2}}. \tag{36}$$

Further, we assert that

$$F^{-1}(T_1 * T_2) = \sqrt{2\pi}F^{-1}(T_1)F^{-1}(T_2) \tag{37}$$

(pointwise multiplication on the right-hand side of (37)). Since the norm in B_k^{ℓ} is given by (24), (b) is clearly a consequence of (36) and (37). To prove the correctness of definition (35), we show that $f(t) = T_{2\tau}(\phi(t+\tau))$ is the element of $C_k^{\ell_1}(\mathbb{R}_t)$ and

$$\| f \|_{C_k^{\ell_1}} \leqslant \| T_2 \|_{C_k^{\ell_2 *}} \| \phi \|_{C_k^{\ell_1+\ell_2}}. \tag{38}$$

It suffices to show that the latter proposition is true in the case when T_2 is finite. Indeed, in the general case T_2 is a limit of the sequence T_{2n}, $n = 1,2,\ldots$, of finite functionals. (38) implies then, that the sequence $T_{2n\tau}(\phi(t + \tau))$ is fundamental in $C_k^{\ell_1}(\mathbb{R}_t)$ and henceforth convergent to some element $f \in C_k^{\ell_1}(\mathbb{R}_t)$. In particular, $T_{2n\tau}(\phi(t + \tau)) \underset{n\to\infty}{\to} f(t)$ for arbitrary fixed $t \in \mathbb{R}$. Thus $f(t) = T_{2\tau}(\phi(t + \tau))$. Consider now the case of finite T_2. Let $\psi_n(\tau) = \psi(\tau/n)$ be the same function as in Lemma 1, satisfying the subsidiary conditions $0 \leqslant \psi(\tau) \leqslant 1$ for all τ. We have

$$T_{2\tau}(\phi(t+\tau)) = T_{2\tau}(\phi(t+\tau)\psi_n(\tau)), \; n \geqslant n_o, \tag{39}$$

where n_o is large enough. We show that the mapping

$$F_n : t \to F_n(t), \; F_n(t)(\tau) = \phi(t+\tau)\psi_n(\tau)$$

is an element of $C_k^{\ell_1}(\mathbb{R}_t, C_k^{\ell_2}(\mathbb{R}_\tau))$, and

$$\| F_n \|_{C_k^{\ell_1}} \leqslant \| T_2 \|_{C_k^{\ell_2 *}} \| \phi \|_{C_k^{\ell_1+\ell_2}} + O(1/n), \text{ as } n \to \infty; \tag{40}$$

this immediately implies our proposition.

Since supp ψ_n is compact, all the derivatives $\dfrac{\partial^{\alpha+\beta}}{\partial t^\alpha \partial \tau^\beta}(\phi(t+\tau)\psi_n(\tau))$ are uniformly continuous for $\alpha \leqslant \ell_1$, $\beta \leqslant \ell_2$, when t lies in a bounded set. It follows that these derivatives may be interpreted as continuous mappings $R \to C(\mathbb{R})$, and thus $t \to F(t)$ is a mapping from $\mathbb{R}$ to $C_k^{\ell_1}(\mathbb{R})$, continuous together with ℓ_2 derivatives. Further,

$$\left\| \frac{\partial^r F(t)}{\partial t^r} \right\|_{C_k^{\ell_1}} = \sup_{\tau\in\mathbb{R}} (1 + |\tau|)^{-k} \sum_{j=0}^{\ell_1} \left| \frac{\partial^{j+r}}{\partial \tau^j \partial t^r} [\phi(t+\tau)\psi_n(\tau)] \right| \leqslant$$

$$\leqslant (1 + |t|)^k \sup_{\tau\in\mathbb{R}} (1 + |t+\tau|)^{-k} \sum_{j=0}^{\ell_1} \sum_{s=0}^{\ell_2} \left| \frac{\partial^{j+s}}{\partial \tau^j \partial t^s} [\phi(t+\tau)\psi_n(\tau)] \right| =$$

$$= (1 + |t|)^k \{ \| \phi \|_{C_k^{\ell_1+\ell_2}} + O(1/n) \}; \tag{41}$$

thus $F \in C_k^{\ell_2}(\mathbb{R}, C_k^{\ell_1}(\mathbb{R}))$ and the required estimate (40) is valid. (We used the inequality

$$1 + |\tau| \geqslant \frac{1 + |t+\tau|}{1 + |t|} \tag{42}$$

here.) Now we have for $T_i \in C_k^{\ell_i}(\mathbb{R})$, $i = 1,2$,

$$F^{-1}(T_1 * T_2) = \frac{1}{\sqrt{(2\pi)}} (T_1 * T_2)(e^{iyt}) =$$

$$= \sqrt{2\pi} \frac{1}{\sqrt{(2\pi)}} T_{1t}\left(\frac{1}{\sqrt{(2\pi)}} T_{2\tau}(e^{i(t+\tau)})\right) =$$

$$= \sqrt{2\pi} \frac{1}{\sqrt{(2\pi)}} T_{1t}\left(e^{iyt} \frac{1}{\sqrt{(2\pi)}} T_{2\tau}(e^{iy\tau})\right) = \sqrt{2\pi} F(T_1) F(T_2). \tag{43}$$

So (37) is valid. Thus (b) is proved. We have shown simultaneously that convolution is an associative and commutative operation (since the same holds for multiplication). Further, the Fourier transform of $y^s e^{i\tau y}$ up to the factor coincides with $\delta^{(s)}(t-\tau)$ (s-th derivative of Dirac δ-function) and thus defines the finite continuous functional on $\tilde{\mathcal{B}}_k^\ell$ for $\ell \geqslant s$. (c) is proved. To prove (d) we note that if $\chi(t)$ were a continuous linear form on $C_k^0(\mathbb{R})$, it would be a continuous linear form on $\mathcal{D}_o(\mathbb{R})$. (Here $\mathcal{D}_o(\mathbb{R})$ is a space of continuous functions on $\mathbb{R}$ with compact support. The sequence $f_n \in \mathcal{D}_o(\mathbb{R})$, $n = 1,2,\ldots$, converges to zero if $\operatorname{supp} f_n \subset K$ for some fixed compact K not depending on n and $f_n(t)$ converges to zero uniformly on K.) We have

$$\chi(y) = \operatorname{sign} y + \chi_1(y), \tag{44}$$

where $\chi_1(y)$ is a finite function with a jump discontinuity at $y = 0$. Then $F(\chi_1)$ is a continuous function and thus defines a continuous functional on $\mathcal{D}_o(\mathbb{R})$. The Fourier transform of sign y is, up to a factor, the distribution v.p.(1/t). (Recall that

$$\text{v.p.}(1/t)(\phi) = \lim_{\varepsilon\to 0} \int_{|t|\geqslant\varepsilon} \frac{\phi(t)}{t}\,dt,\ \phi \in C_o^\infty(\mathbb{R}^1).) \tag{45}$$

Let now $\psi \in C_o^\infty(\mathbb{R})$, $\psi(t) \geqslant 0$, $\psi(t) = 1$ for $|t| \leqslant 1$ be fixed. Let also $\phi(t)$ be a smooth function $\phi(t) \geqslant 0$, $\phi(t) \equiv 0$ for $t \leqslant 0$, $\phi(t) \equiv 1$ for $t \geqslant 1$. Set $\phi_n(t) = \psi(t)\phi(nt)(\ell n\, n)^{-1/2}$. Then we have $\phi_n(t) \in C_o^\infty(\mathbb{R})$, $\phi_n \underset{n\to\infty}{\to} 0$ in $\mathcal{D}_o(\mathbb{R})$. On the other hand

$$\text{v.p.}(1/t)(\phi_n) = \int_0^\infty \psi(t)\,\frac{\phi(nt)}{t}\,dt(\ell n\, n)^{-1/2} \geqslant$$

$$\geqslant (\ell n\, n)^{-1/2} \int_{1/n}^1 \psi(t)\,\frac{\phi(nt)}{t}\,dt = (\ell n\, n)^{-1/2} \int_{1/n}^1 \frac{dt}{t} =$$

$$= (\ell n\ n)^{-1/2}\ell n\ n = (\ell n\ n)^{1/2} \to \infty \text{ as } n \to \infty \tag{46}$$

Thus v.p.(1/t) does not define a continuous functional on $\mathcal{D}_o(\mathbb{R})$. It follows that $\chi \notin \mathcal{B}_k^0$ and the proof of the theorem is complete.

Now we have come near to the basic definition of this item. Let A be a generator of degree k in a Banach space X. If $x \in X$ lies in the domain D_{A^ℓ}, then $e^{iAt}x$ belongs to $C_k^\ell(\mathbb{R},x)$ (see Theorem 5 (a) of item A, Eq. (26)). Thus, if $f \in \mathcal{B}_k^\ell(\mathbb{R})$, the expression

$$f_o(A)x = \frac{1}{\sqrt{(2\pi)}} \int_{-\infty}^\infty F(f)(t)e^{iAt}x\,dt,\ x \in D_{A^\ell} \tag{47}$$

(see notation (9)) is, by Theorem 1, a correctly defined element of X.

<u>Definition 4</u>. Let A be a generator of degree k in X, $f \in \mathcal{B}_k^\ell(\mathbb{R})$. We denote by $f(A) \equiv \mu_A(f)$ the closure of the operator $f_o(A)$, given by equality (47). The operator f(A) is called a function of A with symbol f.

<u>Theorem 3</u>. (a) Definition 4 is correct (i.e., $f_o(A)$ is a closurable linear operator).

(b) If $f \in \mathcal{B}_k^\ell(\mathbb{R})$, $f(A)D_{A^s} \subset D_{A^{s-\ell}}$ for $s \geqslant \ell$, and the following estimate is valid:

$$\|A^{s-\ell}f(A)x\| + \|f(A)x\| \leqslant M_{s\ell}\|f\|_{\mathcal{B}_k^\ell}(\|A^s x\| + \|x\|),\ x \in D_{A^\ell}; \tag{48}$$

in particular f(A) is a bounded operator in X for $f \in \mathcal{B}_k^0(\mathbb{R})$.

(c) If $f \in \mathcal{B}_k^\ell(\mathbb{R})$, each dense linear subset of D_{A^ℓ}, invariant under $\exp(iAt)$, $t \in \mathbb{R}$ is a core of $f(A)$.

(d) The mapping $\mu_A : f \to f(A)$, $f \in \mathcal{B}_k^\infty(\mathbb{R})$ possesses the following algebraic properties:

$$\overline{\mu_A(f)\mu_A(g)} = \mu_A(fg), \text{ or } \overline{f(A)g(A)} = (fg)(A)$$
$$\mu_A(\alpha f + \beta g) = \overline{\alpha\mu_A(f) + \beta\mu_A(g)}, \text{ or } \overline{\alpha f(A) + \beta g(A)} = (\alpha f + \beta g)(A) \tag{49}$$

(the line above the operator denotes its closure),

$$\mu_A(e^{iyt}) = e^{iAt}, \ t \in \mathbb{R}, \ \mu_A(y) = A$$
$$\mu_A((\lambda - y)^{-1}) = R_\lambda(A), \ \mathrm{Im}\lambda \neq 0. \tag{50}$$

Before proving this theorem, we carry out some discussion of the result. The theorem asserts that the image of μ_A is a set in $C(X)$, which forms a commutative and associative algebra with respect to the operations

$$\alpha \cdot B = \begin{cases} \alpha B, & \alpha \neq 0 \\ 0, & \alpha = 0, \end{cases} \qquad B \dotplus C = \overline{B + C}, \ B \circ C = \overline{BC}. \tag{51}$$

Further, μ_A is the homomorphism of algebras continuous in the sense (48) (in particular $\mu_A|_{\mathcal{B}_k^0} : \mathcal{B}_k^0(\mathbb{R}) \to \mathcal{B}(X)$, where $\mathcal{B}(X)$ is the set of continuous linear operators in X, is a Banach algebras' homomorphism). We pay more attention to the latter assertion. Introduce the following norm on D_{A^ℓ}:

$$\|x\|_\ell = \|x\| + \|A^\ell x\|, \ x \in D_{A^\ell}, \ \ell = 1,2,\ldots, \tag{52}$$

and denote by H_A^ℓ the space D_{A^ℓ} supplied with norm (52) (for $\ell = 0$, we set $H_A^0 = X$, $\|x\|_0 = \|x\|$).

<u>Lemma 7</u>. H_A^ℓ is a Banach space. Embeddings $H_A^\ell \to H_A^{\ell-1}$, $\ell = 1,2,\ldots$, are continuous and dense. The operator A^s is continuous from H_A^ℓ to H_A^m for $\ell \geqslant s + m$. The operator $(R_\lambda(A))^s$, $\lambda \in \rho(A)$ is continuous from H_A^ℓ to H_A^m for $\ell \geqslant m - s$. A is a generator of degree k in H_A^ℓ for each ℓ.

<u>Proof</u>. First we prove that H_A^ℓ is a Banach space (i.e., is complete). Recall that A^ℓ is closed (Theorem 5 (b), item A). Let the sequence x_n be fundamental in the norm $\|\cdot\|_\ell$. This means exactly that x_n is fundamental in X (and hence convergent to some $x \in X$) and $A^\ell x_n$ is fundamental in X (and hence convergent to some $y \in X$). Thus $x \in D_{A^\ell}$ and $A^\ell x = y$; $\|x - x_n\|_\ell = \|x - x_n\| + \|A^\ell x_n - y\| \to 0$ as $n \to \infty$, i.e., $x_n \underset{n\to\infty}{\to} x \in H_A^\ell$ in the norm $\|\cdot\|_\ell$.

Next we prove that $A^s : H_A^\ell \to H_A^m$, $\ell \geqslant s + m$ is continuous (in particular for $s = 0$ we obtain that the embeddings $H_A^\ell \to H_A^m$, $\ell \geqslant m$ are continuous).

Clearly $A^s H_A^\ell \subset H_A^m$ for $\ell \geqslant s + m$. Let now $x_n \in H_A^\ell$, $n = 1,2,\ldots$, and $\|x_n\|_\ell \to 0$, $\|A^s x_n - y\|_m \to 0$ as $n \to \infty$. It follows that $\|x_n\| \to 0$, $\|A^s x_n - y\| \to 0$. Since A^s is closed in x (Theorem 5(b), item A), $y = 0$, and we obtain that $A^s : H_A^\ell \to H_A^m$ is closed and everywhere defined. By the closed graph theorem,

A^s is bounded. Now we are able to show that $(R_\lambda(A))^s$ is a bounded operator from H_A^ℓ to $H_A^{\ell+s}$ (the case $m < \ell + s$ follows automatically, since the embeddings $H_A^\ell \to H_A^{\ell-1}$ are continuous as proved already). Indeed, $(A-\lambda)^s D_{H_A^{\ell+s}} =$ $= D_{H_A^\ell}$ for $\lambda \in \rho(A)$ and this operator is continuous. By Banach theorem (on the open mapping) $(R_\lambda(A))^s = ((\lambda - A)^s)^{-1}$ is also continuous. To prove that H_A^ℓ is dense in $H_A^{\ell-1}$ (with respect to the norm $\|\cdot\|_{\ell-1}$), we make use of the following commutative diagram:

$$\begin{array}{ccc} H_A^\ell & \xrightarrow{\text{embedding}} & H_A^{\ell-1} \\ (\lambda - A)^{\ell-1} \downarrow\uparrow (R_\lambda(A))^{\ell-1} & & (R_\lambda(A))^{\ell-1} \uparrow\downarrow (\lambda - A)^{\ell-1} \quad (\lambda \in \rho(A)), \\ H_A^1 = D_A & \xrightarrow{\text{embedding}} & H_A^0 = X \end{array} \tag{53}$$

which gives reduction to the fact that D_A is dense in X (Theorem 1 (a) of item A). Further,

$$\begin{aligned} \|\exp(itA)x\|_\ell &= \|\exp(itA)x\| + \|A^\ell \exp(itA)x\| = \\ &= \|\exp(itA)x\| + \|\exp(itA)A^\ell x\| \leqslant \\ &\leqslant C(1+|t|)^k(\|x\| + \|A^\ell x\|) = C(1+|t|)^k \|x\|_\ell , \end{aligned} \tag{54}$$

so A is a generator of degree k, and the lemma is proved.

<u>Definition 5</u>. The sequence of dense embeddings

$$X = H_A^0 \supset H_A^1 \supset H_A^2 \supset \ldots \supset H_A^\ell \supset \ldots \tag{55}$$

is called a Banach scale* generated by A. (48) means that μ_A realizes a continuous mapping from $\mathcal{B}_k^\ell(\mathbb{R})$ to $\mathcal{B}(H_A^s, H_A^{s-\ell})$ (here $\mathcal{B}(X,Y)$ is the space of bounded linear operators from X to Y). In other words, μ_A is a continuous homomorphism of the algebra $\mathcal{B}_k^\infty(\mathbb{R})$ into the algebra of continuous operators in the Banach scale (55) (we do not give the detailed definitions here; these will be given in Section 3).

<u>Proof of Theorem 3</u>. First of all we notice that if $x \in H_A^s$, $e^{iAt}x =$ $= \mathcal{U}(t)x \in C_k^\ell(\mathbb{R}, H_A^{s-\ell})$ for $s \geqslant \ell$, and

$$\|\mathcal{U}(t)x\|_{C_k^\ell(\mathbb{R}, H_A^{s-\ell})} \leqslant \text{const}\, \|x\|_{H_A^s}. \tag{56}$$

Indeed, this is an easy consequence of Lemma 7 and equality (26) of item A. Thus, if $f \in \mathcal{B}_k^\ell(\mathbb{R})$, $T = 1/\sqrt{2\pi}\mathcal{F}(f)$, $h = T(\mathcal{U}(t)x)$ is a correctly defined element of $H_A^{s-\ell}$ (Theorem 3), and

$$\|h\|_{H_A^{s-\ell}} \leqslant \text{const}\, \|f\|_{\mathcal{B}_k^\ell} \|x\|_{H_A^s}. \tag{57}$$

On the other hand, we have

$$f(A)x = h_1 = T(\mathcal{U}(t)x), \tag{58}$$

*See Section 3 for generalizations.

where the pairing of $C_k^{\ell}(\mathbb{R},X)$ and $C_k^{\ell+}(\mathbb{R})$ stands on the right. To prove that h coincides with h_1 (under the embedding $H_A^{s-\ell} \subset X$), it suffices to prove that $\langle\chi,h\rangle = \langle\chi,h_1\rangle$ for each $\chi \subset X^*$ (recall that $X^* \subset (H_A^{s-\ell})^*$ and is dense since $H_A^{s-\ell}$ is dense in X). But the latter assertion is evident since

$$\langle\chi,h\rangle = \langle\chi,h_1\rangle = T(\langle\chi,\mathcal{U}(t)x\rangle)$$

with usual pairing of $C_k^{\ell+}(\mathbb{R})$ and $C_k^{\ell}(\mathbb{R})$ on the right. Thus (b)) is proved.

Next we show that $R_\lambda(A)f(A)x = f(A)R_\lambda(A)x$, $f \in \mathcal{B}_k^{\ell}(\mathbb{R})$, $x \in D_{A^\ell}$. Indeed, for any $\chi \in X^*$

$$\langle\chi,R_\lambda(A)f(A)x\rangle = \langle R_\lambda(A)^*\chi, f(A)x\rangle =$$
$$= \frac{1}{\sqrt{(2\pi)}} F(f)(\langle R_\lambda(A)^*\chi,\mathcal{U}(t)x\rangle) =$$
$$= \frac{1}{\sqrt{(2\pi)}} F(f)(\langle\chi,R_\lambda(A)(t)x\rangle) = \frac{1}{\sqrt{(2\pi)}} F(f)(\langle\chi,\mathcal{U}(t)R_\lambda(A)\,x\rangle) =$$
$$= \langle\chi,f(A)R_\lambda(A)x\rangle \tag{59}$$

(we used Theorem 1 (c) of item A).

Now let $f \in \mathcal{B}_k^{\ell}(\mathbb{R})$, $x_n \in D_{A^\ell}$, $n = 1,2,\ldots$, $x_n \underset{n\to\infty}{\to} 0$ in X, $y_n = f(A)x_n \underset{n\to\infty}{\to} y$ in X. Set

$$z_n = (R_\lambda(A))^{\ell} y_n \tag{60}$$

for some $\lambda \in \rho(A)$. Then

$$\begin{aligned} \|z_n\| &= \|(R_\lambda(A))^{\ell} f(A)x_n\| = \|f(A)(R_\lambda(A))^{\ell} x_n\| \leqslant \\ &\leqslant \|f(A)\|_{H_A^{\ell}\to X} \|(R_\lambda(A))^{\ell}\|_{X\to H_A^{\ell}} \|x_n\| \to 0 \end{aligned} \tag{61}$$

as $n \to \infty$. Further, $y_n = (\lambda - A)^{\ell} z_n \underset{n\to\infty}{\to} y$. Since $(\lambda - A)^{\ell}$ is closed, then $y = 0$ and (a) is proved.

Let $D \subset D_{A^\ell}$ be a dense linear subset, invariant under $\exp(iAt)$, $t \in \mathbb{R}$, $f \in \mathcal{B}_k^{\ell}$. To prove that D is a core of $f(A)$ we need to show that the domain of closure of restriction of $f(A)$ on D contains D_{A^ℓ}. Let $x \in D_{A^\ell}$. By Theorem 5 (c) of item A, D is a core of A^ℓ, so there exists a sequence $x_n \in D$, $n = 1,2,\ldots$, such that $x_n \underset{n\to\infty}{\to} x$, $A^\ell x_n \underset{n\to\infty}{\to} Ax$. By (48)

$$\|f(A)x_n - f(A)x_m\| \leqslant \text{const}(\|x_n - x_m\| + \|A^\ell x_n - A^\ell x_m\|), \tag{62}$$

so that the sequence $f(A)x_n$ is fundamental and therefore converges to some $y \in X$. It follows that x lies in the domain of closure of $f(A)$ restricted on D. (c) is proved. Set $D = \bigcap_{\ell=1}^{\infty} D_{A^\ell}$. It is invariant under $\exp(iAt)$, $t \in \mathbb{R}$, and we obtain that to prove (49), it is enough to verify that

$$\alpha f_o(A)x + \beta g_o(A)x = (\alpha f + \beta g)_o(A)x, \; x \in D, \tag{63}$$

$$f_o(A)g_o(A)x = (fg)_o(A)x, \; x \in D. \tag{64}$$

(63) is obvious. As for (64), we have $g_o(A)D \subset D$ (see (b)), and we may write for arbitrary $\chi \in X^*$:

$$\langle \chi, (fg)_o(A)x \rangle = \frac{1}{2\pi} (F^{-1}(f)*F^{-1}(g))(\langle \chi, e^{iAt}x \rangle) =$$

$$= \frac{1}{2\pi} F(f)_t(F(g)_\tau(\langle \chi, e^{iAt}e^{iA\tau}x \rangle)) =$$

$$= \frac{1}{\sqrt{(2\pi)}} F^{-1}(f)_t(\frac{1}{\sqrt{(2\pi)}} F^{-1}(g)_\tau \langle (e^{iAt})*\chi, e^{iA\tau}x \rangle) = \tag{65}$$

$$= \frac{1}{\sqrt{(2\pi)}} F(f)_t(\langle (e^{iAt})*\chi, g_o(A)x \rangle) =$$

$$= \frac{1}{\sqrt{(2\pi)}} F(f)(\langle \chi, e^{iAt}g_o(A)x \rangle) = \langle \chi, f_o(A)g_o(A)x \rangle,$$

and (64) is proved.

(50) is now obvious, since $F(e^{iy\tau}) = \sqrt{2\pi}\delta(t - \tau)$, $F(y) = i\sqrt{2\pi}\delta'(t)$ (δ denotes the Dirac delta-function). As for the last of identities (50), it follows from (49) and from the identity $(\lambda - y)(\lambda - y)^{-1} \equiv 1$. The theorem is proved.

At the end of this item we intend to show that, although rather complicated, the symbol space B_k^∞ is nevertheless quite natural for one-operator functional calculus in general Banach spaces. Namely, we present a simple example in which all bounded functions of a given generator A of the degree 0 are of the form $f(A)$, $f \in B_0^0(\mathbb{R})$. Being quite similar to the case $k = 0$, the construction of examples for $k \neq 0$ is omitted here. First we give some extension of the notion of a function of a given operator A, describing the latter in "external" terms.

Definition 6. Let B be a bounded operator in a Banach space X. Let also C be a closed operator in X. We say that B and C commute, if for every $x \in D_C$, $Bx \in D_C$, and

$$BCx = CBx. \tag{66}$$

Example 1. A being a generator, A commutes with e^{iAt}.

Definition 7. Let A be a closed operator in a Banach space X. The closed operator C in X is called a function of A, if every bounded operator B, commuting with A, commutes also with C.

Lemma 8. Let A be a generator of degree k in a Banach space X, $f \in B_k^\ell(\mathbb{R})$. Then $f(A)$ is a function of A in the sense of Definition 7.

Proof. Let B be a bounded operator in X commuting with A. Then B commutes with $\exp(iAt)$. Indeed, for $x \in D_A$ $\exp(iAt)Bx$ and $B\exp(iAt)x$ both satisfy the same Cauchy problem ((3), item A) with $x_o = Bx$, and therefore coincide (item A, Theorem 1 (d)). Next we note that, commuting with A, B commutes with A^ℓ as well. Let $x \in D_{A^\ell}$. We have

$$\begin{aligned}\langle \chi, Bf_o(A)x \rangle &= \langle B^*\chi, f_o(A)x \rangle = \frac{1}{\sqrt{(2\pi)}} F(f)(\langle B^*\chi, \exp(iAt)x \rangle) = \\ &= \frac{1}{\sqrt{(2\pi)}} F(f)(\langle \chi, \exp(iAt)Bx \rangle) = \langle \chi, f_o(A)Bx \rangle, \quad \chi \in X^*,\end{aligned} \tag{67}$$

so $Bf_o(A) = f_o(A)B$. Passing to the closure $f(A)$ of $f_o(A)$, we obtain the desired result. Now we have all we need to construct the next example.

Example 2. Let X be a space of complex-valued continuous functions $f(\tau)$ of the variable $\tau \in \mathbb{R}$ such that $\lim_{|\tau| \to \infty} f(\tau) = 0$, equipped with the usual C-norm:

$$\|f\|_X = \sup_{\tau\in\mathbb{R}} |f(\tau)|. \tag{68}$$

X is a Banach space, and each $f\in X$ is a uniformly continuous function. Indeed, given an $\varepsilon > 0$, choose R such that $|f(\tau)| < \varepsilon/3$ for $|\tau| \geqslant R$. Then $f(\tau)$ is uniformly continuous on $[-R,R]$ so there exists $\delta > 0$ such that $|f(\tau_1) - f(\tau_2)| < \varepsilon/3$ for $\tau_1,\tau_2 \in [-R,R]$, $|\tau_1 - \tau_2| < \delta$. It is easy to verify, however, that $|f(\tau_1) - f(\tau_2)| < \varepsilon$ for any τ_1, τ_2 such that then $|\tau_1 - \tau_2| < \delta$ under these conditions.

Consider now a one-parametric group $\mathcal{U}(t)$ in X defined by

$$(\mathcal{U}(t)f)(\tau) = f(t+\tau), \; f\in X, \; t\in\mathbb{R}. \tag{69}$$

Clearly $\|\mathcal{U}(t)\| = 1$ for all t, $\mathcal{U}(t)$ is strongly continuous since the relation $\|\mathcal{U}(t)f - \mathcal{U}(t_1)f\|_X \underset{t\to t_1}{\to} 0$ is equivalent to uniform continuity of $f(\tau)$. The generator of the group $\mathcal{U}(t)$ is the operator $A = -i(\partial/\partial\tau)$; its domain D_A consists of continuously differentiable functions $\phi(\tau)$, $\tau\in\mathbb{R}$ such that $\phi\in X$ and $\phi'\in X$. Thus A is a generator of degree zero.

Proposition 1. Let C be a bounded function of A in the sense of Definition 7. Then $C = \phi(A)$ for some $\phi \in \mathcal{B}^0_o(\mathbb{R})$.

Proof. C being a function of A, it commutes in particular with $\mathcal{U}(t)$. Let now $f\in X$. We have

$$[Cf](t) = [\mathcal{U}(t)Cf](0) = [C\mathcal{U}(t)f](0) = \chi_C(\mathcal{U}(t)f), \tag{70}$$

where $\chi_C \in X'$ is defined by

$$\chi_C(f) \equiv \langle\chi_C,f\rangle = [Cf](0). \tag{71}$$

Let now $\chi\in X'$. We have

$$\langle\chi,Cf\rangle = \langle\chi_t,\langle\chi_{C\tau},\mathcal{U}(t)f\rangle\rangle = \langle\chi_t,\langle\chi_{C\tau},f(t+\tau)\rangle\rangle \tag{72}$$

(the subscripts t and τ mean the same as in Definition 3). By the Hahn-Banach theorem, χ and χ_C may be extended to linear continuous functions on $C^0_o(\mathbb{R})$, i.e., elements of $C^{0*}_o(\mathbb{R})$. We assert that these extensions may be chosen belonging to $C^{0+}_o(\mathbb{R})$. The convolution of T_1 and T_2 (Definition 3) is then well defined and commutative (see proof of Theorem 2), so that*

$$\langle\chi,Cf\rangle = \chi_{Ct}[\chi_\tau(f(t+\tau))] = \chi_C(\langle\chi,\mathcal{U}(t)f\rangle), \tag{73}$$

and, consequently, $C = \phi(A)$ where $\phi = \sqrt{2\pi}\mathcal{F}^{-1}(\chi_C) \in \mathcal{B}^0_o(\mathbb{R})$. It remains to prove the following:

Lemma 9. Every functional $\chi\in X'$ is the restriction of some element of $C^{0+}_o(\mathbb{R})$.

Proof. Consider the mapping $\psi : \mathbb{R}\to S^1$, $\psi(\tau) = 2\tan^{-1}\tau$ (here S^1 is the unit circle with the angular coordinate $\psi\in(-\pi,\pi]$). Under this mapping X is transferred into the subspace $\tilde{X}$ of $C(S^1)$ (space of continuous functions on S^1), consisting of functions vanishing at $\psi = \pi$. $C^0_o(\mathbb{R})$ is transferred into the space $\tilde{\tilde{X}}$ of bounded functions on S^1, defined and continuous everywhere except the point $\psi = \pi$. Thus we have embeddings

---

* We denote the extensions of χ and χ_C by the same symbols.

$$\tilde{X} \subset C(S^1) \subset \tilde{\tilde{X}}. \tag{74}$$

The norm in each of the three spaces is given by the same formula

$$\| f \| = \sup_{\psi \in (-\pi,\pi)} |f(\psi)|. \tag{75}$$

Finite functionals on $C^0_o(\mathbb{R})$ are clearly transferred into functionals on $\tilde{\tilde{X}}$, whose support does not contain the point $\psi = \pi$. Now let $\tilde{\chi} \in \tilde{X}^*$; by the Hahn-Banach theorem there exists an extension $h \in C^*(S^1)$ of $\tilde{\chi}$. By the Riesz theorem h is a Rhadon measure on S^1. Let $a = h(\{\pi\})$. Then $\tilde{h} = h - a\delta(\psi - \pi)$ is also an extension of $\tilde{\chi}$ and

$$\tilde{h}(Q) = \lim_{n\to\infty} \tilde{h}_n(Q) \equiv \lim_{n\to\infty} \tilde{h}(Q \cap [-\pi + \frac{1}{n}, \pi - \frac{1}{n}]) \tag{76}$$

for each Borel subset Q of S_1. Obviously measure $\tilde{h}_n(Q)$ induces finite functionals $\tilde{\tilde{h}}_n$ on $\tilde{\tilde{X}}$, $\tilde{\tilde{h}} = \lim_{n\to\infty} \tilde{\tilde{h}}_n$ exists in $\tilde{\tilde{X}}^*$, and restriction of $\tilde{\tilde{h}}$ onto $\tilde{X}$ coincides with $\tilde{\chi}$. Lemma 9 is proved.

F. Functions of a Self-Adjoint Operator in a Hilbert Space

Example 2 of the preceding item shows us that we may scarcely hope to enlarge the class $\mathcal{B}^0_k$ of symbols of bounded functions for general generators of degree k in general Banach spaces. It turns out, however, that in the special case (self-adjoint operators in Hilbert spaces) the class of of symbols, which give bounded operators, may be essentially enlarged.

This item may be divided into two sections. In the first section we recall some well-known facts about self-adjoint operators, while the second one is devoted to relations between two approaches to the construction of functional calculus for self-adjoint operators.

To begin with we study in short the spectral measure of a self-adjoint operator A in a Hilbert space H. No proofs will be given since we assume that the reader has already become acquainted with this material (see [13], [71] or any other textbook on functional analysis).

Definition 1. The decomposition of unity in a Hilbert space H is the family $\{E_\lambda\}_{\lambda\in\mathbb{R}}$ of bounded operators in H, possessing the following properties:

(a) For any λ, E_λ is an orthogonal projector in H, i.e.,

$$E^*_\lambda = E_\lambda, \quad E^2_\lambda = E_\lambda; \tag{1}$$

(b) $\{E_\lambda\}$ is a non-decreasing family, i.e.,

$$E_\lambda E_\mu = E_\mu E_\lambda = E_\mu \text{ for } \mu \leqslant \lambda; \tag{2}$$

(c) E_λ is strongly continuous on the left, i.e.,

$$s - \lim_{\mu\to\lambda-0} E_\mu = E_\lambda \text{ (or } \lim_{\mu\to\lambda-0} E_\mu x = E_\lambda x \text{ for all } x \in H); \tag{3}$$

(d) E_λ has strong limits as $\lambda \to \pm\infty$, and these limits are

$$s - \lim_{\lambda\to\infty} E_\lambda = I \text{ (identity operator)}, \quad s - \lim_{\lambda\to-\infty} E_\lambda = 0. \tag{4}$$

It follows from the definition that for any $x,y \in H$, the function

$$E_{x,y}(\lambda) = (E_{\lambda}x,y) \tag{5}$$

is continuous on the left and $dE_{x,y}(\lambda)$ is a correctly defined complex-valued Borel measure on $\mathbb{R}$. We shall use the notation $(dE_{\lambda}x,y)$ for this measure. Let now $\Phi(\lambda)$ be a Borel function of $\lambda \in \mathbb{R}$. Clearly Φ is measurable with respect to all the measures $(dE_{\lambda}x,y)$, $x,y \in H$. Set

$$D_{\Phi} = \{x \in H \mid |\Phi(\lambda)|^2 \text{ is integrable over } \mathbb{R} \text{ with respect to measure } (dE_{\lambda}x, x)\}. \tag{6}$$

Set also

$$(\mu(\Phi)x,y) = \int_{-\infty}^{\infty} \Phi(\lambda)(dE_{\lambda}x,y), \quad x \in D_{\Phi}, \; y \in H. \tag{7}$$

Proposition 1. The equation (7) correctly defines a closed operator $\mu(\Phi)$ with domain D_{Φ}; the following estimate is valid:

$$\|\mu(\Phi)x\|^2 = \int_{-\infty}^{\infty} |\Phi(\lambda)|^2 (dE_{\lambda}x,x). \tag{8}$$

The following notation will be used:

$$\mu(\Phi) = \int_{-\infty}^{\infty} \Phi(\lambda)dE_{\lambda}. \tag{9}$$

Now we may formulate the basic theorem on the structure of self-adjoint operators.

Theorem 1. For any self-adjoint operator A in the Hilbert space H there exists a unique decomposition of unity $E_{\lambda} = E_{\lambda}(A)$ called the spectral family of A, such that

$$A = \int_{-\infty}^{\infty} \lambda dE_{\lambda}. \tag{10}$$

This spectral theorem enables us to construct the functional calculus for a given self-adjoint operator A. Given a Borel function $\Phi(\lambda)$, $\lambda \in \mathbb{R}$, we set

$$\Phi(A) = \int_{-\infty}^{\infty} \Phi(\lambda)dE_{\lambda}(A). \tag{11}$$

The properties of the introduced correspondence $\Phi \to \Phi(A)$ are summarized in the following:

Theorem 2. (a) If $\Phi(\lambda)$ is a bounded function, the operator $\Phi(A)$ is bounded and

$$\|\Phi(A)\| \leqslant \sup_{\lambda \in \sigma(A)} |\Phi(\lambda)|. \tag{12}$$

(b) Moreover, if $|\Phi(\lambda)| \leqslant C(1 + |\lambda|^k)$, $\lambda \in \sigma(A)$, then $D_{A^k} \subset D_{\Phi(A)}$ and

$$\|\Phi(A)x\| \leqslant C_1(\|x\| + \|A^k x\|), \quad x \in D_{A^k}. \tag{13}$$

(c) The following relations are valid:

$$(\Phi(A))^* = \bar{\Phi}(A), \tag{14}$$

$$\Phi(A) \supset \Phi_1(A)\Phi_2(A) \text{ if } \Phi(\lambda) = \Phi_1(\lambda)\Phi_2(\lambda). \tag{15}$$

We have the equality sign in (15) if both $\Phi(\lambda)$ and $\Phi_2(\lambda)$ are bounded on $\sigma(A)$.

$$\Phi(A) \supset \alpha_1\Phi_1(A) + \alpha_2\Phi_2(A) \text{ if } \Phi(\lambda) = \alpha_1\Phi_1(\lambda) + \alpha_2\Phi_2(\lambda) \tag{16}$$

(we use the notation $C_1 \subset C_2$ if $D_{C_1} \subset D_{C_2}$ and $C_1 x = C_2 x$, $x \in D_{C_1}$).

(d) If $\Phi(\lambda) = (\mu - \lambda)^{-1}$, then $\Phi(A) = R_\mu(A)$, $\mathrm{Im}\mu \neq 0$; and $\Phi(A) = A^k$, if $\Phi(\lambda) = \lambda^k$.

Proof. The assertions (a), (c), and (d) are standard, and their proofs may be found, e.g., in [13]. Let $\Phi(\lambda)$ be a Borel function, satisfying $|\Phi(\lambda)| \leqslant C(1 + |\lambda|^k)$, $\lambda \in \sigma(A)$. Then we have for $x \in D_{A^k}$:

$$\begin{aligned} \int_{-\infty}^{\infty} |\Phi(\lambda)|^2 (dE_\lambda x, x) &\leqslant \tilde{C}\left(\int_{-\infty}^{\infty} (dE_\lambda x, x) + \int_{-\infty}^{\infty} \lambda^{2k} (dE_\lambda x, x)\right) \leqslant \\ &\leqslant \tilde{C}(\|x\|^2 + \|A^k x\|^2) < \infty, \end{aligned} \tag{17}$$

and this implies (b) immediately. Thus the theorem is proved.

Next we investigate the relations between the definition of $\Phi(A)$ introduced above and that given in the preceding item (Definition 4 of item D).

Theorem 3. Both definitions coincide for $\Phi \in \mathcal{B}_0^\infty(\mathbb{R})$.

Proof. (a) We show first that $\Phi_t(A) = \exp(itA)$, where $\Phi_t(\lambda) = \exp(it\lambda)$. Indeed, $\|\Phi_t(A)\| \leqslant 1$ for all $t \in \mathbb{R}$. Next, let $x \in D_{A^2}$. We have

$$\begin{aligned} \Phi_{t+\varepsilon}(\lambda) &= e^{i\lambda(t+\varepsilon)} = e^{i\lambda t}(1 + i\lambda\varepsilon) + \varepsilon^2\lambda^2 F_\varepsilon(\lambda) = \\ &= \Phi_t(\lambda) + i\varepsilon\Phi_t(\lambda)\lambda + \varepsilon^2\lambda^2 F_{\varepsilon,t}(\lambda), \end{aligned} \tag{18}$$

where $F_{\varepsilon,t}(\lambda) = \dfrac{e^{i\lambda\varepsilon} - 1 - i\lambda\varepsilon}{\varepsilon^2\lambda^2} e^{i\lambda t}$ is a bounded function.

By Theorem 2 (c) we conclude that

$$\Phi_{t+\varepsilon}(A)x - \Phi_t(A)x = i\varepsilon A(\Phi_t(A)x) + \varepsilon^2 F_{\varepsilon,t}(A)A^2 x \tag{19}$$

and besides $F_{\varepsilon,t}(A)$ is bounded uniformly with respect to ε. Consequently, $\Phi_t(A)x$ is differentiable with respect to t and satisfies the Cauchy problem

$$i\frac{\partial}{\partial t}(\Phi_t(A)x) + A(\Phi_t(A)x) = 0, \quad \Phi_0(A)x = x. \tag{20}$$

By Theorem 1 (d) of item A, $\Phi_t(A)x = \exp(itA)x$. Since D_{A^2} is dense in $\mathcal{H}$ and $\Phi_t(A)$ is bounded, $\Phi_t(A) = \exp(itA)$.

(b) Let now $\Phi \in \mathcal{B}_0^\infty(\mathbb{R})$ be arbitrary. The Fourier transform of Φ, which we denote by $\tilde{\Phi}(t) = F(\Phi)(t)$, lies in $C_0^{\ell+}(\mathbb{R})$ for some natural ℓ. We intend to prove that

$$\int_{-\infty}^{\infty} \Phi(\lambda)(dE_\lambda x, y) = \frac{1}{\sqrt{(2\pi)}}\tilde{\Phi}((\exp(itA)x, y)), \tag{21}$$

$x \in D_{A^\ell}$, $y \in \mathcal{H}$ (here $E_\lambda = E_\lambda(A)$). (21) means that on D_{A^ℓ} both definitions of $\Phi(A)$ give identical results. Since D_{A^ℓ} is a core of $\Phi(A)$ (it is postulated in the former definition and it will be proved below for the latter), this implies the statement of the Theorem. We have

$$\int_{-\infty}^{\infty} \Phi(\lambda)(dE_\lambda x, y) = \int_{-\infty}^{\infty} \frac{1}{\sqrt{(2\pi)}}[\tilde{\Phi}(e^{it\lambda})](dE_\lambda x, y); \tag{22}$$

$$\frac{1}{\sqrt{(2\pi)}}\tilde{\Phi}((\exp(itA)x, y)) = \frac{1}{\sqrt{(2\pi)}}\tilde{\Phi}\left(\int_{-\infty}^{\infty} e^{it\lambda}(dE_\lambda x, y)\right), \tag{23}$$

(23) being valid by part (a) of the proof. Thus we have to verify the possibility of the application of $\tilde{\Phi}$ under the integral sign. Let $\tilde{\Phi}$ be a finite functional. Then applying Lemma 2 of item E, we obtain

$$\tilde{\Phi}\left(\int_{-\infty}^{\infty} e^{it\lambda}(dE_\lambda x,y)\right) = \sum_{j=0}^{\ell} \int_{-R}^{R} d\mu_j(t)\left\{ \frac{\partial^j}{\partial t^j} \int_{-\infty}^{\infty} e^{it\lambda}(dE_\lambda x,y)\right\}, \tag{24}$$

where $d\mu_j$ are the Rhadon measures on $[-R,R]$. Since $x \in D_{A^\ell}$, all the integrals $\int_{-\infty}^{\infty} e^{it\lambda}\lambda^j(dE_\lambda x,y)$, $j \leqslant \ell$, converge absolutely and we can apply the operator $\partial^j/\partial t^j$ under the integral sign. Further, the integrals $\int_{-R}^{R}\int_{-\infty}^{\infty}\lambda^j e^{it\lambda} d\mu_j(t)(dE_\lambda x,y)$ converge absolutely and by the Fubini theorem one may change the order of integration in (24). We obtain that (21) is valid for finite $\tilde{\Phi}$. Now if the sequence $\tilde{\Phi}_n$ of finite functionals converges to $\tilde{\Phi}$ in $C_o^{\ell *}(\mathbb{R}^n)$, $\Phi_n(\lambda) \to \Phi(\lambda)$ pointwise, and $|\Phi_n(\lambda)| \leqslant C(1 + |\lambda|)^\ell$, the constant C then being independent of n (this is a consequence of embedding $B_o^k(\mathbb{R}) \subset C_k^o(\mathbb{R})$, see Lemma 6 of item E). By the Dominated Convergence theorem, $\int_{-\infty}^{\infty}\Phi_n(\lambda)(dE_\lambda x,y) \to \int_{-\infty}^{\infty}\Phi(\lambda)(dE_\lambda x,y)$ as $n \to \infty$. Thus we obtain (21) for $\Phi \in B_o^\ell(\mathbb{R})$. It remains to prove that D_{A^ℓ} is a core of $\int_{-\infty}^{\infty}\Phi(\lambda)dE_\lambda$. But this is obvious. Indeed, let $x \in D_\Phi$. Set

$$x_n = \int_{-n}^{n} dE_\lambda x. \tag{25}$$

Then $x_n \in D_{A^\ell}$, $x_n \overset{n\to\infty}{\to} x$, and

$$\| \Phi(A)x - \Phi(A)x_n\|^2 = \int_{|\lambda| \geqslant n} |\Phi(\lambda)|^2(dE_\lambda x,x) \overset{n\to\infty}{\to} 0, \tag{26}$$

since $\int_{-\infty}^{\infty}|\Phi(\lambda)|^2(dE_\lambda x,x)$ is convergent. The theorem is proved.

In what follows we shall not be interested at all in arbitrary Borel functions as symbols. Only continuous symbols will be used. It follows from Theorem 2 that if $\Phi \in C_\ell^0(\mathbb{R})$, then $D_{\Phi(A)} \supset D_{A^\ell}$ and $\Phi(A)$ is an A^ℓ-bounded operator (i.e., $\|\Phi(A)x\| \leqslant C(\|x\| + \|A^\ell x\|)$). In particular, if $\Phi \in C_o^0(\mathbb{R})$, $\Phi(A)$ is a bounded operator. As Theorem 2 (d) of item E implies, $B_o^0(\mathbb{R})$ is not dense in $C_o^0(\mathbb{R})$. Nevertheless, the mapping $\Phi \to \Phi(A)$ defined here is in a sense the unique extension of that given by Definition 4 of item E. Namely, $B_o^0(\mathbb{R})$ is dense in $C_o^0(\mathbb{R})$ with respect to the locally uniform convergence of uniformly bounded sequences,* and the following theorem is valid:

<u>Theorem 4</u>. Let $\phi_n(y), \phi(y) \in C_o^0(\mathbb{R})$, $n = 1,2,\ldots$, and $\|\phi_n(y)\|_{C_o^0} \leqslant M$ for all n. Let also $\phi_n(y)$ converge to $\phi(y)$ locally uniformly with respect to y as $n \to \infty$. Then $\phi_n(A) \overset{n\to\infty}{\to} \phi(A)$ strongly.

<u>Proof</u>. By Theorem 2 (a) the operator sequence $\{\phi_n(A)\}$ is uniformly bounded. Therefore it is sufficient to prove the strong convergence $\phi_n(A)x \to \phi(A)x$ for $x \in D$, where $D \subset H$ is a dense subset defined by

$$D = \{x \in H \mid x = \int_{-R}^{R} dE_\lambda x,\ R = R(x) \to \infty\}. \tag{27}$$

If $x \in D$, $\phi_n(y) \overset{n\to\infty}{\to} \phi(y)$ uniformly on $[-R(x),R(x)]$. Thus $\|\phi_n(A)x - \phi(A)x\|^2 \leqslant \int_{-R(x)}^{R(x)} |\phi_n(\lambda) - \phi(\lambda)|^2(dE_\lambda x,x) \to 0$ as $n \to \infty$. The theorem is proved.

---

* Really, $B_o^0(\mathbb{R}) \supset C_o^\infty(\mathbb{R})$, and $C_o^\infty(\mathbb{R})$ is dense in $C_o^0(\mathbb{R})$ in the sense that is indicated above.

G. Functions of the Operator, Depending on Parameters

The results of item C are applied here to obtain some theorems on continuity of $f(A(\varepsilon))$ with respect to ε, f being a given symbols, $A(\varepsilon)$ being a family of generators depending on parameter ε and satisfying some continuity conditions. We assume everywhere below that ε is a variable lying in some open subset E of the space $\mathbb{R}^n$, endowed with the induced topology.

Theorem 1. Let $\{A(\varepsilon)\}_{\varepsilon\in E}$ be a uniform family of generators of degree k in a Banach space X. Assume that there exists a dense linear subset $D \subset X$ such that D is a core of $A(\varepsilon)$ for all $\varepsilon \in E$, and $A(\varepsilon)$ is strongly continuous on D (i.e., $A(\varepsilon)x$ is a continuous function of ε for any $x\in D$). Then for any $f \in B_k^\ell(\mathbb{R})$ the operator family $f(A(\varepsilon))$ is strongly continuous with respect to ε on the linear subset

$$D_\ell = \begin{cases} X \text{ if } \ell = 0 \\ D \text{ if } \ell = 1 \\ \left\{ \begin{matrix} x \in D \,|\, A^k(\varepsilon)x \text{ is defined and continuous} \\ \text{with respect to } \varepsilon \text{ for } k \leqslant \ell \end{matrix} \right\} \text{ if } \ell \geqslant 2. \end{cases} \tag{1}$$

Remark 1. In general, D_ℓ, $\ell \geqslant 2$, may be empty. However, $D_\ell \supset D$, if D is invariant under $A(\varepsilon)$ for all ε and $A(\varepsilon) = \Sigma_{j=1}^n \varepsilon_j A_j$. This special case will be of particular interest to us.

Proof of the theorem. Set $U(\varepsilon,t) = \exp(iA(\varepsilon)t)$. We assert that $U(\varepsilon,t)$ is strongly continuous with respect to $(\varepsilon,t)\in E\times\mathbb{R}$. Indeed, $U(\varepsilon,t)x$ is continuous in t, and $U(\varepsilon,t)x \to U(\varepsilon_o,t)x$ as $\varepsilon \to \varepsilon_o$, locally uniformly with respect to t by Theorem 3 of item C. Next, if $x\in D_\ell$, $(-i\partial/\partial t)^k U(\varepsilon,t)x = U(\varepsilon,t)(A(\varepsilon))^k x$ is continuous in (ε,t) for $k \leqslant \ell$.

Now let $f\in B_k^\ell(\mathbb{R})$, $x\in D_\ell$, and let $\tilde{f} = F(f)$ be a finite functional. We obtain, using Lemma 2 of item E:

$$f(A(\varepsilon))x = \sum_{k=0}^{\ell} \int_{\text{supp } \tilde{f}} \left(\frac{\partial^k}{\partial t^k} U(t,\varepsilon)x\right) d\mu_k(t), \tag{2}$$

$d\mu_k(t)$ being the Rhadon measures on supp $\tilde{f}$. The above consideration yields now that $f(A(\varepsilon))x$ continuously depends on ε. If $\tilde{f}$ is not finite, there exists a sequence $f_n \in B_k^\ell(\mathbb{R})$ of symbols with finite Fourier transforms $\tilde{f}_n$, convergent to f in $B_k^\ell(\mathbb{R})$. By Theorem 3 (b) of item E:

$$\| f(A)(\varepsilon))x - f_n(A(\varepsilon))x\| \leqslant C \, \| f - f_n\|_{B_k^\ell} (\| x\| + \| (A(\varepsilon))^\ell x\|). \tag{3}$$

Since $(A(\varepsilon))^\ell x$ is continuous, (3) means that $f_n(A(\varepsilon))x$ converges then to $f(A(\varepsilon))x$ locally uniformly with respect to ε, so that $f(A(\varepsilon))x$ continuously depends on ε. The theorem is proved.

Theorem 2. Let $\{A(\varepsilon)\}_{\varepsilon\in E}$ be a family of self-adjoint operators in the Hilbert space H, which are strongly continuous with respect to and essentially self-adjoint on the dense linear subset $D \subset H$. Then for any $f \in C_o^0(\mathbb{R})$ the operator family $f(A(\varepsilon))$ is strongly continuous in H with respect to ε.

Proof. Since $f(A(\varepsilon))$ is uniformly bounded (the bound is $\| f\|_{C_o^0} = \sup_{\lambda\in\mathbb{R}} |f(\lambda)|$), it is sufficient to prove that $f(A(\varepsilon))x$ is a continuous function of ε for $\chi \in D$. Our considerations are of course local and we may assume that ε lies in a compact subset $E_o \subset E$. Since $A(\varepsilon)\chi$ is continuous, we have

$$C = \sup_{\varepsilon \in E_o} \| A(\varepsilon)x \| < \infty. \tag{4}$$

We wish to obtain the uniform-in-ε approximation of $f(A(\varepsilon))x$ by $f_n(A(\varepsilon))\chi$, where $f_n \in C_o^\infty(\mathbb{R})$. Thus we have to improve the result of Theorem 4 of the preceding item.

<u>Lemma 1</u>. Let $f, g \in C_o^0(\mathbb{R})$. Denote for any $R > 0$:

$$\begin{aligned} M(R) &= \sup_{|\lambda| \geqslant R} |f(\lambda) - g(\lambda)|, \\ m(R) &= \sup_{|\lambda| \leqslant R} |f(\lambda) - g(\lambda)|. \end{aligned} \tag{5}$$

Let A be a self-adjoint operator, $x \in D_A$. Then for any $R > 0$ the following estimate is valid:

$$\| f(A)x - g(A)x \|^2 \leqslant m^2(R) \| x \|^2 + \frac{M^2(R)}{R^2} \| Ax \|^2. \tag{6}$$

<u>Proof</u>. Let $\{E_\lambda\}$ be the spectral family of A. Since $(E_\lambda x, x)$ is a non-decreasing function, we have

$$\int_{\lambda^2 \geqslant R^2} (dE_\lambda x, x) \leqslant \frac{1}{R^2} \int_{\lambda^2 \geqslant R^2} \lambda^2 (dE_\lambda x, x) \leqslant \frac{1}{R^2} \int_{-\infty}^{\infty} \lambda^2 (dE_\lambda x, x) = \frac{\| Ax \|^2}{R^2}. \tag{7}$$

Thus

$$\begin{aligned} \| f(A)x - g(A)x \|^2 &= \int_{-\infty}^{\infty} |f(\lambda) - g(\lambda)|^2 (dE_\lambda x, x) = \\ &= \int_{|\lambda| < R} |f(\lambda) - g(\lambda)|^2 (dE_\lambda x, x) + \int_{|\lambda| > R} |f(\lambda) - g(\lambda)|^2 (dE_\lambda x, x) \leqslant \\ &\leqslant m^2(R) \int_{-\infty}^{\infty} (dE_\lambda x, x) + M^2(R) \int_{|\lambda| > R} (dE_\lambda x, x) \leqslant \\ &\leqslant m^2(R) \| x \|^2 + \frac{M^2(R)}{R^2} \| Ax \|^2. \end{aligned} \tag{8}$$

The lemma is proved. Now we choose the function $f_n(\lambda) \in C_o^\infty(\mathbb{R})$ such that $\| f_n - f \|_{C_o^0} \leqslant 2 \| f \|_{C_o^0}$ and $|f_n(\lambda) - f(\lambda)| \leq (\delta / \sqrt{2} \| x \|)$ for $|\lambda| \leqslant R$, where $R = 2\sqrt{2} C \| f \|_{C_o^0} / \delta$, $\delta = 1/n$. Applying Lemma 1, we easily obtain

$$\| f_n(A(\varepsilon))x - f(A(\varepsilon))x \| \leqslant \frac{1}{n}, \tag{9}$$

so we have reduced the theorem to the case $f \in C_o^\infty(\mathbb{R})$. But $C_o^\infty(\mathbb{R}) \subset \mathcal{B}_o^0$, so we may apply Theorem 1. The theorem is proved.

<u>Remark 2</u>. The reader can easily formulate this theorem for symbols polynomially growing at infinity (similar to the way it was done in Theorem 1).

<u>Remark 3</u>. We also may prove the theorems on differentiability with respect to parameter under some stronger assumptions. However, the expression for the derivative contains functions of several operators, so we delay the proof of these theorems until we develop the suitable techniques.

H. Semigroups in the Pair of Banach Spaces

Dealing with Banach scales we shall sometimes use a generalization of the notion of the generator in a Banach space, which is described below.

Definition 1. A pair of Banach spaces is a collection of two Banach spaces X_1, X_2 and the continuous linear mapping: $j : X_1 \to X_2$.

Definition 2. A strongly continuous tempered group of linear operators in the pair $j : X_1 \to X_2$ is a strongly continuous family $\{U(t)\}_{t\in\mathbb{R}}$ of bounded linear operators $U(t) : X_1 \to X_2$ such that $U(0) = j$,

$$\| U(t)x\|_2 \leqslant C(1 + |t|)^k \| x \|_1, \quad x \in X_1 \tag{1}$$

for some C, k independent of x (k is called the degree of the group);

$$\sum_{\ell=1}^{s} U(t + \tau_\ell)x_\ell = 0 \tag{2}$$

for any $t \in \mathbb{R}$, provided that s, $\tau_\ell \in \mathbb{R}$, $x_\ell \in X_1$, $\ell = 1,\dots,s$ satisfy

$$\sum_{\ell=1}^{s} U(\tau_\ell)x_\ell = 0. \tag{3}$$

Proposition 1. For any fixed $k_1 > k$ there exists an "intermediate" Banach space X_{mid}, a generator A of degree k_1 in X_{mid}, and the pairs $j_1 : : X_1 \to X_{mid}$; $j_2 : X_{mid} \to X_2$ of Banach spaces such that

$$j = j_2 \circ j_1, \quad U(t) = j_2 \circ \exp(iAt) \circ j_1, \quad t \in \mathbb{R}. \tag{4}$$

Proof. Denote by $X_o \subset X_2$ the linear core (i.e., the set of all finite linear combinations) of the elements of the form $U(t)x$, $x \in X_1$, $t \in \mathbb{R}$. For any $t \in \mathbb{R}$ we define the operator $U_o(t) : X_o \to X_o$ by

$$U_o(t)x_o = \sum_{\ell=1}^{s} U(t + \tau_\ell)x_\ell, \quad x_o \in X_o, \tag{5}$$

where $x_o = \Sigma_{\ell=1}^{s} U(\tau_\ell)x_\ell$, $\tau_\ell \in \mathbb{R}$, $x_\ell \in X_1$, $\ell = 1,\dots, s$. The definition (5) is correct since, if x_o has two different representations of the form given above, (5) gives identical results for both of them (see (2) and (3)). The operators $U_o(t)$ form a one-parameter group. Indeed, $U_o(0) = I$ (identical operator in X_o),

$$U_o(t)U_o(\tau)x_o = U_o(t)U_o(\tau) \sum_{\ell=1}^{s} U(\tau_\ell)x_\ell = U_o(t) \sum_{\ell=1}^{s} U(\tau + \tau_\ell)x_\ell =$$

$$= \sum_{\ell=1}^{s} U(t + \tau + \tau_\ell)x_\ell = U_o(t + \tau) \sum_{\ell=1}^{s} U(\tau_\ell)x_\ell = U_o(t + \tau)x_o, \quad x_o \in X_o.$$

We have the commutative triangle of linear mappings:

$$\begin{array}{ccccc} & & X_o & & \\ & \tilde{j}_1 \nearrow & & \searrow \tilde{j}_2 & \\ X_1 & & \xrightarrow{\quad j \quad} & & X_2 \end{array} \tag{6}$$

where $\tilde{j}_2$ is the restriction on X_o of the identity operator in X_2 and $\tilde{j}_1$ is defined since $\mathrm{Im}(j) \subset X_o$. Besides,

$$U(t) = \tilde{j}_2 \circ U_o(t) \circ \tilde{j}_1. \tag{7}$$

Then set

$$\| x_o \|_{mid} = \sup_{t \in \mathbb{R}} \{ (1 + |t|)^{-k_1} \| \tilde{j}_2 \mathcal{U}_o(t) x_o \|_2 \}, \quad x_o \in X_o. \tag{8}$$

We drop $\tilde{j}_2$ below again considering X_o as a subset in X_2. The assertion is that $\| \cdot \|_{mid}$ is a norm function on X_o and

$$\| \tilde{j}_1 x \|_{mid} \leqslant C \| x \|_1, \quad x \in X_1, \tag{9}$$

$$\| \tilde{j}_2 x_o \|_2 \leqslant \| x_o \|_{mid}, \quad x \in X_o. \tag{10}$$

To prove this we note first that $\| \cdot \|_{mid}$ obviously is positive homogeneous and satisfies the triangle inequality. It remains to prove that (9) and (10) are valid and that $\| \cdot \|_{mid}$ does not take the value $+\infty$ (the implication $\| x_o \|_{mid} = 0 \Rightarrow x_o = 0$ follows from (10)). We have

$$\| \sum_{\ell=1}^{s} \mathcal{U}(\tau_\ell) x_\ell \|_{mid} = \sup_{t \in \mathbb{R}} \{ (1 + |t|)^{-k_1} \| \sum_{\ell=1}^{s} \mathcal{U}(t+\tau_\ell) x_\ell \|_2 \} \leqslant$$

$$\leqslant \sup_{t \in \mathbb{R}} \{ (1 + |t|)^{-k_1} \sum_{\ell=1}^{s} \| \mathcal{U}(t + \tau_\ell) x_\ell \|_2 \} \leqslant \tag{11}$$

$$\leqslant C \sum_{\ell=1}^{s} \| x_\ell \|_1 \sup_{t \in \mathbb{R}} \frac{(1 + |t + \tau_\ell|)^{k_1}}{(1 + |t|)^{k_1}} \leqslant C \sum_{\ell=1}^{s} \| x_\ell \|_1 (1 + |\tau_\ell|)^k$$

since $k_1 \geqslant k$. Thus $\| \cdot \|_{mid}$ takes only finite values; setting $s = 1$, $\tau_1 = 0$ in (11), we obtain the inequality (9). (10) is obvious since the expression in curly brackets on the right of (8) is equal to $\| \tilde{j}_2 x_o \|_2$ for $t = 0$.

We assert that $\mathcal{U}_o(t) x_o$ is continuous in the norm $\| \cdot \|_{mid}$ for any $x_o \in X_o$ and that

$$\| \mathcal{U}_o(t) x_o \|_{mid} \leqslant (1 + |t|)^k \| x_o \|_{mid}. \tag{12}$$

Indeed, it suffices to prove that $\mathcal{U}(t)x$ is continuous in the norm $\| \cdot \|_{mid}$ for any $x \in X_1$. We have

$$\| \mathcal{U}(t + \varepsilon)x - \mathcal{U}(t)x \|_{mid} = \sup_{t \in \mathbb{R}} (1 + |\tau|)^{-k_1} \| \mathcal{U}(t + \tau + \varepsilon)x - \mathcal{U}(t + \tau)x \|_2 =$$

$$= \max\{ \sup_{|\tau| \leqslant \mathbb{R}} (1 + |\tau|)^{-k_1} \| \mathcal{U}(t + \tau + \varepsilon)x - \mathcal{U}(t + \tau)x \|_2 +$$

$$+ \sup_{|\tau| \geqslant R} (1 + |\tau|)^{-k_1} \| \mathcal{U}(t + \tau + \varepsilon)x - \mathcal{U}(t + \tau)x \|_2 \} =$$

$$= \max\{ I_1(R,\varepsilon,x) + I_2(R,\varepsilon,x) \}.$$

We estimate first $I_2(R,\varepsilon,x)$:

$$I_2(R,\varepsilon,x) \leqslant C \| x \|_1 (1 + R)^{k-k_1} \sup_{\tau \in \mathbb{R}} \frac{(1 + |t + \tau + \varepsilon|)^k + (1 + |t + \tau|)^k}{(1 + |\tau|)^k} \leqslant$$

$$\leqslant \frac{C_1}{(1 + R)^{k_1 - k}},$$

where C_1 is independent of ε, $|\varepsilon| \leqslant 1$ (but depends on x,t). Thus, given a $\delta > 0$, $I_2(R,\varepsilon,x) < \delta$ for R large enough, $|\varepsilon| < 1$. Fix such R. Then $\mathcal{U}(\lambda)x$ is continuous in the norm $\| \cdot \|_2$ and henceforth uniformly continuous when λ lies in a bounded subset of $\mathbb{R}$. Thus $I_1(R,\varepsilon,x) < \delta$ for ε small enough, and the strong continuity of $\mathcal{U}_o(t)$ in the norm $\| \cdot \|_{mid}$ is proved. The proof of (12) is reduced to direct calculation:

$$\| \mathcal{U}_o(t)x_o \|_{mid} = \sup_{\tau\in\mathbb{R}} \{(1+|\tau|)^{-k_1} \| \mathcal{U}_o(t+\tau)x_o \|_2\} =$$

$$= \sup_{\tau\in\mathbb{R}} \{ \frac{(1+|t+\tau|)^{k_1}}{(1+|\tau|)^{k_1}} (1+|t+\tau|)^{-k_1} \| \mathcal{U}_o(t+\tau)x_o \|_2\} \leqslant$$

$$\leqslant (1+|t|)^{k_1} \sup_{\tau\in\mathbb{R}} \{(1+|t+\tau|)^{-k_1} \| \mathcal{U}_o(t+\tau)x_o \|_2\} = (1+|t|)^{k_1} \|x_o\|_{mid}.$$

Denote now by X_{mid} the completion of X_o in the norm $\|\cdot\|_{mid}$. The closures $\mathcal{U}_o(t)$ of the operators $\mathcal{U}_o(t)$ in X_{mid} form the strongly continuous group of linear operators in X_{mid}, $\overline{\mathcal{U}_o(t)} = \exp(iAt)$, A being a generator of degree k_1 in X_{mid}. The diagram (6) gives rise to the diagram

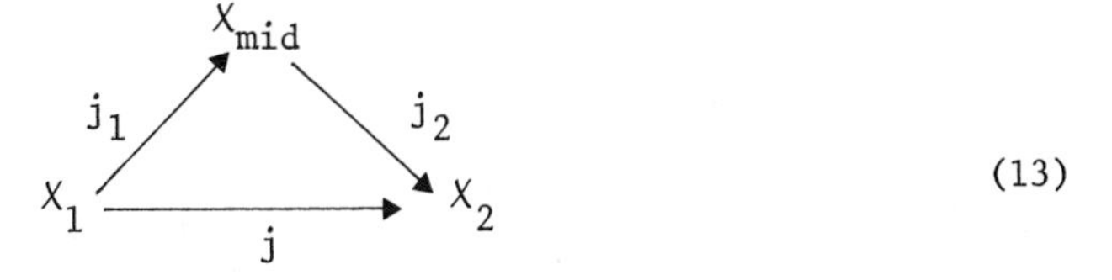

(13)

and (4) is valid. The proposition is proved.

Note. The proof of the above proposition fails for $k_1 = k$. One can prove all the statements except the strong continuity of $\mathcal{U}_o(t)$ (and henceforth $\overline{\mathcal{U}_o(t)}$). There is a simple counterexample:

Counterexample 1. Let $X_1 = \mathbb{C}^1$ (the one-dimensional Banach space), $X_2 = C(\mathbb{R})$ (the space of continuous functions on $\mathbb{R}$ with the usual sup-norm). Let y be a coordinate in $\mathbb{R}$. For any $t\in\mathbb{R}$, $x\in X_1$, we set

$$\mathcal{U}(t)x = x\cdot f_t(y);\; jx = \mathcal{U}(0)x, \tag{14}$$

where

$$f_t(y) = \begin{cases} 0, & y \geqslant t, \\ 1, & y \leqslant t-1/(1+|t|), \\ (1+|t|)(t-y), & 1/(1+|t|) \leqslant y \leqslant t. \end{cases} \tag{15}$$

We have $\|\mathcal{U}(t)\| \leqslant 1$ for all t, and obviously $\mathcal{U}(t)$ is strongly continuous; next, if (3) is valid, we may assume (reducing similar terms if necessary) that $\tau_1 > \tau_2 > \ldots > \tau_s$. All the terms in (3), except the first, vanish for $y\in[\tau_2,\tau_1]$, thus the first one also vanishes identically in this interval and we conclude from (14) and (15) that $x_1 = 0$. By induction, $x_\ell = 0$ for $\ell = 1,\ldots, s$ and (2) is valid. Thus $\mathcal{U}(t)$ satisfies all the conditions of Definition 2 for k = 0. On the other hand, direct computation shows that

$$\sup_{t\in\mathbb{R}} \| \mathcal{U}(t+\varepsilon)x - \mathcal{U}(t)x \|_2 = \|x\|_1 \tag{16}$$

for any $\varepsilon \neq 0$, so that $\mathcal{U}_o(t)$ is not continuous in the norm $\|\cdot\|_{mid}$ for $k_1 = 0$.

The operator A will be called a generator of the group $\mathcal{U}(t)$ (one should bear in mind, however, that the construction of X_{mid} is not unique). For any $f\in \mathcal{B}_k^0(\mathbb{R})$ we define the operator $f_*(A) : X_1 \to X_2$ by means of the relation

$$f_*(A)x = \frac{1}{\sqrt{(2\pi)}} \int_{-\infty}^{\infty} \tilde{f}(t)\mathcal{U}(t)x\,dt,\; x\in X_1 \tag{17}$$

(cf. Definition 4 of item E; the integral in (17) denotes the pairing of $\tilde{f} \in C_k^{0+}(\mathbb{R})$ and $\mathcal{U}(t)x \in C_k^0(\mathbb{R}, \mathcal{X}_2)$).

The following proposition is the immediate consequence of definitions and results obtained in item E.

Proposition 2. (a) $f_*(A)$ is a bounded operator from $\mathcal{X}_1$ to $\mathcal{X}_2$ and

$$\| f_*(A) \| \leqslant \frac{C}{\sqrt{(2\pi)}} \| f \|_{\mathcal{B}_k^0(\mathbb{R})} . \tag{18}$$

(b) If $f \in \mathcal{B}_{k1}^0(\mathbb{R})$, there is a decomposition

$$f_*(A) = j_2 \circ f(A) \circ j_1, \tag{19}$$

where $f(A)$ is a function of the generator A of degree k in the Banach space $\mathcal{X}_{mid}$ (cf. item E). The proof is obvious.

There is an important particular case of the general situation thus described above. Let E be a vector space endowed with the pair of norms $\| \cdot \|_i$, $i = 1,2$ such that $\| x \|_1 \geqslant \| x \|_2$, $x \in E$. Let also $A : E \to E$ be a linear operator such that for any $x_o \in E$, there exists an unique solution of the Cauchy problem

$$i \frac{\partial x(t)}{\partial t} + Ax = 0, \; x(0) = x_o \tag{20}$$

(the derivative in (20) is with respect to $\| \cdot \|_1$), satisfying

$$\| x(t) \|_2 \leqslant C(1 + |t|)^k \| x_o \|_1 . \tag{21}$$

Denoting by $\mathcal{X}_i$, $i = 1,2$ the closure of E with respect to the norm $\| \cdot \|_i$ and by $\mathcal{U}(t) : \mathcal{X}_1 \to \mathcal{X}_2$ the closure of the operator $x_o \to x(t)$, we obtain the strongly continuous tempered group in the pair $(\mathcal{X}_1, \mathcal{X}_2)$ (the mapping j is simply the closure of the identity operator on E). Proposition 1 is now valid for $k_1 = k$, since $x(t)$ is differentiable and therefore continuous in the norm $\| \cdot \|_1$, which properly enables us to prove the strong continuity in $\mathcal{X}_{mid}$ for $k_1 = k$. A slight generalization may be obtained if we consider some wider space instead of $\mathcal{X}_2$ (so that the image $j(\mathcal{X}_1)$ will not be dense in $\mathcal{X}_2$).

3. FUNCTIONS OF SEVERAL OPERATORS

A. Symbol Spaces

Given a n-tuple $A = (A_1, \dots, A_n)$ of generators of degrees k_i, $i = 1, \dots, n$ in a Banach space $\mathcal{X}$, we intend then to define the operator $f(A) \equiv f(\overset{1}{A_1}, \dots, \overset{n}{A_n})$ by means of the Fourier transform

$$f(A)x \equiv f(\overset{1}{A_1}, \dots, \overset{n}{A_n})x =$$

$$= \frac{1}{(2\pi)^{n/2}} \int_{\mathbb{R}^n} F(f)(t_1, \dots, t_n) e^{it_nA_n} \dots e^{it_1A_1} x dt, \tag{1}$$

in complete analogy with the definition, given in the preceding section for functions of a single operator. Thus we have to define the symbol spaces, for which the integral on the right of (1) makes sense under suitable assumptions.

Let $\alpha = (\alpha_1,\ldots,\alpha_n)$, $\beta = (\beta_1,\ldots,\beta_n)$ be multi-indices, $t = (t_1,\ldots,t_n) \in \mathbb{R}^n$. Given a function $\phi : \mathbb{R}^n \to \mathbb{C}$, we set

$$\|\phi\|_{C_\alpha^\beta} = \sup_{t\in\mathbb{R}^n} (1+|t|)^{-\alpha} \sum_{\gamma\leqslant\beta} \left|\frac{\partial^\gamma\phi}{\partial t^\gamma}(t)\right|, \tag{2}$$

where $(1+|t|)^{-\alpha} \overset{\text{def}}{=} (1+|t_1|)^{-\alpha_1}\cdot\ldots\cdot(1+|t_n|)^{-\alpha_n}$, $\gamma \leqslant \beta$ is the abbreviation for $(\gamma_1 \leqslant \beta_1,\ldots,\gamma_n \leqslant \beta_n)$. We denote by $C_\alpha^\beta(\mathbb{R}^n)$ the Banach space of functions $\phi(t)$, $t \in \mathbb{R}^n$, continuous with all the derivatives, which appear on the right of (2), for which the norm (2) is finite.

On the other hand, if N is a non-negative integer, we may also define the space $C_N^\beta(\mathbb{R}^n)$, using the norm

$$\|\phi\|_{C_N^\beta} = \sup_{t\in\mathbb{R}^n} (1+|t|)^{-N} \sum_{\gamma\leqslant\beta} \left|\frac{\partial^\gamma\phi}{\partial t^\gamma}(t)\right|. \tag{3}$$

Here $(1+|t|)^{-N} = (1+(\sum_{j=1}^n t_j^2)^{1/2})^{-N}$.

Following the ideas of Section 2:E, we define the space $C_\alpha^{\beta+}(\mathbb{R}^n)$ (respectively, $C_N^{\beta+}(\mathbb{R}^n)$) as the strong closure of the set of finite continuous functionals over $C_\alpha^\beta(\mathbb{R}^n)$ (respectively, $C_N^\beta(\mathbb{R}^n)$). Next the spaces $B_\alpha^\beta(\mathbb{R}^n)$ and $B_N^\beta(\mathbb{R}^n)$ are defined as the spaces of inverse Fourier transforms of elements of $C_\alpha^{\beta+}(\mathbb{R}^n)$ or $C_N^\beta(\mathbb{R}^n)$, supplied with the norms induced by Fourier transformation.

Both definitions given above are particular cases of the more general situation, consideration of which will also be useful. Let $\{1,\ldots,n\} = I_1 \cup \ldots \cup I_k$ be a partition of the set $\{1,\ldots,n\}$ into the sum of non-empty non-intersecting subsets $I_1,\ldots,I_k$ and let $\nu = (\nu_1,\ldots,\nu_k)$ be multi-indices of the length k, $\beta = (\beta_1,\ldots,\beta_n)$ be multi-indices of the length n. Then we set

$$\|\phi\|_{C^\beta_{I_1,\ldots,I_k,\nu_1,\ldots,\nu_k}} =$$

$$= \sup_{t\in\mathbb{R}^n} (1+|t_{I_1}|)^{-\nu_1} \ldots (1+|t_{I_k}|)^{-\nu_k} \sum_{\gamma\leqslant\beta} \left|\frac{\partial^\gamma\phi}{\partial t^\gamma}(t)\right|, \tag{4}$$

where $|t_{I_j}| = (\sum_{i\in I_j} (t_i)^2)^{1/2}$, and denote by $C^\beta_{I_1,\ldots,I_k,\nu_1,\ldots,\nu_k}(\mathbb{R}^n) \equiv C^\beta_{I,\nu}(\mathbb{R}^n)$ the space of functions $\phi(t)$ continuous with the derivatives of the order $\gamma \leqslant \beta$ such that the expression on the right of (4) is finite. Clearly we come back to $C_\alpha^\beta(\mathbb{R}^n)$ or $C_N^\beta(\mathbb{R}^n)$ respectively in the cases $k = n$ and $k = 1$.

X being a Banach space, the definition of the space $C^\beta_{I,\nu}(\mathbb{R}^n, X)$ of X-valued functions is obvious.

We denote by $C^{\beta+}_{I,\nu}(\mathbb{R}^n)$ the closure of the subspace of finite functionals in $C^{\beta*}_{I,\nu}(\mathbb{R}^n)$. The function $e^{iyt} \equiv e^{i(y_1t_1+\ldots+y_nt_n)}$ belongs to $C^\beta_{I,\nu}(\mathbb{R}^n)$ (recall that the multi-index components are non-negative), so for a given $T \in C^{\beta+}_{I,\nu}(\mathbb{R}^n)$ its inverse Fourier transform may be defined as

$$F^{-1}(T)(y) = (\frac{1}{2\pi})^{n/2} T(e^{iyt}), \tag{5}$$

y being regarded as a parameter on the right of (5). We prove, just as in Lemma 4 of Section 2:E, that $F^{-1}(T)$ is in fact the inverse Fourier transform of T, considered as a tempered distribution, and in particular F^{-1} has no kernel. We denote by $\mathcal{B}^{\beta}_{I,\nu}(\mathbb{R}^n)$ the space of functions of the form $f = F^{-1}(T)$, $T \in C^{\beta+}_{I,\nu}(\mathbb{R}^n)$ and set

$$\| f \|_{\mathcal{B}^{\beta}_{I,\nu}(\mathbb{R}^n)} \overset{\text{def}}{=} \| F(f) \|_{C^{\beta *}_{I,\nu}(\mathbb{R}^n)}, \tag{6}$$

where $F = (F^{-1})^{-1}$ coincides with the (direct) Fourier transform of tempered distributions. (Sometimes we shall use the notation $\mathcal{B}^{\beta}_{\nu}$, C^{β}_{ν}, ... instead of $\mathcal{B}^{\beta}_{I,\nu}$, $C^{\beta}_{I,\nu}$, ... if there will be no confusion; also if multi-indices $\beta,\nu,\dots$, etc. have all their components equal to zero, the corresponding subscript might be dropped.) The spaces $\mathcal{B}^{\beta}_{\alpha}(\mathbb{R}^n)$ and $\mathcal{B}^{\beta}_{N}(\mathbb{R}^n)$ are specializations of $\mathcal{B}^{\beta}_{I,\nu}(\mathbb{R}^n)$. The definition of the latter being somewhat cumbersome, we concentrate our main attention on these specializations. This enables us to avoid obscure calculations with no essential loss of generality.

The elementary properties of the spaces introduced are collected in the following:

<u>Lemma 1</u>. (a) The mapping

$$f(y) \to (1 - \frac{\partial^2}{\partial y^2_{I_1}})^{\nu_1/2} \times \dots \times (1 - \frac{\partial^2}{\partial y^2_{I_k}})^{\nu_k/2} (i + y)^{-\beta} f(y), \tag{7}$$

where

$$\partial^2/\partial y^2_I = \sum_{i \in I} \frac{\partial^2}{\partial y^2_i}\, ; \quad (i + y)^{-\beta} = (i + y_1)^{-\beta_1} \times \dots \times (i + y_n)^{-\beta_n} \tag{8}$$

is the isomorphism of $\mathcal{B}^{\beta}_{I,\nu}$ onto $\mathcal{B}(\mathbb{R}^n) \equiv \mathcal{B}^{0,\dots,0}_{o,\dots,o}(\mathbb{R}^n)$. The mapping

$$f(y) \to (1 - \frac{\partial}{\partial y})^{\alpha} (i + y)^{-\beta} f(y) \tag{9}$$

is the isomorphism of $\mathcal{B}^{\beta}_{\alpha}(\mathbb{R}^n)$ onto $\mathcal{B}(\mathbb{R}^n)$.

(b) There are continuous embeddings

$$\mathcal{B}^{\beta}_{\alpha} \subset \mathcal{B}^{\beta_1}_{\alpha_1},\ \beta \leqslant \beta_1,\ \alpha \geqslant \alpha_1;\ \mathcal{B}^{\beta}_{N} \subset \mathcal{B}^{\beta_1}_{N_1},\ \beta \leqslant \beta_1,\ N \geqslant N_1; \tag{10}$$

$$H^{\alpha + 1/2 + \varepsilon}_{\beta}(\mathbb{R}^n) \subset \mathcal{B}^{\beta}_{\alpha}(\mathbb{R}^n) \subset C^{\alpha}_{\beta}(\mathbb{R}^n) \subset H^{\alpha}_{\beta + 1/2 + \varepsilon}(\mathbb{R}^n), \tag{11}$$

where $\alpha + 1/2 + \varepsilon \overset{\text{def}}{=} (\alpha_1 + 1/2 + \varepsilon, \dots, \alpha_n + 1/2 + \varepsilon)$, and $\beta + 1/2 + \varepsilon$ is defined analogously; $H^{\alpha}_{\beta}(\mathbb{R}^n)$ is the completion of $C^{\infty}_{o}(\mathbb{R}^n)$ with respect to the norm

$$\| \phi \|_{H^{\alpha}_{\beta}} = \| (1 + y^2)^{-\beta/2} (1 - \frac{\partial^2}{\partial y^2})^{\alpha/2} \phi \|_{L^2} \tag{12}$$

(here α,β may have non-integer components).

(c) $\mathcal{B}^{\beta}_{N}(\mathbb{R}^n)$ may be represented as the intersection:

$$B_N^\beta(\mathbb{R}^n) = \bigcap_{|\alpha| \leqslant N} B_\alpha^\beta(\mathbb{R}^n), \tag{13}$$

and the embeddings $B_N^\beta(\mathbb{R}^n) \subset B_\alpha^\beta(\mathbb{R}^n)$, $|\alpha| \leqslant N$ are continuous.

Proof. (a) We assume $k = 1$ for the sake of simplicity. Passing to Fourier transforms, we reduce the desired statement to the verification of whether the mapping

$$f(t) \to (1 + t^2)^{N/2}(1 - \frac{\partial}{\partial t})^{-\beta} f(t)$$

is the isomorphism of $C_N^{\beta+}(\mathbb{R}^n)$ onto $C^+(\mathbb{R}^n) \equiv C_o^{0,\dots,0+}(\mathbb{R}^n)$. The proof of the latter statement does not differ from the one-dimensional case (see Lemma 5 of item E, Section 2). (9) is proved analogously.

(b) The embeddings (10) are obvious, since we pass to Fourier transforms. As for (11), the proof of these embeddings coincides word for word with the proof of Lemma 6 in Section 2:E.

(c) Again we pass to Fourier transforms. It is enough to consider the case $\beta = (0,\dots,0)$, since it follows from (a) that multiplication by $(i + y)^{-\beta}$ isomorphically maps $B_N^\beta(\mathbb{R}^n)$ onto $B_N(\mathbb{R}^n) \equiv B_N^{0,\dots,0}(\mathbb{R}^n)$ and also $B_\alpha^\beta(\mathbb{R}^n)$ onto $B_\alpha(\mathbb{R}^n)$. Thus we need to prove that $C_N^+(\mathbb{R}^n) = \bigcap_{|\alpha| \leqslant N} C_\alpha^+(\mathbb{R}^n)$ and the corresponding embeddings are continuous. In its turn, it suffices to prove that

$$C_N^*(\mathbb{R}^n) = \bigcap_{|\alpha| \leqslant N} C_\alpha^*(\mathbb{R}^n) \tag{14}$$

and the embeddings are continuous, because the subsets of finite functionals on both sides of the latter identity coincide. Once this statement is proved, we obtain by the closed graph theorem that the norm $\| T \|_{C_N^*}$ is equivalent to the norm $||| T ||| = \sum_{|\alpha| < N} \| T \|_{C_\alpha^*}$. Thus if $T_n \to T$ in $C_N^*(\mathbb{R}^n)$, $T_n \to T$ in all $C_\alpha^*(\mathbb{R}^n)$, so $C_N^+(\mathbb{R}^n) \subset \bigcap_{|\alpha| \leqslant N} C_\alpha^+(\mathbb{R}^n)$. Conversely, let $T \in \bigcap_{|\alpha| \leqslant N} C_\alpha^+(\mathbb{R}^n)$. Then $T_n = T\psi(|t|/n) \to T$ in all $C_\alpha^*(\mathbb{R}^n)$, where $\psi \in C_o^\infty(\mathbb{R}^1)$, $\psi(0) = 1$ (cf. Lemma 1 of Section 2:E), so that $T_n \to T$ in $C_N^*(\mathbb{R}^n)$ and conversely $T \in C_N^+(\mathbb{R}^n)$.

Let us prove (14). If $f \in C_N^*(\mathbb{R}^n)$, then $f \dfrac{(1 + t^2)^{N/2}}{(i + t)^\alpha} \in C_\alpha^*(\mathbb{R}^n)$ and henceforth $f \in C_\alpha^*(\mathbb{R}^n)$, since $(i + t)^\alpha / (1 + t^2)^{N/2}$ is bounded for $|\alpha| \leqslant N$. It follows also that $\| f \|_{C_\alpha^*} \leqslant \text{const} \| f \|_{C_N^*}$. On the opposite, if $f \in \bigcap_{|\alpha| \leqslant N} C_\alpha^*(\mathbb{R}^n)$ then $|t|^\alpha f \in C^*(\mathbb{R}^n)$ for $|\alpha| \leqslant n$ (here $|t|^\alpha = \Pi |t_i|^{\alpha_i}$). Since $(1 + t^2)^{N/2} \leqslant \text{const} \cdot \sum_{|\alpha| \leqslant N} |t|^\alpha$, $(1 + t^2)^{N/2} f \in C^*(\mathbb{R}^n)$, and consequently $f \in C_N^*(\mathbb{R}^n)$. The lemma is proved.

We now continue the investigation of the properties of the symbol spaces introduced.

Lemma 2. (a) The pointwise multiplication is a continuous bilinear map from $B_\alpha^{\beta_1}(\mathbb{R}^n) \times B_\alpha^{\beta_2}(\mathbb{R}^n)$ to $B_\alpha^{\beta_1 + \beta_2}(\mathbb{R}^n)$ and from $B_N^{\beta_1}(\mathbb{R}^n) \times B_N^{\beta_2}(\mathbb{R}^n)$ to $B_N^{\beta_1 + \beta_2}(\mathbb{R}^n)$ for any N,α,β_1,β_2 for which the spaces mentioned have been defined.

(b) The function $f(y) = e^{iyt}y^{\gamma} \equiv y_1^{\gamma_1}\dots y_n^{\gamma_n}e^{iy_1t_1}\dots e^{iy_nt_n}$ belongs to $B_{\alpha}^{\beta}(\mathbb{R}^n)$ and to $B_N^{\beta}(\mathbb{R}^n)$ provided that $\gamma \leqslant \beta$ (α and N may be arbitrary). The proof of this lemma coincides exactly with that of Theorem 2 (b) and (d) of Section 2:E and is therefore omitted.

Lemma 3. Let $\alpha = (\alpha_1,\dots,\alpha_{n-1})$, $\tilde{\alpha} = (\alpha,\alpha_n)$, $\beta = (\beta_1,\dots,\beta_{n-1})$, and $\tilde{\beta} = (\beta,\beta_n)$ be multi-indices. The mapping $f(y_1,\dots,y_{n-1}) \to g(y_1,\dots,y_n) \equiv f(y_1,\dots,y_{n-1})$ induces the continuous embedding $B_{\alpha}^{\beta}(\mathbb{R}^{n-1}) \subset B_{\tilde{\alpha}}^{\tilde{\beta}}(\mathbb{R}^n)$.

Proof. Passing to Fourier transforms, we find it necessary to prove $T(t_1,\dots,t_{n-1}) \times \delta(t_n) \in C_{\tilde{\alpha}}^{\tilde{\beta}+}(\mathbb{R}^n)$, where $T(t_1,\dots,t_{n-1}) \in C_{\alpha}^{\beta+}(\mathbb{R}^n)$. But the latter statement is obvious, since the mapping $C_{\tilde{\alpha}}^{\tilde{\beta}}(\mathbb{R}^n) \to C_{\alpha}^{\beta}(\mathbb{R}^{n-1})$: $f(t_1,\dots,t_n) \to f(\underbrace{t_1,\dots,t}_{n-1},0)$ is continuous.

Our next lemma is concerned with the mapping which identifies the pair of arguments of the symbol.

Lemma 4. $f(y_1,\dots,y_n)$ being the function of n arguments, set $[jf] \times (y_1,\dots,y_{n-1}) = f(y_1,\dots,y_{n-1},y_{n-1})$. Thus, jf is a function of $(n-1)$ arguments.

(a) j defines a continuous mapping from $B_N^{\beta}(\mathbb{R}^n)$ to $B_N^{\tilde{\beta}}(\mathbb{R}^{n-1})$, where $\tilde{\beta} = (\beta_1,\dots,\beta_{n-2},\beta_{n-1}+\beta_n)$.

(b) j defines a continuous mapping from $B_{\alpha}^{\beta}(\mathbb{R}^n)$ to $B_{\tilde{\alpha}}^{\tilde{\beta}}(\mathbb{R}^{n-1})$ where $\tilde{\beta}$ is the same as above, while $\tilde{\alpha} = (\alpha_1,\dots,\alpha_{n-2},\min(\alpha_{n-1},\alpha_n))$.

Proof. Consider first the case $\beta = 0$. We define the operator $\mathcal{T}$ by the relation

$$[\mathcal{T}\phi](t_1,\dots,t_n) = \phi(t_1,\dots,t_{n-2},t_{n-1}+t_n), \tag{15}$$

where ϕ is a function depending on $(n-1)$ arguments. Since

$$|t_{n-1}+t_n| \leqslant |t_{n-1}| + |t_n|,$$

$\mathcal{T}$ is a continuous operator from $C_N^0(\mathbb{R}^{n-1})$ to $C_N^0(\mathbb{R}^n)$ and from $C_{\tilde{\alpha}}^0(\mathbb{R}^{n-1})$ to $C_{\alpha}^0(\mathbb{R}^n)$, $\tilde{\alpha}$ being defined in the statement of the lemma. Let f be an element of $B_N^0(\mathbb{R}^n)$ (or $B_{\alpha}^0(\mathbb{R}^n)$). Denote by $T = F(f)$ its Fourier transform and by T_1 the following functional on $C_N^0(\mathbb{R}^{n-1})$ (or $C_{\tilde{\alpha}}^0(\mathbb{R}^{n-1})$), continuous by the above argument:

$$T_1(\phi) = T(\mathcal{T}(\phi)). \tag{16}$$

Calculating its Fourier transform, we obtain

$$\begin{aligned} F^1(T_1) &= \left(\frac{1}{\sqrt{(2\pi)}}\right)^{n-1} T(e^{iy_1t_1}\times\dots\times e^{iy_{n-2}t_{n-2}}e^{iy_{n-1}(t_{n-1}+t_n)}) = \\ &= \sqrt{2\pi}\, jf(y_1,\dots,y_{n-1}), \end{aligned} \tag{17}$$

so the lemma is proved for $\beta = 0$. Let now $\beta \neq 0$. Representing $f(y)$ in the form $f(y) = (i+y_1)^{\beta_1}\dots(i+y_n)^{\beta_n}f_1(y_1,\dots,y_n)$, we then obtain the equality

$$\begin{aligned} &[jf](y_1,\dots,y_{n-1}) = \\ &= (i+y_1)^{\beta_1}\dots(i+y_{n-2})^{\beta_{n-2}}(i+y_{n-1})^{\beta_{n-1}+\beta_n} jf_1(y_1,\dots,y_{n-1}), \end{aligned} \tag{18}$$

which immediately implies, in view of Lemma 1 (a), the desired statement. The proof is now complete.

Next we introduce the difference derivation of symbols, which has applications in numerous formulae of the functional calculus for non-commutative operators. First, let f be a smooth function of a single variable $y \in \mathbb{R}$. We define the difference derivative of f as a function of two variables y_1, y_2, given by the formula

$$\delta f(y_1,y_2) \equiv \frac{\delta f}{\delta y}(y_1,y_2) = \begin{cases} \dfrac{f(y_1) - f(y_2)}{y_1 - y_2}, & y_1 \neq y_2 \\ f'(y_2), & y_1 = y_2. \end{cases} \tag{19}$$

Clearly $\delta f(y_1,y_2)$ is again a smooth function. Difference derivatives of higher order are defined inductively; to obtain $\delta^k f/\delta y^k$ one should fix all the arguments of $\delta^{k-1}f/\delta y^{k-1}$, except one, and apply the operator $\delta/\delta y$ with respect to this argument, according to (19). Thus $\delta^k f/\delta y^k$ is the function of $k + 1$ arguments. It will be shown below that $\delta^k f/\delta y^k$ is a symmetric function of its arguments so that it makes no difference what arguments are fixed in our inductive definition.

Analogously, for a given smooth function $f(y)$, $y = (y_1,\dots,y_n) \in \mathbb{R}^n$ and multi-indices $\alpha = (\alpha_1,\dots,\alpha_n)$, we define the difference derivative of the order α : $\delta^{|\alpha|}f/\delta y^\alpha \equiv (\delta/\delta y_1)^{\alpha_1} \dots (\delta/\delta y_n)^{\alpha_n} f$ which is the function of $n + |\alpha|$ variables, naturally partitioned into n groups, namely, $(y_{(1)},\dots,y_{(n)}) = (y_{11},y_{12},\dots,y_{11+\alpha_1};\dots;y_{n1},y_{n2},\dots,y_{n1+\alpha_n})$ by successive application of the operators $(\delta/\delta y_i)^{\alpha_i}$, $i = 1,\dots,n$ under fixed values of other variables. In what follows we sometimes use different notation for variables; for the sake of convenience we separate the groups of variables by semicolons and the variables inside the groups by commas. We formulate the following lemma only for $n = 1$ in order to clarify the notation. The proof for $n \neq 1$ is identical (up to notation) to that given below.

Lemma 5. $\delta^k f/\delta y^k$ is a symmetrical function of its arguments (i.e., no rearrangement of the arguments affects its value). The following identities are valid:

$$\frac{\delta^k f}{\delta y^k}(y_1,\dots,y_{k+1}) = \int_0^1 d\mu_1 \int_0^{1-\mu_1} d\mu_2 \dots \int_0^{1-\mu_1-\dots-\mu_{k-1}} d\mu_k \times \\ \times f^{(k)}(\mu_1 y_1 + \dots + \mu_k y_k + (1 - \mu_1 - \dots - \mu_k)y_{k+1}) \tag{20}$$

(here $f^{(k)}$ is the k-th derivative of f);

$$\frac{\delta^k f}{\delta y^k}(y_1,\dots,y_{k+1}) = \sum_{j=1}^{k+1} f(y_j) \prod_{\substack{i=1 \\ i\neq j}}^{k} (y_j - y_i)^{-1}, \tag{21}$$

if $y_i \neq y_j$ for $i \neq j$:

$$\frac{\delta^k f}{\delta y^k}(y,\dots,y) = \frac{1}{k!} f^{(k)}(y); \tag{22}$$

$$\frac{\delta^k f}{\delta y^k}(y,x,x,\dots,x) = \frac{1}{(k-1)!}\int_0^1 (1-\tau)^{k-1} f^{(k)}(\tau y + (1-\tau)x)d\tau; \tag{23}$$

$$[(\frac{\partial}{\partial y_1})^{\alpha_1} \dots (\frac{\partial}{\partial y_{k+1}})^{\alpha_{k+1}} \delta^k f(y_1,\dots,y_{k+1})]\Big|_{y_1 = \dots = y_{k+1} = y} =$$

$$= \frac{\alpha_1! \dots \alpha_{k+1}!}{(k+|\alpha|)!} f^{(k+|\alpha|)}(y). \tag{24}$$

Proof. The symmetry of $\delta^k f/\delta y^k$ is a simple consequence of (21). Thus we need only to prove (20) - (24). First we remark that

$$\delta f(y_1,y_2) = \int_0^1 f'(\tau y_1 + (1-\tau)y_2)d\tau. \tag{25}$$

Indeed,

$$\begin{aligned} f(y_1) - f(y_2) &= \int_0^1 \frac{d}{d\tau}[f(y_2 + \tau(y_1 - y_2))]d\tau = \\ &= (y_1 - y_2)\int_0^1 f'(\tau y_1 + (1-\tau)y_2)d\tau, \end{aligned} \tag{26}$$

and we obtain (25) applying the definition (19). Thus (20) is proved for $k = 1$. We proceed now by induction on k. One has

$$\frac{\delta^k f}{\delta y^k}(y_1,\dots,y_{k+1}) =$$

$$= \int_0^1 d\tau \{ \frac{d}{dy} \frac{\delta^{k-1} f}{\delta y^{k-1}}(y,y_3,y_4,\dots,y_{k+1})\}\Big|_{y = \tau y_1 + (1-\tau)y_2} =$$

$$= \int_0^1 d\tau \int_0^1 \mu_2 d\mu_2 \int_0^{1-\mu_2} d\mu_3 \dots \int_0^{1-\mu_2-\dots-\mu_{k-1}} d\mu_k f^{(k)} \times$$

$$\times (\tau\mu_2 y_1 + (1-\tau)\mu_2 y_2 + \mu_3 y_3 + \dots + (1-\mu_2-\dots-\mu_k)y_{k+1}).$$

The following change of variables in the integrand:

$$\nu_1 = \tau\mu_2, \nu_2 = (1-\tau)\mu_2, \nu_3 = \mu_3, \dots, \nu_k = \mu_k; \ \det \frac{\partial(\nu_1,\dots,\nu_k)}{\partial(\tau,\mu_2,\dots,\mu_k)} = \mu_2$$

results in the expression

$$\frac{\delta^k f}{\delta y^k}(y_1,\dots,y_{k+1}) = \int_0^1 d\nu_1 \int_0^{1-\nu_1} d\nu_2 \dots \int_0^{1-\nu_1-\dots-\nu_{k-1}} d\nu_k \times$$

$$\times f^{(k)}(\nu_1 y_1 + \nu_2 y_2 + \dots + \nu_k y_k + (1-\nu_1-\dots-\nu_k)y_{k+1}),$$

which coincides with (20) up to the notation of variables. Thus (20) is proved. Next,

$$\frac{\delta^k f}{\delta y^k}(y,x,x,\dots,x) = \int_0^1 d\mu_1 \int_0^{1-\mu_1} d\mu_2 \dots \int_0^{1-\mu_1-\dots-\mu_{k-1}} d\mu_k \times$$

$$\times f^{(k)}(x(\mu_1 + \dots + \mu_k) + y(1-\mu_1-\dots-\mu_k)).$$

After the unimodular change of variables $\nu_1 = \mu_1 + \dots + \mu_k$, $\nu_2 = \mu_2 + \dots + \mu_k, \dots, \nu_k = \mu_k$, the latter expression takes the form

$$\frac{\delta^k f}{\delta y^k}(y,x,x,\ldots,x) =$$

$$= \int_0^1 d\nu_1 \{ f^{(k)}(\nu_1 x + (1-\nu_1)y) \int_0^{\nu_1} d\nu_2 \int_0^{\nu_2} d\nu_3 \ldots \int_0^{\nu_{k-1}} d\nu_k \} =$$

$$= \int_0^1 f^{(k)}(\nu_1 x + (1-\nu_1)y) \frac{\nu_1^{k-1}}{(k-1)!} d\nu_1,$$

since the multiple integral in curly brackets equals $\nu_1^{k-1}/(k-1)!$ We obtain (23) denoting $\tau = (1 - \nu_1)$. Setting $x = y$ in (23), we come directly to (22), since $\int_0^1 (1-\tau)^{k-1} d\tau = 1/k$. By (20) the expression on the left of (24) equals

$$f^{(k+|\alpha|)}(y) \int_0^1 d\mu_1 \int_0^{1-\mu_1} d\mu_2 \ldots \int_0^{1-\mu_1-\ldots-\mu_{k-1}} d\mu_k \{ \mu_1^{\alpha_1} \ldots \mu_k^{\alpha_k} \times$$
$$\times (1-\mu_1-\ldots-\mu_k)^{\alpha_{k+1}} \} \equiv f^{(k+|\alpha|)}(y) \phi(k,\alpha_1,\ldots,\alpha_{k+1}), \tag{27}$$

where ϕ denotes the multiple integral. After the change of variables $\nu_1 = \mu_1$, $\nu_2 = \mu_1 + \mu_2, \ldots, \nu_k = \mu_1 + \ldots + \mu_k$, we obtain

$$\phi(k,\alpha_1,\ldots,\alpha_{k+1}) = \int_0^1 d\nu_k \{ (1-\nu_k)^{\alpha_{k+1}} \int_0^{\nu_k} d\nu_{k-1} \{ (\nu_k - \nu_{k-1})^{\alpha_k} \times$$
$$\times \int_0^{\nu_{k-1}} d\nu_{k-2} \{ \ldots \int_0^{\nu_2} d\nu_1 \{ (\nu_2 - \nu_1)^{\alpha_2} \nu_1^{\alpha_1} \} \ldots \}\}\}. \tag{28}$$

Compute the inner integral. We then have

$$\int_0^{\nu_2} (\nu_2 - \nu_1)^{\alpha_2} \nu_1^{\alpha_1} d\nu_1 = \nu_2^{1+\alpha_1+\alpha_2} \int_0^1 t^{\alpha_1}(1-t)^{\alpha_2} dt = \nu_2^{1+\alpha_1+\alpha_2} \frac{\alpha_1! \alpha_2!}{(1+\alpha_1+\alpha_2)!}.$$

Thus $\phi(k,\alpha_1,\ldots,\alpha_{k+1}) = \phi(k-1, 1+\alpha_1+\alpha_2, \alpha_3,\ldots,\alpha_{k+1}) \frac{\alpha_1!\alpha_2!}{(1+\alpha_1+\alpha_2)!}$. On the other hand, $\phi(0,\alpha) = 1$ since (24) is clearly valid for $k = 0$. Then applying the obtained relation recursively, we get

$$\phi(k,\alpha_1,\alpha_2,\ldots,\alpha_{k+1}) = \frac{\alpha_1!\alpha_2!}{(1+\alpha_1+\alpha_2)!} \frac{(1+\alpha_1+\alpha_2)!\alpha_3!}{(2+\alpha_1+\alpha_2+\alpha_3)!} \times$$
$$\times \phi(k-2, 2+\alpha_1+\alpha_2+\alpha_3, \alpha_4,\ldots,\alpha_{k+1}) = \frac{\alpha_1!\alpha_2!\ldots\alpha_{k+1}!}{(k+|\alpha|)!}$$

and (24) is proved. It remains to prove (21). For $k = 1$, (21) coincides with definition (19). By induction on k, we have

$$\frac{\delta^k f}{\delta y^k}(y_1,\ldots,y_{k+1}) = \frac{1}{(y_1-y_2)} \{ \sum_{\substack{j=1 \\ j\neq 2}}^{k+1} f(y_j) \prod_{\substack{i=1 \\ i\neq j \\ i\neq 2}}^{k+1} (y_j - y_i)^{-1} -$$

$$- \sum_{j=2}^{k+1} f(y_j) \prod_{\substack{i=2 \\ i\neq j}}^{k+1} (y_j - y_i)^{-1} \} = \frac{1}{y_2 - y_1} \{ \sum_{j=1}^{k+1} (y_j - y_2) f(y_j) \prod_{\substack{i=1 \\ i\neq j}}^{k+1} (y_j - y_i)^{-1} -$$

$$- \sum_{j=1}^{k+1} f(y_j)(y_j - y_1) \prod_{\substack{i=1 \\ i\neq j}}^{k+1} (y_j - y_i)^{-1} \} = \sum_{j=1}^{k+1} f(y_j) \prod_{\substack{i=1 \\ i\neq j}}^{k+1} (y_j - y_i)^{-1}.$$

The proof of the lemma is now complete. In particular, using the difference derivative, we obtain a simple expression for the remainder in the Taylor expansion, as shown in the following:

Lemma 6. There is a representation of the form

$$f(y) = \sum_{k=0}^{N-1} \frac{(y-x)^k}{k!} f^{(k)}(x) + (y-x)^N \frac{\delta^N f}{\delta x^N}(y,x,\dots x) \tag{29}$$

for any N and any smooth function $f(x)$, $x \in \mathbb{R}$.

Proof is carried by induction on N. We come back to the definition for N = 1 and we have

$$\frac{\delta^N f}{\delta x^N}(y,x,\dots,x) = \frac{\delta^N f}{\delta x^N}(x,x,\dots,x) + (y-x)\frac{\delta^{N+1} f}{\delta x^{N+1}}(y,x,\dots,x) =$$

$$= \frac{1}{N!} f^{(N)}(x) + (y-x)\frac{\delta^{N+1} f}{\delta x^{N+1}}(y,x,\dots,x)$$

for any N ((22) has been used), so that the induction step works and the lemma is proved. We give also the n-dimensional formula for N = 1:

Lemma 7. We have

$$f(x_1,\dots,x_n) - f(y_1,\dots,y_n) =$$

$$= \sum_{j=1}^{n} (x_j - y_j) \frac{\delta f}{\delta y_j}(x_2,\dots,x_{j-1};x_j,y_j;y_{j+1},\dots,y_n). \tag{30}$$

The proof is obvious. The analogue of the Leibniz formula is valid for the difference derivative.

Lemma 8. We have

$$\frac{\delta}{\delta y}(fg)(y_1,y_2) = f(y_1)\frac{\delta g}{\delta y}(y_1,y_2) + \frac{\delta f}{\delta y}(y_1,y_2)g(y_2). \tag{31}$$

Proof.

$$\frac{f(y_1)g(y_1) - f(y_2)g(y_2)}{y_1 - y_2} = \frac{f(y_1)(g(y_1)-g(y_2)) + g(y_2)(f(y_1)-f(y_2))}{y_1 - y_2}.$$

Next we show that difference derivation acts in symbol spaces defined above and is continuous there.

Lemma 9. Let $t = (t_1,\dots,t_n) \in \mathbb{R}^n$, $\phi \in C_N(\mathbb{R}^n)$. Consider the function

$$\phi(t_1,\dots,t_{n-1}) = \int_0^{t_{n-1}} \phi(t_1,\dots,t_{n-2},\xi,t_{n-1}-\xi)d\xi. \tag{32}$$

Then $\phi_1 \in C_{N+1}(\mathbb{R}^{n-1})$ and $\|\phi_1\|_{C_{N+1}} \leqslant C\,\|\phi\|_{C_N}$, where C depends only on N and n.

Proof. Clearly $\phi_1(t_1,\dots,t_{n-1})$ is continuous and, moreover, the inequality

$$|\phi_1(t_1,\dots,t_{n-1})| = \left|\int_0^{t_{n-1}} \phi(t_1,\dots,t_{n-2},\xi,t_{n-1}-\xi)d\xi\right| \leqslant$$

$$\leqslant \text{const}\,\|\phi\|_{C_N} \left|\int_0^{t_{n-1}} (1 + t_1^2 + \dots + t_{n-2}^2 + \xi^2 + (t_{n-1}-\xi)^2)^{N/2} d\xi\right| \leqslant$$

$$\leqslant \text{const} \, \| \phi \|_{C_N} \int_0^1 |t_{n-1}| \, (1 + t_1^2 + \ldots + t_{n-2}^2 + t_{n-1}^2 (\xi^2 + (1-\xi)^2))^{N/2} d\xi \leq$$

$$\leqslant \text{const} \, \| \phi \|_{C_N} (1 + t_1^2 + \ldots + t_{n-1}^2)^{N+1/2}$$

holds, which implies the statement of the lemma.

Lemma 10. Difference derivation is a continuous operator from $\mathcal{B}_{N+1}(\mathbb{R}^{n-1})$ to $\mathcal{B}_N(\mathbb{R}^n)$.

Proof. Denote by $I : C_N(\mathbb{R}^n) \to C_{N+1}(\mathbb{R}^{n-1})$ the continuous operator such that $I\phi = \phi_1$, where ϕ_1 is given by (32). It follows that the functional $T_1 : \phi \to T(I\phi)$ belongs to $C_N^+(\mathbb{R}^n)$ provided that $T \in C_{N+1}^+(\mathbb{R}^{n-1})$. Now let $f \in \mathcal{B}_{N+1}(\mathbb{R}^{n-1})$. Set $T = F(f)$. Then

$$F^{-1}(T_1)(y_1,\ldots,y_n) \equiv F^{-1}(T \circ I)(y_1,\ldots,y_n) =$$

$$= \frac{1}{(\sqrt{(2\pi)})^n} T(I(e^{it_1y_1} \cdot \ldots \cdot e^{it_ny_n})) =$$

$$= \frac{1}{(\sqrt{(2\pi)})^n} T(e^{it_1y_1} \cdot \ldots \cdot e^{it_{n-2}y_{n-2}} \int_0^{t_{n-1}} e^{i[\xi y_{n-1} + (t_{n-1} - \xi)y_n]} d\xi) =$$

$$= \frac{1}{(\sqrt{(2\pi)})^n} \frac{-i}{y_{n-1} - y_n} T(e^{it_1y_1} \cdot \ldots \cdot e^{it_{n-2}y_{n-2}} e^{it_{n-1}y_{n-1}} -$$

$$- e^{it_1y_1} \cdot \ldots \cdot e^{it_{n-2}y_{n-2}} e^{it_{n-1}y_n}) = - \frac{i}{\sqrt{(2\pi)}} \frac{\delta f}{\delta y^{n-1}} (y_1,\ldots,y_{n-2};y_{n-1},y_n),$$

and we obtain the statement of the lemma immediately.

Corollary 1. For an arbitrary multi-index $\beta = (\beta_1,\ldots,\beta_{n-1})$, the operator $\delta/\delta y_{n-1}$ is continuous from $\mathcal{B}_{N+1}^{\beta}(\mathbb{R}^{n-1})$ to $\mathcal{B}_N^{\tilde\beta}(\mathbb{R}^n)$, where $\tilde\beta = (\beta_1, \ldots,\beta_{n-1},\beta_{n-1})$.

Proof. Let $f \in \mathcal{B}_{N+1}^{\beta}(\mathbb{R}^{n-1})$. By Lemma 1 (a), $f = (i + y)^{\beta} g$, where $g \in \mathcal{B}_{N+1}$, $\| g \|_{\mathcal{B}_{N+1}} \leqslant \text{const} \, \| f \|_{\mathcal{B}_{N+1}^{\beta}}$. By Lemma 8

$$\frac{\delta f}{\delta y_{n-1}} (y_1,\ldots,y_{n-2};y_{n-1},y_n) = (i + y_1)^{\beta_1} \cdot \ldots \cdot (i + y_{n-2})^{\beta_{n-2}} \times$$

$$\times \{ g(y_1,\ldots,y_{n-1}) \frac{(i + y_{n-1})^{\beta_{n-1}} - (i + y_n)^{\beta_{n-1}}}{y_{n-1} - y_n} +$$

$$+ \frac{\delta g}{\delta y_{n-1}} (y_1,\ldots,y_{n-2};y_{n-1},y_n)(i + y_n)^{\beta_{n-1}} \} =$$

$$= g(y_1,\ldots,y_{n-1})(i + y_1)^{\beta_1} \ldots (i + y_{n-2})^{\beta_{n-2}} \sum_{j=1}^{\beta_{j=1}} (i + y_{n-1})^j (i + y_n)^{\beta_{n-1} - j} +$$

$$+ \frac{\delta g}{\delta y_{n-1}} (y_1,\ldots,y_{n-2};y_{n-1},y_n)(i + y_n)^{\beta_{n-1}} (i + y_1)^{\beta_1} \cdot \ldots \cdot (i + y_{n-2})^{\beta_{n-2}}.$$

We obtain by virtue of Lemmas 2 and 10 that $g \in \mathcal{B}_N^0(\mathbb{R}^n)$, $\frac{\delta g}{\delta y_{n-1}} \in \mathcal{B}_N^0(\mathbb{R}^n)$, and

$(i+y_1)^{\beta_1}\cdot\ldots\cdot(i+y_{n-2})^{\beta_{n-2}}(i+y_{n-1})^{j}(i+y_n)^{\beta_{n-1}-j}\in \mathcal{B}_N^{(\beta_1\beta_2\ldots\beta_{n-2}j\beta_{n-1}-j)}$ $\times(\mathbb{R}^n)$ so that $\frac{\delta f}{\delta y_{n-1}}\in \sum_{j=1}^{\beta_{n-1}} \mathcal{B}_N^{(\beta_1\beta_2\ldots\beta_{n-2}j\beta_{n-1}-j)} \subset \mathcal{B}_N^{\tilde\beta}(\mathbb{R}^n)$ (here Σ denotes the sum of vector spaces, which is not necessarily direct) and, what is more, that $\|\delta f/\delta y_{n-1}\|_{\mathcal{B}_N^{\tilde\beta}} \leqslant \text{const}\,\|f\|_{\mathcal{B}_{N+1}^{\beta}}$. The lemma is proved.

Corollary 2. $\delta/\delta y$ is a continuous operator from $\mathcal{B}_{2N+1}^{\beta}(\mathbb{R})$ to $\mathcal{B}_{N,N}^{(\beta,\beta)}(\mathbb{R}^2)$ (we formulate the statement only for the one-dimensional case, the notations being rather complicated otherwise). Indeed, $\mathcal{B}_{2N}^{(\beta,\beta)}(\mathbb{R}^2)\subset \mathcal{B}_{N,N}^{(\beta,\beta)}(\mathbb{R}^2)$ and the embedding is continuous (Lemma 1 (c)). It remains to apply the above Corollary 1.

To finish with detailed description of symbol spaces, we introduce the space of symbols which plays in a sense the fundamental role in the functional calculus for several non-commuting operators. We denote by $S^\infty(\mathbb{R}^n)$ the following space

$$S^\infty(\mathbb{R}^n) = \bigcup_\beta \bigcap_\alpha \mathcal{B}_\alpha^\beta(\mathbb{R}^n). \tag{33}$$

The embeddings established in Lemma 1 (b) lead us immediately to the following:

Theorem 1. The function $f(y)$ belongs to $S^\infty(\mathbb{R})$ if and only if it is smooth and for some integer $m = m(f)$ and any multi-index $\alpha = (\alpha_1,\ldots,\alpha_n)$ we have

$$\left|\frac{\partial^{|\alpha|}f}{\partial y^\alpha}(y_1,\ldots,y_n)\right| \leqslant C_\alpha(1+|y|)^m; \tag{34}$$

$|y| = (y_1^2+\ldots+y_n^2)^{1/2}$. $S^\infty(\mathbb{R}^n)$ may be represented also in the form

$$S^\infty(\mathbb{R}^n) = \bigcup_\beta \bigcap_\nu \mathcal{B}_{I,\nu}^{\beta}(\mathbb{R}^n) \tag{35}$$

for any partition $I_1\cup\ldots\cup I_k$ of the set $\{1,\ldots,n\}$.

B. Functions of Several Generators in Banach Spaces

We begin with definitions concerned with Feynman ordering of operators. Let X be a Banach space, $(A_1,\ldots,A_n) = A$ be an n-tuple of generators in X, the degree of generator A_i being equal to k_i, $i = 1,\ldots,n$. Set $k = (k_1,\ldots,k_n)$. For any $f\in \mathcal{B}_k(\mathbb{R}^n)$, we define the operator

$$f(\overset{1}{A_1},\ldots,\overset{n}{A_n}) \equiv f(A) \equiv \mu_{\overset{1}{A_1},\ldots,\overset{n}{A_n}}(f) \equiv \mu_A(F) \tag{1}$$

in X by means of the relation

$$\langle h, f(\overset{1}{A_1},\ldots,\overset{n}{A_n})x\rangle \overset{\text{def}}{=} \frac{1}{(2\pi)^{n/2}}\, F(f)(\langle h, e^{it_nA_n}\cdot\ldots\cdot e^{it_1A_1}x\rangle) \tag{2}$$

for any $x\in X$, $h\in X^*$. Here the brackets $\langle\ ,\ \rangle$ denote the pairing between the elements of X^* and X, while $F(f)(\phi(t))$ denotes the value of the functional $F(f)\in C_k^+(\mathbb{R}^n)$ on the element $\phi\in C_k(\mathbb{R}^n)$.

Theorem 1. $f(\overset{1}{A_1},\ldots,\overset{n}{A_n})$ is a correctly defined bounded operator in X and

$$\| f(\overset{1}{A_1},\ldots,\overset{n}{A_n}) \| \leq C \| f \|_{\mathcal{B}_k(\mathbb{R}^n)} \tag{3}$$

for some constant C not depending on f.

Proof. $e^{it_nA_n}\cdot\ldots\cdot e^{it_1A_1}x$ is a continuous X-valued function of $t \in \mathbb{R}^n$ for any $x \in X$. Indeed,

$$e^{it_nA_n}\cdot\ldots\cdot e^{it_1A_1}x - e^{it_n^{(0)}A_n}\cdot\ldots\cdot e^{it_1^{(0)}A_1}x =$$

$$= \sum_{j=1}^{n} e^{it_nA_n}\cdot\ldots\cdot e^{it_{j+1}A_{j+1}}(e^{it_jA_j} - e^{it_j^{(0)}A_j})e^{it_{j-1}^{(0)}A_{j-1}}\cdot\ldots\cdot e^{it_1^{(0)}A_1}x \to 0$$

as $t \to t^{(0)}$ since each $e^{it_jA_j}$ is strongly continuous by definition of a generator and all the factors are bounded when t lies in a compact set.

Since A_i is a generator of degree k_i, $i = 1,\ldots,n$, one has the estimate

$$\| e^{it_nA_n}\cdot\ldots\cdot e^{it_1A_1}x \| \leqslant C(1 + |t_n|)^{k_n}\cdot\ldots\cdot(1 + |t_1|)^{k_1} \| x \|$$

for some constant C. Thus we have proved that $e^{it_nA_n}\cdot\ldots\cdot e^{it_1A_1}x$ is an element of $C_k(\mathbb{R}^n,X)$ and its norm does not exceed $C \| x \|$. It follows that the expression on the right of (2) is defined correctly and $f(\overset{1}{A_1},\ldots,\overset{n}{A_n})$ is in any case a bounded operator from X to X^{**} with the norm not exceeding const $\| f \|_{\mathcal{B}_k}$. It appears however that the range of $f(\overset{1}{A_1},\ldots,\overset{n}{A_n})$ is contained in $X \subset X^{**}$ (see Lemma 1 below); the latter consideration makes the proof complete.

Lemma 1. Let $\phi \in C_\alpha^\beta(\mathbb{R}^n,X)$, $T \in C_\alpha^{\beta+}(\mathbb{R}^n)$. The functional χ on X^* defined by

$$\langle h,\chi \rangle = T(\langle h,\phi \rangle),\ h \in X^*$$

is determined by a (uniquely defined) element $x \in X$. In other words, $T(\phi) \equiv \int T(t)\phi(t)dt$ (the latter notation sometimes will be used if it is convenient) is a correctly defined element of X. The proof of Lemma 1 does not differ from that in the particular case n = 1 (Theorem 1 of Section 2:E).

Unlike the one-dimensional case, one has to make supplementary assumptions in order to define $f(\overset{1}{A_1},\ldots,\overset{n}{A_n})$ for growing symbols. The underlying reason for the distinction between one- and multi-dimensional cases is that, generally speaking, in the latter case the analogue of the scale $\{H_A^\ell\}$ (see Definition 5 of Section 2:E) cannot be defined in such a way that the operators A_i, $i = 1,\ldots,n$, prove to be generators in all of its spaces.

In the following items we deal with the case when such a scale may be constructed or is given a priori; here we give a sketch of another approach which leads us to the consideration of f(A) as an unbounded operator in X. Both approaches are of course equivalent for functions of a single operator.

We introduce the following:

Condition A. There exists a dense linear subset $D \subset X$ such that

(a) D is invariant under $\mathcal{U}(t) = e^{iA_nt_n}\cdot\ldots\cdot e^{iA_1t_1}$ for any $t \in \mathbb{R}^n$.

(b) $\mathcal{U}(t)x$, $x \in D$ is infinitely differentiable with respect to t in the strong sense.

(c) All the derivatives $\mathcal{U}^{(\alpha)}(t)x \equiv (\frac{\partial}{\partial t})^{\alpha}(\mathcal{U}(t)x)$ have the polynomial rate of growth:

$$\| \mathcal{U}^{(\alpha)}(t)x \| \leqslant C(1+|t|)^s, \quad x \in D, \tag{4}$$

where $s = s(\alpha)$, $C = C(x,\alpha)$.

Theorem 2. Let Condition A be satisfied. For every $f \in S^{\infty}(\mathbb{R}^n)$ the equality

$$f(\overset{1}{A}_1,\dots,\overset{n}{A}_n)x = \frac{1}{(\sqrt{(2\pi)})^n} \int F(f)(t)\mathcal{U}(t)x\,dt, \quad x \in D \tag{5}$$

defines $f(\overset{1}{A}_1,\dots,\overset{n}{A}_n)$ as a linear operator with the domain D.

Proof. For some $\beta = (\beta_1,\dots,\beta_n)$, $f \in \bigcap_{N=0}^{\infty} B_N^{\beta}(\mathbb{R}^n)$. In particular, $f \in B_{N_0}^{\beta}(\mathbb{R}^n)$, where $N_0 = \max_{\alpha \leqslant \beta} s(\alpha)$. Condition A guarantees that $\mathcal{U}(t)x \in$ $\in C_{N_0}^{\beta}(\mathbb{R}^n)$. On the other hand, $F(f) \in C_{N_0}^{\beta+}(\mathbb{R}^n)$. Applying Lemma 1, we obtain that $f(\overset{1}{A}_1,\dots,\overset{n}{A}_n)x$ is a correctly defined element of X. The linearity of the operator (5) is obvious, thus the theorem is proved.

Further, we might investigate the closurability of the obtained operators under suitable assumptions. However, we leave this subject and go on with our brief review of the presented approach.

Assume that some of the operators $A_1,\dots,A_n$, namely $A_{j_1},\dots,A_{j_k}$, are not necessarily the generators in X, but for given polynomials $p_1(y),\dots,$ $p_k(y)$, $y \in \mathbb{R}^1$, the operator*

$$\mathcal{U}_{p_1,\dots,p_k}(t_{j_{k+1}},\dots,t_{j_n}) = p_1(\overset{j_1}{A}_{j_1}) \cdot \dots \cdot p_k(\overset{j_k}{A}_{j_k}) \times \exp(i \overset{j_{k+1}}{A}_{j_{k+1}} t_{j_{k+1}}) \dots \exp(i \overset{j_n}{A}_{j_n} t_{j_n}) \tag{6}$$

is defined on the dense linear subset D and Condition A is satisfied for $\mathcal{U}_{p_1,\dots,p_k}(t_{j_{k+1}},\dots,t_{j_n})$ instead of $\mathcal{U}(t)$. Then for any $f \in S^{\infty}(\mathbb{R}^n)$, which has the form

$$F(y_1,\dots,y_n) = p_1(y_{j_1})p_2(y_{j_2}) \dots p_k(y_{j_k})g(y_{j_{k+1}},\dots,y_{j_n}),$$

$g \in S^{\infty}(\mathbb{R}^{n-k})$, we may define the operator $f(\overset{1}{A}_1,\dots,\overset{n}{A}_n)$ by means of the formula

$$f(\overset{1}{A}_1,\dots,\overset{n}{A}_n) = \frac{1}{(\sqrt{(2\pi)})^{n-k}} \int F(g)(t_{j_{k+1}},\dots,t_{j_n}) \times \mathcal{U}_{p_1,\dots,p_k}(t_{j_{k+1}},\dots,t_{j_n})dt_{j_{k+1}} \dots dt_{j_n}, \quad x \in D. \tag{7}$$

In particular, if the above condition is satisfied for any polynomials $p_1,\dots,p_k$ and D does not depend on the choice of polynomials, (7) provides the mapping of the set of functions $f \in C^{\infty}(\mathbb{R}^n)$ polynomial in $(y_{j_1},\dots,y_{j_k})$ into the set of linear operators in X with the domain D.

---

* Here $(j_1,\dots,j_n)$ is the substitution of the set $\{1,\dots,n\}$; the numbers over the operators on the right of (6) have a clear sense (see Introduction to this chapter). Namely, they define the order of the factor; the greater the number over the factor, the further left is its position in the product.

If some of $A_{j_1^1},\ldots,A_{j_{k_n}^n}$ are in fact generators we obtain then two different definitions of $f(A_1, \ldots, A_n)$ for symbols described above. However, it may be shown that both definitions lead to the same result; this statement will be proved (for operators in scales) in the following items.

We finish this item with the definition of Weyl ordering of generators. We introduce the following:

Condition W. The operators $A_1,\ldots,A_n$ are defined on the dense linear subset $D \subset X$. For any $t \in \mathbb{R}^n$ the linear combination $t\cdot A = t_1A_1 + \ldots + t_nA_n$ is a closurable operator and its closure (which will be also denoted by $t\cdot A$) is a generator of degree N in X. Besides,

$$\| e^{i(t\cdot A)} \| \leqslant C(1 + |t|)^N, \tag{8}$$

where C does not depend on t.

This condition being satisfied, we are able to define the Weyl-ordered function of $A_1,\ldots,A_n$ for any $f \in \mathcal{B}_N(\mathbb{R}^n)$. Namely, we set

$$f_W(A) \equiv f_W(A_1,\ldots,A_n) = \frac{1}{(2\pi)^{n/2}} \int_{\mathbb{R}^n} F(f)(t) e^{it\cdot A} dt. \tag{9}$$

The definition (9) is correct and $\| f_W(A) \| \leqslant \text{const} \| f \|_{\mathcal{B}_N}$. To approve it we make use of Corollary 1 of Section 2:C. Since D is a core of $t\cdot A$, the range of $\lambda - tA$ restricted on D is dense in X for all non-real λ and we obtain, by virtue of the mentioned corollary, that e^{itA} is strongly continuous with respect to $t\cdot t$. Now (8) shows that $e^{itA}x \in C_N(\mathbb{R}^n, x)$ for any $x \in X$ and the application of Lemma 1 completes the proof. If also Condition A is satisfied for e^{itA}, we may define $f_W(A_1,\ldots,A_n)$ for symbols $f \in S^\infty(\mathbb{R}^n)$.

C. Linear Operators in Banach Scales

Let Δ be a bidirected set, i.e., a partially ordered set, such that for any $\delta_1,\delta_2 \in \Delta$ there exist the elements $\tilde{\delta},\bar{\delta} \in \Delta$, such that $\bar{\delta} \geqslant \delta_i$, $\tilde{\delta} \leqslant \delta_i$, $i = 1,2$ (in other words both Δ and Δ' are directed sets, where Δ' coincides with Δ as a set and is endowed with the inverse ordering relation).

Definition 1. A Banach scale (over Δ) is a collection of Banach spaces $\{X_\delta\}_{\delta\in\Delta}$ together with linear operators $i_{\delta\delta'} : X_\delta \to X_{\delta'}$ defined for any pair $(\delta,\delta') \in \Delta \times \Delta$ satisfying $\delta \geqslant \delta'$, such that:

(a) All the operators $i_{\delta\delta'}$ are continuous embeddings, and besides $i_{\delta\delta} : X_\delta \to X_\delta$ are identical operators.

(b) For any $\delta,\delta',\delta'' \in \Delta$ such that $\delta \geqslant \delta' \geqslant \delta''$, the diagram

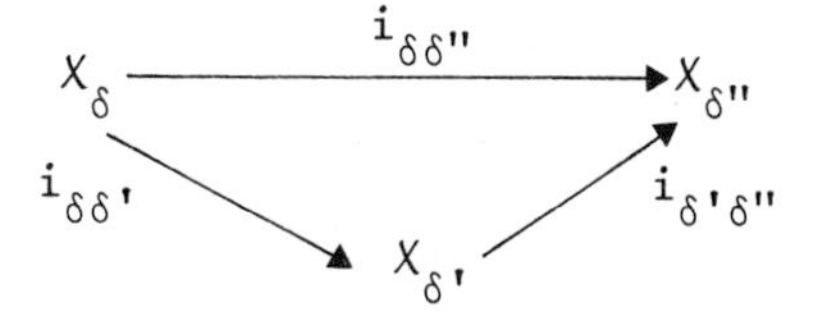

(1)

of Banach spaces and homomorphisms is commutative (i.e., $i_{\delta\delta''} = i_{\delta'\delta''} \circ i_{\delta\delta'}$). A Banach scale is called a dense one if all $i_{\delta\delta'}$ are dense embeddings.

Example 1. $\{W_2^k(\mathbb{R}^n)\}_{k\in\mathbb{R}}$, where $W_2^k(\mathbb{R}^n)$ is the space of Fourier transforms of measurable functions, for which the norm

$$\|\phi\|^2_{W_2^k(\mathbb{R}^n)} = \int_{\mathbb{R}^n} |\tilde{\phi}(x)|^2 (1+x^2)^k dx, \quad (\tilde{\phi} = F(\phi)) \tag{2}$$

is finite, is the well-known Sobolev scale. It is a dense scale.

Example 2. Let Δ be a set of pairs of multi-indices $\delta = (\alpha,\beta)$ of the length n with natural components with the ordering relation $\delta \geqslant \delta' \rightleftarrows (\alpha \geqslant \alpha', \beta \leqslant \beta') \equiv (\alpha_1 \geqslant \alpha_1',\ldots,\alpha_n \geqslant \alpha_n', \beta_1 \leqslant \beta_1',\ldots,\beta_n \leqslant \beta_n')$. The collection of spaces $\{B_\alpha^\beta(\mathbb{R}^n)\}_{(\alpha,\beta)\in\Delta}$ (these spaces were defined in item A) with natural embeddings form a Banach scale over Δ, as follows from Lemma 1 of item A. A. Also $\{H_\alpha^\beta(\mathbb{R}^n)\}_{(\alpha,\beta)\in\Delta}$ is a Banach scale; here the set Δ may be extended to consist of all the pairs of multi-indices with arbitrary real components. The latter scale is dense.

The next proposition follows immediately from the definition.

Proposition 1. Let E be a vector space. Assume that for any element δ of the bidirected set Δ the norm $\|\cdot\|_\delta$ is defined on E and satisfies the condition $\|x\|_\delta \geqslant \|x\|_{\delta'}$ for any $x \in E$, $\delta' \leqslant \delta$. Denote by X_δ the completion E_δ of E in the norm $\|\cdot\|_\delta$ and by $i_{\delta\delta'}$ the closure of the identity operator on E in the corresponding pair of spaces. The collection of these objects forms a dense Banach scale provided that the following compatibility condition is satisfied: if the sequence $\{x_n\}$ of the elements of E converges to 0 in the norm $\|\cdot\|_\delta$ and is fundamental in the norm $\|\cdot\|_{\delta'}$ for some $\delta' \geqslant \delta$, then it converges to 0 in the norm $\|\cdot\|_{\delta'}$ as well.

Notation. The Banach scale will be denoted by $X = \{X_\delta\}_{\delta\in\Delta}$; X_δ will be identified (under the embedding $i_{\delta\delta'}$) with the subset of $X_{\delta'}$ for $\delta \geqslant \delta'$. Consider the union

$$\mathcal{X} = \bigcup_{\delta\in\Delta} X_\delta. \tag{3}$$

$\mathcal{X}$ possesses the structure of a vector space. Indeed, if $x_1, x_2 \in \mathcal{X}$, then $x_i \in X_{\delta_i}$ for some δ_i, $i = 1,2$. Let $\delta \leqslant \delta_1$, $\delta \leqslant \delta_2$ (such δ exists, since Δ is bidirected). Then $x_1, x_2 \in X_\delta$ and we may define their linear combination $ax_1 + bx_2 \in X_\delta \subset \mathcal{X}$. The vector space axioms are obviously valid for such a definition. In applications the vector space $\mathcal{X}$ usually has a natural interpretation. For example, for the scale $\{H_\alpha^\beta(\mathbb{R}^n)\}$ the union (3) is nothing else than the space $S'(\mathbb{R}^n)$ of tempered distributions on $\mathbb{R}^n$. We refer to $\mathcal{X}$ as the total space of the scale. If some linear subspace $E \subset \mathcal{X}$ is dense in any $X_\delta \subset \mathcal{X}$, $\delta \in \Delta$ (and, consequently, we are in the situation of Proposition 1), it will be referred to as the scale base.

We introduce the convergence in the total space $\mathcal{X}$ of the Banach scale $\{X_\delta\}_{\delta\in\Delta}$. Let $x : \Gamma \to \mathcal{X}$, $\gamma \to x_\gamma$ be a generalized sequence in $\mathcal{X}$ (this means nothing more than that Γ is a directed set.)*

Definition 2. $\{x_\gamma\}_{\gamma\in\Gamma}$ converges to $x \in \mathcal{X}$ if and only if there exist $\delta_0 \in \Delta$, $\gamma_0 \in \Gamma$ such that $x_\gamma \in X_{\delta_0}$ for all $\gamma \geqslant \gamma_0$, $x \in X_{\delta_0}$, and $\|x_\gamma - x\|_{\delta_0}$

---

* We shall also speak of $\{X_\gamma\}_{\gamma\in\Gamma}$ as a Γ-sequence for short.

converges to zero (i.e., for any $\varepsilon > 0$ there is $\gamma = \gamma(\varepsilon) \in \Gamma$ such that $\| x_{\gamma'} - x \|_{\delta_0} \leqslant \varepsilon$ for $\gamma' \geqslant \gamma$). In particular, the sequence x_n converges to x if and only if $x_n \in X_\delta$, $x \in X_\delta$ for some $\delta \in \Delta$ and all $n \geqslant n_0$, and $\|x_n - x\|_\delta \to 0$ as $n \to \infty$.

Notation. We write $x_\gamma \to x$ or $\lim_{\gamma\in\Gamma} x_\gamma = x$ if the Γ-sequence x_γ converges to x. We write $x_\gamma \overset{\delta}{\to} x$ if $\|x_\gamma - x\|_\delta \to 0$.

Note. We do not introduce any topology associated with this convergence.

Proposition 2. The convergence introduced above is compatible with linear structure on X (i.e., linear operations are continuous with respect to this convergence). The generalized sequence may have at most one limit.

Proof. Let $x_\gamma \to x$, $\tilde{x}_{\tilde{\gamma}} \to \tilde{x}$, $\gamma \in \Gamma$, $\tilde{\gamma} \in \tilde{\Gamma}$. We prove that $(\Gamma \times \tilde{\Gamma})$-sequence $x_\gamma + \tilde{x}_{\tilde{\gamma}}$ converges to $x + \tilde{x}$. Indeed, let $x_\gamma \overset{\delta}{\to} x$, $\tilde{x}_{\tilde{\gamma}} \overset{\tilde{\delta}}{\to} \tilde{x}$; let also $\delta_1 \leqslant \delta$, $\delta_1 \leqslant \tilde{\delta}$. Then $x_\gamma \overset{\delta_1}{\to} x$, $\tilde{x}_{\tilde{\gamma}} \overset{\delta_1}{\to} \tilde{x}$, so $\|x_\gamma - x\|_{\delta_1} \leqslant \varepsilon/2$, $\|\tilde{x}_{\tilde{\gamma}} - \tilde{x}\|_{\delta_1} \leqslant \varepsilon/2$ for $\gamma \geqslant \gamma(\varepsilon/2)$, $\tilde{\gamma} \geqslant \tilde{\gamma}(\varepsilon/2)$. Thus $\|x_\gamma + \tilde{x}_{\tilde{\gamma}} - x - \tilde{x}\|_{\delta_1} \leqslant \varepsilon$ for these $\gamma, \tilde{\gamma}$, and this means that $x_\gamma + \tilde{x}_{\tilde{\gamma}} \overset{\delta_1}{\to} x + \tilde{x}$. The continuity of multiplication by scalars is proved analogously.

Now if $x_\gamma \to x_1$, $x_\gamma \to x_2$, then $x_\gamma \overset{\delta_1}{\to} x_1$, $x_\gamma \overset{\delta_2}{\to} x_2$ for some δ_1, δ_2, and $x_\gamma \overset{\delta}{\to} x_1$, $x_\gamma \overset{\delta}{\to} x_2$ for $\delta \leqslant \delta_1$, $\delta \leqslant \delta_2$. It follows that $\|x_1 - x_2\|_\delta = 0$, i.e., $x_1 = x_2$. The proposition is proved.

We now expose some results concerned with integration theory to be used in the sequel.

Let $\psi : \mathbb{R}^n \to X$ be a continuous mapping. (This means that ψ transforms convergent generalized sequences into convergent ones. It is easy to see that ψ is continuous if and only if for any $t_0 \in \mathbb{R}^n$ there exists $\delta_0 = \delta_0(t_0) \in \Delta$ such that $\psi(t) \in X_{\delta_0}$ for t lying in some vicinity $U_0 \ni t_0$ and the mapping $\psi : U_0 \to X_{\delta_0}$ is continuous.) Let also the distribution $T \in \mathcal{D}'(\mathbb{R}^n)$ be given. We intend to define, under some conditions, the element $T[\psi] \in X$. Let $\phi \in C_0^\infty(\mathbb{R}^n)$, $\int_{\mathbb{R}^n} \phi(t)dt = 1$. Set $\phi_\varepsilon(t) = \varepsilon^{-n}\phi(t/\varepsilon)$, $\varepsilon \in \mathbb{R}$.

It is well known that

$$T_\varepsilon(t) \equiv (T * \phi_\varepsilon)(t) \equiv \int T(\tau)\phi_\varepsilon(t-\tau)d\tau$$

is a smooth function, and $T_\varepsilon \to T$ in $\mathcal{D}'(\mathbb{R}^n)$ as $\varepsilon \to 0$. Moreover, if T is the distribution of the order k, $T_\varepsilon[f] \overset{\varepsilon\to 0}{\to} T[f]$ for any finite function f with k continuous derivatives. First we define the integral

$$T_\varepsilon[\psi] = \int_{\mathbb{R}^n} T_\varepsilon(t)\psi(t)dt.$$

Let Γ denote the set of finite unions of compact polyhedra in $\mathbb{R}^n$, ordered by the inclusion relation (i.e., $\gamma \leqslant \tilde{\gamma}$ iff $\gamma \subset \tilde{\delta}$). Clearly Γ is a directed set. We set

$$T_\varepsilon[\psi] = \lim_{\gamma\in\Gamma} \int_\gamma T_\varepsilon(t)\psi(t)dt,$$

provided that the limit on the right exists. (The definition of $\int_\gamma T_\varepsilon(t) \times \psi(t)dt$ is obvious since, γ being compact, a standard argument shows that for some $\delta \in \Delta$, ψ is a continuous mapping $\psi : \gamma \to X_\delta$.)

Definition 3. We say that $T[\psi]$ is defined (and equals $x \in X$) for given $T \in \mathcal{D}'(\mathbb{R}^n)$ and continuous, $\psi : \mathbb{R}^n \to X$, if for any $\phi \in C_0^\infty(\mathbb{R}^n)$, $\int_{\mathbb{R}^n}\phi(t)\,dt = 1$, the limit

$$\lim_{\varepsilon\to 0} T_\varepsilon[\psi]$$

exists and equals x.

If ψ lies in the subspace $C_\alpha^\beta(\mathbb{R}^n, X_\delta)$ of the space of all continuous mappings $\psi : \mathbb{R}^n \to X$, and T lies in $C_\alpha^{\beta+}(\mathbb{R}^n)$, we can prove the existence theorem.

Theorem 1. Let $\psi \in C_\alpha^\beta(\mathbb{R}^n, X_\delta) \cdot T \in C_\alpha^{\beta+}(\mathbb{R}^n)$. Then $T[\psi]$ is defined, $T[\psi] \in X_\delta$, and equals the element $T(\psi)$ described in Lemma 1 of item B.

Proof. Let $\phi \in C_0^\infty$, $\int\phi(t)dt = 1$. We assert that $T_\varepsilon = T * \phi_\varepsilon$ lies in $C_\alpha^{\beta+}(\mathbb{R}^n)$. Indeed, the Fourier transform $\mathcal{F}^{-1}(T_\varepsilon) = \mathcal{F}^{-1}(T)\cdot\mathcal{F}^{-1}(\phi_\varepsilon)$ lies in $C_\alpha^{\beta+}(\mathbb{R}^n)$ since $\mathcal{F}^{-1}(T) \in \mathcal{B}_\alpha^\beta(\mathbb{R}^n)$ and $\mathcal{F}^{-1}(\phi_\varepsilon) \in \mathcal{B}_\alpha(\mathbb{R}^n)$ (we have made use of Lemmas 1 and 2 of item A), so that $T_\varepsilon \in C_\alpha^{\beta+}(\mathbb{R}^n)$.

Next, we have $T_\varepsilon(f) = T(\phi_\varepsilon \hat{*} f)$ for any $f \in S(\mathbb{R}^n)$ and henceforth (multi-dimensional analogue of Lemma 3, Section 2:E) for any $f \in C_\alpha^\beta(\mathbb{R}^n)$, where

$$(\phi_\varepsilon \hat{*} f)(t) = \int_{\mathbb{R}^n}\phi_\varepsilon(\tau)f(t+\tau)d\tau.$$

We have

$$\|\phi_\varepsilon \hat{*} f\|_{C_\alpha^\beta} \leqslant C\|f\|_{C_\alpha^\beta};\quad \|\phi_\varepsilon \hat{*} f\|_{C_\alpha^\beta} \leqslant C|\varepsilon|^{-|\beta|}\|f\|_{C_\alpha^0} \tag{5}$$

with the constant C independent of ε, $|\varepsilon| \leqslant 1$.

Let $T_\ell = \psi_\ell T$, where $\psi_\ell \in C_0^\infty(\mathbb{R}^n)$ be a sequence of functionals convergent to T in $C_\alpha^{\beta+}$. We have for $|\varepsilon| \leqslant 1$

$$|T_\varepsilon(f) - T(f)| \leqslant |T_\ell(f) - T(f)| + |T_\ell(\phi_\varepsilon \hat{*} f) - T(\phi_\varepsilon \hat{*} f)| +$$

$$|T_\ell(f - \phi_\varepsilon \hat{*} f)| \leqslant (1+C)\|T_\ell - T\|_{C_\alpha^{\beta*}}\|f\|_{C_\alpha^\beta} +$$

$$+ \|T\|_{C_\alpha^{\beta*}}\|\psi_\ell(f - \phi_\varepsilon \hat{*} f)\|_{C_\alpha^\beta}.$$

Set here $f(t) = f_h(t) = \langle h, \psi(t)\rangle$, where $h \in X_\delta^*$. We assert that $|T_\varepsilon(f_h) - T(f_h)|$ tends to zero uniformly with respect to h, $\|h\| = 1$. Indeed, $\mu > 0$ being given, the first summand is less than $\mu/2$ if we fix $\ell = \ell_0$ such that $\|T_{\ell_0} - T\|_{C_\alpha^{\beta+}} < \frac{\mu}{2}(1+C)^{-1}\{\|\psi\|_{C_\alpha^\beta(\mathbb{R}^n, X_\delta)}\}^{-1}$. Besides, the function $\psi(t)$ is uniformly continuous together with its derivatives up to the order β on any compact subset of $\mathbb{R}^n$, so it turns out that $\psi_{\ell_0}(t)(f_h(t) - (\phi_\varepsilon \hat{*} f_h) \times (t))$ tends to zero uniformly with respect to $t \in \mathbb{R}^n$, $h \in X_\delta^*$, $\|h\| = 1$, together with all the derivatives up to the order β. In other words, $\|\psi_\ell(f_h - \phi_\varepsilon \hat{*} f_h)\|_{C_\alpha^\beta} \to 0$ as $\varepsilon \to 0$ uniformly with respect to h, $\|h\| = 1$, and our assertion is proved once we choose ε small enough. It follows that $\| T_\varepsilon(\psi) - T(\psi) \|_\delta \to 0$ as $\varepsilon \to 0$. It remains to prove that for any fixed ε, $T_\varepsilon[\psi]$ exists and coincides with $T_\varepsilon(\psi)$. Let $\psi_\ell \in C_0^\infty(R^n)$, $0 \leqslant \psi_\ell(t) \leqslant 1$, and $\psi_\ell(t) = 0$ for $|t| \geqslant \ell$. Assume that $\| \psi_n T - T \|_{C_\alpha^{\beta *}} \leqslant m$; then for any two $\gamma, \tilde{\gamma} \subset \Gamma$ such that $\gamma \subset \tilde{\gamma}$ and $\gamma \supset \{t \mid |t| \leqslant \ell + \varepsilon\}$, we have

$$\int_{\tilde{\gamma} \setminus \gamma} | T_\varepsilon(t) | (1 + |t|)^\alpha dt \leqslant C \cdot m |\varepsilon|^{-|\beta|}.$$

Indeed, the function $\dfrac{|T_\varepsilon(t)|}{T_\varepsilon(t)} (1 + |t|)^\alpha$, defined to be zero at the points where $T_\varepsilon(t) = 0$, may be approximated by smooth functions in $L^1(\tilde{\gamma} \setminus \gamma)$, and we obtain that

$$\begin{aligned} \int_{\tilde{\gamma} \setminus \gamma} |T_\varepsilon(t)| (1 + |t|)^\alpha dt &\leqslant \sup_{\substack{\phi \,\ell\, C_0^\infty(\tilde{\gamma} \setminus \gamma) \\ \|\phi\|_{C_\alpha^0} \leqslant 1}} T_\varepsilon(\phi) = \\ &= \sup_{\dots} |T(\phi_\varepsilon * \phi)| = \sup_{\dots} |(\psi_\ell T - T)(\phi_\varepsilon * \phi)| \leqslant \\ &\leqslant C|\varepsilon|^{-|\beta|} \| \psi_n T - T \|_{C_\alpha^{\beta *}} \leqslant C \cdot m |\varepsilon|^{-|\beta|} \end{aligned} \tag{6}$$

by virtue of the second of the inequalities (5).

Since $\| \psi_\ell T - T \|_{C_\beta^{\alpha *}} \to 0$ as $n \to \infty$, (6) implies that the integral

$$\int_{\mathbb{R}^n} T_\varepsilon(t)(1 + |t|)^\alpha dt \tag{7}$$

converges absolutely, so that the limit (4) necessarily exists. Moreover, the convergence of (7) yields that also

$$T_\varepsilon[\psi] = \lim_{n \to \infty} \int_{\mathbb{R}^n} \psi_\ell(t) T_\varepsilon(t) \psi(t) dt.$$

Since the latter expression is simply $T_\varepsilon(\psi)$, the theorem is proved. From now on we do not distinguish between parentheses and square brackets in the expression of $T_\varepsilon(\psi)$.

We come now to the definition of linear operators in Banach scales. Let $X = \{X_\delta\}_{\delta \in \Delta}$, $\mathcal{Y} = \{Y_\sigma\}_{\sigma \in \Sigma}$ be Banach scales, X and Y being their total spaces. A linear operator $A : X \to \mathcal{Y}$ is by definition a linear mapping $A : X \to Y$ of some linear subset $D_A \subset X$ (the domain of A) into Y. If $X = \mathcal{Y}$, we call A the operator in the scale X. To any linear operator $A : X \to \mathcal{Y}$ the collection $\{A_{\delta\sigma}\}_{(\delta,\sigma) \in \Delta \times \Sigma}$ of linear operators $A_{\delta\sigma} : X_\delta \to Y_\sigma$ may be set into correspondence. Namely, we set

$$\begin{aligned} D_{A_{\delta\sigma}} &= \{x \in D_A \mid x \in X_\delta \text{ and } Ax \in Y_\sigma\}, \\ A_{\delta\sigma} x &= Ax, \quad x \in D_{A_{\delta\sigma}}. \end{aligned} \tag{8}$$

Sometimes we drop the subscripts and write A instead of $A_{\delta\sigma}$. We denote by $L(X_\delta, Y_\sigma) \equiv L_{\delta\sigma}(X, \mathcal{Y}) \equiv L_{\delta\sigma}$ the space of linear operators $A : X \to \mathcal{Y}$ such that $A_{\delta\sigma}$ is a bounded-everywhere defined operator and by $C(X_\delta, Y_\sigma) \equiv C_{\delta\sigma}(X, \mathcal{Y}) \equiv$ $\equiv C_{\delta\sigma}$ the space of linear operators $A : X \to \mathcal{Y}$ such that $A_{\delta\sigma}$ is a closed densely defined operator. We call the operator A bounded to the right if

$$A \in L_{right}(X,\mathcal{Y}) \overset{def}{=} \bigcap_{\delta \in \Delta} \bigcup_{\sigma \in \Sigma} L(X_\delta, Y_\sigma) \tag{9}$$

and bounded to the left if

$$A \in L_{left}(X,\mathcal{Y}) \overset{def}{=} \bigcup_{\sigma \in \Sigma} \bigcap_{\delta \in \Delta} L(X_\delta, Y_\sigma). \tag{10}$$

The operators, closed to the right (left), are defined analogously. One can easily see that $A \in L_{right}(X,\mathcal{Y})$ if and only if A is everywhere defined and continuous with respect to convergence introduced in Definition 2. Thus $A : X \to \mathcal{Y}$ is bounded to the right (to the left) if and only if there exists a mapping $\sigma : \Delta \to \Sigma$, $\delta \to \sigma(\delta)$ ($\delta : \Sigma - \Delta$, $\sigma \to \delta(\sigma)$) such that $A_{\delta\sigma(\delta)} :$ $: X_\delta \to Y_{\sigma(\delta)}$ is bounded and everywhere defined (respectively, $A_{\delta(\sigma)\sigma} :$ $: X_{\delta(\sigma)} \to Y_\sigma$ is bounded and everywhere defined).

If A is bounded to the right (to the left) we denote the corresponding mapping by σ_A (respectively, δ_A). Of course, this mapping is not uniquely defined.

In the particular case when $X = \mathcal{Y}$ and Δ is a subset of an Abelian group $\tilde{\Delta}$, the operator $A : X \to X$ will be called a right (left) translator with the step $s \in \tilde{\Delta}$, if A is bounded to the right (to the left) and one may choose $\sigma_A(\delta) = \delta + s$ (respectively, $\delta_A(\sigma) = \sigma - s$). (We assume that $\delta + s \in \Delta$ for any $\delta \in \Delta$ in the first case and that $\sigma - s \in \Delta$ for any $\sigma \in \Delta$ in the second one; of course if $\Delta = \tilde{\Delta}$ the notions of right and left translator with the step s coincide.)

We return to the general case. If $A : X \to \mathcal{Y}$ is a given linear operator we define the following sets of indices:

$$\begin{aligned} &\chi_A = \{(\delta,\sigma) \in \Delta \times \Sigma \mid A \in L(X_\delta, Y_\sigma)\} \\ &\chi_A^C = \{(\delta,\sigma) \in \Delta \times \Sigma \mid A \in C(X_\delta, Y_\sigma)\} \\ &\Delta_A, \Delta_A^C \text{ - projections of } \chi_A \text{ and } \chi_A^C \text{ onto } \Delta \\ &\Sigma_A, \Sigma_A^C \text{ - projections of } \chi_A \text{ and } \chi_A^C \text{ onto } \Sigma. \end{aligned} \tag{11}$$

These sets will be useful below.

<u>Proposition 3</u>. (a) If $(\delta,\sigma) \in \chi_A$, then $(\delta',\sigma') \in \chi_A$ for any $\delta' \geq \delta$, $\sigma' \leq \sigma$.

(b) If $(\delta,\sigma) \in \chi_A^C$, then $(\delta',\sigma') \in \chi_A^C$ for any $\delta' \geq \delta$, $\sigma' \geq \sigma$.

<u>Proof</u>. (a) $A_{\delta'\sigma'} = i_{\sigma\sigma'} A_{\delta\sigma} i_{\delta'\delta}$ is bounded and everywhere defined if the same holds for $A_{\delta\sigma}$.

(b) Let $A \in C(X_\delta, Y_\sigma)$. If $x_n \in D_A$, $n = 1,2,\ldots$, $x_n \overset{\delta'}{\to} x$ and $Ax_n \overset{\sigma'}{\to} y$, then $x_n \overset{\delta}{\to} x$ and $Ax_n \overset{\sigma}{\to} y$, thus $x \in D_A$ and $y = Ax$. This means that $A \in$ $\in C(X_{\delta'}, Y_{\sigma'})$. The proposition is proved.

<u>Proposition 4</u>. Let $\{B_{\delta\sigma}\}_{(\delta,\sigma)\in\chi}$ be a family of linear operators, $B_{\delta\sigma} : X_\delta \to Y_\sigma$ defined for (δ,σ) lying in some subset $\chi \in \Delta \times \Sigma$. The following conditions are equivalent:

(a) There exists a linear operator $A : X \to \mathcal{Y}$ such that $D_{B_{\delta\sigma}} \subset D_A$ for any $(\delta,\sigma) \in \chi$ and $B_{\delta\sigma}x = Ax$ for $x \in D_{B_{\delta\sigma}}$.

(b) For any finite set $\{x_1,\ldots,x_m\}$ of elements of X such that $\sum_{j=1}^{m} x_j = 0$ and $x_j \in D_{B_{\delta_j\sigma_j}}$, $j = 1,\ldots,m$ where $(\delta_j,\sigma_j) \in \chi$, $j = 1,\ldots,m$, we have:

$$\sum_{j=1}^{m} B_{\delta_j\sigma_j}x_j = 0.$$

Proof. (a) $\Rightarrow$ (b). It is evident. (b) $\Rightarrow$ (a). We denote by D_A the linear core of all the sets $D_{B_{\delta\sigma}}$, $(\delta,\sigma) \in \chi$, and set for $x \in D_A$

$$Ax = \sum_{j=1}^{k} B_{\delta_j\sigma_j}x_j \text{ if } x = \sum_{j=1}^{k} x_j,\ x_j \in D_{B_{\delta_j\sigma_j}}. \tag{12}$$

The definition (12) is correct since the condition (b) yields that that $\sum_{j=1}^{k} B_{\delta_j\sigma_j}x_j = \sum_{j=k+1}^{m} B_{\delta_j\sigma_j}x_j$ in the case when the equalities $x = \sum_{j=1}^{k} x_j = \sum_{j=k+1}^{m} x_j$ are two different representations of the element $x \in D_A$. Thus, obviously, the operator A defined by (12) is a linear one. The proposition is proved. The operator constructed in the proof of this proposition is called the (minimal) operator generated by the family $\{B_{\delta\sigma}\}$.

Proposition 5. Let $\{B_{\delta\sigma}\}_{(\delta,\sigma) \in \chi}$ be a family of linear operators $B_{\delta\sigma} : X_\delta \to Y_\sigma$. Assume that:

i) for any $(\delta_1,\sigma_1) \in \chi$, $(\delta_2,\sigma_2) \in \chi$ such that $\delta_1 \geqslant \delta_2$, $\sigma_1 \geqslant \sigma_2$, we have $D_{B_{\delta_1\sigma_1}} \in D_{B_{\delta_2\sigma_2}}$ and $B_{\delta_1\sigma_1}x = B_{\delta_2\sigma_2}x$, $x \in D_{B_{\delta_1\sigma_1}}$;

ii) for any $(\delta_1,\sigma_1) \in \chi$, $(\delta_2,\sigma_2) \in \chi$, there exists a pair $(\delta,\sigma) \in \chi$ such that $\delta \leqslant \delta_i$, $\sigma \leqslant \sigma_i$, $i = 1,2$. Then the conditions (a) and (b) of Proposition 4 are satisfied.

Proof. Let $x_j \in D_{B_{\delta_j\sigma_j}}$, $j = 1,\ldots,m$, $\sum_{j=1}^{m} x_j = 0$. By induction on m we prove, using (ii), that there exists a pair $(\delta,\sigma) \in \chi$ such that $\delta \leqslant \delta_j$, $\sigma \leqslant \sigma_j$, $j = 1,\ldots,m$. Using (i), we obtain

$$\sum B_{\delta_j\sigma_j}x_j = \sum B_{\delta\sigma}x_j = B_{\delta\sigma}\sum x_j = 0,$$

since $B_{\delta\sigma}$ is a linear operator. The proposition is proved.

Proposition 6. Let the family $\{B_{\delta\sigma}\}_{(\delta,\sigma) \in \chi}$ satisfy the conditions of Proposition 5. If all $B_{\delta\sigma}$ are closurable operators, then the family $\{\bar{B}_{\delta\sigma}\}_{(\delta,\sigma) \in \chi}$ also satisfies these conditions.

Proof. Only (ii) should be proved. Let $x \in D_{\bar{B}_{\delta_1\sigma_1}}$, $\bar{B}_{\delta_1\sigma_1}x = y$. This means that there exists a sequence $x_n \in D_{B_{\delta_1\sigma_1}}$, $n = 1,2,\ldots$, such that $x_n \overset{\delta_1}{\to} x$, $B_{\delta_1\sigma_1}x_n \overset{\sigma_1}{\to} y$. But then, according to (i), $x_n \overset{\delta_2}{\to} x$ and $B_{\delta_2\sigma_2}x_n = B_{\delta_1\sigma_1} \overset{\sigma_2}{\to} y$. Thus $x \in D_{\bar{B}_{\delta_2\sigma_2}}$ and $\bar{B}_{\delta_2\sigma_2}x = y$. The proof is complete.

D. Functions of Several Operators in a Banach Scale

Definition 1. A tempered (strongly continuous) group of linear operators in a Banach scale $X = \{X_\delta\}_{\delta\in\Delta}$ is a family $\{\mathcal{U}(t)\}_{t\in\mathbb{R}}$ of linear operators $\mathcal{U}(t)$ in X such that:

(a) The domain $D_{\mathcal{U}(t)} = X$ for all $t \in \mathbb{R}$ and

$$\mathcal{U}(t)\mathcal{U}(\tau)x = \mathcal{U}(t+\tau)x, \ \mathcal{U}(0)x = x, \ x \in X, \ \tau,t \in \mathbb{R}. \tag{1}$$

(b) For any $\delta \in \Delta$ there is a $\sigma \in \Delta$ and conversely for any $\sigma \in \Delta$ there is a $\delta \in \Delta$ such that: (i) $\mathcal{U}(t) \in L(X_\delta, X_\sigma)$ for any $t \in \mathbb{R}$; (ii) $\mathcal{U}(t)$ is strongly continuous in t as an operator from X_δ to X_σ; and (iii) for some $c = c(\delta,\sigma)$ and $k = k(\delta,\sigma)$ the estimate

$$\| \mathcal{U}(t)x \|_\sigma \le c(1 + |t|)^k \| x \|_\delta \tag{2}$$

is valid for any $x \in X_\delta$.

Note. One obtains the definition of a strongly continuous group of linear operators in X omitting the requirement (iii). We denote by $\Lambda_{\mathcal{U}} \subset \subset \Delta \times \Delta$ the set of pairs (δ,σ) for which (i) - (iii) are satisfied.

Definition 2. The generator of a strongly continuous group $\{\mathcal{U}(t)\}$ of linear operators in X is the operator A in X defined by

$$Ax = \lim_{\varepsilon\to 0}(i\varepsilon)^{-1}(\mathcal{U}(\varepsilon)x - x) \equiv -i[\frac{d}{dt}(\mathcal{U}(t)x)]\Big|_{t=0} \tag{3}$$

(the domain $D_A \subset X$ is the set of all elements $x \in X$ for which the limit (3) exists). We write $\mathcal{U}(t) = \exp(iAt)$.

Theorem 1. (a) The domain D_A is dense in X and invariant under $\mathcal{U}(t)$ and besides

$$A\mathcal{U}(t)x = \mathcal{U}(t)Ax = -i\frac{d}{dt}\mathcal{U}(t)x, \ x \in D_A. \tag{4}$$

(b) The operator A is closed (i.e., if Γ-sequence $\{x_\gamma\}$ converges to x and $\{Ax_\gamma\}$ converges to y, then $x \in D_A$ and $Ax = y$).

(c) The Cauchy problem

$$i\frac{dx(t)}{dt} + Ax(t) = 0, \ x\big|_{t=0} = x_0 \in D_A \tag{5}$$

has a unique solution $x(t) = \mathcal{U}(t)x_0$ (thus, there is a one-to-one correspondence between groups and their generators).

(d) For $\mathrm{Im}\lambda \neq 0$ the resolvent $R_\lambda(A) = (\lambda I - A)^{-1}$,

$$R_\lambda(A)x = i\int_0^{\pm\infty} e^{-i\lambda t}\mathcal{U}(t)x\,dt, \ x \in X, \tag{6}$$

is defined (the sign in (6) coincides with the sign of $(-\mathrm{Im}\lambda)$) and moreover $R_\lambda(A) \in L_{right}(X,X) \cap L_{left}(X,X)$, $\chi_{R_\lambda(A)} \supset \Lambda_{\mathcal{U}}$,

(e) $R_\lambda(A)^m \in L(X_\delta, X_\sigma)$ for any m, $(\delta,\sigma) \in \Lambda_{\mathcal{U}}$ and

$$\| R_\lambda(A)^m x \|_\sigma \le c \sum_{j=0}^{k} \frac{k!(m+j-1)!}{j!(k-j)!(m-1)!} |\mathrm{Im}\lambda|^{-m-k} \| x \|_\delta, \tag{7}$$

$$\mathrm{Im}\lambda \neq 0, \ m = 1,2,\ldots$$

are valid for $(\delta,\sigma) \in \Lambda_{\mathcal{U}}$ ($c = c(\delta,\sigma)$, $k = k(\delta,\sigma)$ are the same as in (2)).

(f) There is a representation

$$\exp(iAt)x = \lim_{n\to\infty}(I - \frac{it}{n} A)^{-n}x, \quad x \in X, \tag{8}$$

the convergence in (8) being locally uniform with respect to t.

<u>Proof</u>. We give here a very brief proof of this theorem, since it repeats with slight modifications the proof of Theorem 1 of Section 2:A and of Theorem 1 of Section 2:D.

(a) For $x \in X$ and $\phi \in C_0^\infty(\mathbb{R})$, set

$$x_\phi = \int_{-\infty}^{\infty} \phi(t)\mathcal{U}(t)x dt. \tag{9}$$

The integral (9) makes sense since $x \in X_\delta$ for some δ and consequently $\mathcal{U}(t)x$ is a continuous X_σ-valued function for some σ. The set of elements of the form (9) is dense in X; on the other hand the limit $\lim_{\varepsilon\to 0}(i\varepsilon)^{-1} \times$ $\times (\mathcal{U}(\varepsilon)x_\phi - x_\phi)$ exists and equals $x_{i\phi'}$. Next, if x lies in D_A, then we have

$$\frac{1}{i\varepsilon}(\mathcal{U}(\varepsilon)\mathcal{U}(t)x - \mathcal{U}(t)x) = \mathcal{U}(t)[\frac{1}{i\varepsilon}(\mathcal{U}(\varepsilon)x - x)]. \tag{10}$$

The expression in square brackets tends to Ax in some X_δ, $\mathcal{U}(t) \in L(X_\delta, X_\sigma)$ for some σ, so the right-hand side of (10) tends to $\mathcal{U}(t)Ax$ in X_σ, therefore in X. We obtain $\mathcal{U}(t)x \in D_A$, $A\mathcal{U}(t)x = \mathcal{U}(t)Ax$.

(d) and (e) Set

$$R_m(\lambda,A)x = \frac{i^m}{(m-1)!}\int_0^\infty t^{m-1}e^{-i\lambda t}\mathcal{U}(t)x dt, \quad x \in X, \ \mathrm{Im}\lambda < 0 \tag{11}$$

(the case $\mathrm{Im}\lambda > 0$ is considered in a similar way). The integral (11) is convergent for any $x \in X$, since the estimate (2) is valid. Further, the calculation identical to (14) of Section 2:D shows that $R_m(\lambda,A)$ satisfies the estimates (7). We show next that $R_m(\lambda,A) = (\lambda I - A)^m$. Indeed,

$$\frac{1}{i\varepsilon}(\mathcal{U}(\varepsilon) - I)R_m(\lambda,A)x = \frac{i^{m-1}}{\varepsilon(m-1)!}[\int_\varepsilon^\infty (t-\varepsilon)^{m-1}e^{-i\lambda(t-\varepsilon)}\mathcal{U}(t)x dt -$$

$$- \int_0^\infty t^{m-1}e^{-i\lambda t}\mathcal{U}(t)x dt] = \frac{i^{m-1}}{(m-1)!}[\int_0^\infty \frac{(t-\varepsilon)^{m-1}e^{i\lambda\varepsilon} - t^{m-1}}{\varepsilon}\mathcal{U}(t)x dt -$$

$$- \frac{1}{\varepsilon}\int_0^\varepsilon (t-\varepsilon)^{m-1}e^{-i\lambda(t-\varepsilon)}\mathcal{U}(t)x dt]. \tag{12}$$

The limit on the right as $\varepsilon \to 0$ exists and equals

$$\lambda R_m(\lambda,A)x - R_{m-1}(\lambda,A)x \text{ for } m > 1,$$

$$\lambda R_1(\lambda,A)x - x \text{ for } m = 1.$$

Thus, $R_m(\lambda,A)x \in D_A$ and $(\lambda I - A)R_m(\lambda,A)x = R_{m-1}(\lambda,A)x$ (here we denote $R_0(\lambda,A) \overset{\text{def}}{\equiv} I$). Similarly, $R_m(\lambda,A)(\lambda I - A)x = R_{m-1}(\lambda,A)x$ for $x \in D_A$, so $R_m(\lambda,A) = (\lambda I - A)^m$. Thus (d) and (e) are proved.

(b) $A = \lambda - ((\lambda - A)^{-1})^{-1}$ ($\mathrm{Im}\lambda \neq 0$) and is therefore closed (since $(\lambda - A)^{-1}$ is; we leave details to the reader).

(c) Let $x(t)$ be a solution of (5), and set

$$y(t) = \mathcal{U}(-t)x(t). \tag{13}$$

It is easy to verify that $y(t)$ is differentiable and $dy/dt = 0$, thus $y(t) \equiv$ $\equiv x_o$ and henceforth $x(t) = \mathcal{U}(t)x_o$.

(f) Using (11), we may write for $t > 0$

$$(I - \frac{itA}{n})^{-n}x = (\frac{n}{it})^n(\frac{n}{it} - A)^{-n}x =$$

$$= (\frac{n}{t})^n \frac{1}{(n-1)!} \int_o^\infty \tau^{n-1}e^{-n\tau/t}\mathcal{U}(\tau)x\,d\tau = \tag{14}$$

$$= \frac{1}{(n-1)!} \int_o^\infty \tau^{n-1}e^{-\tau}\mathcal{U}(\frac{t\tau}{n})x\,d\tau.$$

Let $x \in D_{A^2}$. Then

$$\mathcal{U}(\frac{t\tau}{n})x = \mathcal{U}(t)x + t(\frac{\tau - n}{n})\mathcal{U}(t)Ax + t^2(\frac{\tau - n}{n})^2\int_o^1 (1 - \mu) \times$$

$$\times\, \mathcal{U}(t + \mu t(\frac{\tau - n}{n}))A^2x\,d\mu \equiv \mathcal{U}(t)x + [\frac{t\tau}{n} - t]\mathcal{U}(t)Ax + t^2/n^2(\tau - n)^2y(t,\tau) \tag{15}$$

(cf. Lemmas 5 and 6 of item A). For some $\delta \in \Delta$ we have

$$\| y(t,\tau) \|_\delta \leqslant C_1(1 + \frac{\tau}{n})^k \leqslant C_2e^{\tau/n}, \tag{16}$$

where the constants are independent of t if t remains in a compact subset of $\mathbb{R}$. Substituting (15) into (14), we obtain

$$(I - \frac{itA}{n})^{-n}x = I_{n-1}\mathcal{U}(t)x + t(I_n - I_{n-1})\mathcal{U}(t)Ax +$$

$$+ t^2 \frac{1}{n!n} \int_o^\infty \tau^{n-1}(\tau - n)^2e^{-\tau}y(t,\tau)d\tau,$$

where $I_k = \frac{1}{k!} \int_o^\infty \tau^k e^{-\tau} d\tau = 1$.

We have

$$\| \frac{t}{n!n} \int_o^\infty \tau^{n-1}(\tau - n)^2e^{-\tau}y(t,\tau)d\tau \|_\delta \leqslant$$

$$\leqslant C_2t^2 \frac{1}{n!n} \int_o^\infty [\tau^{n+1} - 2n\tau^n + n^2\tau^{n-1}]e^{-\tau(1 - 1/n)} =$$

$$= C_2t^2 \frac{1}{n!n}\{ \frac{(n+1)!}{(1 - 1/n)^{n+2}} - \frac{2n!n}{(1 - 1/n)^{n+1}} + \frac{n^2(n-1)!}{(1 - 1/n)^n} \} =$$

$$= C_2t^2(1 - \frac{1}{n})^{-n}\{(1 + \frac{1}{n})(1 - \frac{1}{n})^{-2} + 1 - 2(1 - \frac{1}{n})^{-1}\} \leqslant C_3t^2/n.$$

It follows that $(I - \frac{itA}{n})^{-n}x \overset{n\to\infty}{\to} \mathcal{U}(t)x$ locally uniformly in t, provided that $x \in D_{A^2}$. Since D_{A^2} is dense in X and direct estimate implies that $(I - \frac{itA}{n})^{-n}x$ are bounded in X uniformly with respect to n,* we conclude that (f) is valid for any $x \in X$.

---

* This means that for any $\delta \in \Delta$ ($\sigma \in \Delta$) there exists $\sigma \in \Delta$ ($\delta \in \Delta$) such that the norms of $(I - itA/n)^{-n}$ as operators from X_δ to X_σ have a common bound.

We introduce, for the sake of convenience, the following notation. If $\mathcal{U}(t)$, $t \in \mathbb{R}^n$ is an operator-valued function, we write $\mathcal{U}(t) \in C_\alpha^\beta(\mathbb{R}^n, X_\delta, X_\sigma)$ if for any $x \in X_\delta$, $\mathcal{U}(t)x \in C_\alpha^\beta(\mathbb{R}^n, X_\sigma)$ and

$$\| \mathcal{U}(t)x \|_{C_\alpha^\beta(\mathbb{R}^n, X_\sigma)} \leqslant C \| x \|_\delta \tag{17}$$

with constant C independent of $x \in X_\delta$. The minimal constant C, for which (17) is satisfied will be denoted by $\| \mathcal{U}(t) \|_{C_\alpha^\beta(\mathbb{R}^n, X_\delta, X_\sigma)}$. Also if $\mathcal{U}(t)$, $t \in \mathbb{R}^n$, is an operator-valued function and also $T \in \mathcal{D}(\mathbb{R}^n)$, we write

$$B = T[\mathcal{U}] \equiv T[\mathcal{U}(t)] \equiv \int_{\mathbb{R}^n} T(t)\mathcal{U}(t)dt \tag{18}$$

for an operator B in the scale X defined by

$$Bx = T[\mathcal{U}(t)x] \tag{19}$$

(the expression on the right of (19) should be treated in the sense of Definition 3 of item C, and the domain D_B consists of all $x \in X$ such that this expression makes sense).

<u>Proposition 1</u>. Let $\mathcal{U}(t) \in C_\alpha^\beta(\mathbb{R}^n, X_\delta, X_\sigma)$, $f \in B_\alpha^\beta(\mathbb{R}^n)$. Then

$$B = (2\pi)^{-n} \int \tilde{f}(t)\mathcal{U}(t)dt \in L(X_\delta, X_\sigma)$$

(here $\tilde{f} = F(f)$ is the Fourier transform of f).

<u>Proof</u>. Let $x \in X_\delta$. Then $\mathcal{U}(t)x \in C_\alpha^\beta(\mathbb{R}^n, X_\sigma)$. Applying Theorem 1 of item C, we obtain $x \in D_B$ and

$$Bx = (2\pi)^{-n/2} F(f)(\mathcal{U}(t)x) \in X_\sigma,$$

the latter pairing being defined in Lemma 1 of item B. From this we can easily obtain

$$\| Bx \|_\sigma \leqslant \text{const} \| f \|_{B_\alpha^\beta} \| \mathcal{U} \|_{C_\alpha^\beta(\mathbb{R}^n, X_\delta, X_\sigma)} \| x \|_\delta. \tag{20}$$

<u>Proposition 2</u>. Let $f(y,z) = f(y_1, \ldots, y_n, z_1, \ldots, z_m) = g_1(y_1, \ldots, y_n) \times g_2(z_1, \ldots, z_m)$, where $g_1 \in B_\alpha^\beta(\mathbb{R}^n)$, $g_2 \in B_\mu^0(\mathbb{R}^m)$. Let also $\mathcal{U}(t_1, \ldots, t_s)$, $V(t_{s+1}, \ldots, t_n)$, $w(\tau_1, \ldots, \tau_m)$ be the operator-valued functions in a Banach scale X such that

$$\mathcal{U} \in C_{\alpha_1, \ldots, \alpha_s}^{\beta_1, \ldots, \beta_s}(\mathbb{R}^s, X_{\delta_1}, X_{\delta_2}),$$

$$V \in C_{\alpha_{s+1}, \ldots, \alpha_n}^{\beta_{s+1}, \ldots, \beta_n}(\mathbb{R}^{n-s}, X_{\delta_3}, X_{\delta_4}), \quad w \in C_\mu^0(\mathbb{R}^m, X_{\delta_2}, X_{\delta_3}).$$

Then for any $x \in X_{\delta_1}$

$$\begin{aligned} &\int_{\mathbb{R}^{m+n}} \tilde{f}(t+\tau)V(t_{s+1}, \ldots, t_n)w(\tau)\mathcal{U}(t_1, \ldots, t_s)x\,dt\,d\tau = \\ &= \int_{\mathbb{R}^n} \tilde{g}_1(t)V(t_{s+1}, \ldots, t_n)\{\int_{\mathbb{R}^m} \tilde{g}_2(\tau)w(\tau)d\tau\}\mathcal{U}(t_1, \ldots, t_s)x\,dt. \end{aligned} \tag{21}$$

Proof. The preceding proposition yields that $\int_{\mathbb{R}^m}\tilde{g}_2(\tau)w(\tau)d\tau \in$ $\in L(X_{\delta_2}, X_{\delta_3})$ so that the right-hand side of (21) is a correctly defined element of X_{δ_4} as well as its left-hand side. Taking an arbitrary element $h \in X^*_{\delta_4}$ and applying it to both sides of (21), we reduce the problem to the verification of the equality

$$[\tilde{g}_1(t)\tilde{g}_2(\tau)](\langle h, V(t_{s+1},\dots,t_n)w(\tau)U(t_1,\dots,t_s)x\rangle) =$$

$$= \tilde{g}_1[\langle h, V(t_{s+1},\dots,t_n)(\int\tilde{g}_2(\tau)w(\tau)d\tau)U(t_1,\dots,t_s)x\rangle].$$

Since the action of the tensor product of distributions may be obtained by successive application of the factors, it is enough to prove that

$$\tilde{g}_2[\langle h, V(t_{s+1},\dots,t_n)w(\tau)U(t_1,\dots,t_s)x\rangle] =$$

$$= \langle h, V(t_{s+1},\dots,t_n)(\int w(\tau)\tilde{g}_2(\tau)d\tau)U(t_1,\dots,t_s)x\rangle$$

for any fixed $t \in \mathbb{R}^n$. Setting $h_1 = V^*(t_{s+1},\dots,t_n)h \in X^*_{\delta_3}$, $x_1 = U(t_1,\dots,t_s)x \in X_{\delta_2}$, we may rewrite the equality in question in the form

$$\tilde{g}_2[\langle h_1, w(\tau)x_1\rangle] = \langle h_1, (\int\tilde{g}_2(\tau)w(\tau))x_1\rangle. \tag{22}$$

But (22) is an immediate consequence of definitions. The proposition is proved.

Now we give the definitions of functions of several operators in a Banach scale $X = \{X_\delta\}_{\delta\in\Delta}$. Let $A_1,\dots,A_n$ be given operators in X. Let $\{j_1,\dots,j_n\}$ be a substitution of the set $\{1,2,\dots,n\}$. We assume that $A_{j_1},\dots,A_{j_k}, k \leqslant n$, are generators in X. Let $f(y_1,\dots,y_n)$ be a function of the form

$$f(y_1,\dots,y_n) = g(y_{j_1},\dots,y_{j_k}) \sum_{\mu_{k+1}+\dots+\mu_n\leqslant N} a_\mu y_{j_{k+1}}^{\mu_{k+1}} \cdots y_{j_n}^{\mu_n}, \tag{23}$$

where $g(y_{j_1},\dots,y_{j_k})$ is a continuous function of tempered growth at infinity (i.e., f is a product of a tempered continuous function in $(y_{j_1},\dots,y_{j_k})$ and a polynomial in $(y_{j_{k+1}},\dots,y_{j_n})$). Consider the expression:

$$U(t_{j_1},\dots,t_{j_k}) = \sum_{\mu_{k+1}+\dots+\mu_n} a_\mu \overset{(j_{k+1})}{A_{j_{k+1}}^{\mu_{k+1}}} \cdots \overset{(j_n)}{A_{j_n}^{\mu_n}} \times$$

$$\times \overset{(j_1)}{\exp(it_{j_1}A_{j_1})}\cdot\dots\cdot\overset{(j_k)}{\exp(it_{j_k}A_{j_k})} \tag{24}$$

which is an operator in X; the number over the operators in (24) denotes the place on which it acts in the product; the operators with lower numbers act before (i.e., they stand closer to the right in the product than the operators with greater numbers). For example,

$$a\overset{(1)}{A}\overset{(3)}{B}\overset{(2)}{\exp(i\tau C)} + b\overset{(5)}{B^2}\overset{(1)}{A^2}\overset{(4)}{\exp(i\tau\mathcal{D})} =$$

$$= aB\exp(i\tau C)A + bB^2\exp(i\tau\mathcal{D})A^2 \tag{25}$$

(here $a,b \in \mathbb{C}$, $A,B,C,\mathcal{D}$ are operators in X). The circles around these numbers will be omitted often.

Definition 3. $f(\overset{1}{A_1},\ldots,\overset{n}{A_n})$ is an operator in X defined by

$$f(\overset{1}{A_1},\ldots,\overset{n}{A_n}) = (2\pi)^{-k/2}\int_{\mathbb{R}^k}\tilde{g}(y_{j_1},\ldots,y_{j_k})\mathcal{U}(t_{j_1},\ldots,t_{j_k})dt_{j_1}\ldots dt_{j_k} \tag{26}$$

(note that the Fourier transform $\tilde{g}(y_{j_1},\ldots,y_{j_k})$ exists since it is of tempered growth).

We shall also denote

$$f(\overset{1}{A_1},\ldots,\overset{n}{A_n}) = f(A) = \mu_A(f) = \mu_{\overset{1}{A_1},\ldots,\overset{n}{A_n}}(f) \tag{27}$$

as it will be more convenient.

Note. The definition is correct only if we fix the substitution $\{j_1,\ldots,j_n\}$ and the number k. For if some of $A_{j_{k+1}},\ldots,A_{j_n}$ were also generators, we might increase k and obtain a different definition of $f(\overset{1}{A_1},\ldots,\overset{n}{A_n})$ for the same f. However, it will be proved below that in cases which are of interest to us, all the variants of definition lead to the same result.

Note. If $k = n$ we obtain the simplest version of the definition:

$$f(\overset{1}{A_1},\ldots,\overset{n}{A_n}) = \\ = (2\pi)^{-n/2}\int_{\mathbb{R}^n}\tilde{f}(t_1,\ldots,t_n)\exp(itA_n)\cdot\ldots\cdot\exp(itA_1)dt_1\ldots dt_n. \tag{28}$$

Example. Let A,C be generators in X, $f(y_1,y_2,y_3) = y_2^2 g(y_1,y_3)$. Then

$$f(\overset{1}{A},\overset{2}{B},\overset{3}{C}) \equiv \overset{2}{B^2}g(\overset{1}{A},\overset{3}{C}) = \frac{1}{2\pi}\int\tilde{g}(t,\tau)\exp(iC\tau)B^2\exp(iAt)dtd\tau.$$

Proposition 3. Let $g(y_{j_1},\ldots,y_{j_k}) \in B_\alpha^\beta(\mathbb{R}^k)$ and $\mathcal{U}(t_{j_1},\ldots,t_{j_k}) \in$ $\in C_\alpha^\beta(\mathbb{R}^k,X_\delta,X_\sigma)$. Then $f(\overset{1}{A_1},\ldots,\overset{n}{A_n}) \in L(X_\delta,X_\sigma)$.

Proof. See Proposition 1.

Proposition 4. Let $A_1,\ldots,A_n$ be generators, $f(y_1,\ldots,y_n) = y_1^{\beta_1}\ldots y_n^{\beta_n}$,

$$\exp(iA_nt_n)\cdot\ldots\cdot\exp(iA_1t_1) \in C_\alpha^\beta(\mathbb{R}^n,X_\delta,X_\sigma) \tag{29}$$

for some α. Then for any $x \in X_\delta$

$$f(\overset{1}{A_1},\ldots,\overset{n}{A_n})x = A_n^{\beta_n}\ldots A_1^{\beta_1}x. \tag{30}$$

Proof. The relation (29) means that for $x \in X_\delta$, $\phi(t) = \exp(iA_nt_n)\ldots\exp(it_1A_1)x$ is a β-times continuously differentiable X_σ-valued function. It is easy to see that its β-th derivative equals

$$\phi^{(\beta)}(t) = i^{|\beta|}\exp(iA_nt_n)A_n^{\beta_n}\cdot\ldots\cdot\exp(iA_1t_1)A_1^{\beta_1}x.$$

The Fourier transform of f equals $\tilde{f} = i^{|\beta|}\delta^{(\beta_n)}(t_n)\ldots\delta^{(\beta_1)}(t_1)$. Applying $h \in X_\sigma^*$ to $\exp(iA_nt_n)\cdot\ldots\cdot\exp(iA_1t_1)x$ and acting by $\tilde{f}$ on the result, we obtain, by virtue of the preceding formula, exactly $\langle h, A_n^{\beta_n} \ldots A_1^{\beta_1} x\rangle$. The proposition is proved.

The above statement is a step to prove the independence of Definition 3 of the choice of k and $\{j_1,\ldots,j_n\}$ (in certain cases). The next step is given by the following theorem, which is also important by itself. First we formulate it in the simplest case:

<u>Theorem 2</u>. Let $A_1,\ldots,A_n$ be generators in X. Let $f(y_1,\ldots,y_n) = g_1(y_1,\ldots,y_s,y_{r+1},\ldots,y_n)g_2(y_{s+1},\ldots,y_r)$, where $r \geqslant s+1$, and also $g_1 \in \in \mathcal{B}^{\beta_1,\ldots,\beta_s,\beta_{r+1},\ldots,\beta_n}_{\alpha_1,\ldots,\alpha_s,\alpha_{r+1},\ldots,\alpha_n}(\mathbb{R}^n)$, $g_2 \in \mathcal{B}^{\beta_{s+1},\ldots,\beta_r}_{\alpha_{s+1},\ldots,\alpha_r}(\mathbb{R}^{r-s})$. Assume that for some $\delta,\delta_1,\delta_2,\sigma$ we have

$$\exp(it_sA_s)\cdot\ldots\cdot\exp(it_1A_1) \in C^{\beta_1,\ldots,\beta_s}_{\alpha_1,\ldots,\alpha_s}(\mathbb{R}^s, X_\delta, X_{\delta_1}),$$

$$\exp(it_rA_r)\cdot\ldots\cdot\exp(it_{s+1}A_{s+1}) \in C^{\beta_{s+1},\ldots,\beta_r}_{\alpha_{s+1},\ldots,\alpha_r}(\mathbb{R}^{r-s}, X_{\delta_1}, X_{\delta_2}), \tag{31}$$

$$\exp(it_nA_n)\cdot\ldots\cdot\exp(it_{r+1}A_{s+1}) \in C^{\beta_{r+1},\ldots,\beta_n}_{\alpha_{r+1},\ldots,\alpha_n}(\mathbb{R}^{n-r}, X_{\delta_2}, X_\sigma).$$

Then for any $x \in X_\delta$

$$f(\overset{1}{A}_1,\ldots,\overset{n}{A}_n)x = \overset{s+1}{C}\; g_1(\overset{1}{A}_1,\ldots,\overset{s}{A}_s,\overset{s+2}{A}_{r+1},\ldots,\overset{n+s+1-r}{A}_n)x, \tag{32}$$

where

$$C = g_2(\overset{1}{A}_{s+1},\ldots,\overset{r-s}{A}_r). \tag{33}$$

<u>Proof</u>. This is an immediate consequence of Proposition 2.

<u>Corollary 1</u>. Under obvious assumptions, we have

$$f(\overset{1}{A}_1,\ldots,\overset{k}{A}_k)g(\overset{k+1}{A}_{k+1},\ldots,\overset{n}{A}_n)x = g(\overset{k+1}{A}_{k+1},\ldots,\overset{n}{A}_n)y, \tag{34}$$

where

$$y = f(\overset{1}{A}_1,\ldots,\overset{k}{A}_k)x. \tag{35}$$

This is a particular case of the above theorem for r = n.

The identities (32) and (34) may be written in a more compact form, if we introduce the useful notation - the so-called "autonomous brackets" ⟦ ⟧. By convention, the expression in these brackets is regarded as a single operator, i.e., the numbers over the operators inside these brackets do not "interact" with the numbers over the operators outside them. The place on which the operator marked by autonomous brackets acts is denoted by the number over the left bracket ⟦ or over the line over the whole operator. Using this notation, we may rewrite (32) and (34) in the following form

$$\begin{aligned} &g_1(\overset{1}{A}_1,\ldots,\overset{s}{A}_s,\overset{r+1}{A}_{r+1},\ldots,\overset{n}{A}_n)g_2(\overset{s+1}{A}_{s+1},\ldots,\overset{r}{A}_r)x = \\ &= \overset{s+1}{\llbracket}\, g_2(\overset{1}{A}_{s+1},\ldots,\overset{r-s}{A}_r)\rrbracket\, g_1(\overset{1}{A}_1,\ldots,\overset{s}{A}_s,\overset{s+2}{A}_{r+1},\ldots,\overset{n+s+1-r}{A}_n)x, \end{aligned} \tag{36}$$

$$f(\overset{1}{A}_1,\dots,\overset{k}{A}_k)g(\overset{k+1}{A}_{k+1},\dots,\overset{n}{A}_n)x =$$

$$= \overset{1}{[\![} f(\overset{1}{A}_1,\dots,\overset{k}{A}_k) \overset{2}{]\!]\,[\![} g(\overset{1}{A}_{k+1},\dots,\overset{n-k}{A}_n)]\!] x \equiv \tag{37}$$

$$\equiv [\![g(\overset{1}{A}_{k+1},\dots,\overset{n-k}{A}_n)]\!]\,[\![f(\overset{1}{A}_1,\dots,\overset{k}{A}_k)]\!] x.$$

The same theorem is valid if some of the operators $A_1,\dots,A_n$ are not the generators, but the symbols in question are polynomial in variables, corresponding to those operators. We do not write the analogue of the condition (31) for this case since it would take too much space. On the other hand these conditions are obvious and the reader may write them himself, if desired. Combining the proved theorem with the preceding proposition, we conclude that under obvious conditions the definition of $f(\overset{1}{A}_1,\dots,\overset{n}{A}_n)$ does not depend on k and $\{j_1,\dots,j_n\}$.

Our next theorem is concerned with the case when two of the operators $A_1,\dots,A_n$ coincide and no other operators act between them. Again we give the formulation only for the case when all the operators are generators in X.

<u>Theorem 3 ("shifting indices" theorem)</u>. Let $f(y_1,\dots,y_n) \in \mathcal{B}_\alpha^\beta(\mathbb{R}^n)$, $A_1,\dots,A_s = A_{s+1},\dots,A_n$ be generators in X,

$$\begin{gathered}\exp(it_nA_n)\cdot\ldots\cdot\exp(it_{s+1}A_{s+1})\exp(it_sA_s)\ \ldots\ \times\\ \times\ \exp(it_1A_1) \in C_\alpha^\beta(\mathbb{R}^n,X_\delta,X_\sigma).\end{gathered} \tag{38}$$

Then for any $x \in X_\delta$, we have

$$f(\overset{1}{A}_1,\dots,\overset{n}{A}_n)x = g(\overset{1}{A}_1,\dots,\overset{s}{A}_s,\overset{s+1}{A}_{s+2},\dots,\overset{n-1}{A}_n)x, \tag{39}$$

where $g(y_1,\dots,y_s,y_{s+2},\dots,y_n) = f(y_1,\dots,y_s,y_s,y_{s+2},\dots,y_n)$.

<u>Note</u>. We adopt the convention that the right-hand side of (39) will be written in the form $f(\overset{1}{A}_1,\dots,\overset{s}{A}_s,\overset{s}{A}_s,\overset{s+1}{A}_{s+2},\dots,\overset{n-1}{A}_n)x$.

<u>Proof</u>. By Lemma 4 of item A, $g \in \mathcal{B}_{\tilde\alpha}^{\tilde\beta}(\mathbb{R}^{n-1})$, where

$$\tilde\alpha = (\alpha_1,\dots,\alpha_{s-1},\min\{\alpha_s,\alpha_{s+1}\},\alpha_{s+2},\dots,\alpha_n),$$

$$\tilde\beta = (\beta_1,\dots,\beta_{s-1},\beta_s+\beta_{s+1},\beta_{s+2},\dots,\beta_n).$$

We assert that $\exp(it_nA_n)\cdot\ldots\cdot\exp(it_{s+2}A_{s+2})\exp(it_sA_s)\cdot\ldots\cdot\exp(it_1A_1) \in$ $\in C_{\tilde\alpha}^{\tilde\beta}(\mathbb{R}^{n-1},X_\delta,X_\sigma)$ so that the right-hand side of (39) is well defined. Indeed, it is easy to prove that if a X_σ-valued function $F(t,\tau)$ satisfies $F(t+x,\tau-x) \equiv F(t,\tau)$, is (β_s,β_{s+1})-times continuously differentiable, and satisfies the estimates

$$\left|\frac{\partial^{\gamma_1+\gamma_2}}{\partial t^{\gamma_1}\partial t^{\gamma_2}} F(t,\tau)\right| \leqslant C(1+|t|)^{\alpha_s}(1+|\tau|)^{\alpha_{s+1}},\ \gamma_1 \leqslant \beta_s,\ \gamma_2 \leqslant \beta_{s+1}, \tag{40}$$

then $f(t) \equiv F(t,t)$ is $(\beta_s+\beta_{s+1})$-times continuously differentiable and

$$\left|\frac{\partial^\gamma f(t)}{\partial t^\gamma}\right| \leqslant C(1+|t|)^{\min\{\alpha_s,\alpha_{s+1}\}}, \ \gamma \leqslant \beta_s + \beta_{s+1}. \tag{41}$$

Next, taking any $h \in X^*_\delta$, $x \in X_\delta$, we may write

$$F(f)(\langle h, \exp(it_n A_n)\cdot\ldots\cdot\exp(i(t_s + t_{s+1})A_s)\cdot\ldots\cdot\exp(it_1 A_1)x\rangle) =$$

$$= F(g)(\langle h, \exp(it_{n-1}A_n)\cdot\ldots\cdot\exp(it_s A_s)\cdot\ldots\cdot\exp(it_1 A_1)x\rangle).$$

(This connection between actions of $F(f)$ and $F(g)$ on test functions may be easily ascertained from the proof of Lemma 4 of item A.)

We now summarize all the theorems proved and give their application to the case when all the operators involved are translators and all the symbols considered belong to S^∞.

<u>Theorem 4</u>. Let all the operators $A_1,\ldots,A_n$ be right translators in X. Then for any symbols $f,g,h \in S^\infty$, polynomial in the variables corresponding to that of operators $A_1,\ldots,A_n$ which are not generators, we have:

(a) $f(\overset{1}{A_1},\ldots,\overset{n}{A_n})$ is the right translator in X independent of the choice of k and $\{j_1,\ldots,j_k\}$ in Definition 3.

(b)
$$f(\overset{1}{A_1},\ldots,\overset{s}{A_s},\overset{r+1}{A_{r+1}},\ldots,\overset{n}{A_n})g(\overset{s+1}{A_{s+1}},\ldots,\overset{r}{A_r}) =$$
$$= f(\overset{1}{A_1},\ldots,\overset{s}{A_s},\overset{r+1}{A_{r+1}},\ldots,\overset{n}{A_n}) [\![g(\overset{1}{A_{s+1}},\ldots,\overset{r-s}{A_r})]\!]. \tag{42}$$

(c)
$$h(\overset{1}{A_1},\ldots,\overset{s}{A_s},\overset{s+1}{A_{s+1}},\ldots,\overset{n}{A_n}) = h(\overset{1}{A_1},\ldots,\overset{s}{A_s},\overset{s}{A_s},\ldots,\overset{n-1}{A_n}) \tag{43}$$

if $A_s = A_{s+1}$.

The same statements are valid for left translators with the only difference that they are valid not on all $x \in X$, but for $x \in X_\delta$, where δ is sufficiently large ($\delta \geqslant \delta_0$, where δ_0 depends on the operators and symbols involved).

<u>Proof</u>. Using the definitions of S^∞, generators, and right translators, we may always choose for given δ such α, β, σ that the requirements of the above theorems are satisfied. The case is similar for left translators.

4. REGULAR REPRESENTATIONS

A. Definition and the Main Property

Let X be a Banach scale, $A_1,\ldots,A_n$ be an n-tuple of operators. For the sake of simplicity we assume throughout this item that all these operators are generators in X. We consider also the scale $\mathcal{B} = \{\mathcal{B}^\beta_\alpha(\mathbb{R}^n)\}$ (see example 2 of Section 3:C). There is a mapping μ_A, described in Definition 3 of Section 3:D, which sets into correspondence an operator $f(A_1, \ldots,A_n)$ in X to any element $f \in \mathcal{B}$.

<u>Definition 1</u>. The left regular representation of the n-tuple $A = (A_1,\ldots,A_n)$ is the n-tuple $L = (L_1\ldots,L_n)$ of the operators in the scale $\mathcal{B}$ such that the following identity is valid:

$$A_i \mu_A(f) = \mu_A(L_i f), \ i = 1,\dots,n \tag{1}$$

for any $f \in \mathcal{B}$.

Definition 2. The right regular representation of the n-tuple $A = (A_1,\dots,A_n)$ is the n-tuple $R = (R_1,\dots,R_n)$ of the operators in the scale $\mathcal{B}$ such that

$$\mu_A(f) A_i = \mu_A(R_i f), \ i = 1,\dots,n \tag{2}$$

for any $f \in \mathcal{B}$.

Of course, regular representations do not exist, in the general case, for an arbitrary n-tuple A of generators in X. The existence of a regular representation yields a system of relations satisfied by this tuple. This system has the form

$$A_i A_j x = f_{ij}(\overset{1}{A}_1,\dots,\overset{n}{A}_n)x, \ x \in D_{A_i A_j}, i,j = 1,\dots,n, \tag{3}$$

where

$$f_{ij}(y) = L_i(y_j). \tag{4}$$

Conversely, if the operators $A_1,\dots,A_n$ satisfy certain algebraic relations and some functional-analytic restrictions, the regular representations exist. These questions will be considered in the sequel; here we assume that the regular representation exists and investigate the consequences of this fact. Moreover, we restrict ourselves to consideration of the case when the left regular representation $L = (L_1,\dots,L_n)$ exists.

Theorem 1. Assume that the operators $L_1,\dots,L_n$ are generators in $\mathcal{B}$. Let also

$$\begin{aligned} \mathcal{U}_A(t) &\equiv \exp(it_n A_n)\cdot\ldots\cdot\exp(it_1 A_1) \in \\ &\in C^{\beta}_{\alpha}(\mathbb{R}^n, X_\delta, X_{\delta_1}) \cap C^{\nu}_{\mu}(\mathbb{R}^n, X_{\delta_1}, X_\sigma) \cap C^{\eta}_{\rho}(\mathbb{R}^n, X_\delta, X_\sigma); \end{aligned} \tag{5}$$

$$\begin{aligned} \mathcal{U}_L(t) &\equiv \exp(it_n A_n)\cdot\ldots\cdot\exp(it_1 L_1) \in \\ &\in C^{\nu}_{\mu}(\mathbb{R}^n, \mathcal{B}^{\beta}_{\alpha}(\mathbb{R}^n), \mathcal{B}^{\eta}_{\rho}(\mathbb{R}^n)). \end{aligned} \tag{6}$$

Then for any $x \in X_\delta$, $g \in \mathcal{B}^{\beta}_{\alpha}(\mathbb{R}^n)$, $f \in \mathcal{B}^{\nu}_{\mu}(\mathbb{R}^n)$, we have

$$[\![f(\overset{1}{A}_1,\dots,\overset{n}{A}_n)]\!] [\![g(\overset{1}{A}_1,\dots,\overset{n}{A}_n)]\!] x = (f(\overset{1}{L}_1,\dots,\overset{n}{L}_n)g)(\overset{1}{A}_1,\dots,\overset{n}{A}_n)x \tag{7}$$

or, in another form,

$$\mu_A(f) \circ \mu_A(g)x = \mu_A(\mu_L(f)(g))x. \tag{8}$$

In other words, the knowledge of regular representation enables us to write the product of two functions of $(\overset{1}{A}_1,\dots,\overset{n}{A}_n)$ in the form of some new function of $(\overset{1}{A}_1,\dots,\overset{n}{A}_n)$.

Proof. Fix some $x \in X_\delta$. The mapping $m_x : g \to \mu_A(g)x$ is a continuous linear mapping from $\mathcal{B}^{\beta}_{\alpha}(\mathbb{R}^n)$ to X_{δ_1} and from $\mathcal{B}^{\eta}_{\rho}(\mathbb{R}^n)$ to X_σ. It is enough to prove that

$$m_x[\mathcal{U}_L(t)g] = \mathcal{U}_A(t)m_x[g], \ g \in \mathcal{B}^{\beta}_{\alpha}(\mathbb{R}^n). \tag{9}$$

Indeed, applying to (9) the distribution $(2\pi)^{-n/2}F(f)$, we obtain the desired result. We prove (9) by induction on k, setting

$$\mathcal{U}_{Lk}(t) = \mathcal{U}_L(t_1,\ldots,t_k,0,\ldots,0),$$
$$\mathcal{U}_{Ak}(t) = \mathcal{U}_A(t_1,\ldots,t_k,0,\ldots,0) \tag{10}$$

and proving that

$$m_x[\mathcal{U}_{Lk}(t)g] = \mathcal{U}_{Ak}(t)m_x[g], \quad k = 0,1,\ldots,n. \tag{11}$$

(11) is evident for $k = 0$, so we must only show how the induction step works. One should distinguish between two cases:

(a) $\nu_k \neq 0$. In this case both sides of (11) are differentiable with respect to t_k X_σ-valued functions and

$$\frac{\partial}{\partial t_k} m_x[\mathcal{U}_{Lk}(t)g] = m_x[Lk\mathcal{U}_{Lk}(t)g] = iA_k m_x[\mathcal{U}_{Lk}(t)g] \tag{12}$$

by (1),

$$\frac{\partial}{\partial t_k} \mathcal{U}_{Ak}(t)m_x[g] = iA_k\mathcal{U}_{Ak}(t)m_x[g]. \tag{13}$$

(11) is valid for $t_k = 0$ by the induction assumption, so we may use Theorem 1 (c) of Section 3:D and obtain (11) for any $t_k \in \mathbb{R}$.

(b) $\nu_k = 0$. We use the same idea as above but the technique is now more complicated. Let $t_1,\ldots,t_{k-1}$ be fixed. We introduce the norm $\|\ \|_{mid}$ on $\mathcal{B}_\alpha^\beta(\mathbb{R}^n)$, setting

$$\|g\|_{mid} = \sup_{t_k\in\mathbb{R}} (1+|t_k|)^{-\mu_{k-1}} \|\mathcal{U}_{Lk}(t)g\|_{\mathcal{B}_\rho^\eta(\mathbb{R}^n)}, \tag{14}$$

and denote by $\mathcal{B}_{mid}$ the completion of $\mathcal{B}_\alpha^\beta$ with respect to the norm $\|\ \|_{mid}$. The results of Section 2:H are applicable to the case; here the mapping j (see Definition 1 of Section 2:H) is given by

$$j = \mathcal{U}_L(t_1,\ldots,t_{k-1},0,\ldots,0). \tag{15}$$

By Proposition 1 of Section 2:H, $\mathcal{U}_{Lk}(t)$ gives rise to a strongly continuous group $w(t_k)$ in $\mathcal{B}_{mid}$ and

$$\mathcal{U}_{Lk}(t) = j_2 w(t_k) j_1, \tag{16}$$

where $j_1 : \mathcal{B}_\alpha^\beta(\mathbb{R}^n) \to \mathcal{B}_{mid}$ and $j_2 : \mathcal{B}_{mid} \to \mathcal{B}_\rho^\eta(\mathbb{R}^n)$ are continuous mappings. Define the mapping $M_x : \mathcal{B}_{mid} \to X_\sigma$ by setting

$$M_x(g) = m_x \circ j_2(g). \tag{17}$$

Clearly M_x is a continuous linear mapping. Denote by $\tilde{L}$ the generator of the group $w(\tau)$. We intend to show that

$$M_x\tilde{L}\tilde{g} = A_k M_x\tilde{g} \tag{18}$$

for any $\tilde{g} \in D_{\tilde{L}}$. If only (18) were proved, it follows that $M_x w(t_k) = \exp(itA_k)M_x$ and consequently,

$$m_x \mathcal{U}_{Lk}(t)g = m_x j_2 w(t_k) j_1 g = M_x w(t_k) j_1 g =$$
$$= \exp(it_k A_k) m_x j_2 j_1 g \equiv \exp(it_k A_k) m_x \mathcal{U}_{L(k-1)} g = \mathcal{U}_{Ak}(t) m_x g \tag{19}$$

for any $g \in \mathcal{B}_\alpha^\beta(\mathbb{R}^n)$, as desired. We prove (18) in the following way:

Let $\tilde{g} \in D_{\tilde{L}}$, $h = j_2\tilde{g}$. We claim that $h \in D_{Lk}$ and

$$Lkh = j_2\tilde{L}\tilde{g}. \tag{20}$$

It is enough to show that $\exp(it_k L_k) j_2(\tilde{g}) = j_2(w(t_k)\tilde{g})$. The latter identity is valid on the subset $j_1(\mathcal{B}_\alpha^\beta)$ dense in $\mathcal{B}_{mid}$ and therefore valid for any $g \in \mathcal{B}_{mid}$ ($\exp(it_k L_k)$ is continuous from $\mathcal{B}_\rho^\eta(\mathbb{R}^n)$ to some $\mathcal{B}_{\rho'}^{\eta'}(\mathbb{R}^n)$ and therefore closed in $\mathcal{B}_\rho^\eta(\mathbb{R}^n)$). From (20) we obtain

$$M_x\tilde{L}\tilde{g} = m_x j_2 \tilde{L}\tilde{g} = m_x L_k j_2 \tilde{g} = A_k m_x j_2 \tilde{g} = A_k M_x \tilde{g}, \tag{21}$$

i.e., (18) is valid. The theorem is therefore proved.

B. Agreement Conditions

It turns out that in general even the most simple algebraic properties are not valid for functions of several generators. Thus we have to impose some additional conditions on the mutual behavior of generators (or their groups) in order to obtain substantial functional calculus. First of all we present a striking example of pathological behavior.

Example 1. There exist two self-adjoint operators, A_1 and A_2, in the Hilbert space H, such that

(a) For some dense linear subset $D \subset D_{A_1} \cap D_{A_2}$ the restrictions $A_1|_D$ and $A_2|_D$ are essentially self-adjoint.

(b) A_1 commutes with A_2 on D,

$$A_1A_2x = A_2A_1x, \; x \in D. \tag{1}$$

(c) Nevertheless, the corresponding groups $\exp(iA_1t)$ and $\exp(iA_2t)$ do not commute, contradicting the expectations based on formal algebraic calculations.

The construction is rather simple. Consider the Riemannian surface Γ of the multi-valued analytic function $\sqrt{z} = \sqrt{x + iy}$. x and y may be considered as coordinates on Γ, each point (x,y), $x^2 + y^2 \neq 0$ corresponding exactly to two points of Γ. Hence the Lebesgue measure $\mu = dxdy$ is defined on Γ, and we may define the Hilbert space $L^2(\Gamma,\mu)$ of square-summable functions on Γ. Next, for each point (x,y) such that $y \neq 0$ we may uniquely define its shift $g_x^t(x,y) = (x + t, y)$, $t \in \mathbb{R}$ by the requirement that $g_x^t(x,y)$ be a continuous curve on Γ. If $y = 0$ this definition fails when $x(x + t) \leqslant 0$. However, the subset $\{y = 0\} \subset \Gamma$ has a zero measure, and we are able to define the following group of unitary operators in $L^2(\Gamma)$:

$$\mathcal{U}_1(t)f = (g_x^t)^* f, \; f \in L^2(\Gamma), \; t \in \mathbb{R}. \tag{2}$$

The generator of the semigroup $\mathcal{U}_1(t)$ will be denoted A_1. Let $D = C_0^\infty(\Gamma)$. Obviously

$$A_1 f = -i\,\frac{\partial f}{\partial x}, \; f \in C_0^\infty(\Gamma). \tag{3}$$

Proposition 1. $C_o^\infty(\Gamma)$ is a core of A_1.

Proof. Consider the subspace $D_1 \subset C_o^\infty(\Gamma)$ defined by

$$D_1 = \{\phi \in C_o^\infty(\Gamma) \mid y \neq 0 \text{ on supp } \phi\}. \tag{4}$$

D_1 is dense in $L^2(\Gamma)$ and invariant under $U_1(t)$. Thus D_1 is a core of A_1 by Theorem 3 of Section 2:A. All the more, $C_o^\infty(\Gamma) \supset D_1$ is a core of A_1. The proposition is proved.

Quite analogously we define the shift $g_y^t(x,y)$ and the corresponding group $U_2(t) = (iA_2t)$:

$$U_2(t)f = (g_y^t)^* f; \tag{5}$$

$$A_2 f = -i \frac{\partial f}{\partial y}, \quad f \in C_o^\infty(\Gamma). \tag{6}$$

Thus, A_1 and A_2 are essentially self-adjoint operators defined on $C_o^\infty(\Gamma)$. They commute on $C_o^\infty(\Gamma)$, since $\partial^2 f/\partial x \partial y = \partial^2 f/\partial y \partial x$. On the other hand, $\exp(iA_1t)$ and $\exp(iA_2t)$ do not commute. Indeed, let $f \in C_o^\infty(\Gamma)$ be a function with support in the small neighborhood of the point (1,1) on one of the sheets of Γ. Then the function

$$\tilde{f} = U_1(-2)U_2(-2)U_1(2)U_2(2)f \tag{7}$$

has its support in the neighborhood of the same point on another sheet of Γ and does not therefore coincide with f. It is easy to see that in the discussed example even the commutability $A_1 \exp(iA_2t) = \exp(iA_2t)A_1$ does not take place. Indeed, the domain D_{A1} is not invariant under $\exp(iA_2t)$; moreover, for $f \in C_o^\infty(\Gamma)$, $\exp(iA_2t)f$ may be jump discontinuous in the x-axis direction for suitable values of t.

After this consideration we proceed to the discussion of positive results which guarantee us the required commutability. It appears to be most convenient to represent them in the form given below, since this particular form admits easy applications to commutation relations.

Let X,Y be Banach spaces, A,C be the generators of strongly continuous semigroups in X and Y respectively. Let also $B : X \to Y$ be a closed densely defined linear operator. We are looking for conditions under which $B \exp(iAt) = \exp(iCt)A$ (i.e., D_B is invariant under $\exp(iAt)$ and $B \exp(iAt)x = \exp(iCt)bx$). The above counterexample shows that besides the algebraic condition "BA = CB on a suitable dense subset," we have to require some additional functional-analytic conditions. We shall speak about the latter ones as of agreement conditions.

Theorem 1. Assume that there exists a core $D \subset D_B$ of the operator B such that for any $x \in D_o = \{x \in D \mid x \in D_A \text{ and } Ax \in D\}$, we have $Bx \in D_C$ and $BAx = CBx$. Then

$$\exp(iAt)x \in D_B, \quad B\exp(iAt)x = \exp(iCt)Bx, \quad x \in D_B, \tag{8}$$

provided that at least one of the following agreement conditions is then satisfied:

(a) D_o is also the core of B, D is invariant under $\exp(iAt)$, and $y(t) = B\exp(iAt)x$ is a continuous Y-valued function of t for $x \in D$.

(b) D is invariant under $R_\lambda(A)$ for $\mathrm{Im}\lambda < -\omega_o$, where ω_o is the maximum of the types of semigroups $\exp(iAt)$ and $\exp(iCt)$.

Conversely, let (8) be satisfied. Then the linear subset $D = D_B$ satisfies all the mentioned properties.

Proof. Assume first that (b) is satisfied. Let $\mathrm{Im}\lambda < -\omega_o$. Since D is invariant under $R_\lambda(A)$, we have $R_\lambda(A)x \in D_o$ for $x \in D$, so that $(C - \lambda) \times BR_\lambda(A)x = B(A - \lambda)R_\lambda(A)x = Bx$, $x \in D$, or

$$BR_\lambda(A)x = R_\lambda(C)Bx, \ x \in D, \ \mathrm{Im}\lambda < -\omega_o. \tag{9}$$

Set $x_\mu = (-i\mu R_{-i\mu}(A))^{[\mu t]}x$, $\mu \geq \omega_o$; $x \in D$ and $t \geq 0$ being fixed. Since D is invariant under any power of $R_\lambda(A)$ as well, we obtain by successive application of (9) that

$$y_\mu \equiv Bx_\mu = (-i\mu R_{-i\mu}(C))^{[\mu t]}Bx.$$

By Theorem 1 (f) of Section 2:A, $x_\mu \to \exp(iAt)x$ and $y_\mu \to \exp(iCt)Bx$ as $\mu \to \infty$. Since B is closed, $\exp(iAt)x \in D_B$ and $B\exp(iAt)x = \exp(iCt)Bx$. Thus (8) is proved for $x \in D$. Now let $x \in D_B$ be arbitrary. Since D is a core of B, there exists a sequence $x_n \in D$ B-convergent to x (i.e., $x_n \to x$ and $Bx_n \to Bx$). Set $y_n = \exp(iAt)x_n$. Then $y_n \to \exp(iAt)x$ and $By_n = B\exp(iAt)x_n = \exp(iCt)Bx_n \to \exp(iCt)Bx$. Thus (8) is proved, since B is a closed operator.

Assume now that (a) is satisfied. It is sufficient to prove (8) for $x \in D_o$ and then to repeat the above argument. We note first that D_o is invariant under $\exp(iAt)$. Indeed, let $x \in D_o$. Then $\exp(iAt)x \in D$ (since D is invariant) and $A\exp(iAt)x = \exp(iAt)Ax \in D$, since $Ax \in D$, so that $\exp(iAt)x \in D_o$. To prove (8) for $x \in D_o$ consider the expression $[B\mathcal{U}(t+\varepsilon)x - B\mathcal{U}(t)x]/\varepsilon$, where $\mathcal{U}(t) = \exp(iAt)$. We have

$$\frac{1}{\varepsilon}[B\mathcal{U}(t+\varepsilon)x - B\mathcal{U}(t)x] = iB\int_0^1 \mathcal{U}(t+\lambda\varepsilon)Axd\lambda.$$

$Ax \in D$, therefore $B\mathcal{U}(t + \lambda\varepsilon)Ax$ is a continuous function of λ. Since B is a closed operator, we may apply B under the integration sign, obtaining

$$\frac{1}{\varepsilon}[B\mathcal{U}(t+\varepsilon)x - B\mathcal{U}(t)x] = i\int_0^1 B\mathcal{U}(t+\lambda\varepsilon)Axd\lambda.$$

Using the continuity of the integrand again, we conclude that

$$\frac{1}{\varepsilon}[B\mathcal{U}(t+\varepsilon)x - B\mathcal{U}(t)x] \overset{\varepsilon\to 0}{\to} iB\mathcal{U}(t)Ax = iBA\mathcal{U}(t)x = iCB\mathcal{U}(t)x, \ x \in D_o$$

(the last equality is valid due to invariance of D_o under $\mathcal{U}(t)$). Thus we have proved that for $x \in D_o$, $y(t) = B\mathcal{U}(t)x$ is differentiable and satisfies the Cauchy problem

$$i\frac{dy(t)}{dt} + Cy(t) = 0, \ y(0) = Bx. \tag{10}$$

Since C is a generator of a strongly continuous semigroup, the solution of (10) is defined uniquely (Theorem 1 (d) of Section 2:A), so $y(t) = \exp(iCt) \times Bx$, and (8) is proved for $x \in D_o$.

Assume now that (8) is satisfied and set $D = D_B$. Then D_B is invariant under $\exp(iAt)$ by assumption, $B\exp(iAt)x$ is obviously continuous for $x \in D_B$ since it equals $\exp(iCt)Bx$.

Next we prove (b). We have

$$R_\lambda(A)x = i\int_0^\infty e^{-i\lambda t}\exp(iAt)xdt, \ \mathrm{Im}\lambda < -\omega_o.$$

For $x \in D_B$, the integral

$$i\int_0^\infty Be^{-i\lambda t}\exp(iAt)x\,dt = i\int_0^\infty e^{-i\lambda t}\exp(iCt)Bx\,dt = R_\lambda(C)Bx$$

converges absolutely and has a continuous integrand; since B is closed we conclude that $R_\lambda(A)x \in D_B$, so D_B is $R_\lambda(A)$-invariant. Moreover, (9) is valid for $x \in D_B$. We mention that $D_o = \{x \mid x = R_\lambda(A)y \text{ for some } y \in D_B\}$. Indeed, if $x = R_\lambda(A)y$, then $Ax = (A - \lambda)x + \lambda x = -y + \lambda x \in D_B$, so $x \in D_o$. Conversely, if $x \in D_o$, then $x = R_\lambda(A)y$, where $y = \lambda x - Ax \in D_B$. Now we are able to prove that $BAx = CBx$ for $x \in D_o$. Let $x = R_\lambda(A)y$, $y \in D_B$. Then by (9)

$$BAx = B(A-\lambda)x + \lambda Bx = -By + \lambda BR_\lambda(A)y =$$

$$= By + \lambda R_\lambda(C)By = (\lambda R_\lambda(C) - I)By = CR_\lambda(C)By = CBR_\lambda(A)y = CBx,$$

as required. It remains to prove that D_o is a core of B. For $x \in D_B$, set $x_\mu = -i\mu R_{-i\mu}(A)x \in D_o$. We claim that x_μ is B-convergent to x as $\mu \to \infty$. Indeed, by Theorem 1 (f) of Section 2:A

$$(-i\mu R_{-i\mu}(A))^{[\mu t]}x \to \exp(iAt)x, \quad \mu \to \infty,$$

uniformly with respect to $t \in [0,T]$ for any $T < \infty$. Set $t = 1/\mu$. Thus, we obtain

$$\| -i\mu R_{-i\mu}(A)x - \exp(\tfrac{i}{\mu} A)x \| \overset{\mu\to\infty}{\to} 0,$$

$$\| \exp(\tfrac{i}{\mu} A)x - x \| \overset{\mu\to\infty}{\to} 0,$$

so that $-i\mu R_{-i\mu}(A)x = x_\mu \overset{\mu\to\infty}{\to} x$. On the other hand

$$y_\mu \equiv Bx_\mu = -i\mu BR_{-i\mu}(A)x = -i\mu R_{-i\mu}(C)Bx,$$

and the same argument shows that $y_\mu \overset{\mu\to\infty}{\to} Bx$. Thus D_o is a core of B and the theorem is proved. The reformulation of the theorem for the case of strongly continuous groups is obvious.

We employ Theorem 1 to prove the modification of the Krein-Shikhvatov theorem, concerned with strongly continuous representations of Lie groups in Banach spaces. First of all we recall some notions and facts from the theory of Lie groups and their representations. We omit the proofs since they may be found in standard textbooks.

Let Γ be a n-dimensional Lie algebra over $\mathbb{R}$ with the basis $a_1,\ldots,a_n$, so that

$$[a_i,a_j] = \sum_{k=1}^n \lambda_{ij}^k a_k, \quad i,j = 1,\ldots,n, \tag{11}$$

where λ_{ij}^k are the structure constants of Γ (with respect to the basis $a_1, \ldots, a_n$). Let G be a Lie group with Lie algebra Γ. We shall use the canonical coordinates of the second genus in the neighborhood of the neutral element $e \in G$: the coordinate tuple $x = (x_1,\ldots,x_n)$ lying in the neighborhood of zero in $\mathbb{R}^n$ corresponds to the element $g(x) \in G$ equal to

$$g(x) = g_n(x_n)g_{n-1}(x_{n-1}) \ldots g_1(x_1), \tag{12}$$

where $g_i(t)$ is the one-parametric subgroup of G, corresponding to $a_i \in \Gamma$ (i.e., $g_i(0) = e$, $g_i(t + \tau) \equiv g_i(t)g_i(\tau)$, and a_i is a tangent vector of the curve $g_i(t)$ for $t = 0$). The composition law in the coordinate system $(x_1,\dots,x_n)$ has the form

$$g(x)g(y) = g(\psi(x,y)), \tag{13}$$

where ψ is a smooth mapping of the neighborhood of the origin in $\mathbb{R}^n \times \mathbb{R}^n$ into $\mathbb{R}^n$, $\psi(y,0) = \psi(0,y) \equiv y$. It is easy to calculate the derivative $(\partial\psi/\partial x) \times \times (x,y)$. Since our consideration is local, we may assume, by the Ado theorem, that Γ is realized as a matrix Lie algebra, and the vicinity of e in G as that in a matrix Lie group. Thereafter $g(x)$ has the form

$$g(x) = e^{x_n a_n} e^{x_{n-1} a_{n-1}} \cdot \dots \cdot e^{x_1 a_1}, \tag{14}$$

where e^{xa} is the usual matrix exponent. We calculate now the derivative $(\partial/\partial\psi)g(\psi)$. Later we substitute $\psi = x$ and $\psi = \psi(x,y)$ into the obtained expression. We have

$$\frac{\partial}{\partial\psi_j} g(\psi) = e^{\psi_n a_n} e^{\psi_{n-1} a_{n-1}} \cdot \dots \cdot e^{\psi_j a_j} a_j e^{\psi_{j-1} a_{j-1}} \cdot \dots \cdot e^{\psi_1 a_1}. \tag{15}$$

Using the fact that for given elements a,b of a matrix Lie algebra

$$e^{tb} a e^{-tb} = e^{t\,ad_b}(a), \tag{16}$$

where ad_b is an operator of commutation with b in this algebra: $ad_b = [b,a]$, we obtain

$$\frac{\partial}{\partial\psi_j} g(\psi) = [e^{\psi_n ad_{a_n}} \cdot \dots \cdot e^{\psi_j ad_{a_j}}(a_j)]g(\psi). \tag{17}$$

To simplify the expression in square brackets we note that in the basis $(a_1,\dots,a_n)$ the operator ad_{a_s} has the matrix Λ_s with the elements $(\Lambda_s)_{pq} = \lambda^p_{sq}$ and consequently the operator $\exp(\psi_n ad_{a_n}) \cdot \dots \cdot \exp(\psi_j \Lambda_j)$ is represented by the matrix $\exp(\psi_n\Lambda_n) \cdot \dots \cdot \exp(\psi_j\Lambda_j)$. Thus

$$\begin{aligned} \frac{\partial}{\partial\psi_j} g(\psi) &= \sum_{p=1}^{n} [\exp(\psi_n\Lambda_n) \cdot \dots \cdot \exp(\psi_j\Lambda_j)]_{pj} a_p g(\psi) = \\ &= \Big(\sum_{p=1}^{n} B_{pj}(\psi) a_p\Big) g(\psi), \end{aligned} \tag{18}$$

where $B(\psi) = B(\psi_1,\dots,\psi_n)$ is the matrix with the elements

$$B_{pq}(\psi) = [\exp(\psi_n\Lambda_n) \cdot \dots \cdot \exp(\psi_q\Lambda_q)]_{pq}. \tag{19}$$

In particular, $B(0) = I$ (identity matrix), so that the inverse matrix $C(\psi) = B^{-1}(\psi)$ is defined when ψ is close to zero.

Calculating the derivative with respect to x on both sides of (13), we obtain

$$\frac{\partial}{\partial x_i} g(\psi(x,y)) = \sum_{p,j=1}^{n} \frac{\partial\psi_j}{\partial x_i} B_{pj}(\psi) a_p g(\psi(x,y)), \tag{20}$$

$$\begin{aligned} \frac{\partial}{\partial x_i} g(\psi(x,y)) &= \frac{\partial}{\partial x_i}(g(x)g(y)) = \sum_{p=1}^{n} B_{pi}(x) a_p g(x)g(y) = \\ &= \sum_{p=1}^{n} B_{pi}(x) a_p g(\psi(x,y)). \end{aligned} \tag{21}$$

The matrices a_p, $p = 1,\ldots,n$ are linearly independent and so are $a_p g(\psi(x,y))$ since $g(\psi)$ is invertible. Thus the comparison of (20) and (21) yields

$$B_{pi}(x) = \sum_{j=1}^{n} B_{pj}(\psi) \frac{\partial \psi_j}{\partial x_i}, \tag{22}$$

from where

$$\frac{\partial \psi_j}{\partial x_i}(x,y) = \sum_{k=1}^{n} C_{jk}(\psi(x,y))B_{ki}(x) \tag{23}$$

or simply $\frac{\partial \psi}{\partial x} = C(\psi)B(x) = B^{-1}(\psi)B(x)$. Recall the definition of the representation of G in a Banach space $\mathcal{X}$.

Definition 1. A (strongly continuous) representation of the Lie group G in the Banach space $\mathcal{X}$ is a function $g \to T(g)$ on G, whose values are bounded linear operators in $\mathcal{X}$, such that T is continuous in the strong sense, $T(0) = I$ and $T(g_1)T(g_2) = T(g_1 g_2)$ for any $g_1, g_2 \in G$. In particular, for any element $a \in \Gamma$ of the corresponding Lie algebra, the family $T_a(t) = T(g_a(t))$, $t \in \mathbb{R}$, where $g_a(t)$ is a one-parametric subgroup of G, corresponding to a, is a strongly continuous group of bounded linear operators in $\mathcal{X}$ (Definition 1 of Section 2:A) and therefore has the form $T_a(t) = \exp(itA)$, where A is its generator. The operator $A = A(a)$ will be called the generator of representation T, corresponding to $a \in \Gamma$.

The theorem given below enables us to construct representations of Lie groups starting from representations of their Lie algebras.

Theorem 2. Let Γ be a Lie algebra, given by relations (11), and G be a corresponding connected simply connected Lie group. Assume that $A_1,\ldots,A_n$ are the generators of strongly continuous groups of bounded linear operators in the Banach space $\mathcal{X}$; $D \subset \mathcal{X}$ is a dense linear subset such that

(a) $D \subset D_{A_1} \cap \ldots \cap D_{A_n}$ and is invariant under operators A_j, their resolvents $R_\lambda(A_j)$, and groups e^{itA_j}, $j = 1,\ldots,n$.

(b) The operators iA_j form the representation of Γ in the space D, i.e.,

$$[A_j, A_k]h = -i \sum_{s=1}^{n} \lambda_{jk}^{s} A_s h, \quad h \in D, \; j,k = 1,\ldots,n. \tag{24}$$

Then there exists a representation T of the group G in $\mathcal{X}$, such that A_j are its generators, $A_j = A(a_j)$.

Proof. Since G is connected and simply connected, it is enough to construct $T(g)$ for g lying in the neighborhood of unity, namely in the neighborhood which is covered by canonical system of coordinates. For such g, we set

$$T(g(x)) \equiv T(x) = e^{iA_n x_n} \cdot \ldots \cdot e^{iA_1 x_1}. \tag{25}$$

$T(x)$ is a bounded strongly continuous function, and besides $T_{a_i}(t) = \exp \times$ $\times\, (iA_i t)$ is a strongly continuous group with the generator A_i, so all we need is to prove that the operators $T(x)$ satisfy the group law in the vicinity of zero, i.e.,

$$T(x)T(y) = T(\psi(x,y)) \tag{26}$$

for x,y small enough. It is sufficient to prove that (26) holds on the dense subset $D \subset \mathcal{X}$. Note that D is a core of each A_i by Lemma 2 of Section 2:A.

Lemma 1. Let $h \in D$, we have

$$e^{iA_s t}A_k h = \sum_{p=1}^{n} [\exp(t\Lambda_s)]_{pk} A_p e^{iA_s t}h, \quad s,k = 1,\dots,n. \tag{27}$$

Proof. Consider the Banach space $Y = \mathcal{X} \times \mathbb{C}^n = \underbrace{\mathcal{X} \oplus \mathcal{X} \oplus \dots \oplus \mathcal{X}}_{n \text{ summands}}$. The operators in Y may be represented as matrices with operators in $\mathcal{X}$ as their elements. We introduce the operator $B : \mathcal{X} \to Y$ which is the closure* from D of the operator $h \to (A_1 h,\dots,A_n h) \in Y$, and the operator $C_s : Y \to Y$, which has the form (Λ_s' denotes the transpose of Λ_s):

$$C_s = A_s \otimes I + iI \otimes \Lambda_s' = \tag{28}$$

$$= \begin{pmatrix} A_s & & 0 \\ & \ddots & \\ 0 & & A_s \end{pmatrix} + i \begin{pmatrix} \lambda_{s1}^1 I & \dots & \lambda_{s1}^n I \\ \dots & \dots & \dots \\ \lambda_{sn}^1 I & \dots & \lambda_{sn}^n I \end{pmatrix}, \quad D_{C_s} = D_{A_s} \oplus \dots \oplus D_{A_s}.$$

C_s is a generator of the strongly continuous group $\exp(iC_s t)$ in the space Y. Indeed, direct computation shows that we should set

$$\exp(iC_s t) = \exp(iA_s t) \otimes \exp(-\Lambda_s' t). \tag{29}$$

Further, we have by (24), using also the antisymmetry of λ_{jk}^s with respect to lower indices,

$$BA_s h = (A_1 A_s h,\dots,A_n A_s h) = (A_s A_1 h,\dots,A_s A_n h) -$$
$$- i\Big(\sum_{\ell=1}^{n} \lambda_{1s}^{\ell} A_\ell h,\dots, \sum_{\ell=1}^{n} \lambda_{ns}^{\ell} A_\ell h\Big) = (A_s \otimes I)Bh + i(I \times \Lambda_s')Bh = C_s Bh, \quad h \in D. \tag{30}$$

Since D is invariant under resolvent $R_\lambda(A_s)$, we may apply Theorem 1 and obtain that

$$B \exp(iA_s t)h = \exp(iC_s t)Bh, \quad h \in D_B \supset D. \tag{31}$$

The latter identity may be written in the form

$$A_p \exp(iA_s t)h = \exp(iA_s t) \sum_{k=1}^{n} [\exp(-\Lambda_s' t)]_{pk} A_k h =$$
$$= \exp(iA_s t) \sum_{k=1}^{n} [\exp(-\Lambda_s t)]_{kp} A_k h. \tag{32}$$

(32) is equivalent to (27) since $\exp(-\Lambda_s t)$ is the inverse matrix to $\exp(\Lambda_s t)$. The lemma is thereby proved.

Lemma 2. For any $h \in D$, $T(\lambda)h$ is differentiable with respect to $\lambda \in \mathbb{R}^n$ and

$$-i \frac{\partial}{\partial \lambda_j} T(\lambda)h = \sum_{p=1}^{n} B_{pj}(\lambda) A_p T(\lambda)h. \tag{33}$$

Proof. Set $h(\lambda) = T(\lambda)h$. For any $\varepsilon = (\varepsilon_1,\dots,\varepsilon_n) \in \mathbb{R}^n$, we have

---

* B is closurable since $h_n \to 0$, $Bh_n \to \tilde{h} = (\tilde{h}_1,\dots,\tilde{h}_n)$ implies $A_i h_n \to \tilde{h}_i$, $i = 1,\dots,n$, so that $\tilde{h}_i = 0$, since A_i are closed operators.

$$h(\lambda+\varepsilon)-h(\lambda) = \sum_{j=1}^{n} [h(\lambda_1,\ldots,\lambda_{j-1},\lambda_j+\varepsilon_j,\ldots,\lambda_n+\varepsilon_n) -$$

$$- h(\lambda_1,\ldots,\lambda_j,\lambda_{j+1}+\varepsilon_{j+1},\ldots,\lambda_n+\varepsilon_n)] = i\sum_{j=1}^{n}\varepsilon_j e^{iA_n(\lambda_n+\varepsilon_n)}\cdot\ldots\cdot\times$$

$$\times\, e^{iA_{j+1}(\lambda_{j+1}+\varepsilon_{j+1})}\int_0^1 d\tau e^{iA_j(\lambda_j+\tau\varepsilon_j)}A_j e^{iA_{j-1}\lambda_{j-1}}\cdot\ldots\cdot e^{iA_1\lambda_1}h =$$

$$= i\sum_{j=1}^{n}\varepsilon_j e^{iA_n\lambda_n}\cdot\ldots\cdot e^{iA_j\lambda_j}A_j e^{iA_{j-1}\lambda_{j-1}}\cdot\ldots\cdot e^{iA_1\lambda_1}h + o(\|\varepsilon\|),$$

when $\|\varepsilon\| = (\varepsilon_1^2 + \ldots + \varepsilon_n^2)^{1/2} \to 0$; we have used the strong continuity of $\exp(iA_n\lambda_n)\cdot\ldots\cdot\exp(iA_j\lambda_j)$ and the invariance of D under $\exp(iA_i t)$. Thus $h(\lambda)$ is differentiable and

$$-i\frac{\partial}{\partial\lambda_j}h(\lambda) = e^{iA_n\lambda_n}\cdot\ldots\cdot e^{iA_j\lambda_j}A_j e^{iA_{j-1}\lambda_{j-1}}\cdot\ldots\cdot e^{iA_1\lambda_1}h. \tag{34}$$

Successive application of Lemma 1 yields

$$-i\frac{\partial}{\partial\lambda_j}h(\lambda) = \sum_{p=1}^{n}[\exp(\lambda_n\Lambda_n)\cdot\ldots\cdot\exp(\lambda_j\Lambda_j)]_{pj}A_p h(\lambda).$$

The lemma is proved. For $h \in D$, set now

$$h_1(x,y) = T(x)T(y)h,\quad h_2(x,y) = T(\psi(x,y))h. \tag{35}$$

We have $h_1(0,y) = h_2(0,y) = T(y)h$. By Lemma 2,

$$-i\frac{\partial h_1}{\partial x_j}(x,y) = \sum_{p=1}^{n}B_{pj}(x)A_p h_1(x,y), \tag{36}$$

$$-i\frac{\partial h_2}{\partial x_j}(x,y) = \sum_{s=1}^{n}\frac{\partial\psi_s(x,y)}{\partial x_j}\sum_{p=1}^{n}B_{ps}(\psi(x,y))A_p h_2(x,y) = \tag{37}$$

$$= \sum_{p=1}^{n}B_{pj}(x)A_p h_2(x,y),$$

since (22) is valid. We note that

$$B_{pj}(x_1,\ldots,x_j,0,\ldots,0) = [\exp(x_j\Lambda_j)]_{pj} = \delta_{pj}, \tag{38}$$

since the j-th column of Λ_j consists of zeros ($\lambda_{jj}^p \equiv 0$), so we have

$$-i\frac{\partial h_i}{\partial x_j}(x_1,\ldots,x_j,0,\ldots,0,y) = A_j h_i(x_1,\ldots,x_j,0,\ldots,0,y),$$
$$i = 1,2,\; j = 1,\ldots,n. \tag{39}$$

Now we may prove that $h_1(x_1,\ldots,x_j,0,\ldots,0,y) = h_2(x_1,\ldots,x_j,0,\ldots,0,y)$, $j = 0,\ldots,n$ by induction on j. It is valid for $j = 0$ and, if it is valid for $j = j_o$, we have that for $j = j_o + 1$, both h_1 and h_2 satisfy identical Cauchy data at $x_j = 0$ for equation (39). Application of Theorem 1 (d) of Section 1:A yields the desired result. Thus $T(x)T(y)h = T(\psi(x,y))h$. The theorem is proved.

C. Scales Generated by the Tuple of Unbounded Operators

Now we proceed to the investigation of more concrete scales of the type arising when the problems concerned with asymptotics are considered. The general results may be essentially improved for such scales, and many more detailed theorems may be stated. In this item we introduce different types of scales associated with a given collection of closed operators in a Banach space.

Let X be a Banach space with the norm $\|\cdot\|$, $E \subset X$ be a dense linear subset. Assume also that closed operators $B_1, \ldots, B_m$ in X are given such that E is the core of each of these operators and E is invariant under these operators.

(a) Let $\rho = (\rho_1, \ldots, \rho_m)$ be a m-tuple of positive integers. Consider the following norm defined on E:

$$\|x\|_k \equiv \|x\|_{\rho,B,k} = \sum_{0 \leqslant \langle\rho,\nu\rangle \leqslant k} \|B_{\nu_1} \cdot \ldots \cdot B_{\nu_s} x\|, \quad x \in E, \; k = 1,2,\ldots . \tag{1}$$

Here the number of elements s in the sequence $\{\nu_i\}$ of integers, satisfying $1 \leqslant \nu_i \leqslant m$, is not fixed: the empty product $(s = 0)$ is by convention the identity operator, $\langle\rho,\nu\rangle$ is defined by $\langle\rho,\nu\rangle = \rho_{\nu_1} + \rho_{\nu_2} + \ldots + \rho_{\nu_s}$ (the number of entries of B_j in the product is counted with the "weight" ρ_j); the sum is taken over all the sequences $(\nu_1, \ldots, \nu_s)$ satisfying then the enumerated conditions. In particular, when $\rho_1 = \ldots = \rho_m = 1$, we obtain the definition of the norm

$$\|x\|_{B,k} = \sum_{0 \leqslant s \leqslant k} \|B_{\nu_1} \cdot \ldots \cdot B_{\nu_s} x\|, \quad x \in E. \tag{2}$$

We denote by $H^k_{B,\rho}$ the completion of E with respect to the norm (1) (the notation H^k_B will be used in the particular case of the norm (2)). It is clear that $H^0_{B,\rho} = X$. Next, we have the inequality

$$\|x\|_{\rho,B,k} \leqslant \|x\|_{\rho,B,s}, \quad k \leqslant s$$

so that the identical operator on E is extended to continuous operators $i_{sk} : H^s_{B,\rho} \to H^k_{B,\rho}$ for $s \geqslant k$, and we have $i_{sk} \circ i_{rs} = i_{rk}$, $r \geqslant s \geqslant k$.

(b) For an arbitrary multi-index $\alpha = (\alpha_1, \ldots, \alpha_m)$ with non-negative integer components, we define the norm

$$\|x\|_{B,\alpha} = \sum_{\alpha(\nu) \leqslant \alpha} \|B_{\nu_1} \cdot \ldots \cdot B_{\nu_s} x\|, \tag{3}$$

where $\alpha(\nu) = (\alpha_1(\nu), \ldots, \alpha_m(\nu))$, $\alpha_i(\nu)$ denotes the number of entries of B_i in the product $B_{\nu_1} \cdot \ldots \cdot B_{\nu_s}$. The completion of E with respect to the norm (3) is denoted by H^α_B. Again we have continuous operators $i_{\alpha\beta} : H^\alpha_B \to H^\beta_B$ defined for $\beta \leqslant \alpha$, which are the closures of the identical operator defined on E.

Theorem 1. $H_{B,\rho} = \{H^k_{B,\rho}\}$ and $\tilde{H}_B = \{H^\alpha_B\}$ are Banach scales. The operators B_i, $i = 1, \ldots, m$, are left translators in these scales and moreover $\delta_{B_i}(k) = k + \rho_i$ for the scale $H_{B,\rho}$, $\delta_{B_i}(\alpha) = (\alpha_1, \ldots, \alpha_{i-1}, \alpha_i + 1, \alpha_{i+1}, \ldots, \alpha_n)$ for the scale $\tilde{H}_B$. In other words, B_i is the left translator in the scales defined, and its step is $-\rho_i$ for the scale $H_{B,\rho}$ and $(0, \ldots, 0, -1, 0, \ldots, 0)$ for the scale $\mathcal{H}_B$.

Proof. (a) Consider first the spaces $H^k_{B,\rho}$. The estimate $\| B_i x \|_{k-\rho_i} \leqslant C \| x \|_k$ follows immediately from the definition. We may write

$$\| x \|_k \sim \| x \|_{k-1} + \sum_{i:\ \rho_i \leqslant k} \| B_i x \|_{k-\rho_i}, \quad x \in E,\ k \geqslant 1 \tag{4}$$

(here $\sim$ denotes the equivalence of the norms). We prove by induction on k that $i_{k,k-1} : H^k_{B,\rho} \to H^{k-1}_{B,\rho}$ has no kernel and that B_j are closurable from E in the space $H^k_{B,\rho}$. It is valid for $k = 0$, since B_j are closed in $X = H^0_{B,\rho}$ by assumption. Assume that the statement is proved for all $k < k_0$. Set $k = k_0$. We prove first that $i_{k,k-1}$ has no kernel. Let $x_n \in E$ be a sequence convergent in $H^k_{B,\rho}$ to some element x (we write $x_n \overset{k}{\to} x$ for short) and let $x_n \overset{k-1}{\to} 0$. Then $x_n \overset{k-\rho_i}{\to} 0$, since $\rho_i \geqslant 1$ for all i. Next, from (4) we obtain that the sequence $B_i x_n$ is fundamental in $H^{k-\rho_i}_{B,\rho}$, so $B_i x_n \overset{k-\rho_i}{\to} y_i \in H^{k-\rho_i}_{B,\rho}$. Since by the induction assumption B_i is closurable in $H^{k-\rho_i}_{B,\rho}$, $y_i = 0$ for all i. Using (4) again, we obtain $\| x_n \|_k \to 0$, i.e., $x = 0$. Thus the triviality of $\operatorname{Ker}(i_{k,k-1})$ is shown. It follows that $\operatorname{Ker}(i_{ks}) = \{0\}$ for all $s \leqslant k$. Next, we show that B_j is closurable in $H^k_{B,\rho}$ for any j. Let $x_n \in E$, $x \overset{k}{\to} 0$, $B_j x_n \overset{k}{\to} y \in H^k_{B,\rho}$. It follows that $x_n \overset{k-1}{\to} 0$, and $B_j x_n \overset{k-1}{\to} i_{k,k-1} y$. Since B_j is closurable in $H^{k-1}_{B,\rho}$, we obtain $i_{k,k-1} y = 0$, therefore $y = 0$ since $i_{k,k-1}$ has the trivial kernel.

(b) The case of spaces H^α_β is considered in a similar way. We mention only that instead of (4) we have

$$\| x \|_{B,\alpha} \sim \sum_{i:\ \alpha_i > 0} \left(\| x \|_{B,(\alpha_1,\dots,\alpha_i-1,\dots,\alpha_n)} + \| B_i x \|_{B,(\alpha_1,\dots,\alpha_i-1,\dots,\alpha_n)} \right), \quad |\alpha| > 0, \tag{5}$$

and the proof proceeds by induction on $|\alpha|$ instead of k. The theorem is proved.

Now we concentrate our attention on the scale $H_{B,\rho}$. Let $A : E \to E$ be a given linear operator. We are seeking the conditions under which A gives rise to a translator in the scale $H_{B,\rho}$. These conditions are rather simple, however. Introduce first a convenient terminology. Given a product $B_{\nu_1} B_{\nu_2} \dots B_{\nu_s}$, we call the number $\langle \rho, \nu \rangle = \rho_{\nu_1} + \dots + \rho_{\nu_s}$ the length of this product. Similarly, $\rho_{\nu_1} + \dots + \rho_{\nu_s}$ is called the length of the commutator $K_\nu(A) = [B_{\nu_1}[B_{\nu_2} \dots [B_{\nu_s}, A] \dots]]$. We adopt the convention that A itself is the commutator of the length zero.

Theorem 2. Assume that there exists a function $\phi : \mathbb{Z}_+ \cup \{0\} \to \mathbb{Z}_+ \cup \{0\}$ (where $\mathbb{Z}_+$ is the set of positive integers), such that for any $r \in \mathbb{Z}_+ \cup \{0\}$ and any commutator $K_\nu(A)$ of the length r, the following estimate is valid:

$$\| K_\nu(A) x \|_0 \leqslant C \| x \|_{\phi(r)}, \quad x \in E \tag{6}$$

with the constant C depending only on $r = \langle \rho, \nu \rangle$. Then A induces a left translator in the scale $H_{B,\rho}$. More precisely, set

$$\psi(k) = \max_{0 \leqslant r \leq k} (k + \phi(r) - r), \quad k \in \mathbb{Z}_+ \cup \{0\}. \tag{7}$$

Then

$$\| Ax \|_k \leqslant \text{const} \| x \|_{\psi(k)}, \quad x \in E, \tag{8}$$

so that A extends to a bounded linear operator from $H^{\psi(k)}_{B,\rho}$ to $H^k_{B,\rho}$.*

Proof. We have

$$\| Ax \|_k = \sum_{0 \leqslant \langle\rho,\nu\rangle \leqslant k} \| B_{\nu_1} B_{\nu_2} \dots B_{\nu_s} Ax \|_o, \quad x \in E.$$

We intend to show that $B_{\nu_1} \dots B_{\nu_s} Ax$ is the linear combination of the terms having the form $K_\mu(A) B_{\varepsilon_1} B_{\varepsilon_2} \dots B_{\varepsilon_t} x$, where $\langle\rho,\mu\rangle + \langle\rho,\varepsilon\rangle = \langle\rho,\nu\rangle$. Once it has been proved, we might estimate these terms, using (6)

$$\| K_\mu(A) B_{\varepsilon_1} \dots B_{\varepsilon_t} x \|_o \leqslant C \| B_{\varepsilon_1} \dots B_{\varepsilon_t} x \|_{\phi(\langle\rho,\mu\rangle)} \leqslant$$

$$\leqslant C \| x \|_{\langle\rho,\varepsilon\rangle + \phi(\langle\rho,\mu\rangle)} = C \| x \|_{\langle\rho,\nu\rangle + \phi(r) - r} \leqslant$$

$$\leqslant C \| x \|_{k + \phi(r) - r} \leqslant C \| x \|_{\psi(k)}$$

(here $r = \langle\rho,\mu\rangle$, C denotes different constants in different places). This estimate yields (8) immediately. We prove the required representation of $B_{\nu_1} \dots B_{\nu_s} Ax$ by induction on s. Assume that $B_{\nu_2} \dots B_{\nu_s} Ax$ is a linear combination of the terms $K_\mu(A) B_{\varepsilon_1} \cdot \dots \cdot B_{\varepsilon_t} x$ with $\langle\rho,\mu\rangle + \langle\rho,\varepsilon\rangle = \langle\rho,\nu\rangle - \rho_{\nu_1}$. We have

$$B_{\nu_1} K_\mu(A) B_{\varepsilon_1} \cdot \dots \cdot B_{\varepsilon_t} x =$$

$$= K_\mu(A) B_{\varepsilon_1} \cdot \dots \cdot B_{\varepsilon_t} x + [B_{\nu_1}, K_\mu(A)] B_{\varepsilon_1} \cdot \dots \cdot B_{\varepsilon_t} x.$$

Since both terms on the right have the length $\langle\rho,\nu\rangle$, we obtain the required representation. The theorem is thereby proved.

Now we turn our attention to the case when the given operators satisfy Lie commutation relations. Let $A_1, \dots, A_n$ be operators, defined on a dense invariant subset E of a Banach space X and closurable in X (we use the same notation for their closures). Assume that $A_1, \dots, A_n$ satisfy Lie commutation relations on E:

$$[A_j, A_k] x = -i \sum_{s=1}^{n} \lambda^s_{jk} A_s x, \quad x \in E, \ j,k = 1, \dots, n, \tag{9}$$

where $\lambda^s_{jk} \in \mathbb{R}$ are the structure constants. We choose some number $m \leqslant n$ and consider the scale $H_{(A_1,\dots,A_m)} = \{H^k_{(A_1,\dots,A_m)}\}$, generated by the tuple $(A_1, \dots, A_m)$. Let L denote the Lie algebra consisting of all linear combinations of $(A_1, \dots, A_n)$, and L_1 denote the subspace of L generated by $(A_1, \dots, A_m)$. We set

---

* It is not difficult to construct a non-decreasing function $\psi^*(k)$ defined for sufficiently large k, such that A extends to a bounded operator from $H^k_{B,\rho}$ to $H^{\psi^*(k)}_{B,\rho}$. Namely, for $k \geqslant \psi(0)$ set $\psi^*(k) = \max\{r \mid \psi(r) \leqslant k\}$. Then clearly ψ^* is non-decreasing and A is bounded in the mentioned pair of spaces since there is a decomposition $H^k_{B,\rho} \overset{i_{k\psi(\psi^*(k))}}{\to} H^{\psi(\psi^*(k))}_{B,\rho} \overset{A}{\to} H^{\psi^*(k)}_{B,\rho}$.

$$L_2 = L_1 + [L_1, L_1],\ L_3 = L_2 + [L_1, L_2], \ldots,\ L_s = L_{s-1} + [L_1, L_{s-1}], \ldots \quad (10)$$

We have the non-decreasing sequence of subspaces $L_1 \subset L_2 \subset \ldots \subset L_s \subset \ldots$ of L. This sequence becomes stable for some $s = s_o$, i.e., $L_{s_o} = L_{s_o+1} = \ldots = \tilde{L}_1$, where L_1 is the Lie subalgebra in L generated by $(A_1, \ldots, A_m)$.

Proposition 1. The elements of $\tilde{L}_1$ are left translators in $H_{(A_1,\ldots,A_m)}$. More precisely, if $B = \sum_{j=1}^{n} \alpha_j A_j \in L_s$, then

$$\| Bx \|_k \leqslant C_k \| x \|_{k+s} \left(\sum_{j=1}^{n} |\alpha_j| \right), \quad (11)$$

where C_k depends only on k.

Proof. The elements of L_s are sums of products of the length $\leqslant s$ of the elements $A_1, \ldots, A_m$. Let $C_1, \ldots, C_r$ be a basis of L_s. Then $B = \sum_{\nu=1}^{r} \beta_r C_r$ and besides $\sum_{\nu=1}^{r} |\beta_r| \leqslant \text{const} \sum_{j=1}^{n} |\alpha_j|$, since all the norms in a finite dimensional vector space L_s are equivalent. Thus (11) turns out to be an immediate consequence of the norm definition (2).

Assume now that the operators $A_1, \ldots, A_n$ satisfy the conditions of Theorem 2 of the preceding item (E plays the role of the subset D mentioned there). In particular, the operators $A_1, \ldots, A_n$ (and all their real linear combinations) are generators in X. We should like to find out whether they are also generators in $H_{(A_1,\ldots,A_m)}$. For this we shall make use of the identity

$$A_j \exp(iBt)x = \exp(iBt) \sum_{k=1}^{n} [\exp(-\Lambda t)]_{kj} A_k x, \quad x \in E. \quad (12)$$

Here $B = \sum_{\ell=1}^{n} \alpha_\ell A_\ell$, $\alpha_\ell \in \mathbb{R}$, $\Lambda = \sum_{\ell=1}^{n} \alpha_\ell A_\ell$ is the corresponding matrix of the associated representation of L. The identity (12) was proved in item B (see (32) of item B) for basis elements of a Lie algebra, but it is easy to see that a posteriori this proof is valid for arbitrary B of the above form.

Theorem 3. Let $L_B \subset L$ be a minimal invariant subspace of the matrix* Λ, containing L_1. If $L_B \subset \tilde{L}$, B is a generator in the scale $H_{(A_1,\ldots,A_m)}$ (in particular, A_j itself is a generator in the scale $H_{(A_1,\ldots,A_m)}$ for $j = 1, \ldots, m$). If $L_B \subset L_s$, then the estimates

$$\| \exp(iBt)x \|_k \leqslant C_k P_B(t) (P_\Lambda(t))^k \| x \|_{sk}, \quad x \in H^{sk}_{(A_1,\ldots,A_m)} \quad (13)$$

are valid, where $P_\Lambda(t)$ is the norm of $\exp(-\Lambda t)$, $P_B(t)$ is the norm of the operator $\exp(iBt)$ in the space X.

Proof. We proceed by induction on k. Assume that the statement is proved for $k = k_o$. Set $k = k_o + 1$. It is enough to prove the estimate (13) for $x \in E$

---

* $A_1, \ldots, A_n$ is the basis of L and the action of an $(n \times n)$-matrix Λ on L is defined as the action on the coordinates with respect to this basis. Thus, $\Lambda A_j = \sum_k \Lambda_{kj} A_k = \sum_{\ell,k} \alpha_\ell \lambda^k_{\ell j} A_k = i \sum_\ell \alpha_\ell [A_\ell, A_j] = i[B, A_j]$, i.e., Λ is the matrix of the operator iad_B.

$$\| \exp(iBt)x \|_k = \| \exp(iBt)x \|_{k_o} + \sum_{j=1}^{m} \| A_j \exp(iBt)x \|_{k_o} =$$

$$= \| \exp(iBt)x \|_{k_o} + \sum_{j=1}^{m} \| \exp(iBt) \sum_{k=1}^{n} [\exp(-\Lambda t)]_{kj} A_k x \|_{k_o} \leqslant \tag{14}$$

$$\leqslant C_{k_o} P_B(t)(P_\Lambda(t))^{k_o} (\| x \|_{k_o s} + \sum_{j=1}^{m} \| \sum_{k=1}^{n} [\exp(-\Lambda t)]_{kj} A_k x \|_{k_o s}), \ x \in E,$$

by the induction assumption. We claim that $\sum_{k=1}^{n} [\exp(-\Lambda t)]_{kj} A_k \in L_B$ for any $j \in \{1,\ldots,m\}$. Indeed, the latter sum is the result of action of the operator $\exp(-\Lambda t)$ on the element $A_j \in L_1$. We have

$$\exp(-\Lambda t)A_j = \sum_{r=0}^{\infty} 1/r!(-t)^r \Lambda^r A_j$$

and for any r, $\Lambda^r A_j \in L_B$ since $L_B \ni A_j$ and L_B is invariant under Λ. Applying Proposition 1 to the right-hand side of (14), we come to the estimate

$$\| \exp(iBt)x \|_k \leqslant C_{k_o} P_B(t)(P_\Lambda(t))^{k_o} (\| x \|_{k_o s} + \text{const} \| x \|_{k_o s+s}) \times$$

$$\times \sum_{j=1}^{m} \sum_{k=1}^{n} |[\exp(-\Lambda t)]_{kj}| \leqslant \text{const} \cdot P_B(t)(P_\Lambda(t))^{k_o+1} \| x \|_{(k_o+1)s}.$$

The theorem is proved.

We consider now the case of nilpotent Lie algebras. Assume that the algebra L is nilpotent. We assume that the basis $A_1,\ldots,A_n$ of L is chosen in such a way that $\lambda_{jk}^s = 0$ for $s \leqslant k$. In other words, all the matrices Λ_j of associated representation of L are strictly lower-triangular matrices. An important example is given by the so-called stratified Lie algebras. A Lie algebra L is a stratified Lie algebra if $L = \bigoplus_{r=1}^{N} L^r$ (the direct sum of linear subspaces) and

$$[L^r, L^s] \subset L^{r+s} \quad (L^{r+s} \overset{\text{def}}{=} \{0\}, \text{ if } r+s > N), \tag{15}$$

and besides L^1 generates the whole algebra.

If we choose the basis $A_1,\ldots,A_m$ in L^1 and extend it to the basis $A_1, \ldots, A_n$ of the whole algebra, such that $A_j \in L^{r(j)}$, where $r(j)$ is a non-decreasing function, this basis will satisfy the condition formulated above. We mention that in this case the spaces defined by (10) have the form

$$L_r = \bigoplus_{s \leqslant r} L^s. \tag{16}$$

In particular, $L_N = L$.

Theorem 4. Let L be a nilpotent algebra and let $A_1,\ldots,A_n$ be the generators of tempered semigroups in X, satisfying the conditions of Theorem 2 of item B. Then $A_1,\ldots,A_n$ generate tempered semigroups in the scales $H_{A,\rho}$ for any $\rho = \rho_1,\ldots,\rho_n$.

Proof. We consider the case $\rho_1 = \ldots = \rho_n = 1$; the general case is considered in an analogous way (the simple generalization of Theorem 3 is required). Applying Theorem 3, we obtain the desired result. Indeed, all the matrices Λ_j are nilpotent, and so $\| \exp(-\Lambda t) \|$ has not more than polynomial growth at infinity.

Theorem 5. Let L be a stratified Lie algebra $L = \bigoplus_{r=1}^{N} L^r$, satisfying the conditions of Theorem 2 of item B. Let $A_1,\ldots,A_n$ be the basis in L constructed above, and assume that $\exp(itA_1),\ldots,\exp(itA_n)$ are semigroups of tempered growth in X. Then $A_1,\ldots,A_n$ generate tempered semigroups in the scale $H_{(A_1,\ldots,A_m)}$.

Proof. This immediately follows from Theorem 3.

Note. The statement of Theorem 4 is valid for any choice of the basis in L. The special choice of basis given will be used in the following items to construct a regular representation.

D. Regular Representation of a Nilpotent Lie Algebra

In this item we give the explicit expression for regular representation of the nilpotent Lie algebra and prove that these operators are generators in the scale of symbol spaces.

Let Γ be a real n-dimensional Lie algebra with the basis $a_1,\ldots,a_n$ and commutation relations

$$[a_i,a_j] = \sum_{k=1}^{n} \lambda_{ij}^{k} a_k, \quad i,j = 1,\ldots,n. \tag{1}$$

We suppose that Γ is nilpotent and that all the matrices Λ_i,

$$(\Lambda_i)_{kj} = \lambda_{ij}^{k}, \quad i,k,j = 1,\ldots,n, \tag{2}$$

of the associated representation are strictly lower-triangular, i.e., $\lambda_{ij}^{k} = 0$ for $j \geqslant k$. Equivalently, if we denote by $\Gamma_\ell \subset \Gamma$ the linear subspace with the basis $a_\ell, a_{\ell+1}, \ldots, a_n$, then

$$[\Gamma_\ell,\Gamma] \subseteq \Gamma_{\ell+1}, \quad \ell = 1,\ldots,n \tag{3}$$

($\Gamma_{n+1} = 0$ by convention).

Let X be a Banach space, $A_1,\ldots,A_n$ be the generators of tempered semigroups in X, satisfying the conditions of Theorem 2, item B. Thus, A_i are the generators of the strongly continuous representation of the Lie group G correspondent to Γ. By Theorems 4 and 5, item C, these operators are generators of tempered groups in the scales $H_{A,\rho}$ and in $H_{(A_1,\ldots,A_m)}$ (the latter assertion is valid if Γ is a stratified Lie algebra generated by $(A_1,\ldots,A_m)$).

The total space of these scales is X, endowed with the convergence, induced by the norm $\|\cdot\|_X$; thus for any $f \in B$ (the total space of the scale $\{B_\alpha^\beta(\mathbb{R}^n)\}$) the operator $f(\overset{1}{A_1},\ldots,\overset{n}{A_n})$ is defined as

$$f(\overset{1}{A_1},\ldots,\overset{n}{A_n})x = \left(\frac{1}{2\pi}\right)^n F(f)[\exp(it_nA_n)\cdot\ldots\cdot\exp(it_1A_1)x], \quad x \in X, \tag{4}$$

the expression on the right is understood in the sense of regularization given in Definition 3 of Section 3:C, where the convergence is that in the Banach space X.

In order to obtain the operators $L_1,\ldots,L_n$ of left regular representation for the operators $A_1,\ldots,A_n$, we make some preliminary calculations.

Lemma 1. For any $x \in D$, we have

$$A_j \exp(it_n A_n)\cdot\ldots\cdot\exp(it_1 A_1)x = \tag{5}$$
$$= \tilde{L}_j(\exp(it_n A_n)\cdot\ldots\cdot\exp(it_1 A_1)x),\ j = 1,\ldots,n,$$

where

$$\tilde{L}_j = -i \sum_{k=j}^{n} P_{jk}(t_{j+1},\ldots,t_{k-1}) \frac{\partial}{\partial t_k}, \tag{6}$$

where P_{jk} are polynomials with real coefficients, $P_{kk} = 1$ for all k, and p_{jj+1} are constants.

Proof. By Lemma 2 of the item B we have for any $x \in D$

$$-i \frac{\partial}{\partial t_j} U(t)x = \sum_p B_{pj}(t) A_p U(t)x, \tag{7}$$

where $U(t) = \exp(it_n A_n)\cdot\ldots\cdot\exp(it_1 A_1)$,

$$B_{pq}(t) = [e^{t_n \Lambda_n}\cdot\ldots\cdot e^{t_q \Lambda_q}]_{pq}. \tag{8}$$

The lower-triangularity of the matrices Λ_j and the antisymmetry of the structure constants λ^k_{ij} with respect to the lower indices yields that in fact the matrix Λ_j has the form:

$$\Lambda_j = \left(\begin{array}{c|c} 0 & 0 \\ \hline & \begin{matrix} 0 & 0 \\ & \ddots \\ * & 0 \end{matrix} \end{array}\right) \tag{9}$$

where only the elements marked by * in the lower right block of the size $(n - j) \times (n - j)$ may be non-zero. Consequently, the matrix $\exp(t\Lambda_j)$ has the form

$$\exp(t\Lambda_j) = \left(\begin{array}{c|c} \begin{matrix} 1 & & 0 \\ & 1 & \\ 0 & & 1 \end{matrix} & 0 \\ \hline 0 & \begin{matrix} 1 & & 0 \\ & 1 & \\ * & & 1 \end{matrix} \end{array}\right) = \left(\begin{array}{c|c} E_j & 0 \\ \hline 0 & M_{n-j}(t) \end{array}\right) \tag{10}$$

(E_j is the unit matrix of the size $j \times j$, $M_{n-j}(t)$ is the polynomial in t of the order $\leqslant n - j - 1$ with coefficients which are the real matrices of the size $(n - j) \times (n - j)$:

$$M_{n-j}(t) = E_{n-j} + M^{(2)}_{n-j} t + \ldots + M^{(n-j-1)}_{n-j} t^{n-j-1}, \tag{11}$$

where all $M^{(s)}_{n-j}$ are strictly lower-triangular matrices. It follows that $B(t) = (B_{pq}(t))^n_{p,q=1}$ is a lower-triangular matrix with ones on the main diagonal,

$$B(t) = E + R(t), \tag{12}$$

where $R(t)$ is strictly lower-triangular, and $R_{pq}(t)$ is a polynomial in $(t_{q+1}, \ldots, t_{p-1})$ (constant if $p = q + 1$). Thus the matrix $B(t)^{-1}$ exists and

$$B(t)^{-1} = E - R(t) + R(t)^2 - \ldots + (-1)^{n-1} R(t)^{n-1} \tag{13}$$

is lower-triangular and the element $B(t)^{-1}_{pq}$ is a polynomial in $(t_{q+1},\ldots, t_{p-1})$, constant for $p = q + 1$ (the latter statement is easily proved by induction). Setting $P_{jk} = B(t)^{-1}_{kj}$, we obtain the statement of the lemma.

Lemma 2. Let $x \in D$, $F(f) \in S(\mathbb{R}^n)$. Then

$$A_j [\![f(\overset{1}{A}_1,\ldots,\overset{n}{A}_n)]\!] x = (L_j f)(\overset{1}{A}_1,\ldots,\overset{n}{A}_n)x, \tag{14}$$

where

$$\begin{aligned} L_j &= \sum_{k=j}^{n} y_k P_{jk}\left(-i\frac{\partial}{\partial y_{j+1}},\ldots,-i\frac{\partial}{\partial y_{k-1}}\right) = \\ &= y_j + \sum_{k=j+1}^{n} y_k P_{jk}\left(-i\frac{\partial}{\partial y_{j+1}},\ldots,-i\frac{\partial}{\partial y_{k-1}}\right). \end{aligned} \tag{15}$$

Proof. Under assumptions of the lemma, we have

$$\begin{aligned} &A_j [\![f(\overset{1}{A}_1,\ldots,\overset{n}{A}_n)]\!] x = \\ &= A_j (2\pi)^{-n/2}\int F(f)(t)e^{it_nA_n}\cdot\ldots\cdot e^{it_1A_1}x\,dt = \\ &= (2\pi)^{-n/2}\int F(f)A_j e^{it_nA_n}\cdot\ldots\cdot e^{it_1A_1}x\,dt = \\ &= (2\pi)^{-n/2}\int F(f)(t)\tilde{L}_j(e^{it_nA_n}\cdot\ldots\cdot e^{it_1A_1}x)dt = \\ &= (2\pi)^{-n/2}\int ({}^t\tilde{L}_j(F(f)(t))e^{it_nA_n}\cdot\ldots\cdot e^{it_1A_1}x)dt = \\ &= (2\pi)^{-n/2}\int F(L_jf)(t)e^{it_nA_n}\cdot\ldots\cdot e^{it_1A_1}x\,dt = \\ &= (L_jf)(\overset{1}{A}_1,\ldots,\overset{n}{A}_n)x. \end{aligned} \tag{16}$$

Here ${}^t\tilde{L}_j = -\tilde{L}_j$ is the transpose (in L_2) of $\tilde{L}_j$. The lemma is proved.

Theorem 1. The operators L_j given by (15) give the left regular representation of the tuple $A = (A_1,\ldots,A_n)$.

Proof. Passing to the limit in Lemma 2, we conclude that its statement is valid for any $x \in X$. Now let $f \in B$, $x \in D_{f(\overset{1}{A}_1,\ldots,\overset{n}{A}_n)} \cap D_{(L_jf)(\overset{1}{A}_1,\ldots,\overset{n}{A}_n)}$. Then it follows from Definition 3 of Section 3:C that there exists a sequence f_m convergent to f in $S'(\mathbb{R}^n)$ such that $f_m(\overset{1}{A}_1,\ldots,\overset{n}{A}_n)x$ converges to $f(\overset{1}{A}_1,\ldots,\overset{n}{A}_n)x$, $(L_jf_m)(\overset{1}{A}_1,\ldots,\overset{n}{A}_n)x$ converges to $(L_jf)(\overset{1}{A}_1,\ldots,\overset{n}{A}_n)x$, and $F(f_m) \in S(\mathbb{R}^n)$, $m = 1,2,3,\ldots$. We have

$$(L_jf_m)(\overset{1}{A}_1,\ldots,\overset{n}{A}_n)x = A_j [\![f_m(\overset{1}{A}_1,\ldots,\overset{n}{A}_n)]\!] x, \quad m = 1,2,3,\ldots .$$

Since A_j is a closed operator, we may pass to the limit and obtain

$$A_j [\![f(\overset{1}{A}_1,\ldots,\overset{n}{A}_n)]\!] x = (L_jf)(\overset{1}{A}_1,\ldots,\overset{n}{A}_n)x. \tag{17}$$

The theorem is proved.

Now we give the proof of the analogous theorem in the arbitrary scale $X = \{X_\delta\}_{\delta\in\Delta}$.

Theorem 2. Let $A_1,\ldots,A_n$ be generators and right translators in the scale X. Assume that the operators $(iA_1),\ldots,(iA_n)$ form a representation of the Lie algebra Γ. Then the operators (15) form the left regular representation of the tuple $(A_1,\ldots,A_n)$.

Proof. We need only to show that the statement of Lemma 1 is valid for any $x \in X$ under the conditions of the theorem. Then the same takes place for Lemma 2, and the proof goes as in Theorem 1. Next, it is sufficient to prove that Lemma 1 of item B takes place for any $h \in X$ in the situation considered.

Set $Y = X \oplus \ldots \oplus X$ (the direct sum of n copies of the space X endowed with evident convergence). Set also

$$B\otimes = A_1 x \oplus \ldots \oplus A_n x. \tag{18}$$

$B : X \to Y$ is a continuous operator. Besides, we have

$$BA_j = (A_j \otimes I - iI \oplus \Lambda_j)B, \; j = 1,\ldots,n. \tag{19}$$

Since B is continuous, it follows that

$$B\exp(iA_j t) = \exp\{i(A_j \otimes I - iI \otimes \Lambda_j)t\}B \tag{20}$$

(cf. the analogous argument in the proof of Theorem 1 of item A). Thus the desired statement is proved, and Theorem 2 follows.

We establish now the important property of the operators $L = (L_1,\ldots, L_n)$ of the left regular representation of the nilpotent Lie algebras.

Theorem 3. The operators L_j, $j = 1,\ldots,n$ given by (15) are generators of tempered groups in the scale $\{B^{\beta}_{\alpha}(\mathbb{R}^n)\}$.

Proof. In view of embeddings established in Lemma 1 (b) of Section 3:A, it is enough to prove that L_j are the generators of tempered groups in the scale $\{H^{\beta}_{\alpha}(\mathbb{R}^n)\}$. Passing to Fourier transforms, we come to the problem of establishing that $\tilde{L}_j$, $j = 1,\ldots,n$ given by (6) are generators in the scale $\{H^{\beta}_{\alpha}(\mathbb{R}^n)\}$. The operators $\exp(i\tau\tilde{L}_j)$ may be calculated explicitly.

Lemma 3. For any function $\phi(t) \in S'(\mathbb{R}^n)$, we have

$$\exp(i\tau\tilde{L}_j)\phi(t) = \phi(T_{(j)}(t,\tau)), \tag{21}$$

where $T_{(j)}(t,\tau)$ is defined as the solution of the system of ordinary differential equations

$$\dot{T}^k_{(j)} = P_{jk}(T^{j+1}_{(j)},\ldots,T^{k-1}_{(j)}); \; k = 1, \ldots, n \tag{22}$$

(here we set $P_{jk} \equiv 0$ for $k < j$).

Proof. Indeed, $\tilde{L}_j$ is a vector field, and $F(t,\tau) = \exp(i\tau\tilde{L}_j)\phi(t)$ satisfies the linear partial differential equation of the first order

$$i\frac{\partial}{\partial\tau}F + \tilde{L}_j F = 0. \tag{23}$$

Solving (23) by the characteristics method, we come to the system (22). Next, since the coefficients of the vector field L_j have the special form (6), the solution of (22) may be given more explicitly. Namely,

$$T^k_{(j)}(t,\tau) = t_k, \; k < j$$
$$T^k_{(j)}(t,\tau) = t_k + Q_{jk}(t_{j+1},\ldots,t_{k-1},\tau), \tag{24}$$

where Q_{jk} are polynomials. Thus, the Jacobian

$$\det \frac{\partial T_{(j)}}{\partial t}(t,\tau) \equiv 1 \tag{25}$$

for all τ, t, so that $\exp(i\tau\tilde{L}_j)$ is a unitary group in $L_2(\mathbb{R}^n)$, and all the derivatives of $T^k_{(j)}$ have the polynomial estimate of growth with respect to all the variables. These properties yield the statement of the theorem immediately. Theorem 3 is proved.

E. Pseudo-Differential Operators in Spaces of Smooth Functions

In the sequel (namely, in the chapters concerned with construction of asymptotics), we shall need the developed calculus of pseudo-differential operators defined in spaces of smooth functions (which will be symbol spaces). The results presented below contain somewhat more information than we could obtain directly from general theorems proved in this chapter and are derived in a slightly different way. Our aim is to expose results which arise in asymptotics theory.

We introduce the scale $\{H^k_\ell\}_{k,\ell\in\mathbb{R}}$. The space $H^k_\ell(\mathbb{R}^n)$ is the completion of $S(\mathbb{R}^n)$ with respect to the norm

$$\|u\|_{H^k_\ell} = \left(\int (1+|x|^2)^{-\ell}\,|(1-\Delta)^{k/2}u(x)|^2 dx\right)^{1/2}. \tag{1}$$

Consider the algebra of smooth functions of tempered growth

$$H(\mathbb{R}^n) = \bigcup_\ell \bigcap_k H^k_\ell(\mathbb{R}^n). \tag{2}$$

This is an algebra with unity over the field C. When it will not cause confusion, the argument $(\mathbb{R}^n)$ will be omitted from the notation.

It is clear that $H(\mathbb{R}^n)$ is nothing else but the space $S^\infty(\mathbb{R}^n)$ introduced in Section 3:A. We also could make use of the spaces $H^\beta_\alpha(\mathbb{R}^n)$ instead of $H^k_\ell(\mathbb{R}^n)$, but the latter turn out to be more convenient.

We give the definition of convergence in H. A generalized sequence of functions $\{f_m\} \subset H$ converges to zero, if $\exists \ell\, \forall k : \|f_m\|_{H^k_\ell} \to 0$. The convergence introduced is compatible with algebraic operations in H and also separable (i.e., the generalized sequence may have at most one limit). The proof is almost the same as in Proposition 2 of Section 3:C.

Consider the algebra $\mathcal{L}H \equiv \mathcal{L}H(\mathbb{R}^n)$ of all continuous linear operators in $H(\mathbb{R}^n)$. We let $\mathcal{L}(B_1 \to B_2)$ denote the Banach space of all continuous linear mappings of a Banach space B_1 into a Banach space B_2.

<u>Lemma 1</u>. The algebra $\mathcal{L}H$ is of the form

$$\mathcal{L}H = \bigcap_\ell \bigcup_r \bigcap_s \bigcup_k \mathcal{L}(H^k_\ell \to H^s_r). \tag{3}$$

The equality (3) should be considered in the sense that each operator $T \in \mathcal{L}H$ has the following property: $\forall \ell \exists r\; \forall s \exists k$ such that T can be extended to a continuous linear mapping from H_ℓ^k into H_r^s. Since $S(\mathbb{R}^n)$ is dense in all the spaces of the scale $\{H_\ell^k\}$, this extension is unique, so there will be no confusion.

Definition 1. The elements of the algebra LH will be called operators (in the space H).

Proof of Lemma 1. Let $T \in LH$. Assume that T does not belong to the right-hand side of (3). This means that for some fixed $\forall r \, \exists_{s=s(r)} \forall k$, T may not be extended to a continuous linear operator from H_ℓ^k to $H_r^{s(r)}$. Consider the directed set Δ of triplets (r,m,ε), where $r \in \mathbb{R}$, $m \in \mathbb{Z}_+$ and $\varepsilon \in \mathbb{R}_+ \setminus \{0\}$ with the ordering relation defined by $(r,m,\varepsilon) \leqslant (r',m',\varepsilon')$ if $m \leqslant m'$ and $\varepsilon \geqslant \varepsilon'$. Since $H_\ell^k \subset H_\ell^{k'}$ for $k \geqslant k'$ and this embedding is continuous, the above argument yields that for any $(r,m,\varepsilon) \in \Delta$, we may find $f = f_{rm\varepsilon} \in H$ with the properties:

(a) $f_{rm\varepsilon} \in H_\ell^m$;

(b) $\| f_{rm\varepsilon} \|_{H_\ell^k} \leqslant \varepsilon$ for all integers k such that $0 \leqslant k \leqslant m$;

(c) $\| Tf_{rm\varepsilon} \|_{H_r^{s(r)}} \geqslant 1$ or $Tf_{rm\varepsilon} \notin H_r^{s(r)}$.

The generalized sequence $\{f_{rm\varepsilon}\}_{(rm\varepsilon)\in\Delta}$ converges to zero. Indeed, for any $k \in \mathbb{R}$, we have

$$\| f_{rm\varepsilon} \|_{H_\ell^k} \leqslant C\delta \tag{4}$$

for $\varepsilon \leqslant \delta$, $m \geqslant k_o$, where $k_o \in \mathbb{Z}_+$ is any fixed element greater than k, C is the norm of embedding operator $H_\ell^{k_o} \subset H_\ell^k$. On the other hand, (c) means that $Tf_{rm\varepsilon}$ is not convergent to zero so T is not continuous. The obtained contradiction proves the lemma since the inclusion of the right-hand side of (3) into LH is evident. There is a natural convergence in LH: the generalized sequence $\{T_m\} \subset LH$ converges to zero if $\forall \ell \, \exists r \, \forall s \, \exists k$ so that $T_m \to 0$ in $L(H_\ell^k \to H_r^s)$. Since the space H is not covered by theorems of Section 3, we need to construct the functional calculus in H independently.

Definition 2. If in the algebra LH we are given a one-parameter multiplicative group $\{e_t, t \in \mathbb{R}\}$ which is differentiable and grows slowly as $|t| \to \infty$, that is, $\forall \ell \, \exists r \, \forall s \, \exists k \, \exists p > 0 \, \forall t \in \mathbb{R}$, we have $\| e_t \|_{H_\ell^k \to H_r^s} \leqslant C(1 + |t|)^p$, then the operator

$$A = -i \left. \frac{de_t}{dt} \right|_{t=0} \tag{5}$$

will be called a generator in H. We denote $e_t = e_t(A)$.

Theorem 1. The operators of multiplication by x_j and differentiation $\hat{p}_j = -ih(\partial/\partial x_j)$ are generators in $H(\mathbb{R}^n)$.* Let ϕ be a smooth real function on $\mathbb{R}^n$, all of whose derivatives are bounded. Then for an arbitrary $j = 1, \ldots, n$ the operator $\hat{p}_j + \phi(x)$ is a generator in $H(\mathbb{R}^n)$.

Proof. The groups generated by the operators enumerated in the theorem have the form

* Here $h \neq 0$ is a real parameter.

$$e_t(x_j)f(x) = e^{itx_j}f(x), \quad f \in H, \tag{6}$$

(the multiplication operator)

$$e_t(\hat{p}_j)f(x) = f(x_1,\ldots,x_{j-1},x_j + ht, x_{j+1},\ldots,x_n), \quad f \in H, \tag{7}$$

(the "shift" operator)

$$e_t(\hat{p}_j + \phi(x))f(x) = \\ = \exp\{i\int_0^t \phi(x_1,\ldots,x_j + \tau h,\ldots,x_n)d\tau\} e_t(\hat{p}_j)f(x), \quad f \in H. \tag{8}$$

The required estimates follow immediately from these explicit formulae. The theorem is proved.

Let $A_1,\ldots,A_m$ be generators in $H(\mathbb{R}^n)$, and let $\pi_1,\ldots,\pi_m$ be real numbers such that $\pi_j \neq \pi_k$ for $j \neq k$. The numbers can be written in increasing order: $\pi_{j_1} < \ldots < \pi_{j_m}$. We consider the product of the groups corresponding to $A_1,\ldots,A_m$ arranged in the same order, and apply the Fourier transform of the function f to this product:

$$f(\overset{\pi_1}{A_1},\ldots,\overset{\pi_m}{A_m}) = (2\pi)^{-m/2}\mathcal{F}(f)(e_{t_{j_m}}(A_{j_m})\cdot\ldots\cdot e_{t_{j_j}}(A_{j_1})). \tag{9}$$

Lemma 2. The mapping $f \to (\overset{\pi_1}{A_1},\ldots,\overset{\pi_m}{A_m})$ is continuous and linear from $H(\mathbb{R}^m)$ into $\mathcal{L}H(\mathbb{R}^n)$. In other words, $\forall q\ \forall \ell\ \exists r\ \forall s\ \exists k\ \exists p\ \exists C \geqslant 0$, $\forall f \in H_q^p(\mathbb{R}^m)$ we have the estimate

$$\| f(\overset{\pi_1}{A_1},\ldots,\overset{\pi_m}{A_m}) \|_{H_\ell^k(\mathbb{R}^n) \to H_r^s(\mathbb{R}^n)} \leqslant C \| f \|_{H_q^p(\mathbb{R}^m)}. \tag{10}$$

Proof. As in Lemma 1, we show that the system of estimates (10) is equivalent to the continuity of the mapping $f \to f(\overset{\pi_1}{A_1},\ldots,\overset{\pi_m}{A_m})$. Next, we prove that $\forall q\ \forall \ell\ \exists r\ \forall s\ \exists k\ \exists p$:

$$e_{t_{j_m}}(A_{j_m})\cdot\ldots\cdot e_{t_{j_1}}(A_{j_1}) \in H_p^q(\mathbb{R}^m, \mathcal{L}(H_\ell^k(\mathbb{R}^n) \to H_r^s(\mathbb{R}^n))) \tag{11}$$

(we denote by $H_p^q(\mathbb{R}^m,\mathcal{B})$, where $\mathcal{B}$ is a Banach space, the space of $\mathcal{B}$-valued functions on $\mathbb{R}^n$ with finite norm $\| \ \|_{H_p^q}$). Once (11) is proved, the statement of the lemma is obtained immediately; indeed, it is an easy exercise to verify that the Fourier transform maps $H_q^p(\mathbb{R}^m)$ onto $H_{-p}^{-q}(\mathbb{R}^m)$ continuously and that the continuous pairing $H_{-p}^{-q}(\mathbb{R}^m) \times H_p^q(\mathbb{R}^m,\mathcal{B}) \to \mathcal{B}$ (which arises from L_2 scalar product) is well defined. To prove (11), consider the derivative of the group product of the order $\alpha = \alpha_1,\ldots,\alpha_m$. It has the form

$$A_{j_m}^{\alpha_{j_m}} e_{t_{j_m}}(A_{j_m})\cdot\ldots\cdot A_{j_1}^{\alpha_{j_1}} e_{t_{j_1}}(A_{j_1}) \equiv B_1(t)\cdot\ldots\cdot B_r(t), \tag{12}$$

where $r = m + |\alpha|$, and $B_i(t)$, $i = 1,\ldots,r$ are operators satisfying the property

$$\forall \ell\ \exists r\ \forall s\ \exists k\ \exists p\ \exists C : \| B(t) \|_{H_\ell^k \to H_r^s} \leqslant C(1 + |t|)^p. \tag{13}$$

Thus all we need is to prove that the product of the operators satisfying (13) also satisfies (13); then (11) will follow immediately. Let $B_i(t)$, $i = 1,2$, satisfy (13). This may be rewritten in the more compact form

$$\| B_i(t) \|_{H_\ell^{k_i(\ell,s)} \to H^s_{r_i(\ell)}} \leqslant C_i(\ell,s)(1+|t|)^{p_i(\ell,s)}. \tag{14}$$

Set $B(t) = B_1(t)B_2(t)$. Then

$$\| B(t) \|_{H_\ell^{k(\ell,s)} \to H^s_{r(\ell)}} \leqslant C(\ell,s)(1+|t|)^{p(\ell,s)}, \tag{15}$$

where $r(\ell) = r_1(r_2(\ell))$, $k(\ell,s) = k_2(\ell,k_1(r_2(\ell),s))$, $C(\ell,s) = C_1(r_2(\ell),s) \times \times C_2(\ell,k_1(r_2(\ell),s))$, $p(\ell,s) = p_1(r_2(\ell),s) + p_2(\ell,k_1(r_2(\ell),s))$. The lemma is proved.

Definition 3. The mapping $f \to f(\overset{\pi_1}{A_1},\ldots,\overset{\pi_m}{A_m})$ will be called the ordered quantization corresponding to the ordered set of operators which are $\overset{\pi_1}{A_1},\ldots,\overset{\pi_m}{A_m}$. The numbers $\pi_1,\ldots,\pi_m$ indicate the order of the action of the operators in this set; an operator with a lower index acts first. The function f is called the symbol of the operator $f(\overset{\pi_1}{A_1},\ldots,\overset{\pi_m}{A_m})$.

Now we come to the investigation of pseudo-differential operators in $H(\mathbb{R}^n)$. Let $f \in H(\mathbb{R}^n_x \times \mathbb{R}^n_p)$.

Definition 4. An operator

$$f(\overset{2}{x},-ih\overset{1}{\frac{\partial}{\partial x}}) = f(\overset{2}{x_1},\ldots,\overset{2}{x_n},-ih\overset{1}{\frac{\partial}{\partial x_1}},\ldots,-ih\overset{1}{\frac{\partial}{\partial x_n}}) \tag{16}$$

is called a $1/h$ pseudo-differential operator. (Here we put coinciding numbers over the operators which commute with each other; it is easy to see that the notation is correct.)

We consider operators depending on the parameter h since this will be useful in the asymptotic theory and we should like to present the dependence on h more explicitly.

Lemma 3. The operator (16) acts according to the formula

$$f(\overset{2}{x},-ih\overset{1}{\frac{\partial}{\partial x}})u(x) = \overset{-1/h}{F}_{p\to x} f(x,p) F^{1/h}_{y\to p} u(y), \quad u \in S(\mathbb{R}^n), \tag{17}$$

where

$$\begin{aligned} F^{1/h}_{y\to p} u &= (2\pi h)^{-n/2} \int e^{-(i/h)\langle y,p\rangle} u(y)\,dy, \\ \overset{-1/h}{F}_{p\to x} v &= (2\pi h)^{-n/2} \int e^{(i/h)\langle x,p\rangle} v(p)\,dp \end{aligned} \tag{18}$$

are the $1/h$-Fourier transform and its inverse, respectively. Lemma 3 is proved by standard calculations using formulae (6) and (7).

Consider the function $e_p(x) \equiv \exp\{(i/h)xp\}$, which for each $p \in \mathbb{R}^n$ belongs to $H(\mathbb{R}^n)$. Let T be an arbitrary operator in $LH(\mathbb{R}^n)$. We associate with it a function of the variables $p,x \in \mathbb{R}^n$ by means of the formula

$$\mathrm{Smb}\{T\} \overset{\mathrm{def}}{=} e_{-p}(x)(Te_p)(x). \tag{19}$$

Definition 5. The function defined by (19) is called the symbol of the operator T. We note that if

$$T = f(\overset{2}{x}, -ih \overset{1}{\frac{\partial}{\partial x}})$$

is a 1/h pseudo-differential operator, then $\mathrm{Smb}\{T\} = f$ (it follows from Lemma 3). It is natural to assume that an analogous assertion holds in the general case:

$$T = \mathrm{Smb}\{T\}(\overset{2}{x}, -ih \overset{1}{\frac{\partial}{\partial x}}). \tag{20}$$

(20) is valid in the case $\mathrm{Smb}\{T\} \in H(\mathbb{R}^n_x \times \mathbb{R}^n_p)$. Indeed, the operator $T - \mathrm{Smb}\{T\}(\overset{2}{x}, -ih \overset{1}{\frac{\partial}{\partial x}})$ annuls the function $e_p(x)$ for any $p \in \mathbb{R}^n$, and the set of linear combinations of functions $e_p(x)$ is dense in $H(\mathbb{R}^n)$. But $\mathrm{Smb}\{T\}$ belongs, in general, to a broader algebra of functions. We will describe it next.

Let $\underline{H}^{s_1,s_2}_{r_1,r_2}(\mathbb{R}^n_x \times \mathbb{R}^n_p)$ denote the completion of $C^\infty_0(\mathbb{R}^n_x \times \mathbb{R}^n_p)$ with respect to the norm:

$$\| f \|_{H^{s_1,s_2}_{r_1,r_2}} = (\iint (1+|p|^2)^{-r_2}(1+|x|^2)^{-r_1} |(1-\Delta_p)^{s_2/2}(1-\Delta_x)^{s_1/2} \times \times f(x,p)|^2 dp dx)^{1/2}. \tag{21}$$

We set

$$\mathcal{H}(\mathbb{R}^n_x \times \mathbb{R}^n_p) \overset{\mathrm{def}}{=} \bigcap_{s_2} \bigcup_{r_1} \bigcap_{s_1} \bigcup_{r_2} H^{s_1,s_2}_{r_1,r_2}(\mathbb{R}^n_x \times \mathbb{R}^n_p). \tag{22}$$

In this algebra of functions we introduce a natural convergence by analogy with the above constructions. Clearly $H(\mathbb{R}^n_x \times \mathbb{R}^n_p) \subset \mathcal{H}(\mathbb{R}^n_x \times \mathbb{R}^n_p)$, where the embedding is continuous.

First suppose that $T \in \mathcal{L}H(\mathbb{R}^n)$. It is easy to show that the function $\mathrm{Smb}\{T\}$, defined by (19), belongs to the algebra $\mathcal{H}(\mathbb{R}^n_x \times \mathbb{R}^n_p)$. It turns out that each element in $\mathcal{H}(\mathbb{R}^n_x \times \mathbb{R}^n_p)$ can be regarded as the symbol of some 1/h pseudo-differential operator.

Theorem 2. The mapping

$$f \to f(\overset{2}{x}, -ih \frac{\partial}{\partial x})$$

can be extended to a linear homeomorphism

$$\mu : \mathcal{H}(\mathbb{R}^n_x \times \mathbb{R}^n_p) \to \mathcal{L}H(\mathbb{R}^n) \tag{23}$$

of the algebra of symbols onto the algebra of operators. The inverse mapping is given by the formula $\mu^{-1}(T) = \mathrm{Smb}\{T\}$.

Proof. Since for $f \in L_2(\mathbb{R}^n_x \times \mathbb{R}^n_p)$ the operator

$$\hat{f} \equiv f(\overset{2}{x}, -ih \overset{1}{\frac{\partial}{\partial x}})$$

is a Hilbert-Schmidt operator in $L^2(\mathbb{R}^n)$, we have

$$\| \hat{f} \|_{L^2(\mathbb{R}^n) \to L^2(\mathbb{R}^n)} \leqslant [\mathrm{Tr}(\hat{f} * \hat{f})]^{1/2} = \frac{1}{(2\pi h)^{n/2}} \| f \|_{L^2(\mathbb{R}^n_x \times \mathbb{R}^n_p)}. \tag{24}$$

As a consequence of this estimate in the case of a scale $\{H^k\}$ we have the inequality

$$\| \hat{f} \|_{H_\ell^{r+k}(\mathbb{R}^n) \to H_{\ell+s}^r(\mathbb{R}^n)} \leqslant C_{r,\ell,s,k} \| f \|_{H_{s,k}^{r,\ell}(\mathbb{R}_x^n \times \mathbb{R}_p^n)}, \tag{25}$$

which holds for all $f \in H_{s,k}^{r,\ell}$.

In $H(\mathbb{R}_x^n \times \mathbb{R}_p^n)$ we consider the convergence which is induced by $\mathcal{H}(\mathbb{R}_x^n \times \mathbb{R}_p^n)$. $H(\mathbb{R}_x^n \times \mathbb{R}_p^n)$ is a dense subset of $\mathcal{H}(\mathbb{R}_x^n \times \mathbb{R}_p^n)$, and (25) yields that $\mu : \mathcal{H} \to \mathcal{L}H(\mathbb{R}^n)$ (defined on this dense subset) is continuous. Since $\mathcal{L}H(\mathbb{R}^n)$ is a complete space (the verification of this fact is any easy exercise), the mapping μ can be extended to a continuous linear mapping between these spaces.

We will show that μ is a homeomorphism; that is, the inverse mapping $\mu^{-1} = \mathrm{Smb} : \mathcal{L}H \to \mathcal{H}$ is continuous. To do this we establish the following estimate on the symbol of an operator T in the algebra $\mathcal{L}H(\mathbb{R}^n)$: $\forall \ell \in 2\mathbb{Z}_+$ $\exists s \in \mathbb{R}$ $\forall r \in 2\mathbb{Z}_+$ $\exists k \in \mathbb{R}$:

$$\| \mathrm{Smb}\{T\} \|_{H_{s+\ell,r-k}^{r,\ell}(\mathbb{R}_x^n \times \mathbb{R}_p^n)} \leqslant C_{r,\ell,s,k}^{(1)} \| T \|_{H_{r+n}^{k+n}(\mathbb{R}^n) \to H_s^r(\mathbb{R}^n)}. \tag{26}$$

Here $2\mathbb{Z}_+$ is the set of non-negative even integers. This estimate is equivalent to the assertion of continuity of the mapping $\mathrm{Smb} : \mathcal{L}H \to \mathcal{H}$.

We now prove (26). Since $T \in \mathcal{L}H$, we know that

$$\forall \ell \in 2\mathbb{Z}_+ \ \exists s \in \mathbb{R} \ \forall r \in 2\mathbb{Z}_+ \ \exists k \in \mathbb{R} : \| T \|_{H_{r+n}^{k+n} \to H_s^r} < \infty.$$

We set $\mathrm{Smb}\{T\} = f$; then

$$T = f(\overset{2}{x}, -ih \overset{1}{\frac{\partial}{\partial x}}) = \hat{f}.$$

From (24) we find that

$$\| f \|^2_{H_{s+\ell,r-k}^{r,\ell}} = (2\pi h)^n \mathrm{Tr}(\hat{f}_1, \hat{f}_1^*),$$

where

$$f_1(x,p) = (1 + |x|^2)^{-(r+s)/2}(1 + |p|^2)^{(k-r)/2}(1 - \Delta_x)^{r/2}(1 - \Delta_p)^{\ell/2} f(x,p).$$

Consequently

$$\| f \|_{H_{s+\ell,r-k}^{r,\ell}} \leqslant C_{\ell,r,k,s}^{(2)} \| \hat{f}_1 \circ (1 - \Delta)^{n/2} \circ (1 + |x|^2)^{n/2} \|_{L^2 \to L^2}.$$

In the scale $\{H_\ell^k\}$ we have to estimate the operator $\hat{f}_1$ in terms of the norm of $\hat{f}$. This is easy to do by using the formula which connects the symbol f_1 to the symbol f (we choose $r, \ell \in 2\mathbb{Z}_+$ so that f_1 is expressed in terms of the derivatives of f of integer order). As a result we obtain

$$\| f \|_{H_{s+\ell,r-k}^{r,\ell}} \leqslant C_{\ell,r,k,s}^{(1)} \| \hat{f} \|_{H_{\ell+n}^{k+n} \to H_s^r}.$$

The theorem is proved.

By virtue of Theorem 2 the linear structure and the structure of the convergence in $H(\mathbb{R}^n_x \times \mathbb{R}^n_p)$ and in $LH(\mathbb{R}^n)$ are isomorphic. But the algebraic structures of these spaces are distinct. By using the homeomorphism μ we can transfer from LH to H the non-commutative multiplication $*$, with respect to which μ is an isomorphism of algebras.

Definition 6. The product

$$f * g = \mathrm{Smb}\{[f(\overset{2}{x},-ih\overset{1}{\frac{\partial}{\partial x}})] \circ [g(\overset{2}{x},-ih\overset{1}{\frac{\partial}{\partial x}})]\} \tag{27}$$

is called the twisted product of the functions f and g.

We consider the algebra $LH(\mathbb{R}^n_x \times \mathbb{R}^n_p)$ of continuous mappings from $H(\mathbb{R}^n_x \times \mathbb{R}^n_p)$ into itself. We introduce the mapping $\nu : LH \to H$ which associates with each continuous operator Q in H the function $\nu(Q) \overset{\mathrm{def}}{=} Q(1)$, where 1 is the identity in H. Thus the mapping ν applies the operator to the identity.

We note that, analogously to (3),

$$LH(\mathbb{R}^n_x \times \mathbb{R}^n_p) = \bigcap_{s_2} \bigcup_{k_2} \bigcap_{\ell_1} \bigcup_{r_1} \bigcap_{s_1} \bigcup_{k_1} \bigcap_{\ell_2} \bigcup_{r_2} L(H^{k_1,k_2}_{\ell_1,\ell_2} \to H^{s_1,s_2}_{r_1,r_2}). \tag{28}$$

Therefore we can define a homeomorphism

$$j : LH(\mathbb{R}^n) \to LH(\mathbb{R}^n_x \times \mathbb{R}^n_p) \tag{29}$$

by means of the formula

$$j(T) = [e^{-(i/h)xp}]T_x[e^{(i/h)xp}], \tag{30}$$

where T_x is the operator T acting with respect to the variable x in $H(\mathbb{R}^n_x \times \mathbb{R}^n_p)$, and $[e^{ixp/h}]$ is the operator of multiplication by the exponential. From our definitions, we obtain the following assertion.

Lemma 4. (a) The following diagram of an algebra homeomorphism is commutative:

$$\begin{array}{ccc} & \overset{\mu}{\nearrow} & LH(\mathbb{R}^n) \\ H(\mathbb{R}^n_x \times \mathbb{R}^n_p) & & \downarrow j \\ & \underset{\nu}{\nwarrow} & LH(\mathbb{R}^n_x \times \mathbb{R}^n_p) \end{array}$$

Here H is regarded as an algebra with the twisted product $*$.

(b) We have the formulas

$$j\mu(f) = f(\overset{2}{x},p - ih\overset{1}{\frac{\partial}{\partial x}}), \quad (f * g)(x,p) = [j\mu(f)]g(x,p). \tag{31}$$

These formulas give us a method of calculating the symbol of a composition of operators in the algebra $LH(\mathbb{R}^n)$. The proof is obvious.

F. Some Estimates for Functions of a Tuple of Non-Commuting Self-Adjoint Operators

In this item we consider estimates for functions of a n-tuple $A_1,\dots,A_n$ of self-adjoint operators in a Hilbert space for which a regular representation exists. As it was shown in this chapter, an operator $f(\overset{1}{A_1},\dots,\overset{n}{A_n})$ is bounded without any additional assumptions, if $f \in B_0(\mathbb{R}^n)$. However, in applications this requirement is often too restrictive; in particular, a smooth function $f(y_1,\dots,y_n)$ homogeneous of degree zero for large $|y|$, does not necessarily belong to $\mathcal{B}_0$.

It is well known that pseudo-differential operators with symbols from $S^m_{\rho,\delta}(\mathbb{R}^n \times \mathbb{R}^n)$ are bounded in the space $L^2(\mathbb{R}^n)$ for $m = 0$ (see, for example, [30,45]). The boundedness of $f(\overset{1}{A}_1,\dots,\overset{n}{A}_n)$ for symbols of a functional class wider than $\beta_0(\mathbb{R}^n)$ follows in this and many other cases from:

(a) the existence of non-trivial commutation relations;
(b) the self-adjointness of operators $A_1,\dots,A_n$.

We assume throughout this item that the left regular representation exists. In terms of this representation we formulate a sufficient condition, under which the operators with symbols from $S^m_{\rho,0}(\mathbb{R}^{n-k} \times \mathbb{R}^k)$ are bounded in the scale of spaces, generated by operators $(A_1,\dots,A_k)$.

Being combined with the methods of [52] and [39], the results of this item yield the proof of boundedness of operators $f(\overset{1}{A}_1,\dots,\overset{n}{A}_n)$ with oscillating symbols of the form $f = \exp(iS)\phi$. However, we do not dwell on these questions.

We establish some estimates for functions of tuples of Feynman-ordered non-commuting self-adjoint operators in a Hilbert space. Further, by saying that self-adjoint operators A and B commute, we mean in fact that their spectral families commute:

$$E_\lambda(A)E_\mu(B) - E_\mu(B)E_\lambda(A) = 0, \quad \lambda,\mu \in \mathbb{R}. \tag{1}$$

Our basic aim is to establish the estimates for functions of tuples, for which a regular representation exists. However, we first prove some preliminary results.

Let H be a Hilbert space and $A_1,\dots,A_n$ be (unbounded) self-adjoint operators in H. From now on we assume that the intersection $D_{A_1} \cap \dots \cap D_{A_n}$ of domains of operators $A_1,\dots,A_n$ contains a linear subset D, which is dense in H and invariant under operators A_j and corresponding unitary groups $\exp(itA_j)$, $j = 1,\dots,n$; further, we assume that for any sequence of indices $j_1,\dots,j_m \in \{1,\dots,n\}$ the product of groups

$$\exp(it_m A_{j_m}) \times \dots \times \exp(it_1 A_{j_1})$$

is infinitely differentiable with respect to $t \in \mathbb{R}^m$ in the strong sense on the domain D, and all its derivatives grow no faster than the powers of t:

$$\left\| \frac{\partial^{|\alpha|}}{\partial t^\alpha} (\exp(it_m A_{j_m}) \times \dots \times \exp(it_1 A_{j_1})u) \right\| \leqslant C(1 + |t|)^N, \quad u \in D, \tag{2}$$

where N may depend on $|\alpha|$, and C depends on $|\alpha|$ and on the choice of $u \in D$. Under these conditions the formula

$$f(A) \equiv f(\overset{1}{A}_1,\dots,\overset{n}{A}_n) = \langle \tilde{f}, Q_A \rangle, \tag{3}$$

where

$$\tilde{f}(t) = (2\pi)^{-n} \int f(\xi) e^{-it\xi} d\xi, \quad t,\xi \in \mathbb{R}^n, \tag{4}$$

is the Fourier transform of the function f,

$$Q_A \equiv Q_A(t) = \exp(it_n A_n) \times \dots \times \exp(it_1 A_1), \tag{5}$$

and the brackets $\langle , \rangle$ denote the pairing of a distribution with a test function in the weak topology on D, defines for any $f \in S^{\infty}(\mathbb{R}^n)$ a linear operator $f(A)$ on a dense linear subset $D_{\infty} \supset D$, invariant under all operators $f(A)$, $f \in S^{\infty}(\mathbb{R}^n)$. The operator (3) is correctly defined and bounded (for any self-adjoint operators $A_1, \ldots, A_n$) if $f \in \mathcal{B}_0(\mathbb{R}^n)$, i.e., f belongs to the space of Fourier transforms of the limits of finite functionals over the space $C(\mathbb{R}^n)$ of continuous functions bounded in $\mathbb{R}^n$ with the norm $\| f \|_C = \sup_{y \in \mathbb{R}^n} f(y)$ (see Section 3 of this Chapter).

Under these assumptions operator (3) can be rewritten as a multiple Stiltjes integral

$$f(\overset{1}{A_1}, \ldots, \overset{n}{A_n}) = f(\lambda_1, \ldots, \lambda_n) dE_{\lambda_n}(A_n) \times \ldots \times dE_{\lambda_1}(A_1), \tag{6}$$

where $dE_{\lambda}(A_i)$ is the spectral measure of operator A_i; the integral is considered in the sense of weak convergence in H. Formula (6) makes sense for a wider class of symbols than formula (3); however, one should remember that for $f \notin S^{\infty}(\mathbb{R}^n)$, operator (6) need not necessarily be densely defined.

Theorem 1. Let operators $A_{k+1}, \ldots, A_n$ be pairwise commutative, and the function $f(y_1, \ldots, y_n)$ satisfy the estimates

$$\left| \frac{\partial^{\alpha_1 + \ldots + \alpha_k}}{\partial y_1^{\alpha_1} \ldots \partial y_k^{\alpha_k}} f(y_1, \ldots, y_n) \right| \leqslant C(1 + |y_1| + |y_2|)^{-k-\varepsilon} \tag{7}$$

for some $\varepsilon > 0$ and all $\alpha = (\alpha_1, \ldots, \alpha_k)$ with $|\alpha| = \alpha_1 + \ldots + \alpha_k \leqslant k + 1$. Then the operator $f(\overset{1}{A_1}, \ldots, \overset{n}{A_n})$ is bounded in H (and hence can be extended by continuity on the whole space H).

We divide the proof into several lemmas:

Lemma 1. Let self-adjoint operators $B_1, \ldots, B_m$ be pairwise commutative in a Hilbert space H, and let the sequence $g_n(x)$ of continuous bounded functions in $x = (x_1, \ldots, x_m) \in \mathbb{R}^m$ be uniformly bounded (i.e., $|g_n(x)| \leqslant M$ for all n, x) and converge to the function $g(x)$ as $n \to \infty$ locally uniformly with respect to $x \in \mathbb{R}^m$. Then the sequence of operators $g_n(B_1, \ldots, B_m)$ converges to the operator $g(B_1, \ldots, B_n)$ as $n \to \infty$ in the strong sense.

Proof. First of all since spectral measures $dE_{\lambda_1}(B_1), \ldots, dE_{\lambda_m}(B_m)$ commute, we have the estimate

$$\| g_n(B_1, \ldots, B_m) \| \leqslant \sup_{x \in \mathbb{R}^m} |g_n(x)| \leqslant M; \tag{8}$$

hence it is sufficient to verify the convergence on some dense subset of H. We choose this subset to consist of the following vectors

$$u = E_{\Delta_1}(B_1) E_{\Delta_2}(B_2) \ldots E_{\Delta_m}(B_m) v, \quad v \in H, \tag{9}$$

where $\Delta_1, \ldots, \Delta_m$ are compact Borel subsets of the real axis. If u has the form (9), then

$$\| g_n(B_1, \ldots, B_m) u - g(B_1, \ldots, B_m) u \| \leqslant \sup_{x \in \Delta_1 \times \ldots \times \Delta_m} |g_n(x) - g(x)| \tag{10}$$

and, consequently, $g_n(B_1, \ldots, B_m) u \to g(B_1, \ldots, B_m) u$. The lemma is proved.

Lemma 2. Under conditions of Theorem 1 the Fourier transform $\tilde{\phi}_v(t_1, \ldots, t_k)$ of the H-valued function

$$\phi_v(y_1,\ldots,y_k) = \bar{f}(y_1,\ldots,y_k,A_{k+1},\ldots,A_n)v, \quad v \in H, \tag{11}$$

is a continuous function and

$$\int_{\mathbb{R}^k} \| \tilde{\phi}_v(t_1,\ldots,t_k) \| dt_1 \ldots dt_k \leqslant C \| v \|, \tag{12}$$

where the constant C is independent of v.

Proof. Since $f(y_1,\ldots,y_n)$ is continuous and bounded with respect to all the variables, Lemma 1 yields by estimate (7) that $\phi_v(y_1,\ldots,y_k)$ is a continuous summable function; hence the Fourier transform exists:

$$\tilde{\phi}_v(t_1,\ldots,t_k) = (2\pi)^{-k} \int \phi_v(y_1,\ldots,y_k) e^{-ity} dy_1 \ldots dy_k, \tag{13}$$

and it is a bounded continuous H-valued function. Further,

$$t^\alpha \tilde{\phi}_v(t_1,\ldots,t_k) = (2\pi)^{-k} \int [(-i \frac{\partial}{\partial y})^\alpha \phi_v(y_1,\ldots,y_k)] e^{-ity} dy_1 \ldots dy_k \tag{14}$$

for any multi-index α with $|\alpha| \leqslant k + 1$ (the differentiability of $\phi_v(y_1,\ldots, y_k)$ follows from Lemma 1), hence the norm of $t^\alpha \tilde{\phi}_v$ is bounded for $|\alpha| \leqslant k + 1$ and can be estimated by const $\| v \|$. Now we have

$$\| (1 + t^2)^{(k+1)/2} \tilde{\phi}_v(t) \| \leqslant \text{const} \sum_{|\alpha| \leqslant k+1} \| t^\alpha \tilde{\phi}_v(t) \| \leqslant C \| v \|, \tag{15}$$

which immediately yields (12).

Lemma 3. Under conditions of the theorem, the formula holds:

$$(f(\overset{1}{A}_1,\ldots,\overset{n}{A}_n)u,v) = \int_{\mathbb{R}^k} (\exp(it_k A_k) \times \ldots \times \exp(it_1 A_1)u, \tilde{\phi}_v(t_1,\ldots,t_k)) dt_1 \ldots dt_k. \tag{16}$$

(Here the brackets $(\cdot,\cdot)$ denote the scalar product in H.)

Proof. First let $f \in S^\infty(\mathbb{R}^n)$. We have

$$\begin{aligned}
(f(\overset{1}{A}_1,\ldots,\overset{n}{A}_n)u,v) &= \int_{\mathbb{R}^n} \tilde{f}(t_1,\ldots,t_n)(\exp(it_n A_n) \times \ldots \times \exp(it_1 A_1) \times \\
&\times u,v) dt_1 \ldots dt_n = \int_{\mathbb{R}^k \times \mathbb{R}^{n-k}} \tilde{f}(t_1,\ldots,t_n)(\exp(it_k A_k) \times \ldots \times \exp(it_1 A_1) \times \\
&\times u, \exp(it_{k+1} A_{k+1}) \times \ldots \times \exp(it_n A_n)v) dt_1 \ldots dt_n = \\
&= \int_{\mathbb{R}^k} (\exp(it_k A_k) \times \ldots \times \exp(it_1 A_1)u, \int_{\mathbb{R}^{n-k}} \bar{\tilde{f}}(t_1,\ldots,t_n) \times \\
&\times \exp(it_{k+1} A_{k+1}) \times \ldots \times \exp(it_n A_n) v dt_{k+1} \ldots dt_n) dt_1 \ldots dt_k = \\
&= \int_{\mathbb{R}^k} (\exp(it_k A_k) \times \ldots \times \exp(it_1 A_1)u, \tilde{\phi}_v(t_1,\ldots,t_k)) dt_1 \ldots dt_k,
\end{aligned} \tag{17}$$

Q.E.D.

Now if $f \notin S^\infty(\mathbb{R}^n)$ but the estimates (7) hold, we can approximate f by means of a sequence $f_s \in S^\infty(\mathbb{R}^n)$, $s = 1,2,\ldots$, so that the difference $f - f_s$ satisfies estimates (7) with the constant C_s tending to zero. Passing to the limit as $s \to \infty$, we obtain the desired statement. Then combining with (16), we obtain the statement of the theorem.

One could prove this theorem in a completely different way, based directly upon the representation (6).

Let now the tuple $A_1,\dots,A_n$ of self-adjoint operators in a Hilbert space $\mathcal{H}$ be given, satisfying the conditions formulated at the beginning of this item, and let, moreover, the following conditions be satisfied:

(a) operators $A_{k+1},\dots,A_n$ commute with each other;

(b) the left regular representation $L_1,\dots,L_n$ of the tuple $(\overset{1}{A}_1,\dots,\overset{n}{A}_n)$ exists in $S^\infty(\mathbb{R}^n)$, the operators $L_1,\dots,L_n$ being generators in $S^\infty(\mathbb{R}^n)$.

In other words, according to definitions of item A we have for any two symbols $f,g \in S^\infty(\mathbb{R}^n)$:

$$[\![f(\overset{1}{A}_1,\dots,\overset{n}{A}_n)]\!] [\![g(\overset{1}{A}_1,\dots,\overset{n}{A}_n)]\!] = h(\overset{1}{A}_1,\dots,\overset{n}{A}_n) \tag{18}$$

on $\mathcal{D}$, where

$$h(y) = f(\overset{1}{L}_1,\dots,\overset{n}{L}_n)(g(y)), \tag{19}$$

and $L_1,\dots,L_n$ are given linear operators in $S^\infty(\mathbb{R}^n)$, the operator $f(\overset{1}{L}_1,\dots,\overset{n}{L}_n)$ is defined in S^∞ by the formula

$$f(\overset{1}{L}_1,\dots,\overset{n}{L}_n) = \int \tilde{f}(t_1,\dots,t_n)\exp(iL_nt_n)\dots\exp(iL_1t_1). \tag{20}$$

Here the convergence is understood in the weak sense. Condition (b) is satisfied, for example, if the operators $A_1,\dots,A_n$ form a nilpotent Lie algebra and satisfy the agreement conditions (see item B and item D of this section).

Let $y = (y',y'')$, $y' = (y_1,\dots,y_k)$, $y'' = (y_{k+1},\dots,y_n)$. Analogous notations will be used for multi-indices $\alpha = (\alpha_1,\dots,\alpha_n)$. By $S^m_\rho(\mathbb{R}^k,\mathbb{R}^{n-k}) \equiv$ $\equiv S^m_\rho$, where $m \in \mathbb{R}$, $\rho \geqslant 0$, we denote the space of functions $f(y) \in C^\infty(\mathbb{R}^n)$ such that

$$\left| \frac{\partial^{|\alpha|} f(y)}{\partial y^\alpha} \right| \leqslant C_\alpha (1 + |y'|)^{m-\rho|\alpha'|}, \quad |\alpha| = 0,1,2,\dots, \tag{21}$$

and by G^m_ρ we denote the set of operators $f(\overset{1}{A}_1,\dots,\overset{n}{A}_n)$ with symbols $f \in S^m_\rho$. We introduce the following condition on representation operators:

<u>Condition (ρ)</u>. For any $m \in \mathbb{R}$, $f \in S^m_\rho$ and any interchange $\pi = (\pi_1,\dots,\pi_n)$ of the numbers $(1,\dots,n)$, the operator $f(\overset{\pi_1}{L}_1,\dots,\overset{\pi_n}{L}_n)$ is a pseudo-differential operator in $S^\infty(\mathbb{R}^n)$,

$$f(\overset{\pi_1}{L}_1,\dots,\overset{\pi_n}{L}_n) = \mathcal{H}(\overset{2}{y},-i\overset{1}{\frac{\partial}{\partial y}}), \tag{22}$$

with symbol $\mathcal{H}(y,\xi)$ satisfying the following conditions:

$$\mathcal{H}(y,0) - f(y) \in S^{m-\rho}_\rho, \tag{23}$$

$$\left| \frac{\partial^{|\alpha+\beta|}\mathcal{H}}{\partial y^\alpha \partial \xi^\beta}(y,\xi) \right| \leqslant C_{\alpha\beta}(\Phi(y,\xi))^{m-\rho(|\alpha'|+|\beta''|)}, \tag{24}$$

where $\Phi(y,\xi)$ is a function in $\mathbb{R}^{2n}$, such that for some $N_0 \geqslant 0$ the following inequalities hold:

$$1 \leqslant \Phi(y,\xi) \leqslant C(1+|y'|)(1+|\xi|)^{N_o}, \qquad (25)$$

$$1/\Phi(y,\xi) \leqslant C(1+|y'|)^{-1}(1+|\xi|)^{N_o} \qquad (26)$$

with constant C independent of (y,ξ).

Theorem 2. Under conditions mentioned above:

(a) $\bigcup_{m\in\mathbb{R}} G_\rho^m$ is an algebra with filtration. More precisely, if $f_i \in S_\rho^{m_i}$, i = 1,2, then

$$[\![f_1(\overset{1}{A}_1,\ldots,\overset{n}{A}_n)]\!] [\![f_2(\overset{1}{A}_1,\ldots,\overset{n}{A}_n)]\!] = g(\overset{1}{A}_1,\ldots,\overset{n}{A}_n), \qquad (27)$$

where $g \in S_\rho^{m_1+m_2}$. Moreover,

$$f_1(y)f_2(y) - g(y) \in S_\rho^{m_1+m_2-\rho}. \qquad (28)$$

(b) For any non-negative integer m, denote by $\mathcal{H}^m$ the completion of $\mathcal{D}$ in the norm $\|\cdot\|_m$, which is given by the formula

$$\|u\|_o = \|u\|, \qquad (29)$$

where $\|\cdot\|$ is the norm in $\mathcal{H}$;

$$\|u\|_m = \|u\|_{m-1} + \sum_{j=1}^{k} \|A_j u\|_{m-1}, \quad m = 1,2,\ldots \qquad (30)$$

(see also item C of this section).

Then each operator $f(\overset{1}{A}_1,\ldots,\overset{n}{A}_n) \in G_\rho^m$ can be extended to a continuous operator from $\mathcal{H}^s$ into $\mathcal{H}^{s+[-m]}$ (here $[-m]$ is the integer part of the number $-m$) for $s \geqslant \max(-[-m],0)$. In particular, the operators with symbols from S_ρ^0 are bounded in $\mathcal{H}$.

Proof of Theorem 2. First we prove the following:

Proposition 1. Let $f_i \in S_\rho^{m_i}$, i = 1,2; $(\pi_1,\ldots,\pi_n)$ be an interchange of numbers $(1,\ldots,n)$. Then

$$g \overset{\text{def}}{=} f_1(\overset{\pi_1}{L}_1,\ldots,\overset{\pi_n}{L}_n)(f_2(y_1,\ldots,y_n)) \in S_\rho^{m_1+m_2} \qquad (31)$$

and

$$g(y) - f_1(y)f_2(y) \in S_\rho^{m_1+m_2-\rho}. \qquad (32)$$

Item (a) of the theorem is a special case of Proposition 1.

Proof of Proposition 1. Obviously, (32) implies (31); to prove (32), we have by Condition (ρ)

$$g(y) = \mathcal{H}(\overset{2}{y},-i\overset{1}{\frac{\partial}{\partial y}})f_2(y) = \mathcal{H}(y,0)f_2(y) -$$

$$- i\int_o^1 d\tau\{\langle \mathcal{H}_\xi(\overset{2}{y},-i\tau\overset{1}{\frac{\partial}{\partial y}}),f_{2y}(y)\rangle\} =$$

$$= f_1(y)f_2(y) - i\int_o^1 d\tau\{\langle \mathcal{H}_\xi(\overset{2}{y},-i\tau\overset{1}{\frac{\partial}{\partial y}}),f_{2y}(y)\rangle\} + \psi(y) \equiv f_1(y)f_2(y) - iG(y) + \psi(y),$$

where $\psi(y) \in S_\rho^{m_1 + m_2 - \rho}$ by virtue of (23); the brackets $\langle\cdot,\cdot\rangle$ denote summing with respect to the coordinate number from 1 to n. We estimate the derivatives of the function G(y). Set $\varepsilon_j = 0$, $j \in \{1,\dots,k\}$, $\varepsilon_j = 1$, $j \in \{k+1,\dots,n\}$.

By (24) we obtain

$$\left| \frac{\partial^{|\alpha+\beta|}}{\partial y^\alpha \partial \xi^\beta} \times H_{\xi_j}(y,\tau\xi) \right| \leqslant C(\Phi(y,\xi))^{m_1 - \rho(|\alpha'| + |\beta''| + \varepsilon_j)} \tag{34}$$

uniformly in $\tau \in [0,1]$ (here and subsequently the letter C denotes different constants). Using (25) and (26) and the inequality $\Phi(y,\xi)^{-1} \leqslant 1$, which follows from (25), we can obtain the following inequality from (34)

$$\begin{aligned} &\left| \frac{\partial^{|\alpha+\beta|}}{\partial y^\alpha \partial \xi^\beta} H_{\xi_j}(y,\tau\xi) \right| \leqslant \\ &\leqslant C(1+|y'|)^{m_1 - \rho\varepsilon_j - \rho|\alpha'|}(1+|\xi|)^{N_0|m_1 - \rho\varepsilon_j - \rho|\alpha'||} \end{aligned} \tag{35}$$

with the constant C independent of $\tau \in [0,1]$. We set

$$F_{\alpha,N,j}(y,\xi) = \left[\int_0^1 \frac{\partial^{|\alpha|}}{\partial y^\alpha} H_{\xi_j}(y,\tau\xi)d\tau\right](1+\xi^2)^{-N}. \tag{36}$$

For N large enough and all $\alpha \leqslant \alpha_0$ (where α_0 is any fixed multi-index) and $|\beta| = 1,2,\dots,$ we have

$$\left| \frac{\partial^{|\beta|}}{\partial \xi^\beta} F_{\alpha,N,j}(y,\xi) \right| \leqslant C(1+|y'|)^{m_1 - \rho\varepsilon_j - \rho|\alpha'|}(1+|\xi|)^{-n-1}. \tag{37}$$

Estimate (37) yields that the Fourier transform $\hat{F}_{\alpha,N,j}(y,\eta)$ with respect to variables ξ is summable and

$$\int |\hat{F}_{\alpha,N,j}(y,\eta)|d\eta \leqslant C(1+|y'|)^{m - \rho\varepsilon_j - \rho|\alpha'|}. \tag{38}$$

We have

$$\begin{aligned} G^{(\alpha_0)}(y) &= \sum_{j=1}^{N} \sum_{\beta+\gamma=\alpha_0} \frac{\alpha_0!}{\beta!\gamma!} F_{\beta,N,j}\left(\overset{2}{y}, -i\frac{\overset{1}{\partial}}{\partial y}\right) \times (1-\Delta)^N f_{2y_j}^{(\gamma)}(y) = \\ &= \frac{1}{(2\pi)^n} \sum_{\beta+\gamma=\alpha_0} \sum_{j=1}^{N} \frac{\alpha_0!}{\beta!\gamma!} \int_{\mathbb{R}^n} \hat{F}_{\beta,N,j}(y,y-N)(1-\eta)^N f_{2y_j}^{(\gamma)}(\eta)d\eta. \end{aligned} \tag{39}$$

From (38) and the fact that $f_2 \in S_\rho^{m_2}$, it follows that each term in the sum (39) can be estimated by

$$C(1+|y'|)^{m_1 + m_2 - \rho - \rho|\beta'| - \rho|\gamma'|} = C(1+|y'|)^{m_1 + m_2 - \rho - \rho|\alpha_0'|}. \tag{40}$$

Thus the inclusion $G(y) \in S_\rho^{m_1 + m_2 - \rho}$ and the Proposition 1 are proved.

Now we prove the validity of statement (b) of Theorem 2. First of all we establish the boundedness of operators with symbols from S_ρ^0. Let $f \in S_\rho^0$. We construct a symbol $g \in S_\rho^0$ that

$$(f(A))^* f(A) + (g(A))^* g(A) = M^2 + \psi(A), \tag{41}$$

where $M = 1 + \sup_{y\in\mathbb{R}^n} |f(y)|$, $\psi(y) \in S_\rho^{-k-1}$; the asterisk denotes the conjugation in H. (41) immediately implies that the operator f(A) is bounded in H, since $\psi(A)$ is bounded by Theorem 1. In the case of pseudo-differential operators the idea to use identities of the form (41) to prove the boundedness was proposed by Kumano-Go.

Since operators $A_1,\dots,A_n$ are self-adjoint, the operator conjugate to $f(A)$ is given on $\mathcal{D}$ by the equality

$$(f(A))^* = \bar{f}(\overset{n}{A_1},\dots,\overset{1}{A_n}), \tag{42}$$

where $\bar{f}$ is a complex-conjugate symbol, and the numbers over operators in $\bar{f}$ are set in the inverse order. If $g \in S^0_\rho$, then by Proposition 1 we have

$$(f(A))^* f(A) + (g(A))^* g(A) = (|f|^2 + |g|^2)(A) + \psi(A), \quad \psi \in S^{-\rho}_\rho. \tag{43}$$

We solve equation (41) by the method of successive approximations. Set

$$g_o(y) = \sqrt{M^2 - |f(y)|}. \tag{44}$$

It is evident that $g_o \in S^0_\rho$ and we have (for the sake of brevity, we denote by capital letters the operators, which correspond to symbols that are denoted by small letters):

$$F^*F + G^*_oG_o = M^2 + \Psi_o, \quad \psi_o \in S^{-\rho}_\rho. \tag{45}$$

Now we set for $j = 1,2,\dots,$

$$g_j(y) = -\frac{1}{2}\operatorname{Re}\psi_{j-1}(y)/g_{j-1}(y) + g_{j-1}(y) \equiv r_j(y) + g_{j-1}(y), \tag{46}$$

where ψ_{j-1} should satisfy the conditions:

$$F^*F + G^*_{j-1}G_{j-1} = M^2 + \Psi_{j-1}; \tag{47}$$

$$\psi_j \in S^{-\rho(j+1)}_\rho, \quad \operatorname{Im}\psi_j \in S^{-\rho(j+2)}_\rho. \tag{48}$$

It turns out that such a choice of ψ_j is always possible. Really, (47) yields that $\Psi_j = \psi_j(A)$ is symmetric on $\mathcal{D}$. Setting $f_1 = \psi_j(y_1,\dots,y_n)$, $\pi_\ell = n + 1 - \ell$, $\ell = 1,\dots,n$, $f_2 = 1$, we obtain by Proposition 1, that

$$\psi_j(A) = \bar{\psi}_j(\overset{n}{A_1},\dots,\overset{1}{A_n}) = \bar{\psi}_j(\overset{1}{A_1},\dots,\overset{n}{A_n}) + \chi(\overset{1}{A_1},\dots,\overset{n}{A_n}),$$

where $\chi \in S^{-\rho(j+2)}_\rho$, if $\psi_j \in S^{-\rho(j+1)}_\rho$. Set

$$\psi'_j(y) = \frac{1}{2}[\psi_j(y) + \bar{\psi}_j(y) + \chi(y)]; \tag{49}$$

then $\psi'_j(A) = \psi_j(A)$ and ψ'_j satisfies both conditions in (48). Thus it is sufficient to construct symbols ψ_j satisfying the first condition in (48), and the validity of the second condition may then be provided.

We proceed by induction on j. Let symbols g_{j-1}, ψ_{j-1} satisfying the induction proposition be already constructed. Note that the symbols g_{j-1}, r_j are real ones. Using (46), we obtain:

$$\begin{aligned} F^*F + G^*_jG_j &= F^*F + G^*_{j-1}G_{j-1} + G^*_{j-1}R_j + R^*_jG_{j-1} + R^*_jR_j = \\ &= M^2 + \Psi_{j-1} + G^*_{j-1}R_j + R^*_jG_{j-1} + R^*_jR_j, \end{aligned} \tag{50}$$

where $r_j \in S_\rho^{-\rho j}$, $g_{j-1} \in S_\rho^0$, $\bar{g}_{j-1}r_j + \bar{r}_j g_{j-1} = 2g_{j-1}r_j = -\psi_{j-1}$, and hence by Proposition 1, we have

$$F^*F + G_j^*G_j = M^2 + \psi_j,\ \psi_j \in S_\rho^{-\rho(j+1)}, \tag{51}$$

and the induction step is carried out.

Thus we have shown already that the operators with symbols from S_ρ^0 are bounded. Now we prove that if $F \in G_\rho^m$ (m being integer), then $F : H^s \to H^{s-m}$ is a continuous operator. We have:

$$\text{(a)} \quad A_j \in G_\rho^1,\ j = 1,\dots,k; \tag{52}$$

$$\text{(b)} \quad A_j : H^s \to H^{s-1} \tag{53}$$

are continuous operators for $j = 1,\dots,k$. Then the implication holds:

$$\text{(c)} \quad F \in G_\rho^m \Rightarrow [A_j, F] \equiv A_jF - FA_j \in G_\rho^{m-\rho},\ j = 1,\dots,k; \tag{54}$$

(d) if $F \in G_\rho^0$, then F is bounded in H^0,

$$\text{(e)} \quad G_\rho^m G_\rho^{m'} \subset G_\rho^{m+m'}. \tag{55}$$

The boundedness of an operator $F \in G_\rho^k$ from H^s into H^{s-k} follows from (a) - (e) by standard argument analogous to that carried out in item C, and hence omitted here. The theorem is proved.

It should be noted that the verification of Condition (ρ) in the general case may be rather difficult. This condition, however, holds in the case of usual pseudo-differential operators, $n = 2k$, $A_{j+k} = x_j$, $A_j = -i\frac{\partial}{\partial x_j}$ in $L_2(\mathbb{R}^k)$; $L_j = y_j - i\frac{\partial}{\partial y_{k+j}}$, $L_{k+j} = y_{k+j}$, $j = 1,\dots,k$;

$$H(y,\xi) = f(y_1 + \xi_1,\dots,y_k + \xi_k, y_{k+1},\dots,y_n).$$

One may set $\Phi(y,\xi) = 1 + |y' + \xi'|$, $N_0 = 1$. In this case estimates (25) and (26) with $N_0 = 1$ follow from the Peetre inequality. Also the following assertion, to be used in Section 1 of Chapter 4, is evident.

Corollary 1. Let function $f(y^{(1)}, y^{(2)}, y^{(3)})$, $y^{(i)} \in \mathbb{R}^n$, $i = 1,2,3$, satisfy the estimates

$$\left| \frac{\partial^{|\beta+\gamma+\delta|} f}{\partial y^{(1)\beta} \partial y^{(2)\gamma} \partial y^{(3)\delta}} \right| \leqslant C_{\beta\gamma\delta}\left(1 + \sqrt{(y^{(1)})^2 + (y^{(2)})^2}\right)^{-|\beta| - |\gamma|},$$
$$|\beta| + |\gamma| + |\delta| = 0,1,2,\dots. \tag{56}$$

Then the operator $f(-i\overset{1}{\frac{\partial}{\partial \xi}}, \overset{2}{\xi}, \overset{2}{\xi})$ is bounded in $L_2(\mathbb{R}_\xi^n)$ and the upper estimate of its norm is completely defined by the constants $C_{\beta\gamma\delta}$.

III
Asymptotic solutions for pseudo-differential equations

1. THE CANONICAL OPERATOR ON A LAGRANGIAN MANIFOLD IN $\mathbb{R}^{2n}$

A. WKB-Approximations

We introduce first some basic notions related to the problems considered. Let $x = (x_1,\dots,x_n)$ be coordinates in $\mathbb{R}^n$, and let

$$H(\overset{2}{x},-ih\overset{1}{\frac{\partial}{\partial x}},h) \equiv H(\overset{2}{x_1},\dots,\overset{2}{x_n},-ih\overset{1}{\frac{\partial}{\partial x_1}},\dots,-ih\overset{1}{\frac{\partial}{\partial x_n}},h) \equiv \hat{H} \tag{1}$$

be a $1/h$ pseudo-differential operator ($1/h$-PDO in the sequel) in $\mathbb{R}^n$. The definition of a $1/h$ pseudo-differential operator was given in Section 4:E of Chapter 2. We require throughout the chapter that the symbol $H(q,p,h)$, $(q,p) = (q_1,\dots,q_n,p_1,\dots,p_n) \in \mathbb{R}^{2n}$ of the $1/h$-PDO be a smooth function of the parameter $h \in [0,1]$ with values in $S^\infty(\mathbb{R}^n)$, i.e.,

$$\left| \frac{\partial^{|\alpha|+|\beta|+k} H(q,p,h)}{\partial q^\alpha \partial p^\beta \partial h^k} \right| \leqslant C_{\alpha\beta k}(1+|q|+|p|)^m, \quad h \in [0,1], \; p,q \in \mathbb{R}^n. \tag{2}$$

The function $H_0(q,p,h) \overset{\text{def}}{=} H(q,p,0)$ is called the principal symbol (or the Hamiltonian) of the $1/h$-PDO.

A $1/h$ pseudo-differential equation ($1/h$-PDE in the sequel) is the equation of the form

$$H(\overset{2}{x},-ih\overset{1}{\frac{\partial}{\partial x}},h)u(x,h) = v(x,h), \tag{3}$$

where $v(x,h)$ is a given function. If u,v are vector functions and also $H(\overset{2}{x},-ih\overset{1}{\frac{\partial}{\partial x}},h) = \| H_{ij}(\overset{2}{x},-ih\overset{1}{\frac{\partial}{\partial x}},h) \|$ is a matrix of $1/h$-PDO's, then (3) is a system of $1/h$-PDE's.

The typical example of the $1/h$-PDE is the well-known quantum-mechanical Schrödinger equation

$$-ih\frac{\partial\psi}{\partial t} + H(\overset{2}{x},-ih\overset{1}{\frac{\partial}{\partial x}},h)\psi = 0, \tag{4}$$

where $H(\overset{2}{x}, -ih\overset{1}{\frac{\partial}{\partial x}}, h) = -\frac{h^2}{2m}\Delta + V(x)$ is the energy operator of the quantum system under consideration, and $\psi = \psi(x,t,h)$ is its wave function. We introduce the problem of seeking the approximate solution of the general 1/h-PDE of the form (4) as $h \to 0$; such approximate solutions are known in quantum mechanics as quasi-classical or WKB-approximations (named after Wentzel, Kramers and Brillouin).

If the operator H were an ordinary differential operator with constant coefficients, $H = H(-ih\frac{d}{dx})$, $H(p) = \sum_{k=0}^{N} C_k p^k$ being a polynomial in the variable $p \in \mathbb{R}$, the equation (4) would have special solutions of the form

$$\psi(x,t,h) = e^{(i/h)S(x,t)}, \tag{5}$$

where $S(x,t) = ax + bt$ is a linear function in x and t, and besides $b + \sum_{k=0}^{N} C_k a^k = 0$, i.e., S(x,t) satisfies the equation

$$\frac{\partial S}{\partial t} + H(\frac{\partial S}{\partial x}) = 0. \tag{6}$$

As an analogy to this case we seek the approximate solutions of (4) in the form

$$\psi(x,t,h) = e^{(i/h)S(x,t)}\phi(x,t,h), \tag{7}$$

where $\phi(x,t,h)$ depends on h smoothly; for instance,

$$\phi(x,t,h) = \sum_{k=0}^{N} (-ih)^k \phi_k(x,t) \tag{8}$$

is a polynomial in h, S(x,t) is a real-valued function. After substitution of the expression (7) into equation (4), we come to the expansion of the form:

$$\begin{aligned} -ih\frac{\partial\psi}{\partial t} + H(\overset{2}{x}, -ih\overset{1}{\frac{\partial}{\partial x}}, h)\psi = e^{(i/h)S(x,t)}\{[\frac{\partial S}{\partial t} + H_o(x, \frac{\partial S}{\partial x})]\phi(x,t,h) + \\ + (-ih)[\frac{\partial}{\partial t} + \frac{\partial H_o}{\partial p}(x, \frac{\partial S}{\partial x})\frac{\partial}{\partial x} + \frac{1}{2}\operatorname{tr}(\frac{\partial^2 H_o}{\partial p^2}(x, \frac{\partial S}{\partial x})\frac{\partial^2 S}{\partial x^2}) + i\frac{\partial H}{\partial h}(x, \frac{\partial S}{\partial x}, 0)] \times \\ \times \phi(x,t,h) + \sum_{j=2}^{N+1}(-ih)^j P_j(\overset{2}{x}, -i\overset{1}{\frac{\partial}{\partial x}}, t)\phi(x,t,h)\} + h^{N+2}e^{(i/h)S(x,t)}f(x,t,h), \end{aligned} \tag{9}$$

where $P_j(x, -i\frac{\partial}{\partial x}, t)$ are differential operators with smooth coefficients, independent of h; f(x,t,h) is smooth with respect to all the variables. The reader can easily verify the validity of expansion (9) for the case when $H(\overset{2}{x}, -ih\overset{1}{\frac{\partial}{\partial x}})$ is a differential operator*; this will be enough to grasp the idea, see item B for exact formulations in the general case.

---

* Hint: prove the identity

$$e^{(-i/h)S(x,t)}H(\overset{2}{x}, -ih\overset{1}{\frac{\partial}{\partial x}}, h)e^{(i/h)S(x,t)} = H(\overset{2}{x}, \overset{1}{[\![}\frac{\partial S}{\partial x}, -ih\frac{\partial}{\partial x}]\!], h),$$

and obtain the expansion of its right-hand side in powers of h.

To obtain the approximate solution we require that the coefficients of the powers of h within the curly brackets be equal to zero. This requirement yields the system of equations:

$$\frac{\partial S}{\partial t} + H_o(x, \frac{\partial S}{\partial x}) = 0, \tag{10}$$

$$\frac{\partial \phi_o}{\partial t} + H_{op}(x, \frac{\partial S}{\partial x}) \frac{\partial \phi_o}{\partial x} + F(x,t)\phi_o = 0, \tag{11}$$

$$\frac{\partial \phi_k}{\partial t} + H_{op}(x, \frac{\partial S}{\partial x}) \frac{\partial \phi_k}{\partial x} + F(x,t)\phi_k = -\sum_{j=0}^{k-1} P_{k-j+1}(\overset{2}{x}, -i\overset{1}{\frac{\partial}{\partial x}}, t)\phi_j, \; k = 1,\ldots,N, \tag{12}$$

where

$$F(x,t) = i\frac{\partial H}{\partial h}(x, \frac{\partial S}{\partial x}, 0) - \frac{1}{2}\operatorname{tr}(\frac{\partial^2 H_o}{\partial p^2}(x, \frac{\partial S}{\partial x})\frac{\partial^2 S}{\partial x^2}). \tag{13}$$

Once the system (10) - (12) is satisfied, we have

$$-ih\frac{\partial \psi}{\partial t} + H(\overset{2}{x}, -ih\overset{1}{\frac{\partial}{\partial x}}, h)\psi = r_N(x,t,h) \equiv h^{N+2}e^{(i/h)S(x,t)}g(x,t,h), \tag{14}$$

where g(x,t,h) is a smooth function. Assume that the Hamiltonian $H_o(q,p)$ is real-valued. Then we are able to construct the solution of (10) - (12). Prescribe for the sake of definiteness the Cauchy data

$$\psi(x,0,h) = e^{(i/h)S_o(x)}\phi_o(x,h), \; \phi_o(x,h) = \sum_{k=0}^{N}(-ih)^k\phi_{ok}(x) \tag{15}$$

for the unknown function ψ in the equation (4) (we assume $\operatorname{Im}S_o \equiv 0$). First of all, we find the solution of the equation (10) with initial data $S(x,0) = S_o(x)$. (10) is a well-known, in classical mechanics, Hamilton-Jacobi equation (it is often called the characteristic equation for (4)), and it may be solved by the method of bicharacteristics. Namely, consider the system of ordinary differential equations (the Hamiltonian system generated by H_o):

$$\dot{q} = \frac{\partial H_o}{\partial p}(q,p), \quad \dot{p} = -\frac{\partial H_o}{\partial q}(q,p). \tag{16}$$

(The solutions of the system (16) are known as the bicharacteristics of the equation (4).) Denote by $(q(q_o,t),p(q_o,t))$ the solution of (16), corresponding to the initial data

$$q|_{t=0} = q_o, \quad p|_{t=0} = \frac{\partial S}{\partial x}(q_o), \tag{17}$$

and consider the function

$$W(q_o,t) = S_o(q_o) + \int_0^t [p(q_o,\tau)\frac{\partial H_o}{\partial p}(q(q_o,\tau),p(q_o,\tau)) - H_o(q(q_o,\tau),p(q_o,\tau))]d\tau. \tag{18}$$

Consider also the system of equations

$$q(q_o,t) = x \in \mathbb{R}^n. \tag{19}$$

Since $\frac{\partial q}{\partial q_o}(q_o,0) = E$, (19) has a smooth solution $q_o = q_o(x,t)$ defined for (x,t) belonging to some open set $\mathcal{U}_o \subset \mathbb{R}^{n+1}$, containing the hyperplane $\{t = 0\}$ and satisfying the property: if $(x,t) \in \mathcal{U}_o$, then $(x,\tau) \in \mathcal{U}_o$ for any τ between 0 and t. For a moment we restrict ourselves to consideration of the points $(x,t) \in \mathcal{U}_o$ only. We set

$$S(x,t) = W(q_o(x, t), t). \tag{20}$$

The function (20) satisfies (10) and the prescribed initial condition (see [2]).

Next we turn to equations (11) and (12). Introducing the new unknown functions

$$\chi_k(q_o,t) = \phi_k(q(q_o,t),t), \tag{21}$$

we may rewrite (11) and (12) in the form

$$\frac{\partial \chi_k}{\partial t} + F(q(q_o,t),t)\chi_k = -\sum_{j=0}^{k-1} \tilde{P}_{k-j+1}(\overset{2}{q_o}, -i\overset{1}{\frac{\partial}{\partial q_o}}, t)\chi_j,\ k = 0,1,\ldots,N, \tag{22}$$

where $\tilde{P}_{k-j+1}$ are some new differential operators, and the sum on the right is meant to be zero for $k = 0$. The equation (22) for χ_k is called the transport equation and is an ordinary differential equation along the bicharacteristics. We may successively resolve the system (22) by means of elementary integration:

$$\chi_k(q_o,t) = e^{-\int_0^t F(q(q_o,\tau),\tau)d\tau} \phi_{ok}(q_o) - \int_0^t e^{-\int_\tau^t F(q(q_o,\tau'),\tau')d\tau'} \times$$
$$\times \sum_{j=0}^{k-1} \tilde{P}_{k-j+1}(\overset{2}{q_o}, -i\overset{1}{\frac{\partial}{\partial q_o}}, \tau)\chi_j(q_o,\tau)d\tau,\ k = 0,\ldots,N. \tag{23}$$

Now make some conclusions. First of all, the WKB-approximation in the considered situation enables us to obtain for arbitrary N the "approximate solution" of the equation (4), which satisfies the Cauchy data (15) and satisfies the equation up to the remainder $r_N(x,t,h)$ (14) which may be estimated in the following way:

$$\left|(-ih\frac{\partial}{\partial x})^\alpha(-ih\frac{\partial}{\partial t})^\beta r_N(x,t,h)\right| \leqslant C_{\alpha\beta}(K)h^{N+2},$$
$$|\alpha| + \beta = 0,1,2,\ldots,\quad (x,t) \in K \tag{24}$$

for any compact subset $K \subset \mathcal{U}_o$. We write $r_N(x,t,h) = O(h^{N+2})$ in $\mathcal{U}_o$ for the function $r_N(x,t,h)$ satisfying the estimates (24),* and call any function $\psi(x,t,h)$ satisfying (15) and such that

$$-ih\frac{\partial\psi}{\partial t} + \hat{H}\psi = O(h^N) \text{ in the domain } \mathcal{U} \subset \mathbb{R}^{n+1}, \tag{25}$$

the asymptotic solution modulo h^N in $\mathcal{U}$ of the problem (4), (15). And secondly, the WKB-approximation in the presented form does not provide the asymptotic solution for all (x,t) even if the solution of the Hamiltonian system (16) exists for all t. Indeed, the simple example (one-dimensional harmonic oscillator - $H(q,p,h) = \frac{1}{2}(q^2 + p^2)$) shows that the solutions of system (10) - (12) have singularities in the points of the caustic (i.e., in the points where the Jacobian $\det(\partial q/\partial q_o)$ vanishes). On the other hand, it is known that in the same example the exact solution with finite Cauchy data exists for all (x,t) and has no singularities. Thus we should like to improve the method in order to obtain global asymptotic solutions, provided that the Hamiltonian flow is globally defined. (We note, however,

---

* The definition of $O(h^N)$ will be refined in the subsequent item.

that additional conditions should be imposed if non-finite Cauchy data are to be considered.)

Such improvement becomes possible as soon as we broaden the class of functions within the asymptotic solution constructed. Let $I \subset \{1,\dots,n\} \equiv$ $\equiv [n]$ be some subset $\bar{I} = \{1,\dots,n\} \setminus I$; we consider the WKB-approximation in mixed coordinate-momenta representation:

$$\psi(x,t,h) = \bar{F}^{1/h}_{\xi_{\bar{I}} \to x_{\bar{I}}} [e^{(i/h)S_I(x_I,\xi_{\bar{I}},t)} \phi_I(x_I,\xi_{\bar{I}},t,h)]; \tag{26}$$

here $S_I(x_I,\xi_{\bar{I}},t)$ is a real-valued function, $\phi_I(x_I,\xi_{\bar{I}},t,h) = \sum_{k=0}^{N} (-ih)^k \phi_{Ik} \times$ $\times (x_{\bar{I}},\xi_{\bar{I}},t)$. The representation of the asymptotic solution in the form (7) is the particular case of (26), when $I = \emptyset$. Substituting (26) into (4), we obtain the system of equations (see theorem on commutation in item B; however, the result is not unexpected since the 1/h-Fourier transformation transfers differentiation into multiplication by coordinates and vice versa):

$$\frac{\partial S_I}{\partial t} + H_o(x_I, -\frac{\partial S_I}{\partial \xi_{\bar{I}}}, \frac{\partial S_I}{\partial x_I}, \xi_{\bar{I}}) = 0, \tag{27}$$

$$\frac{\partial \phi_{Ik}}{\partial t} + H_{op_I}(x_I, -\frac{\partial S_I}{\partial \xi_{\bar{I}}}, \frac{\partial S_I}{\partial x_I}, \xi_{\bar{I}}) \frac{\partial \phi_{Ik}}{\partial x_I} - H_{ox_{\bar{I}}}(x_I, -\frac{\partial S_I}{\partial \xi_{\bar{I}}}, \tag{28}$$

$$\frac{\partial S_I}{\partial x_I}, \xi_{\bar{I}}) \frac{\partial \phi_{Ik}}{\partial \xi_{\bar{I}}} + F_I(x_I,\xi_{\bar{I}}, t)\phi_{Ik} = -\sum_{j=0}^{k-1} P_{k-j+1}(\overset{2}{x_I}, \overset{2}{\xi_{\bar{I}}}, -i\overset{1}{\frac{\partial}{\partial x_I}}, -i\overset{1}{\frac{\partial}{\partial \xi_{\bar{I}}}})$$

$$\phi_{Ij}, k = 0,\dots,N,$$

analogous to (10) - (12). If the support of ϕ_I lies in the set $\{|\xi_{\bar{I}}| < R\}$ and the system (27) - (28) is satisfied, we obtain

$$-ih \frac{\partial \psi}{\partial t} + H(\overset{2}{x}, -i\overset{1}{\frac{\partial}{\partial x}}, h)\psi = r_{IN}(x,t,h) \equiv$$

$$\equiv h^{N+2} \bar{F}^{1/h}_{\xi_{\bar{I}} \to x_{\bar{I}}} \{e^{(i/h)S_I(x_I,\xi_{\bar{I}},t)} g_I(x_I,\xi_{\bar{I}},t,h)\} = O(h^{N+2-(|\bar{I}|/2)}) \tag{29}$$

(the last equality may be established by direct estimate). The fact that the accuracy is less than in (14) (the factor $h^{-|\bar{I}|/2}$) is not essential since N may be chosen as large as we desire. (Note also that this factor disappears in L^2-estimates.)

Solving (27) by virtue of the method of bicharacteristics, we come exactly to Hamiltonian system (16), while the equation (28) for ϕ_{Ik} is reduced to an ordinary differential equation along the bicharacteristics. It turns out that the global asymptotic solution of the problem (4) and (15) may be written as the linear combination of functions of the form (26) with every possible I; all these functions correspond to the same family of bicharacteristics, defined by (16) and (17), and are combined together by means of partition of unity. To prove the formulated assertion, one should show that, primarily, the solutions of the form (26) may serve as continuations of each other and, secondly, that for any point lying on the bicharacteristic of our family such as $I \subset \{1,\dots,n\}$ may be found so that the asymptotic solution of the form (26) may be defined in the neighborhood of this point. The latter property is guaranteed by a lemma on local coordinates (see next item). As for the former one, we give here the proof of it in the simplest case. Before doing this, we present a theorem on asymptotics of integrals with rapidly oscillating integrands.

Let $\Phi(x,\omega)$, $x \in \mathbb{R}^n$, $\omega \in \mathbb{R}^m$ be a smooth real-valued function.

Definition 1. The point $(x_0,\omega_0) \in \mathbb{R}^n \times \mathbb{R}^m$ is a stationary point of Φ (with respect to the variables x), if

$$\frac{\partial \Phi}{\partial x}(x_0,\omega_0) = 0. \tag{30}$$

The stationary point (x_0,ω_0) is non-degenerate if the determinant of the matrix

$$\frac{\partial^2 \Phi}{\partial x^2}(x,\omega) = \left\| \begin{matrix} \frac{\partial^2 \Phi}{\partial x_1 \partial x_1} & \cdots & \frac{\partial^2 \Phi}{\partial x_1 \partial x_n} \\ \cdots & \cdots & \cdots \\ \frac{\partial^2 \Phi}{\partial x_n \partial x_1} & \cdots & \frac{\partial^2 \Phi}{\partial x_n \partial x_n} \end{matrix} \right\| \tag{31}$$

is not equal to zero at (x_0,ω_0).

Note that if (x_0,ω_0) is a non-degenerate stationary point of Φ then by the implicit function theorem the equation $\frac{\partial \Phi}{\partial x}(x,\omega) = 0$ defines in the vicinity of (x_0,ω_0) the unique smooth function $x = x(\omega)$. Definition 1 also makes sense if the dependence of Φ on ω is only continuous (the function $x(\omega)$ is continuous in this case).

Consider the integral

$$I[\phi](\omega,h) = \frac{e^{i(\pi n/4)}}{(2\pi h)^{n/2}} \int_{\mathbb{R}^n} e^{(i/h)\Phi(x,\omega)} \phi(x,\omega)dx, \tag{32}$$

where $\Phi(x,\omega)$, $\phi(x,\omega)$ are functions, continuous in ω together with all their derivatives with respect to x, $\mathrm{Im}\Phi(x,\omega) \equiv 0$, $\phi(x,\omega)$ vanishes outside some ball $|x| < R(k)$ provided that ω belongs to a compact set K. The function Φ is called the phase of the integral (32), and the function ϕ is called its amplitude.

Theorem 1. (Stationary phase method).

(a) Assume that the phase Φ has no stationary points on the support supp ϕ. Then $I[\phi](\omega,h)$ decays faster than any power of h as $h \to 0$ locally uniformly with respect to ω. If Φ,ϕ depend on ω smoothly, the same is valid for all the derivatives of $I[\phi]$ with respect to ω.

(b) Assume that there exists a non-degenerate stationary point $x = x(\omega)$ of the phase Φ and no other stationary points of Φ lie on the support of ϕ. Then the integral (32) has for any N an asymptotic expansion

$$I[\phi](\omega,h) = e^{(i/h)\Phi(x(\omega),\omega)} \left(\det\left[-\frac{\partial^2 \Phi}{\partial x^2}(x(\omega),\omega)\right]\right)^{-1/2} \times$$
$$\times \left\{\phi(x(\omega),\omega) + \sum_{k=1}^{N-1} (-ih)^k (V_k[\Phi]\phi)(x(\omega),\omega)\right\} + h^N R_N[\Phi,\phi](\omega), \tag{33}$$

where $V_k[\Phi]$ are independent of h differential operators in x of the order $\leqslant 2k$ with coefficients dependent on the derivatives up to the order $2k + 2$ of the phase Φ with respect to x taken in the stationary point; the branch of the square root is fixed by the following choice of the argument of $\det\left[-\frac{\partial^2 \Phi}{\partial x^2}\right]$:

$$\arg \det\left[-\frac{\partial^2\Phi}{\partial x^2}(x(\omega),\omega)\right] = \sum_{k=1}^{n} \arg \lambda_k, \tag{34}$$

where $-3\pi/2 < \arg \lambda_k < \pi/2$, $\lambda_1,\dots,\lambda_n$ are the eigenvalues of the matrix $-\frac{\partial^2\Phi}{\partial x^2}(x(\omega),\omega)$, counted with their multiplicities. The remainder $R_N = R_N[\Phi,\phi](\omega)$ has the estimate: for any compact subset $K \subset \mathbb{R}^m$

$$\left|\left(-ih\frac{\partial}{\partial x}\right)^{\alpha} R_N\right| \leqslant C_{k,\alpha} \tag{35}$$

((35) is valid for $|\alpha| = 0$ if Φ,ϕ depend on ω continuously and for any α if Φ depend on ω smoothly).

Consider now the WKB-approximation of the form (26) and show that under certain conditions its asymptotic expansion due to Theorem 1 results in the approximation (7). We have

$$\psi(x,t,h) = \frac{e^{i(\pi/4)|\bar{I}|}}{(2\pi h)^{|\bar{I}|/2}} \int_{\mathbb{R}^{|\bar{I}|}} e^{(i/h)[\xi_{\bar{I}}x_{\bar{I}} + S_I(x_I,\xi_{\bar{I}},t)]} \phi_I(x_I,\xi_{\bar{I}},t,h)\,d\xi_{\bar{I}}. \tag{36}$$

The stationary point of the phase of integral (26) is given by the equation

$$x_{\bar{I}} + \frac{\partial S_I}{\partial \xi_{\bar{I}}}(x_I,\xi_{\bar{I}},t) = 0, \tag{37}$$

while the matrix of second derivatives of the phase with respect to $\xi_{\bar{I}}$ coincides with that of S_I. Assume that the solution $\xi_{\bar{I}} = \xi_{\bar{I}}(x,t)$ of the equation (37) exists and is unique on supp ϕ_I, and that $\det\left[\frac{\partial^2 S_I}{\partial\xi_{\bar{I}}\partial\xi_{\bar{I}}}(x_I,\xi_{\bar{I}} \times (x,t),t)\right] \neq 0$. Then by Theorem 1, we obtain

$$\begin{aligned}\psi(x,t,h) &= e^{(i/h)[S_I(x_I,\xi_{\bar{I}}(x,t),t) + x_{\bar{I}}\xi_{\bar{I}}(x,t)]}\phi(x,t,h) + O(h^N) \equiv \\ &\equiv e^{(i/h)S(x,t)}\phi(x,t,h) + O(h^N),\end{aligned} \tag{38}$$

where

$$S(x,t) = S_I(x_I,\xi_{\bar{I}}(x,t),t) + x_{\bar{I}}\xi_{\bar{I}}(x,t), \tag{39}$$

and the explicit form of the function $\phi(x,t,h)$ may be obtained from (33). The equations (37) and (39) show that the functions S and S_I are the Legendre transforms [2] of each other. It follows, in particular, that the equations

$$q_{\bar{I}} = -\frac{\partial S_I}{\partial \xi_{\bar{I}}}(q_I,p_{\bar{I}},t), \quad p_I = \frac{\partial S_I}{\partial x_I}(q_I,p_{\bar{I}},t) \tag{40}$$

define the same set of points $(p,q) \in \mathbb{R}^{2n}$ as the equation

$$p = \frac{\partial S}{\partial x}(q,t). \tag{41}$$

In other words, we see, taking into consideration the connection between solutions of Hamilton-Jacobi equations and bicharacteristics, that S and S_I correspond to the same family of solutions of (16).

Our considerations in this item have a preliminary character, so we do not give further details here; in particular, we pay no attention to how the amplitude ϕ is transformed, although this matter is rather interesting and leads to some topological conditions. We hope that the motivations for construction of global asymptotic solutions have become clearer, and thus we move on to detailed formulations

B. The Canonical Operator on a Lagrangian Manifold in $\mathbb{R}^{2n}$

The canonical operator is a global version of the WKB-approximations to solutions of 1/h-PDE's, described in the preceding item. As we have already seen, the WKB-approximations of the form

$$\psi(x,h) = e^{(i/h)S(x)}\phi(x,h), \tag{1}$$

and

$$\psi_I(x,h) = F^{-1/h}_{\xi_{\bar{I}} \to x_{\bar{I}}}\{e^{(i/h)S_1(x_I,\xi_{\bar{I}})}\phi_1(x_I,\xi_{\bar{I}},h)\}, \tag{2}$$

where ϕ,ϕ_1 are polynomials in h; S,S_1 are real-valued, define functions that are coincident (up to a unimodular factor) modulo $O(h^N)$ as $h \to 0$, provided that the systems of equations

$$p = \frac{\partial S}{\partial x}(q) \tag{3}$$

and

$$p_I = \frac{\partial S_I}{\partial x_I}(q_I,p_{\bar{I}}), \quad q_{\bar{I}} = -\frac{\partial S_I}{\partial \xi_{\bar{I}}}(q_I,p_{\bar{I}}), \tag{4}$$

$q,p \in \mathbb{R}^n$, define the same submanifold in the space $\mathbb{R}^{2n} = \mathbb{R}^n_q \times \mathbb{R}^n_p$ and, besides the functions ϕ and ϕ_1 are properly related. The submanifold $L \subset \mathbb{R}^{2n}$, given by (3) or (4), is isotropic with respect to the differential form

$$\omega^2 = \sum_{j=1}^{n} dp_j \wedge dq_j \tag{5}$$

in $\mathbb{R}^{2n}$ (i.e., the restriction of ω^2 on the tangent space of L vanishes identically and, as will be shown below, any isotropic submanifold L of dimension n may be locally represented in the form (4) for a suitable $I \subset [n]$. These observations lead to the idea of representing the global asymptotic solutions of 1/h-PDE's in the form of linear combinations of functions, having the form (2) with various $I \subset [n]$ and corresponding to the same isotropic submanifold L. The canonical operator is the realization of this idea; roughly speaking, the canonical operator acts on functions defined on L and transforms them into functions on $\mathbb{R}^n_x$, dependent on the parameter h; for functions on L with support small enough, the result has the form (2).

It turns out that the condition on $\psi = K\phi$ (where K is the canonical operator on L) to be an asymptotic solution to the 1/h-PDE $\hat{H}\psi = 0$ is reduced to geometric conditions on L plus the transport equation for ϕ (the reader has become acquainted briefly with a local variant of these latter conditions in item A). It turns out also that, in general, there is a hindrance to the existence of the canonical operator on a given isotropic submanifold $L \subset \mathbb{R}^n$, namely, some element of the cohomology group $H^1(L,\mathbb{R})$. The requirement that this cohomology class be zero is known as the "quantization condition" and leads, in particular, to asymptotics of eigenvalues for 1/h-PDO's.

We come now to exact definitions and formulations.

Parameters. In the subsequent exposition we deal with various objects, depending on parameters, such as smooth functions, mappings of manifolds, vector fields, differential forms, etc. Usually these parameters will be denoted by $\omega = (\omega_1,\ldots,\omega_m)$ and vary in some simply connected open subset $\Omega \subset \mathbb{R}^m$ or, more generally, on a simply connected manifold Ω (possibly with a non-empty border $\partial\Omega$). If M is some manifold, we denote by $C^{\infty,0}_{\alpha,\omega} \equiv C^{\infty,0}_{\alpha,\omega}(M\times\Omega)$ the space of (complex-valued) functions $f(\alpha,\omega)$, $\alpha \in M$, $\omega \in \Omega$, continuously dependent on ω in C^∞-topology on M. (Equivalently, this means that $f(\alpha,\omega)$ is infinitely smooth in α and continuous in ω with all its derivatives with respect to α.) As a rule, we assume that the components of all our mappings, forms, vector fields, etc. in local coordinates belong to the space $C^{\infty,0}_{\alpha,\omega}$; all the exceptions to this convention are mentioned explicitly.

The convenient notation will often be used: if the function, mapping, etc. $f(\alpha,\omega)$ depends on the arguments α,ω, we write $f_\omega(\alpha) \overset{\text{def}}{=} f(\alpha,\omega)$.

We present the implicit function theorem for $C^{\infty,0}_{\alpha,\omega}$-mappings.

Theorem 1. Let $f_i(\alpha_1,\ldots,\alpha_n,\omega_1,\ldots,\omega_m) \equiv f_i(\alpha,\omega) \in C^{\infty,0}_{\alpha,\omega}$, $i = 1,\ldots,k \leqslant n$, and assume that

$$f_1(\alpha,\omega) = \ldots = f_k(\alpha,\omega) = 0 \tag{6}$$

for $(\alpha,\omega) = (\alpha^{(0)},\omega^{(0)})$ and the Jacobian

$$\frac{D(f_1,\ldots,f_k)}{D(\alpha_1,\ldots,\alpha_k)} = \det\begin{pmatrix} \frac{\partial f_1}{\partial\alpha_1} & \cdots & \frac{\partial f_1}{\partial\alpha_k} \\ \cdot & \cdot\,\cdot\,\cdot & \cdot \\ \frac{\partial f_k}{\partial\alpha_1} & \cdots & \frac{\partial f_k}{\partial\alpha_k}\end{pmatrix} \tag{7}$$

is not equal to zero at $(\alpha^{(0)},\omega^{(0)})$. Then the system of equations (6) has a unique solution $(\alpha_1(\alpha_{k+1},\ldots,\alpha_n,\omega),\ldots,\alpha_k(\alpha_{k+1},\ldots,\alpha_n,\omega))$ in some neighborhood of $(\alpha^{(0)},\omega^{(0)})$ and $\alpha_i(\alpha_{k+1},\ldots,\alpha_n,\omega) \in C^{\infty,0}_{(\alpha_{k+1},\ldots,\alpha_n),\omega}$.

The space $\hat{H}_{h,\ell oc}(\mathbb{R}^n \times \tilde{\Omega})$. First of all, we define the space to which our approximate solutions should belong. Assume that a set $\tilde{\Omega} \subset \Omega \times (0,1]$ of pairs (ω,h) is given (this set arises naturally from the quantization conditions, as we shall see soon). For given $h \in (0,1]$, we shall denote by $\Omega_h \subset \Omega$ the set of $\omega \in \Omega$ such that $(\omega,h) \in \tilde{\Omega}$.

Definition 1. $\hat{H}_{h,\ell oc}(\mathbb{R}^n \times \tilde{\Omega})$ is the space of functions $f(x,\omega,h)$, defined for $x \in \mathbb{R}^n$, $(\omega,h) \in \tilde{\Omega}$ smooth in x and satisfying the condition: for any function $\phi \in C^{\infty,0}_{x,\omega}(\mathbb{R}^n \times \Omega)$ with compact support $\operatorname{supp}\phi \subset \mathbb{R}^n \times \Omega$, the expression

$$\begin{aligned}\|\phi f\|_k &= \sup_{(\omega,h)\in\tilde{\Omega}} \|(\phi f)(\cdot,\omega,h)\|_{k,h} \equiv \\ &\equiv \sup_{(\omega,h)\in\tilde{\Omega}} \int_{\mathbb{R}^n} (\overline{\phi f})(x,\omega,h)(1-h^2\Delta_x)^k(\phi f)(x,\omega,h)\,dx\end{aligned} \tag{8}$$

is finite for any integer $k \geqslant 0$ (here $\Delta_x = \sum_{j=1}^n \frac{\partial^2}{\partial x_j^2}$ is a Laplacian). It is obvious that if $\|f\|_k$ is finite, then $\|\phi f\|_k$ is finite for any $\phi \in C^{\infty,0}_{x,\omega} \times (\mathbb{R}^n \times \Omega)$.

For any real N we write $f = \hat{O}(h^N)$ if $f/h^N \in \hat{H}_{h,\ell oc} \equiv \hat{H}_{h,\ell oc}(\mathbb{R}^n \times \tilde{\Omega})$. Also we write $f = O(h^N)$ if for any multi-index $\beta = (\beta_1,\ldots,\beta_n)$ and any compact $K \subset \mathbb{R}^n \times \tilde{\Omega}$, the estimate is valid:

$$\left| \frac{\partial^{|\beta|} f}{\partial x^\beta}(x,\omega,h) \right| \leqslant C_{K,\beta} h^{N-|\beta|}, \quad (x,\omega,h) \in K. \tag{9}$$

We say that $f = \hat{O}(h^\infty)$ if $f = \hat{O}(h^N)$ for any N.

Proposition 1. (a) The following implications are valid:

$$f = O(h^N) \Rightarrow f = \hat{O}(h^N); \quad f = \hat{O}(h^N) \Rightarrow f = O(h^{N-n/2}). \tag{10}$$

(b) For any compact set $K \subset \mathbb{R}^n \times \tilde{\Omega}$, we have

$$C_{K,k} |||f|||_k \leqslant \|f\|_k \leqslant C_k |||f|||_k, \qquad k = 0,1,2,\ldots, \quad f \in \hat{H}_{h,loc}, \operatorname{supp} f \subset K, \tag{11}$$

where $C_{K,k}, C_k > 0$, C_k does not depend on K; the right-hand inequality remains valid for any function f with finite $|||f|||_k$,

$$|||f|||_k = \sup_{(\omega,h)\in\tilde{\Omega}} |||f|||_{k,h} = \sup_{(\omega,h)\in\tilde{\Omega}} \int_{\mathbb{R}^n} \bar{f}(x,\omega,h) \times (1 + x^2 - h^2\Delta_x)^k f(x,\omega,h)dx. \tag{12}$$

The norm (12) is often more convenient than (8), since (12) is invariant under 1/h-Fourier transformation in any group of variables.

We introduce the notion of support modulo $\hat{O}(h^\infty)$ of the function $f \in \hat{H}_{h,\ell oc}(\mathbb{R}^n \times \tilde{\Omega})$, in a way analogous to the notion of singular support in the theory of distributions.

Definition 2. Let $f(x,\omega,h) \in \hat{H}_{h,\ell oc}(\mathbb{R}^n \times \tilde{\Omega})$. The support modulo $\hat{O}(h^\infty)$ of f is the minimal closed subset $K \subset \mathbb{R}^n \times \Omega$, such that for any function $\phi \in C^{\infty,0}_{x,\omega}(\mathbb{R}^n \times \Omega)$ satisfying $\operatorname{supp} \phi \cap K = \emptyset$, we have $\phi f = \hat{O}(h^\infty)$. The notation: $K = \operatorname{sing\,supp} f$.

We do not consider the singular support in the usual sense here, so there should be no confusion while using this notation for Lagrangian manifolds in $\mathbb{R}^{2n}$ and canonical coverings.

In the space $\mathbb{R}^{2n} = \mathbb{R}^{2n}_{(q,p)}$ consider the differential 2-form:

$$\omega^2 = \sum_{j=1}^{n} dp_j \wedge dq_j = dp \wedge dq. \tag{13}$$

Obviously, ω^2 is a closed non-degenerate 2-form, i.e., a symplectic structure in $\mathbb{R}^{2n}$ (see [2]).

Definition 3. Let L be a smooth manifold, and let $i : L \to \mathbb{R}^{2n}$ be an immersion. The pair (L,i) is called a Lagrangian manifold if for any $\alpha \in L$, $i_*(T_\alpha L)$ is a Lagrangian subspace in $T_{i(\alpha)}\mathbb{R}^{2n}$, i.e., the restriction of ω^2 on $i_*(T_\alpha L)$ is a zero 2-form.

Proposition 2. (L,i) is a Lagrangian manifold if and only if

$$i^*\omega^2 \equiv 0. \tag{14}$$

If (L,i) is a Lagrangian manifold, then $\dim L \leqslant n$.

The proof of (14) is obvious. As for the latter statement, take any $\alpha \in L$ and set $\Lambda = i_*(T_\alpha L)$. The condition $\omega^2|_\Lambda = 0$ is equivalent to saying Λ is orthogonal to $J\Lambda$, where J is the matrix of the form ω^2; thus $\dim \Lambda + \dim J\Lambda \leqslant 2n$; since J is non-degenerate, $\dim J\Lambda = \dim \Lambda$ and thus $\dim \Lambda \leqslant n$. The proposition is proved.

Thus (L and $i(L)$ may be identified locally and even globally if i is a proper embedding), the notion of a Lagrangian manifold is synonymous to that of a manifold isotropic with respect to the symplectic structure ω^2; the term "Lagrangian manifold" was proposed by one of the authors in the book [50]. In the sequel we deal only with Lagrangian manifolds of maximal dimension n. Such manifolds exist, as is shown by the following:

Example 1. Let $S = S(x)$ be a smooth real-valued function defined on the open subset $L \subset \mathbb{R}^n_x$. We define the embedding $i : L \to \mathbb{R}^{2n}_{q,p}$ by

$$q(x) = x, \; p(x) = \frac{\partial S}{\partial x}(x), \; x \in L. \tag{15}$$

Since $i^*pdq = dS$, we have $i^*\omega^2 = di^*pdq = ddS = 0$, so that (15) defines a Lagrangian submanifold in $\mathbb{R}^{2n}$. The definition of Lagrangian manifold, depending on parameters $\omega \in \Omega$, may be given in a natural way:

Definition 4. The pair $(L \times \Omega, i : L \times \Omega \to \mathbb{R}^{2n})$ is a Lagrangian manifold (depending on parameters), if $i \in C^{\infty,0}_{\alpha,\omega}(L \times \Omega)$ and for any $\omega \in \Omega$, (L, i_ω) is a Lagrangian manifold. We write $i(\alpha,\omega) = (q(\alpha,\omega), p(\alpha,\omega))$.

Our next aim is to cover $L \times \Omega$ by open sets, such that the restriction of i on each of these sets could be described by equations of the type (4).

For each $I \subset [n]$, we define the mapping $\pi_I : L \times \Omega \to \mathbb{R}^n_{(x_I,\xi_{\bar{I}})} \times \Omega$,

$$\pi_I : (\alpha,\omega) \to (x_I, \xi_{\bar{I}}, \omega) \equiv (q_I(\alpha,\omega), p_{\bar{I}}(\alpha(\omega),\omega)). \tag{16}$$

Recall that we have adopted the convention: if $I = [n]$ the subscript I may be omitted; in particular, $\pi_{[n]}$ is denoted by π.

Definition 5. The canonical covering of the Lagrangian manifold $(L \times \Omega, i)$ is the set $\{(\mathcal{U},I)\} = \{(\mathcal{U}_j, I_j)\}_{j\in J}$ of pairs $(\mathcal{U},I)$, where $\mathcal{U} \in L \times \Omega$ is an open subset and $I \subset [n]$ is some subset, such that:

(a) $\{\mathcal{U}_j\}_{j\in J}$ is a locally finite covering of $L \times \Omega$, and $\mathcal{U}_j$ are shrinkable sets.

(b) The projections $\tilde{\mathcal{U}}_j$ of $\mathcal{U}_j$ onto L and all finite intersections $\tilde{\mathcal{U}}_{j_1\ldots j_s} = \tilde{\mathcal{U}}_{j_1} \cap \ldots \cap \tilde{\mathcal{U}}_{j_s}$ are shrinkable sets.

(c) The restriction of the mapping π_{I_j} on $\mathcal{U}_j$ is a homeomorphism of $\mathcal{U}_j$

onto some open subset $V_j \subset \mathbb{R}^n_{(x_I,\xi_{\bar I})} \times \Omega$, $I = I_j$, for any $j \in J$, and the inverse mapping $\pi^{-1}_{I_j}$ belongs to $C^{\infty,0}_{(x_I,\xi_{\bar I}),\omega}(V_j)$.

The canonical atlas corresponding to a canonical covering $\{(U_j I_j)\}_{j\in J}$ is the set $\{(U_j,\pi_{I_j})\}_{j\in J}$ (here the restriction of the mapping π_{I_j} on U_j is denoted by the same letter).

The pair (U_j,π_{I_j}) is called a canonical chart on $L \times \Omega$, non-singular or singular, depending on whether $I_j = [n]$ or not; $(x_{I_j},\xi_{\bar I_j},\omega)$ are called the canonical coordinates in U_j. We write

$$q = q(x_{I_j},\xi_{\bar I_j},\omega) \equiv (x_{I_j},q_{\bar I_j}(x_{I_j},\xi_{\bar I_j},\omega),$$

$$p = p(x_{I_j},\xi_{\bar I_j},\omega) \equiv (p_{I_j}(x_{I_j},\xi_{\bar I_j},\omega),\xi_{\bar I_j}), \quad (x_I,\xi_{\bar I_j},\omega) \in V_j$$

for the expression $i \circ \pi^{-1}_{I_j}$ of the functions $i(\alpha,\omega) = (q(\alpha,\omega),p(\alpha,\omega))$ in the canonical coordinates of the chart (U_j,π_{I_j}).

Theorem 2. The canonical covering exists for any Lagrangian manifold $(L \times \Omega, i)$.

Proof. We make use of the following:

Lemma 1. (lemma on local coordinates) For any $(\alpha^{(0)},\omega^{(0)}) \in L \times \Omega$, there exists a subset $I \subset [n]$ and a neighborhood $U \subset L \times \Omega$ of the point $(\alpha^{(0)},\omega^{(0)})$ such that $(\alpha,\omega) \to \pi_I(\alpha,\omega)$ is a coordinate system in U.*

Indeed, the Lagrangian subspace $i_{\omega *}(T_{\alpha(0)}L)$ is projected isomorphically on the coordinate Lagrangian subspace $\{p_I = 0, q_{\bar I} = 0\}$ for some $I \subset [n]$ (see [2], [52] for detailed proof). Thus, the Jacobian $\det \frac{\partial(q_I,p_{\bar I})}{\partial\alpha}(\alpha,\omega) \neq 0$ (here $\alpha = (\alpha_1,\dots,\alpha_n)$ is any system of local coordinates on L). It remains to apply Theorem 1.

By Lemma 1 we may find the covering of $L \times \Omega$ by open sets U_σ in which the coordinate systems of the prescribed type are valid. It may be assumed that all U_σ have the form $U_\sigma = U'_\sigma \times U''_\sigma$, $U'_\sigma \subset L$, $U''_\sigma \subset \Omega$, where the sets U'_σ and all their intersections are shrinkable, and also U''_σ are shrinkable (it is enough to choose these sets as small enough open balls in some Riemannian metric). Passing to a locally finite subcovering, we obtain the desired results. The theorem is proved.

The point $(\alpha^{(0)},\omega^{(0)}) \in L \times \Omega$ will be called non-singular if we may choose $I = [n]$ (in Lemma 1), and singular in the opposite case. The caustic $\Sigma \subset \mathbb{R}^n_x \times \Omega$ is by definition the set of projections $\pi(\alpha,\omega)$ of singular points $(\alpha,\omega) \in L \times \Omega$ (recall that we drop the subscript I when $\bar I = \emptyset$).

---

* We do not require the differentiability with respect to ω in the definition of the coordinate system.

The canonical operator: preliminary definition and discussion. Let a Lagrangian manifold $(L \times \Omega, i : L \times \Omega \to \mathbb{R}^{2n})$ be given. We define some functional spaces on $L \times \Omega$. Recall that the continuous mapping $r : A \to B$ of topological spaces is called proper if the pre-image $r^{-1}(K)$ is a compact subset in A for any compact subset $K \subset B$.

Definition 6. $A(L \times \Omega, i)$ is a subspace of $C^{\infty,0}_{\alpha,\omega}(L \times \Omega)$ consisting of all functions ϕ such that the restriction $\pi|_{\mathrm{supp}\,\phi}$ of the mapping π on the support of ϕ is a proper mapping. $A[h](L \times \Omega, i)$ is a space of polynomials in the parameter $h \in [0,1]$ with coefficients from $A(L \times \Omega, i)$. $A^N[h](L \times \Omega, i)$ is a subspace of $A[h](L \times \Omega, i)$ consisting of polynomials of the order $\leqslant N$.

Proposition 3. All the spaces described in Definition 6 are $C^{\infty,0}_{\alpha,\omega} \times$ $\times\,(L \times \Omega)$-modules, invariant under differential (in α) operators on $L \times \Omega$ with $C^{\infty,0}_{\alpha,\omega}$-coefficients. The space $A^N[h](L \times \Omega, i)$ is naturally isomorphic to the quotient space $A[h](L \times \Omega, i)/I^{N+1}[h](L \times \Omega, i)$, where $I^{N+1}[h](L \times \Omega, i)$ $= h^{N+1}A[h](L \times \Omega, i)$.

The proof of this statement is obvious. Let $\mathcal{U} \subset L \times \Omega$ be an open subset. We denote by $A(\mathcal{U}, i)$, $A[h](\mathcal{U}, i)$, $A^N[h](\mathcal{U}, i)$, $I^{N+1}[h](\mathcal{U}, i)$ the subspaces of the spaces enumerated, consisting of functions with supports lying in $\mathcal{U}$. For the sake of brevity, we use in the sequel a shortened notation for defined spaces: A, $A[h]$, $A^N[h]$, $I^{N+1}[h]$, $A_{\mathcal{U}}$, $A_{\mathcal{U}}[h]$, $A^N_{\mathcal{U}}[h]$, $I^{N+1}_{\mathcal{U}}[h]$. The canonical operator

$$K : A[h] \to \hat{H}_{h,\ell oc}(\mathbb{R}^n \times \tilde{\Omega}) \tag{17}$$

is defined in the following way. The set $\tilde{\Omega}$ depends on some cohomology classes on $L \times \Omega$ and will be defined later. We choose a canonical atlas $\{(\mathcal{U}_j, \pi_{I_j})\}_{j \in J}$ of $L \times \Omega$ and $C^{\infty,0}_{\alpha,\omega}$-partition of unity* $\{e_j\}_{j \in J}$ on $L \times \Omega$ subordinate to the covering $\{\mathcal{U}_j\}_{j \in J}$. We set

$$[K\phi](x,\omega,h) = \sum_{j \in J} [K_j(e_j\phi)](x,\omega,h), \quad \phi \in A[h], \quad (x,\omega,h) \in \mathbb{R}^n \times \tilde{\Omega}, \tag{18}$$

where

$$K_j : A_{\mathcal{U}_j}[h] \to \hat{H}_{h,\ell oc}(\mathbb{R}^n \times \Omega \times (0,1]) \tag{19}$$

is an elementary operator corresponding to the chart $(\mathcal{U}_j, \pi_{I_j})$ to be defined later in this item, and the sum (18) is locally finite.

It turns out that for any j_1, j_2, $\mathcal{U}_{j_1} \cap \mathcal{U}_{j_2} \neq \emptyset$, we have

---

*This means that $e_j = e_j(\alpha,\omega) \in C^{\infty,0}_{\alpha,\omega}(L \times \Omega)$, supp $e_j \subset \mathcal{U}_j$ for any j, and $\sum_{j \in J} e_j \equiv 1$ on $L \times \Omega$ (note that the sum is locally finite). Since by our assumptions $L \times \Omega$ is a manifold, the existence of unity partitions is then guaranteed by standard theorems.

$$K_{j_1}\phi = K_{j_2}\phi + \hat{O}(h), \quad \phi \in A_{U_{j_1} \cap U_{j_2}}[h] \tag{20}$$

(but (20) is not valid in general for $\tilde{\Omega} = \Omega \times (0,1]$).

The property (20) enables us when seeking the solution of the 1/h-PDE $\hat{H}\psi = 0$ to glue the transport operators modulo $I^{-1}[h]$ into a global operator on $L \times \Omega$ having the form of a vector field plus a function. The subsequent terms of the expansion may be taken into account by means of the regular theory of perturbations.

Also (20) implies that the operator

$$K^{(0)} : A^0[h] \equiv A \to \hat{H}_{h,\ell oc}(\mathbb{R}^n \times \tilde{\Omega})/\mathrm{mod}\ \hat{O}(h) \tag{21}$$

in the quotient spaces, induced by K, does not depend on the choice of the canonical atlas and the subordinate partition of unity.

Elementary canonical operators. Two difficulties might arise if we made an attempt to define the action of the operator K_j on the function ϕ by formula (2), where $I = I_j$, S_1 corresponds to our Lagrangian manifold, and ϕ_1 is the expression of ϕ in canonical coordinates. Primarily, the integral might diverge if the set U_j is not precompact. Secondly, one may pick out from the stationary phase method (Theorem 1, item A), that (20) could be held only if ϕ were multiplied by the square root of the Jacobian of the canonical coordinates changed under the transition from the chart $(U_{j_1}, \pi_{I_{j_1}})$ to the chart $(U_{j_2}, \pi_{I_{j_2}})$, i.e., if ϕ were not a function but a section of a certain line bundle over $L \times \Omega$. The first difficulty is eliminated by the virtue of suitable cut-off functions. As for the second one, there are different ways to avoid this difficulty. One may either consider the space of line bundle sections on $L \times \Omega$ instead of a function space (just as L. Hörmander did in his definition of Fourier integral operators [29]) or identify these two spaces by means of some simplification of the arising line bundles. We have chosen the second way here; the simplification is "inserted" into the definition of the elementary canonical operator, so that there is no need to describe the bundle at all. The adopted variant of exposition is a traditional one in the canonical operator theory ([50], [52], [54], etc.). In order to fulfill our intention, we should fix some non-degenerate smooth measure μ on L, dependent on parameters $\omega \in \Omega$.

Definition 7. A measure on the Lagrangian manifold $(L \times \Omega, i)$ is a smooth non-vanishing differential n-form on L, dependent on parameters $\omega \in \Omega$ such that its density in any system of local coordinates* $(\alpha_1, \ldots, \alpha_n, \omega)$ on $L \times \Omega$ belongs to $C^{\infty,0}_{\alpha,\omega}$.

If μ is a measure on $(L \times \Omega, i)$ and $(\alpha_1, \ldots, \alpha_n, \omega)$ is a system of local coordinates, we denote the density of μ in these coordinates by $D\mu/D\alpha$:

$$\mu = \frac{D\mu}{D\alpha}(\alpha,\omega)d\alpha_1 \wedge d\alpha_2 \wedge \ldots \wedge d\alpha_n. \tag{22}$$

If $(\alpha'_1, \ldots, \alpha'_n, \omega)$ is another coordinate system, then obviously

$$\frac{D\mu}{D\alpha'} = \frac{D\mu}{D\alpha} \cdot \frac{D\alpha}{D\alpha'}. \tag{23}$$

---

* $(\alpha_1, \ldots, \alpha_n)$ may depend on ω in such a way that $\alpha_1(\alpha,\omega)\ldots\alpha_n(\alpha,\omega) \in C^{\infty,0}_{\alpha,\omega}$; recall that ω are parameters so that, for instance, the exterior differentiation d acts only with respect to the variables α, parameters ω being fixed.

Thus, the requirement of Definition 7 is that $D\mu/D\alpha \in C^{\infty,0}_{\alpha,\omega}$ and does not take zero values; by (23) these conditions do not depend on the choice of the coordinate system.

In the particular case, when $(x_I, \xi_{\bar{I}}, \omega)$ is a canonical coordinate system on $L \times \Omega$, we denote the density of the measure μ by μ_I:

$$\mu_I(x_I, \xi_{\bar{I}}, \omega) = \left(\frac{D\mu}{D(x_I, \xi_{\bar{I}})} \cdot \pi_I^{-1}\right)(x_I, \xi_{\bar{I}}, \omega). \tag{24}$$

Measures on the Lagrangian manifold $(L \times \Omega, i)$ always exist, as is shown by the following:

Proposition 4. The differential form

$$\sigma = i^*(d(q_1 - ip_1) \wedge \ldots \wedge d(q_n - ip_n)) \tag{25}$$

is a measure on the Lagrangian manifold $(L \times \Omega, i)$.

Proof. Let $(\alpha^{(0)}, \omega^{(0)}) \in L \times \Omega$ be an arbitrary point. By Lemma 1, for some $I \subset [n]$, $(q_I(\alpha,\omega), p_{\bar{I}}(\alpha,\omega), \omega)$ is a system of local coordinates in the vicinity of $(\alpha^{(0)}, \omega^{(0)})$. Consider the following linear transformation of the space $\mathbb{R}^{2n}$:

$$(q,p) \to (q',p') = S_I(q,p); \quad q' = (q_I, p_{\bar{I}}); \quad p' = (p_I; -q_{\bar{I}}). \tag{26}$$

S_I is a canonical transformation, i.e., $S_I^*\omega^2 = \omega^2$. Moreover,

$$d(q_1 - ip_1) \wedge \ldots \wedge d(q_n - ip_n) =$$
$$= (-i)^{|\bar{I}|} S_I^*\{d(q_1' - ip_1') \wedge \ldots \wedge d(q_n' - ip_n')\}. \tag{27}$$

Set $i' = S_I \circ i$. Since S_I preserves the symplectic structure, $(L \times \Omega, i')$ is a Lagrangian manifold, the point $(\alpha^{(0)}, \omega^{(0)})$ is now non-singular, and we have

$$\sigma_I(\alpha,\omega) \equiv \frac{D\sigma}{D(x_I, \xi_{\bar{I}})} = (-i)^{|\bar{I}|} \frac{D\sigma'}{Dx'},$$
$$\sigma' = i'^*\{d(q_1' - ip_1') \wedge \ldots \wedge d(q_n' - ip_n')\}. \tag{28}$$

Thus, we have reduced the situation to the non-singular case (from this point we drop the primes). We have

$$\frac{D\sigma}{Dx} = \det \frac{\partial(q_1 - ip_1, \ldots, q_n - ip_n)}{\partial(x_1, \ldots, x_n)} \equiv$$
$$\equiv \det \frac{\partial(x_1 - ip_1(x,\omega), \ldots, x_n - ip_n(x,\omega))}{\partial(x_1, \ldots, x_n)} = \det\left(E - i\frac{\partial p(x,\omega)}{\partial x}\right). \tag{29}$$

In the canonical coordinates (x,ω) we have

$$0 = i^*\omega^2 = \sum_{\ell,r} \frac{\partial p_\ell(x,\omega)}{\partial x_r} dx_r \wedge dx_\ell = \sum_{\ell < r} \left\{\frac{\partial p_\ell}{\partial x_r} - \frac{\partial p_r}{\partial x_\ell}\right\} dx_r \wedge dx_\ell; \tag{30}$$

thus $\frac{\partial p(x,\omega)}{\partial x}$ is a symmetric matrix with real entries.

Lemma 2. Let C be a symmetric $(n \times n)$-matrix, $C = C_1 + iC_2$, where C_1 and C_2 are matrices with real entries, and C_2 is positive semi-definite. Then the spectrum of C lies in the upper half-plane. The proof of this lemma is carried out in [52].

Applying this lemma to the matrix $\frac{\partial p}{\partial x}$ (which, however, has a zero imaginary part), we obtain that the spectrum of $E - i\frac{\partial p(x,\omega)}{\partial x}$ lies in the half-plane $\{\mathrm{Re}\lambda \geqslant 1\}$. Thus $\frac{D\sigma}{Dx} \neq 0$ and Proposition 4 is proved.

The measure σ is uniquely determined by the immersion i and will be called the canonical measure on $(L \times \Omega, i)$. This measure is complex-valued; obviously the real-valued measure on $(L \times \Omega, i)$ exists if and only if L is an orientable manifold. If μ is any measure on $(L \times \Omega, i)$, it differs from σ by a factor, which is a non-vanishing element of $C^{\infty,0}_{\alpha,\omega}(L \times \Omega)$. This factor will be denoted by

$$f_\mu(\alpha,\omega) \equiv \frac{D\mu}{D(q - ip)} \overset{\mathrm{def}}{=} \mu/\sigma. \tag{31}$$

We come to the definition of phase functions, analogous to S, S_1 in (1) and (2) in the parametric case. Let $I \subset [n]$. Consider the differential 1-form:

$$\omega^1_I = p_I dq_I - q_{\bar{I}} dp_{\bar{I}} \tag{32}$$

in $\mathbb{R}^{2n}$. Obviously $d\omega^1_I = \omega^2$, hence the form $i^*\omega^1_I$ is a closed 1-form on L (dependent on parameters ω). Consider the Pfaff equation

$$dW_I = i^*\omega^1_I. \tag{33}$$

For any connected simply connected open subset $\tilde{\mathcal{U}} \subset L$, we may thus obtain a solution of (33), belonging to $C^{\infty,0}_{\alpha,\omega}(\tilde{\mathcal{U}} \times \Omega)$ and a unique modulo additive term which is dependent only on ω. If I, $I \subset [n]$ and W_I, W_J are the solutions of the corresponding Pfaff equations in $\tilde{\mathcal{U}} \times \Omega$, direct computation yields that

$$W_I(\alpha,\omega) + p_{\bar{I}}(\alpha,\omega)q_{\bar{I}}(\alpha,\omega) = W_J(\alpha,\omega) + p_{\bar{J}}(\alpha,\omega)q_{\bar{J}}(\alpha,\omega) + a(\omega), \tag{34}$$

where $a(\omega)$ is some continuous function.

Let now $(\mathcal{U}, \pi_I : \mathcal{U} \to \mathcal{V})$ be a canonical chart on L.

Definition 8. An action, or a phase function in the chart $(\mathcal{U}, \pi_I)$, is the function defined in $\mathcal{V}$

$$S(x_I, \xi_{\bar{I}}, \omega) = (W_I \circ \pi_I^{-1})(x_I, \xi_{\bar{I}}, \omega), \tag{35}$$

where W_I is a solution of (33) in $\tilde{\mathcal{U}} \times \Omega$ (recall that $\tilde{\mathcal{U}}$ is the projection of $\mathcal{U}$ onto L). Thus, action in the chart is defined modulo a continuous function of ω. From (33) immediately follows:

Proposition 5. If $S(x_I, \xi_{\bar{I}}, \omega)$ is the action in the chart $(\mathcal{U}, \pi_I)$, then the restriction $i|_{\mathcal{U}}$ is given in canonical coordinates by the system of equations

$$q(x_I, \xi_{\bar{I}}, \omega) = (x_I, -\frac{\partial S}{\partial \xi_{\bar{I}}}(x_I, \xi_{\bar{I}}, \omega)); \; p(x_I, \xi_{\bar{I}}, \omega) = (\frac{\partial S}{\partial x_I}(x_I, \xi_{\bar{I}}, \omega), \xi_{\bar{I}}). \tag{36}$$

At last, we define the notion of the cut-off function in the chart $(\mathcal{U},\pi_I)$, which enables us to define the elementary canonical operator.

Definition 9. A cut-off function in the canonical chart $(\mathcal{U},\pi_I : \mathcal{U} \to \mathcal{V})$ is a function $\chi(x,\xi_{\bar{I}},\omega) \equiv \chi(x_I,x_{\bar{I}},\xi_{\bar{I}},\omega)$ defined in $\mathbb{R}^{|\bar{I}|}_{x_{\bar{I}}} \times \mathcal{V}$ and satisfying the conditions:

(a) $\chi(x,\xi_{\bar{I}},\omega) \in C^{\infty,0}_{(x,\xi_{\bar{I}}),\omega}(\mathbb{R}^{|\bar{I}|}_{x_{\bar{I}}},\mathcal{V})$ and $\chi(x,\xi_{\bar{I}},\omega) \equiv 1$ in the neighborhood of the set

$$\{(x,\xi_{\bar{I}},\omega) \in \mathbb{R}^{|\bar{I}|}_{x_{\bar{I}}} \times \mathcal{V} \,|\, x_{\bar{I}} = q_{\bar{I}}(x_I,\xi_{\bar{I}},\omega)\}. \tag{37}$$

(b) For any compact set $K \subset \mathbb{R}^n_x \times \Omega$ and any function $\phi \in A_{\mathcal{U}}[h]$, there exists a number R such that the conditions $(x,\omega) \in K$, $(x_I,\xi_{\bar{I}},\omega) \in \text{supp} \times (\phi \circ \pi_I^{-1}) = \pi_I(\text{supp}\ \phi)$, $(x,\xi_{\bar{I}},\omega) \in \text{supp}\ \chi$ yield that $|\xi_{\bar{I}}| < R$.

Proposition 6. The cut-off function χ exists.

The proof will be given below, in Proposition 8, where the stronger statement is formulated. Let the following objects be fixed: a measure μ on the Lagrangian manifold $(L \times \Omega,i)$; a canonical chart $(\mathcal{U},\pi_I : \mathcal{U} \to \mathcal{V})$ on $L \times \Omega$; an action $S(x_I,\xi_{\bar{I}},\omega)$ in the chart $(\mathcal{U},\pi_I)$; a continuous branch $\arg \mu_I \times (x_I,\xi_{\bar{I}},\omega)$ of the argument of $\mu_I(x_I,\xi_{\bar{I}},\omega)$ in $\mathcal{V}$ (such a branch exists since $\mathcal{V}$ is simply connected and μ_I does not vanish, and since $\mathcal{V}$ is connected any two branches differ by $2\pi k$, k being an integer); and a cut-off function $\chi(x,\xi_{\bar{I}},\omega)$.

Definition 10. An elementary canonical operator in the chart $(\mathcal{U},\pi_I)$ is an operator

$$K_{e\ell} : A_{\mathcal{U}}[h] \to \hat{H}_{h,\ell oc}(\mathbb{R}^n \times \Omega \times (0,1]) \tag{38}$$

defined by the formula:*

$$[K_{e\ell}\phi](x,\omega,h) = F^{1/h}_{\xi_{\bar{I}}\to x_{\bar{I}}}\{\chi(x,\xi_{\bar{I}},\omega)e^{(i/h)S(x_I,\xi_{\bar{I}},\omega)}\sqrt{\mu_I(x_I,\xi_{\bar{I}},\omega)}(\phi \circ \pi_I^{-1}) \times$$

$$\times (x_I,\xi_{\bar{I}},\omega,h)\} \equiv \frac{e^{i(\pi/4)|\bar{I}|}}{(2\pi h)^{|\bar{I}|/2}}\int_{\mathbb{R}^{|\bar{I}|}} d\xi_{\bar{I}}\{\chi(x,\xi_{\bar{I}},\omega) \times \tag{39}$$

$$\times e^{(i/h)[S(x_I,\xi_{\bar{I}},\omega) + x_{\bar{I}}\xi_{\bar{I}}]}\sqrt{\mu_I(x_I,\xi_{\bar{I}},\omega)}(\phi \circ \pi_I^{-1})(x_I,\xi_{\bar{I}},\omega,h)\}.$$

Note that for $(x,\omega) \in K$ (a compact subset in $\mathbb{R}^n \times \Omega$) and $\phi \in A_{\mathcal{U}}[h]$, the integration in (39) is in fact over the finite region $|\xi_{\bar{I}}| < R = R(K,\phi)$. Also it is clear that $K_{e\ell}$ is a linear operator.

---

* In (39) $\sqrt{\mu_I}$ is a continuous branch of the square root defined by $\sqrt{\mu_I} = |\mu_I|^{1/2} \times \exp(\frac{1}{2}\arg \mu_I)$, where the branch $\arg \mu_I$ was fixed above.

Theorem 3. (a) Definition (10) is correct (i.e., the operator $K_{e\ell}$ really acts in the space (38));

(b) For $\phi \in A_{\mathcal{U}}[h]$, $K_{e\ell}\phi$ does not depend modulo $\hat{O}(h^\infty)$ on the choice of the cut-off function χ;

(c) The support modulo $\hat{O}(h^\infty)$ of $K_{e\ell}\phi$ is contained in the set $\pi(\mathrm{supp}\ \phi)$.

Proof. We start from (b). Let χ_1, χ_2 be cut-off functions, $\theta = \chi_1 - \chi_2$. Then the integrand in

$$I(x,\omega,h) = \frac{e^{i(\pi/4)|\bar{I}|}}{(2\pi h)^{|\bar{I}|/2}} \int_{\mathbb{R}^{|\bar{I}|}} d\xi_{\bar{I}} \{\theta(x,\xi_{\bar{I}},\omega) e^{(i/h)[S(x_I,\xi_{\bar{I}},\omega) + x_{\bar{I}}\xi_{\bar{I}}]} \times \tag{40}$$
$$\times [\sqrt{\mu_I}(\phi \circ \pi_I^{-1})](x_I,\xi_{\bar{I}},\omega,h)\}$$

vanishes in the neighborhood of stationary points of the phase $\Phi(x,\xi_{\bar{I}},\omega) = S(x_I,\xi_{\bar{I}},\omega) + x_{\bar{I}}\xi_{\bar{I}}$ (see Proposition 5 and Definition 9). We may rewrite (40) in the form

$$I(x,\omega,h) = h^{N + [|\bar{I}|+1/2] - (|\bar{I}|/2)} \int_{\mathbb{R}^{|\bar{I}|}} d\xi_{\bar{I}} \{a(x_I,\xi_{\bar{I}},\omega,h) \times \tag{41}$$
$$\times (L^{N+[|\bar{I}|+1/2]} e^{(i/h)\Phi(x,\xi_{\bar{I}},\omega)})\},$$

where $a(x_I,\xi_{\bar{I}},\omega,h)$ is a polynomial in h, smooth with respect to $(x_I,\xi_{\bar{I}})$, $a(x_I,\xi_{\bar{I}},\omega,h) = 0$ for $|\xi_{\bar{I}}| > R(K)$ when $(x,\omega) \in K$, where K is any compact in $\mathbb{R}^n \times \Omega$, and

$$L = -i\left|\frac{\partial\Phi}{\partial\xi_{\bar{I}}}\right|^{-2} \frac{\partial\Phi}{\partial\xi_{\bar{I}}} \frac{\partial}{\partial\xi_{\bar{I}}}, \quad Le^{(i/h)\Phi} = h^{-1}e^{(i/h)\Phi} \tag{42}$$

is a differential operator with coefficients non-singular on supp a. Integrating by parts $N + [\frac{|\bar{I}|+1}{2}]$ times we obtain immediately that $I(x,\omega,h) = O(h^N)$. Since N is arbitrary, (b) is proved. Similarly, (c) is valid since if $\psi \in C^{\infty,0}_{x,\omega}(\mathbb{R}^n \times \Omega)$, $\mathrm{supp}\ \psi \cap \pi(\mathrm{supp}\ \phi) = \phi$, then we may set $\theta(x,\xi_{\bar{I}},\omega) = \psi(x)\chi(x,\xi_{\bar{I}},\omega)$ in (40) and then proceed as above. It remains to prove (a). Let $\psi \in C^{\infty,0}_{x,\omega}(\mathbb{R}^n \times \Omega)$, $K = \mathrm{supp}\ \psi$ be compact. By similar argument we obtain that modulo $\hat{O}(h^\infty)$

$$\psi[K_{e\ell}\phi] = \psi \bar{F}^{1/h}_{\xi_{\bar{I}} \to x_{\bar{I}}} \{\chi(\xi_{\bar{I}}) e^{(i/h)S(x_I,\xi_{\bar{I}},\omega)} [(\phi \circ \pi_I^{-1}) \times \tag{43}$$
$$\times \sqrt{\mu}](x_I,\xi_{\bar{I}},\omega,h)\},$$

where $\chi(\xi_{\bar{I}}) \in C_0^\infty(\mathbb{R}^{|\bar{I}|})$, $\chi(\xi_{\bar{I}}) = 0$ for $|\xi_{\bar{I}}| > R(K,\phi)$. Using Proposition 1, we obtain that $\psi[K_{e\ell}\phi] = \psi\hat{O}(h^0) = \hat{O}(h^0) \subset \hat{H}_{h,\ell oc}(\mathbb{R}^n \times \Omega \times (0,1])$. The theorem is proved.

The global canonical operator and quantization conditions. Now we may present the complete construction of the canonical operator, outlined some pages earlier. Assume that a Lagrangian manifold $(L \times \Omega, i)$ is given. With no loss of generality, we assume that L is connected (otherwise one should consider connected components of L) and we recall that Ω is assumed to be connected and simply connected. We fix:

i) some measure μ on $(L \times \Omega, i)$;

ii) some canonical atlas $\{(\mathcal{U}_j, \pi_{I_j} : \mathcal{U}_j \to \mathcal{V}_j)\}_{j \in \mathcal{J}}$ of $(L \times \Omega, i)$ and some $C^{\infty,0}_{\alpha,\omega}$-partition of the unity, $\{e_j\}_{j \in \mathcal{J}}$ on $L \times \Omega$, subordinate to the covering $\{\mathcal{U}_j\}_{j \in \mathcal{J}}$;

iii) the family of cut-off functions $\{\chi_j(x, \xi_{\bar{I}_j}, \omega)\}_{j \in \mathcal{J}}$ in the canonical charts, satisfying the condition: for any compact set $K \subset \mathbb{R}^n \times \Omega$ and any function $\phi \in \mathcal{A}[h]$, $\chi_j(x, \xi_{\bar{I}_j}, \omega) = 0$ when $(x, \omega) \in K$ and $(x_{I_j}, \xi_{\bar{I}_j}, \omega) \in \pi_{I_j}(\operatorname{supp} \phi \cap \mathcal{U}_j)$ for almost all (i.e., for all except some finite subset) $j \in \mathcal{J}$; the existence of such a family is proved in Proposition 8 below and such a family will be called concordant with the canonical atlas;

iv) the point $\alpha^{(0)} \in \Omega$ which will be called the initial point;

v) some continuous branch $\arg \frac{D\mu}{D\sigma}(\alpha^{(0)}, \omega)$ of the argument of $D\mu/D\sigma$ for $\alpha^{(0)}$ fixed, $\omega \in \Omega$ (such a branch exists since Ω is assumed to be simply connected).

To determine uniquely the elementary canonical operators K_j corresponding to the canonical charts $(\mathcal{U}_j, \pi_{I_j})$ according to Definition 10, we have only to fix the choice of the phase function $S_j(x_I, \xi_{\bar{I}}, \omega)$ and of a continuous branch of $\arg \mu_{I_j}(x_{I_j}, \xi_{\bar{I}_j}, \omega)$ in $\mathcal{V}_j$ for each $j \in \mathcal{J}$. We perform this in the following way. For each $j \in \mathcal{J}$, we choose a point $\alpha^{(j)} \in \tilde{\mathcal{U}}_j$, which will be called the central point of $\mathcal{U}_j$ and a differentiable path

$$\gamma_j : [0,1] \to L, \ \gamma_j(0) = \alpha^{(0)}, \ \gamma_j(1) = \alpha^{(j)}. \tag{44}$$

<u>Definition 11.</u> We set

$$W_j(\alpha, \omega) = \int_{\gamma_j} i^* \omega^1 + \int_{\alpha^{(j)}}^{\alpha} i^* \omega^1 - p_{\bar{I}_j}(\alpha, \omega) q_{\bar{I}_j}(\alpha, \omega), \ \alpha \in \tilde{\mathcal{U}}_j, \tag{45}$$

$$\arg_j \frac{D\mu}{D\sigma}(\alpha, \omega) = \arg \frac{D\mu}{D\sigma}(\alpha^{(0)}, \omega) + \int_{\gamma_j} d(\arg \frac{D\mu}{D\sigma}) + \int_{\alpha^{(j)}}^{\alpha} d(\arg \frac{D\mu}{D\sigma}), \ \alpha \in \tilde{\mathcal{U}}_j \tag{46}$$

for any $\omega \in \Omega$; $\int_{\alpha^{(j)}}^{\alpha}$ is taken over any path lying in $\tilde{\mathcal{U}}_j$ and connecting the points $\alpha^{(j)}$ and α (note that the form $d(\arg \frac{D\mu}{D\sigma})$ is defined correctly, since various branches of $\arg \frac{D\mu}{D\sigma}$ differ by a constant)

$$\begin{aligned} \arg \mu_{I_j}(x_{I_j}, \xi_{\bar{I}_j}, \omega) &= [(\arg_j \frac{D\mu}{D\sigma}) \circ \pi_{I_j}^{-1}](x_{I_j}, \xi_{\bar{I}_j}, \omega) + \\ &+ \arg \sigma_{I_j}(x_{I_j}, \xi_{\bar{I}_j}, \omega), \ (x_{I_j}, \xi_{\bar{I}_j}, \omega) \in \mathcal{V}_j, \end{aligned} \tag{47}$$

where the branch of $\arg \sigma_{I_j}(x_{I_j}, \xi_{\bar{I}_j}, \omega)$ is chosen in the following special way:

$$\arg \sigma_{I_j}(x_{I_j},\xi_{\bar{I}_j},\omega) = \sum_{k=1}^{n} \arg \lambda_k - \frac{\pi}{2}|\bar{I}_j|, \tag{48}$$

where $-\frac{\pi}{2} < \arg \lambda_k < \frac{\pi}{2}$, λ_k are the eigenvalues of the matrix $(\partial(q_{I_j} - ip_{I_j}, p_{\bar{I}_j} + iq_{\bar{I}_j}))/\partial(q_{I_j}, p_{\bar{I}_j})$, counted with their multiplicities. We also set

$$S_j(x_{I_j},\xi_{\bar{I}_j},\omega) = (W_j \circ \pi_{I_j}^{-1})(x_{I_j},\xi_{\bar{I}_j},\omega), \quad (x_{I_j},\xi_{\bar{I}_j},\omega) \in V_j. \tag{49}$$

Proposition 7. The above definition is correct.

Proof. We have to show that (a) W_j given by (45) satisfies the Pfaff equation (see Definition 8); (b) λ_k used in (48) lie in the right half-plane; and (c) (47) really gives some branch of the argument $\arg \mu_{I_j}(x_{I_j}, \xi_{\bar{I}_j},\omega)$.

(a) Since $\tilde{U}_j$ is shrinkable and $i^*\omega^1$ is closed, (45) is a correctly defined expression ((46) is as well). Next, we have

$$dW_j = i^*\omega^1 - d(i^* p_{\bar{I}_j} q_{\bar{I}_j}) = i^*(pdq - p_{\bar{I}_j} dq_{\bar{I}_j} - q_{\bar{I}_j} dp_{\bar{I}_j}) =$$

$$= i^*(p_{I_j} dq_{I_j} - q_{\bar{I}_j} dp_{\bar{I}_j}) = i^*\omega^1_I$$

as desired.

(b) Under the canonical transformation S_{I_j} (26), the matrix $\dfrac{\partial(q_{I_j} - ip_{I_j}, p_{\bar{I}_j} + iq_{\bar{I}_j})}{\partial(q_{I_j}, p_{\bar{I}_j})}$ becomes $\dfrac{\partial(q-ip)}{\partial q}$, and we have shown already in the proof of Proposition 4 that the spectrum of the latter matrix lies in the right half-plane. Next,

$$\sigma_{I_j}(x_{I_j},\xi_{\bar{I}_j},\omega) = (-i)^{|\bar{I}_j|} \det \frac{\partial(q_{I_j} - ip_{I_j}, p_{\bar{I}_j} + iq_{\bar{I}_j})}{\partial(q_{I_j}, p_{\bar{I}_j})} \tag{50}$$

so that (48) really gives some branch of the argument $\arg \sigma_{I_j}$.

(c) This follows from the fact that

$$\mu_{I_j} = (\frac{D\mu}{D\sigma} \circ \pi_{I_j}^{-1}) \circ \sigma_{I_j}. \tag{51}$$

The proposition is proved.

We define now the elementary canonical operator K_j, corresponding to the canonical chart (U_j, π_{I_j}) by virtue of Definition 10, and the pre-canonical operator

$$\overset{0}{K} : A[h] \to \hat{H}_{h,loc}(\mathbb{R}^n \times \Omega \times (0,1]) \tag{52}$$

by virtue of the formula

$$\overset{0}{K}\phi = \sum_{j\in J} K_j e_j \phi \equiv \sum_{j\in J} \bar{F}^{1/h}_{\xi_{\bar{I}_j} \to x_{\bar{I}_j}} \{\chi_j(x,\xi_{\bar{I}_j},\omega) \times$$

$$\times\, e^{(i/h)S_j(x_{I_j},\xi_{\bar{I}_j},\omega)} \sqrt{\mu_{I_j}(x_{I_j},\xi_{\bar{I}_j},\omega)}\,(e_j\phi \circ \pi^{-1}_{I_j})(x_{I_j},\xi_{\bar{I}_j},\omega,h)\} \tag{53}$$

(recall once more that the argument of the square root in (53) is assumed to be equal to one half of the expression (47)).

Definition 12. The canonical operator

$$K : A[h] \to \hat{H}_{h,\ell oc}(\mathbb{R}^n \times \tilde{\Omega}) \tag{54}$$

is a composition of the pre-canonical operator $\overset{0}{K}$ with the natural restriction map $\hat{H}_{h,\ell oc}(\mathbb{R}^n \times \Omega \times (0,1]) \to \hat{H}_{h,\ell oc}(\mathbb{R}^n \times \tilde{\Omega})$. Here the set $\tilde{\Omega} \subset \Omega \times$ $\times (0,1]$ is selected by quantization conditions as given below:

Definition 13. (a) The pair $(\omega,h) \in \Omega \times (0,1]$ satisfies the quantization condition if the cohomology class

$$\Theta(\omega,h) \equiv \frac{1}{h}[i^*\omega^1] + \frac{1}{2}[d \arg \frac{D\mu}{D\sigma}] \in H^1(L,\mathbb{R}) \tag{55}$$

is trivial modulo 2π, i.e., for any 1-cycle γ on L we have

$$<\Theta(\omega,h),[\gamma]> \equiv \frac{1}{h}\oint_\gamma i^*\omega^1 + \frac{1}{2}\operatorname*{var}_{\gamma} \arg \frac{D\mu}{D\sigma} = 2\pi k \tag{56}$$

with some integer $k = k(\gamma)$ (here $\operatorname*{var}_{\gamma} \arg \frac{D\mu}{D\sigma} \equiv \oint_\gamma d \arg \frac{D\mu}{D\sigma}$ is the variation of $\arg \frac{D\mu}{D\sigma}$ along γ).

(b) $\tilde{\Omega} \subsetneq \Omega \times (0,1]$ is the set of all pairs (ω,h), satisfying the quantization condition.

Note. It is clear that it is enough to verify (56) for some collection $\{\gamma_s\}$ of cycles such that $\{[\gamma_s]\}$ is a base of the homology group $H_1(L,\mathbb{R})$. The set Ω is selected by the system of equations:

$$< \Theta(\omega,h),[\gamma_s]> \equiv 0 (\mathrm{mod}\ 2\pi) \tag{57}$$

which turns out to be a finite system in most of the applications. Also the observation that the second summand in (55) does not depend in fact on (ω,h) is useful. Indeed, Ω is assumed to be connected, $<[d \arg \frac{D\mu}{D\sigma}],[\gamma]>$ depends on parameters continuously and may take only discrete series of values, multiple to 2π, therefore being a constant. Thus, (57) may be reduced to a system

$$\frac{1}{2\pi h}\oint_{\gamma_s} i^*\omega(pdq) = \begin{cases} k \\ k + 1/2 \end{cases}, \tag{58}$$

depending on whether $\operatorname*{var}_{\gamma_s} \arg \frac{D\mu}{D\sigma}$ is a multiple of 4π or not. (58) is the well-known quantum-mechanical quantization condition (where parameters ω usually include energy and other physical characteristics of the considered system). For further information on quantization conditions see item D of the current section.

Proposition 8. (a) There exists a family of cut-off functions $\{\chi_j \times \times (x,\xi_{\bar{I}_j},\omega)\}_{j\in J}$ concordant with the given canonical atlas.

(b) For the concordant family of cut-off functions the right-hand side of the equality (53) is a locally finite sum (thus, the definition of the pre-canonical operator is correct).

Proof. (a) Let $\chi \in C_0^\infty(\mathbb{R})$ be a function, satisfying the conditions: $\chi(z) = 1$ for $|z| \leqslant 1/2$ and $\chi(z) = 0$ for $|z| \geqslant 1$. We define the cut-off function $\chi_j(x,\xi_{\bar{I}_j},\omega)$ on $\mathbb{R}^{|\bar{I}_j|}_{x_{\bar{I}_j}} \times V_j$ by

$$\chi_j(x,\xi_{\bar{I}_j},\omega) = \chi(|x_{\bar{I}_j} - q_{\bar{I}_j}(x_{I_j},\xi_{\bar{I}_j},\omega)|^2), \tag{59}$$

where $|\cdot|^2$ is the square of the usual Euclidean norm in $\mathbb{R}^{|\bar{I}|}$.

We claim that the function (59) satisfies the conditions of Definition 9 and that the family $\{\chi_j\}_{j\in J}$ is concordant with the canonical atlas. Indeed, let $\phi \in A[h]$. If A denotes the support supp ϕ, the restriction $\pi|_A : A \to \mathbb{R}^n_x \times \Omega$ is a proper mapping. Now let $K \subset \mathbb{R}^n_x \times \Omega$ be a fixed compact set. We denote by $K_1 \subset \mathbb{R}^n_x \times \Omega$ the compact set, consisting of the points $(x,\omega) \in \mathbb{R}^n_x \times \Omega$ such that $\mathrm{dist}(x,x') \leqslant 1$ for some $(x',\omega) \in K$ (the distance is defined in terms of the usual Euclidean norm in $\mathbb{R}^n_x$). Assume that $x \in K$, $(x,\xi_{\bar{I}_j},\omega) \in \mathrm{supp}\,\chi_j$ and $(\alpha,\omega) = \pi^{-1}_{\bar{I}_j}(x_{I_j},\xi_{\bar{I}_j},\omega) \in A$. If follows from the definition of χ_j that $(x_{I_j},q_{\bar{I}_j}(x_{I_j},\xi_{\bar{I}_j},\omega),\omega) \equiv \pi(\alpha,\omega) \in K_1$, so that (α,ω) belongs to a compact set $\pi^{-1}(K_1) \cap A$. Since the canonical covering is locally finite, we conclude that for at most a finite number of elements $j \in J$ all the inclusions mentioned above may be valid, i.e., for $x \in K$ only a finite number of terms on the right-hand side of (53) may be non-zero. Further, from the above arguments it follows that $(x_{I_j},\xi_{\bar{I}_j},\omega)$ belongs to a compact set $\pi_{I_J}(\pi^{-1}(K_1) \cap A)$, so that the estimate $|\xi_{\bar{I}_j}| < R$ for some R large enough is valid, i.e., the conditions of Definition 9 are satisfied. The proposition is thereby proved.

Now we come directly to the comparison of elementary canonical operators on the intersections of the canonical charts which will give us the foundation of the introduced quantization conditions.

Theorem 4. Let U_j, U_k be any pair of elements of the canonical covering with non-empty intersection $U_{jk} = U_j \cap U_k$. There exist:

i) numbers $C^{(1)}_{jk}(\omega) \in \mathbb{R}$ and $C^{(2)}_{jk} \in \pi\mathbb{Z}$, where $C^{(1)}_{jk}(\omega)$ continuously depend on ω,

ii) differential operators V^s_{jk}, $s=0,1,2,\ldots$, on L of the order $\leqslant 2s$ with the coefficients independent of h defined in the intersection U_{jk} and belonging to $C^{\infty,0}_{\alpha,\omega}(U_{jk})$,

such that for any $\phi \in A_{U_{ij}}[h]$ and any natural N the equality holds:

$$K_k\phi = \exp\{ \frac{i}{h} c^{(1)}_{jk}(\omega) + ic^{(2)}_{jk}\} K_j V^{(N)}_{jk}\phi + \hat{O}(h^{N+1}), \tag{60}$$

where

$$V^{(N)}_{jk} \equiv V^{(N)}_{jk}[h] = \sum_{s=0}^{N} (-ih)^s V^s_{jk}; \tag{61}$$

$\hat{O}(h^{N+1})$ is meant in the space $H_{h,loc}(\mathbb{R}^n \times \Omega \times (0,1])$. $c^{(1)}_{jk}$ and V^s_{jk} are uniquely defined by these conditions, while $c^{(2)}_{jk}$ is defined uniquely modulo the multiple of 2π.

Theorem 5. The numbers $c^{(1)}_{jk}(\omega)$ and $c^{(2)}_{jk}$ and the operators $V^{(N)}_{jk}$ introduced in the preceding theorem satisfy the following properties:

(a) $V^{(N)}_{jj}$ is the identity operator, $V^{(N)}_{jj} = 1$ for any j and N. For any non-empty intersection $\mathcal{U}_{jk}$, V^0_{jk} is the identity operator defined in this intersection, $V^0_{jk} = 1$. If $\mathcal{U}_{jk\ell} = \mathcal{U}_j \cap \mathcal{U}_k \cap \mathcal{U}_\ell$ is non-empty, we have

$$V^{(\infty)}_{jk} \circ V^{(\infty)}_{k\ell} = V^{(\infty)}_{j\ell} \tag{62}$$

(the latter identity is understood as the equality of formal power series in h with the coefficients which are operators in $A_{\mathcal{U}_{jk\ell}}$).

(b) We have (under suitable choice of $c^{(2)}_{jk}$)

$$c^{(1)}_{jk} + c^{(1)}_{k\ell} - c^{(1)}_{j\ell} = 0, \quad c^{(2)}_{jk} + c^{(2)}_{k\ell} - c^{(2)}_{j\ell} = 0, \tag{63}$$

provided that the intersection $\mathcal{U}_{jk\ell}$ is non-empty. Also

$$c^{(1)}_{kk} = c^{(2)}_{kk} = 0 \tag{64}$$

for any $k \in J$.

(c) There are explicit formulas for $c^{(1)}_{jk}(\omega)$, $c^{(2)}_{jk}$. Consider any path $\gamma_{jk} : [0,1] \to L$ such that $\gamma_{jk}(0) = \alpha^{(j)}$, $\gamma_{jk}(1) = \alpha^{(k)}$, and besides, $\gamma_{jk}([0,1/2]) \subset \tilde{\mathcal{U}}_j$ and $\gamma_{jk}([1/2,1]) \subset \tilde{\mathcal{U}}_k$. Then

$$c^{(1)}_{jk}(\omega) = \int_{\gamma_k} i^*\omega^1 - \int_{\gamma_j} i^*\omega^1 - \int_{\gamma_{jk}} i^*\omega^1, \tag{65}$$

$$c^{(2)}_{jk} = \frac{1}{2}\left[\arg_k \frac{D\mu}{D\sigma}(\alpha^{(k)},\omega) - \arg_j \frac{D\mu}{D\sigma}(\alpha^{(j)},\omega) - \underset{\gamma_{jk}}{\mathrm{var}} \arg \frac{D\mu}{D\sigma}(\alpha,\omega)\right] \tag{66}$$

(see Definition 11 for W_k, W_j, $\arg_k \frac{D\mu}{D\sigma}$, $\arg_j \frac{D\mu}{D\sigma}$).

(d) Formulas (65) and (66) enable us to extend the definition of $c^{(1)}_{jk}(\omega)$, $c^{(2)}_{jk}$ onto the set of all pairs (j,k) such that $\tilde{\mathcal{U}}_j \cap \tilde{\mathcal{U}}_k \neq \emptyset$, thus preserving the properties (63) and (64). Therefore $c^{(1)}_{jk}(\omega)$ and $c^{(2)}_{jk}$ are the 1-cocycles of the covering $\{\tilde{\mathcal{U}}_j\}_{j\in J}$ of L and therefore define the cohomology classses $c^{(1)}(\omega)$, $c^{(2)} \in H^1(L,\mathbb{R})$. We have

$$c^{(1)}(\omega) = [i^*\omega^1], \quad c^{(2)} = \frac{1}{2}\left[d \arg \frac{D\mu}{D\sigma}\right]. \tag{67}$$

Theorem 6. (a) We have

$$\frac{1}{h} C_{jk}^{(1)}(\omega) + C_{jk}^{(2)} = 2\pi m \text{ for all } \mathcal{U}_{jk} \neq 0 \tag{68}$$

for some $m \in \mathbb{Z}$ if and only if the quantization condition is satisfied for the pair $(\omega,h) \in \Omega \times (0,1]$.

(b) The canonical operator K, defined by (54): (i) does not depend on the choice of the central points $\alpha^{(j)}$ and paths γ_j; (ii) modulo $\hat{O}(h^\infty)$ does not depend on the choice of the family of the cut-off functions χ_j, concordant with the canonical covering; and (iii) modulo $\hat{O}(h)$ does not depend on the choice of the partition of unity and of the canonical covering.

(c) Thus, the operator

$$K^{(0)} : A \equiv A^0[h] \to \hat{H}_{h,\ell oc}(\mathbb{R}^n \times \tilde{\Omega})/\text{mod } \hat{O}(h) \tag{69}$$

in quotient spaces depends only on the Lagrangian manifold $(L \times \Omega, i)$, measure μ, the initial point $\alpha^{(0)}$, and the prescribed value of arg $\frac{D\mu}{D\sigma}(\alpha^{(0)},\omega^{(0)})$ for some $\omega^{(0)} \in \Omega$.

Note. One might introduce on the right-hand side of (45) the auxiliary additive term, depending only on $\omega \in \Omega$. Once it has been done, the canonical operator will depend on the choice of this term as well.

Proof of Theorems 4, 5 and 6. First of all, we prove that the identity (60) is valid with $C_{jk}^{(1)}$ and $C_{jk}^{(2)}$ given by (65) and (66), respectively.

It is enough to prove (60) for functions ϕ with a compact support. Indeed, if $\psi \in C_{x,\omega}^{\infty,0}(\mathbb{R}^n \times \Omega)$ has a compact support, then $\psi K_k \phi = \psi K_k \phi_1$, where $\phi_1 \in A_{\mathcal{U}_{jk}}[h]$ has a compact support and also $\psi K_j V_{jk}^{(N)} \phi = \psi K_j V_{jk}^{(N)} \phi_1$ (these facts are simple consequences of the definition of the cut-off functions). Next, if the support of ϕ is compact, then

$$K_k \phi = F^{1/h}_{\xi_{\bar{I}_k} \to x_{\bar{I}_k}} \{ e^{(i/h) S_k(x_{I_k},\xi_{\bar{I}_k},\omega)} \sqrt{\mu_{I_k}(x_{I_k},\xi_{\bar{I}_k},\omega)} \times$$
$$\times (\phi \circ \pi_{I_k}^{-1})(x_{I_k},\xi_{\bar{I}_k},\omega,h)\} + \hat{O}(h^\infty), \tag{70}$$

and the analogous formula is valid for the right-hand side of (60). At last, for $\psi \in C_{x,\omega}^{\infty,0}(\mathbb{R}^n \times \Omega)$ with the compact support

$$\| \psi f \|_{s,h} \leqslant \text{const} \| f \|_{s,h} \leqslant \text{const} |\!|\!| f |\!|\!|_{s,h}, \tag{71}$$

the latter inequality being valid by Proposition 1 (b), the constants in (71) depend only on s and ψ. Thus, we have gotten rid of cut-off functions of any sort and, using the invariance of the norm $|\!|\!| \cdot |\!|\!|$ under the 1/h-Fourier transform, we may reduce our problem to the following one:

Given a function $\phi \in A_{\mathcal{U}_{jk}}[h]$ with a compact support, one should verify that

$$F^{1/h}_{x_{\bar{I}_j} \to \xi_{\bar{I}_j}} \bar{F}^{1/h}_{\xi_{\bar{I}_k} \to x_{\bar{I}_k}} \{ e^{(i/h)S_k(x_{I_k}, \xi_{\bar{I}_k}, \omega)} \sqrt{\mu_{I_k}(x_{I_k}, \xi_{\bar{I}_k}, \omega)} \times$$

$$\times (\phi \circ \pi^{-1}_{I_k})(x_{I_k}, \xi_{\bar{I}_k}, \omega, h)\} = \exp\{ \frac{i}{h} C^{(1)}_{jk}(\omega) + iC^{(2)}_{jk}\} \times$$

$$\times e^{(i/h)S_j(x_{I_j}, \xi_{\bar{I}_j}, \omega)} \sqrt{\mu_{I_j}(x_{I_j}, \xi_{\bar{I}_j}, \omega)}((V^{(N)}_{jk}\phi) \circ \pi^{-1}_{I_j}) \times \tag{72}$$

$$\times (x_{I_j}, \xi_{\bar{I}_j}, \omega, h) + R_{N+1}(x_{I_j}, \xi_{\bar{I}_j}, \omega, h),$$

where the remainder R_{N+1} has the estimate

$$|||R_{N+1}|||_{s,h} \leqslant C_s h^{N+1}, \quad s = 0,1,2,\dots . \tag{73}$$

The validity of expansion (72) will be proved by means of the stationary phase method (see Theorem 1 of item A). First of all, we make reduction to the case $I_j = \emptyset$. To perform this, consider the canonical transformation S_{I_j} defined by (26) and set

$$(x', \xi') = S_{I_j}(x, \xi) \equiv (x_{I_j}, \xi_{\bar{I}_j}, \xi_{I_j}, -x_{\bar{I}_j});$$

$$(q', p') = S_{I_j}(q, p) \equiv (q_{I_j}, p_{\bar{I}_j}, p_{I_j}, -q_{\bar{I}_j}). \tag{74}$$

Thus, we define in fact the new Lagrangian manifold $(L \times \Omega, i' : (\alpha, \omega) \to \to (S_{I_j}(q(\alpha,\omega), p(\alpha,\omega)), \omega))$. In the "primed" variables we have now:

$$I'_j = \emptyset, \ \bar{I}'_k = (\bar{I}_j \setminus \bar{I}_k) \cup (\bar{I}_k \setminus \bar{I}_j); \tag{75}$$

$$F^{1/h}_{x_{\bar{I}_j} \to \xi_{\bar{I}_j}} \bar{F}^{1/h}_{\xi_{\bar{I}_k} \to x_{\bar{I}_k}} = e^{-i(\pi/2)|\bar{I}_j \setminus \bar{I}_k|} \bar{F}^{1/h}_{\xi'_{\bar{I}_k} \to x'_{\bar{I}_k}}; \tag{76}$$

$$\arg \mu_{I_j}(x_{I_j}, \xi_{\bar{I}_j}, \omega) = \arg \mu'(x', \omega) - \arg \sigma'(x', \omega) +$$

$$+ \arg \sigma_{I_j}(x_{I_j}, \xi_{\bar{I}_j}, \omega) = \arg \mu'(x', \omega) - \frac{\pi}{2} |\bar{I}_j|; \tag{77}$$

$$\arg \mu_{I_k}(x_{I_k}, \xi_{\bar{I}_k}, \omega) = \arg \mu'_{I'_k}(x'_{I'_k}, \xi'_{\bar{I}'_k}, \omega) -$$

$$- \arg \mu'_{I'_k}(x'_{I'_k}, \xi'_{\bar{I}'_k}, \omega) + \arg \sigma_{I_k}(x_{I_k}, \xi_{\bar{I}_k}, \omega) =$$

$$= \arg \mu'_{I'_k}(x'_{I'_k}, \xi'_{\bar{I}_k}, \omega) - \frac{\pi}{2} |\bar{I}_k| + \frac{\pi}{2} |\bar{I}'_k| \tag{78}$$

(we easily obtain (77) - (78) from (47) - (48); cf. the proof of Proposition 7),

$$|\mu_{I_j}(x_{I_j}, \xi_{\bar{I}_j}, \omega)| = |\mu'(x', \omega)|,$$

$$|\mu_{I_k}(x_{I_k},\xi_{\bar{I}_k},\omega)| = |\mu'_{I'_k}(x'_{I'_k},\xi'_{\bar{I}_k},\omega)|,$$
$$S_k(x_{I_k},\xi_{\bar{I}_k},\omega) = S'_k(x'_{I'_k},\xi'_{\bar{I}'_k},\omega), \tag{79}$$
$$c^{(1)\prime}_{jk}(\omega) = c^{(1)}_{jk}(\omega),\ c^{(2)\prime}_{jk} = c^{(2)}_{jk},$$

etc. Since

$$-\frac{\pi}{2}|\bar{I}_j\setminus\bar{I}_k| - \frac{\pi}{4}|\bar{I}_k| + \frac{\pi}{4}|\bar{I}'_k| =$$
$$= -\frac{\pi}{2}|\bar{I}_j\setminus\bar{I}_k| - \frac{\pi}{4}|\bar{I}_k| + \frac{\pi}{4}|\bar{I}_j\setminus\bar{I}_k| + \frac{\pi}{4}|\bar{I}_k - \bar{I}_j| = \tag{80}$$
$$= -\frac{\pi}{4}(|\bar{I}_j\setminus\bar{I}_k| + |\bar{I}_k| - |\bar{I}_k\setminus\bar{I}_j|) = -\frac{\pi}{4}|\bar{I}_j|,$$

(72) becomes in new variables (we omit the primes now):

$$F^{1/h}_{\xi_{\bar{I}_k}\to x_{\bar{I}_k}}\{e^{(i/h)S_k(x_{I_k},\xi_{\bar{I}_k},\omega)}\sqrt{\mu_{I_k}(x_{I_k},\xi_{\bar{I}_k},\omega)}(\phi\circ\pi^{-1}_{I_k})(x_{I_k},\xi_{\bar{I}_k},\omega,h)\} =$$
$$= \exp\{\frac{i}{h}c^{(1)}_{jk}(\omega) + ic^{(2)}_{jk}\}e^{(i/h)S_j(x,\omega)}\sqrt{\mu(x,\omega)}\times \tag{81}$$
$$\times((V^{(N)}_{jk}\phi)\circ\pi^{-1})(x,\omega,h) + R_{N+1}(x,\omega,h),$$

where R_{N+1} is expected to satisfy (73). The left-hand side of (81) may be rewritten in the form

$$I(x,\omega,h) = \frac{e^{i(\pi/4)|\bar{I}_k|}}{(2\pi h)^{|\bar{I}_k|/2}}\int e^{(i/h)[S_k(x_{I_k},\xi_{\bar{I}_k},\omega)+x_{\bar{I}_k}\xi_{\bar{I}_k}]}\times$$
$$\times\sqrt{\mu_{I_k}(x_{I_k},\xi_{\bar{I}_k},\omega)}(\phi\circ\pi^{-1}_{I_k})(x_{I_k},\xi_{\bar{I}_k},\omega,h)d\xi_{\bar{I}_k}; \tag{82}$$

the integrand in (82) has a compact support. We apply Theorem 1 of item A. The equations of the stationary point $\xi_{\bar{I}_k} = \xi_{\bar{I}_k}(x,\omega)$ are

$$x_{\bar{I}_k} + \frac{\partial S_k}{\partial\xi_{\bar{I}_k}}(x_{I_k},\xi_{\bar{I}_k},\omega) = 0; \tag{83}$$

by Proposition 5 they are equivalent to equations of the Lagrangian manifold in the chart $(\mathcal{U}_k,\pi_{I_k})$:

$$x_{\bar{I}_k} = q_{\bar{I}_k}(x_{I_k},\xi_{\bar{I}_k},\omega). \tag{84}$$

Since $\operatorname{supp}\phi\subset\mathcal{U}_k\cap\mathcal{U}_j$, the equation (84) then has a unique solution on $\operatorname{supp}\phi\circ\pi^{-1}_{I_k}$; this solution is given by equations of the Lagrangian manifold in the chart $(\mathcal{U}_j,\pi)$:

$$\xi_{\bar{I}_k} = p_{\bar{I}_k}(x,\omega). \tag{85}$$

Moreover, on $\operatorname{supp}\phi\circ\pi^{-1}_{I_k}$, we have

$$\det\left(-\frac{\partial^2 S_k}{\partial \xi^2_{\bar{I}_k}}\right) = \det \frac{\partial q_{\bar{I}_k}}{\partial \xi_{\bar{I}_k}} = \det \frac{\partial q}{\partial (x_{I_k}, \xi_{\bar{I}_k})} \neq 0. \tag{86}$$

Next, we obtain, using (45) and (65), that

$$S_k(x_{I_k}, p_{\bar{I}_k}(x,\omega),\omega) + x_{\bar{I}_k} p_{\bar{I}_k}(x,\omega) = (W_k \circ \pi^{-1})(x,\omega) + x_{\bar{I}_k} p_{\bar{I}_k}(x,\omega) = \tag{87}$$

$$= ((W_k + i^*(p_{\bar{I}_k} q_{\bar{I}_k})) \circ \pi^{-1})(x,\omega) = \{[(\int_{\gamma_j} + \int_{\alpha(j)}^{\alpha}) i^* \omega_1] \circ \pi^{-1}\}(x,\omega) =$$

$$= \{[(\int_{\gamma_j} + \int_{\gamma_{jk}} - \int_{\gamma_k} + \int_{\gamma k} + \int_{\alpha(k)}^{\alpha}) i^* \omega] \circ \pi^{-1}\}(x,\omega) = C_{jk}^{(1)}(\omega) + S_j(x,\omega).$$

Thus, we obtain

$$I(x,\omega,h) = e^{(i/h)[C_{jk}^{(1)}(\omega) + S_j(x,\omega)]} \{[\det \frac{\partial q}{\partial (x_{I_k}, \xi_{\bar{I}_k})}]^{-1/2} \times$$

$$\times \sum_{s=0}^{N} (-ih)^s V_s [\sqrt{\mu_{I_k}(x_{I_k}, \xi_{\bar{I}_k}, \omega)} (\phi \circ \pi_{I_k}^{-1})(x_{I_k}, \xi_{\bar{I}_k}, \omega)]\} |_{\xi_{\bar{I}_k} = p_{\bar{I}_k}(x,\omega)} +$$

$$+ R_{N+1}(x,\omega,h), \tag{88}$$

where the remainder satisfies the required estimates, V_s are differential (in $\xi_{\bar{I}_k}$) operators of the order $\leqslant 2s$, and $V_0 = 1$. Denoting

$$V_{jk}^s = (\sqrt{\mu_{I_k}})^{-1} V_s \sqrt{\mu_{I_k}}, \quad V_{jk}^{(N)} = \sum_{s=0}^{N-1} (-ih)^s V_{jk}^s \tag{89}$$

(we regard $(x_{I_k}, \xi_{\bar{I}_k}, \omega)$ as the coordinates on $L \times \Omega$ in formula (89)), we may rewrite (88) in the form

$$I(x,\omega,h) = e^{(i/h)[C_{jk}^{(1)}(\omega) + S_j(x,\omega)]} \{[\det \frac{\partial q}{\partial (x_{I_k}, \xi_{\bar{I}_k})}]^{-1/2} \times$$

$$\times \sqrt{\mu_{I_k}(x_{I_k}, \xi_{I_k}, \omega)} ((V_{jk}^{(N)} \phi) \circ \pi_{I_k}^{-1})(x_{I_k}, \xi_{\bar{I}_k}, \omega)\} |_{\xi_{\bar{I}_k} = p_{\bar{I}_k}(x,\omega)} + \tag{90}$$

$$+ R_{N+1}(x,\omega,h).$$

In (89) and (90) the argument

$$\arg \det \frac{\partial q}{\partial (x_{I_k}, \xi_{\bar{I}_k})} = \sum_{m=0}^{n} \arg \lambda_m, \quad -\frac{3\pi}{2} < \arg \lambda_m \leqslant \frac{\pi}{2}, \tag{91}$$

where λ_m are the eigenvalues of the matrix $\frac{\partial q}{\partial (x_{I_k}, \xi_{\bar{I}_k})}$. We have

$$\mu_{I_k}(x_{I_k}, \xi_{\bar{I}_k}, \omega) (\det \frac{\partial q}{\partial (x_{I_k}, \xi_{\bar{I}_k})})^{-1} = \frac{D\mu}{D(x_{I_k}, \xi_{\bar{I}_k})} \frac{D(x_{I_k}, \xi_{\bar{I}_k})}{Dx} = \frac{D\mu}{Dx} \tag{92}$$

(our notations are not completely pure, but it seems that confusion is unlikely to occur). Thus, to prove (60), it remains to show that for our choice of the arguments we have

$$\arg \mu(x,\omega) - \arg \mu_{I_k}(x_{I_k},\xi_{\bar{I}_k},\omega) - \arg\det \frac{\partial q}{\partial(x_{I_k},\xi_{\bar{I}_k})} \equiv -2C^{(2)}_{jk} \pmod{4\pi} \quad (93)$$

(note that modulo 2π (93) is obviously valid). To verify (93), we rewrite its left-hand side in the form

$$\arg\mu - \arg\mu_{I_k} - \arg\det\frac{\partial q}{\partial(x_{I_k},\xi_{\bar{I}_k})} =$$

$$= \int_{\gamma_j} d(\arg\frac{D\mu}{D\sigma}) + \int_{\alpha^{(j)}}^{\alpha} d(\arg\frac{D\mu}{D\sigma}) - \int_{\gamma_k} d(\arg\frac{D\mu}{D\sigma}) - \int_{\alpha^{(k)}}^{\alpha} d(\arg\frac{D\mu}{D\sigma}) + \quad (94)$$

$$+ \arg\sigma - \arg\sigma_{I_k} - \arg\det\frac{\partial q}{\partial(x_{I_k},\xi_{\bar{I}_k})}$$

(here $\alpha \in \tilde{U}_j \cap \tilde{U}_k$, $(x,\xi) = (q(\alpha,\omega),p(\alpha,\omega))$). Since $(\int_{\alpha^{(j)}}^{\alpha} - \int_{\alpha^{(k)}}^{\alpha})d(\arg \times \frac{D\mu}{D\sigma}) = \int_{\gamma_{jk}} d(\arg\frac{D\mu}{D\sigma})$, (93) turns out to be equivalent to

$$\arg\sigma - \arg\sigma_{I_k} - \arg\det\frac{\partial q}{\partial(x_{I_k},\xi_{\bar{I}_k})} \equiv 0 \pmod{4\pi}. \quad (95)$$

To prove that (95) holds, we fix any point $(\alpha,\omega) \in U_{jk}$ and consider the matrix-valued function (we write I instead of I_k thereafter):

$$A(t,\tau) = \frac{\partial(q - ip)}{\partial(\alpha_1,\ldots,\alpha_n)} \cdot \left[\frac{\partial(q_I - i\tau p_I, (q_{\bar{I}} - i\tau p_{\bar{I}})\cos t + (p_{\bar{I}} + i\tau q_{\bar{I}})\sin t)}{\partial(\alpha_1,\ldots,\alpha_n)}\right]^{-1}. \quad (96)$$

In (96) $(\alpha_1,\ldots,\alpha_n)$ is an arbitrary coordinate system on L in the neighborhood of the point α; $A(t,\tau)$ does not depend on the choice of this cooordinate system. We set also

$$J(t,\tau) = \det A(t,\tau). \quad (97)$$

Since σ is a non-degenerate measure, it is clear that $J(t,\tau) \neq 0$ for any (t,τ) for which it is defined. We assert that $A(t,\tau)$ is defined for $\tau > 0$. Indeed, we have

$$\frac{\partial(q_I - i\tau p_I, (q_{\bar{I}} - i\tau p_{\bar{I}})\cos t + (p_{\bar{I}} + i\tau q_{\bar{I}})\sin t)}{\partial(\alpha_1,\ldots,\alpha_n)} = \frac{\partial(\tilde{q} - i\tau\tilde{p})}{\partial(\alpha_1,\ldots,\alpha_n)}, \quad (98)$$

where

$$\tilde{q} = (q_I, q_{\bar{I}}\cos t + p_{\bar{I}}\sin t), \quad \tilde{p} = (p_I, p_{\bar{I}}\cos t - q_{\bar{I}}\sin t). \quad (99)$$

The transformation $(q,p) \to (\tilde{q},\tilde{p})$ of $\mathbb{R}^{2n}$, given by (99) is a canonical one, i.e., it preserves the form ω^2. Thus, $\tilde{i} : L \times \Omega \to \mathbb{R}^{2n}$, $\tilde{i}(\alpha,\omega) = (\tilde{q}(\alpha,\omega), \tilde{p}(\alpha,\omega))$ is a Lagrangian manifold. It is enough to prove that the form $\tilde{\sigma} = \tilde{i}^*(d(\tilde{q}_1 - i\tau\tilde{p}_1) \wedge \ldots \wedge d(\tilde{q}_n - i\tau\tilde{p}_n))$ is non-degenerate. Make a change of variables: $\tilde{\tilde{q}} = \tilde{q}$, $\tilde{\tilde{p}} = \tau\tilde{p}$. Since then $\tilde{\omega}^2 = d\tilde{p} \wedge d\tilde{q} = \tau^{-1} d\tilde{\tilde{p}} \wedge d\tilde{\tilde{q}}$, the mapping $\tilde{\tilde{i}} : (\alpha,\omega) \to (\tilde{\tilde{q}}(\alpha,\omega), \tilde{\tilde{p}}(\alpha,\omega))$ is also a Lagrangian manifold and it suffices to apply Proposition 4. Next, we have obviously

$$J(0,0) = \det\frac{\partial(q - ip)}{\partial x} = \sigma, \quad (100)$$

$$J(\frac{\pi}{2},0) = \det\frac{\partial(q - ip)}{\partial(x_I,\xi_{\bar{I}})} = \sigma_I, \quad (101)$$

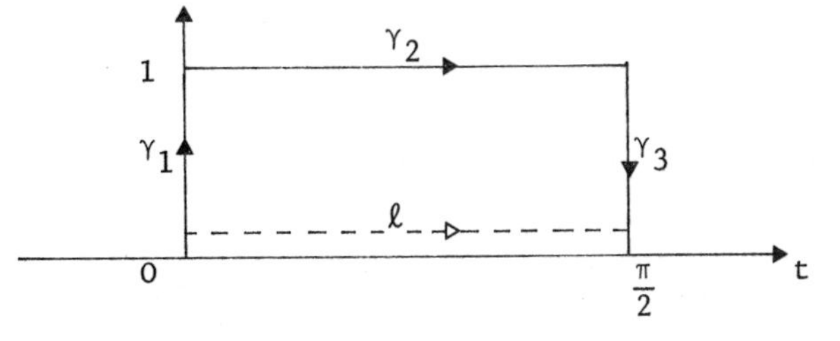

Fig. 1.

$$J(t,1) = \det \frac{\partial(q-ip)}{\partial(\alpha_1,\ldots,\alpha_n)} \cdot \left[\det \frac{\partial(q_I - ip_I, (q_{\bar I} - ip_{\bar I})e^{it})}{\partial(\alpha_1,\ldots,\alpha_n)}\right]^{-1} = e^{-i|\bar I|t}. \quad (102)$$

Consider a connected simply connected domain $\Gamma \subset \mathbb{R}^2_{t,\tau}$ such that Γ contains the half-plane $\tau > 0$, the points $(0,0)$ and $(\frac{\pi}{2},0)$, and $J(t,\tau)$ is defined in Γ (the existence of such a domain follows from the above consideration). The continuous branch $\arg J(t,\tau)$ of the argument of the Jacobian $J(t,\tau)$ in the domain Γ exists and may be fixed by fixing its value in any point of Γ. We fix the choice of it by setting

$$\arg J(0,0) = \arg \sigma \quad (103)$$

((100) implies that the definition (103) is correct). We assert that

$$\arg J(\frac{\pi}{2},0) = \arg \sigma_I. \quad (104)$$

To prove (104), consider in the domain Γ the contour $\gamma = \gamma_1 + \gamma_2 + \gamma_3$, shown in Figure 1. Obviously,

$$\arg J(\frac{\pi}{2},0) = \arg J(0,0) + \underset{\gamma}{\operatorname{var}} \arg J(t,\tau) =$$
$$= \arg \sigma + \underset{\gamma_1}{\operatorname{var}} \arg J(t,\tau) + \underset{\gamma_2}{\operatorname{var}} J(t,\tau) + \underset{\gamma_3}{\operatorname{var}} \arg J(t,\tau). \quad (105)$$

From (102) we obtain immediately

$$\underset{\gamma_2}{\operatorname{var}} \arg J(t,\tau) = -\frac{\pi}{2}|\bar I|. \quad (106)$$

Next, $\underset{\gamma_3}{\operatorname{var}} \arg J(t,\tau) = \underset{\gamma_3}{\operatorname{var}} \arg \tilde J(t,\tau)$, where

$$\tilde J(\frac{\pi}{2},\tau) = e^{i(\pi/2)|\bar I|} J(\frac{\pi}{2},\tau) =$$
$$= \det \frac{\partial(q_I - ip_I, p_{\bar I} + iq_{\bar I})}{\partial(\alpha_1,\ldots,\alpha_n)} \cdot \left[\det \frac{\partial(q_I - i\tau p_I, p_{\bar I} + i\tau q_{\bar I})}{\partial(\alpha_1,\ldots,\alpha_n)}\right]^{-1}. \quad (107)$$

We have also

$$J(0,\tau) = \det \frac{\partial(q-ip)}{\partial\alpha} \left[\det \frac{\partial(q-i\tau p)}{\partial\alpha}\right]^{-1}. \quad (108)$$

We intend to show that

$$\underset{\gamma_1}{\operatorname{var}} \arg J(t,\tau) = -\sum_{m=1}^{n} \arg \lambda_m,$$
$$\underset{\gamma_3}{\operatorname{var}} \arg \tilde J(t,\tau) = \sum_{m=1}^{n} \arg \mu_m, \quad (109)$$

where λ_m, μ_m are the eigenvalues of the matrices $\frac{\partial(q-ip)}{\partial x}$, $\frac{\partial(q_I - ip_I, p_{\bar{I}} + iq_{\bar{I}})}{\partial(x_I, \xi_{\bar{I}})}$, respectively, the values of their arguments being taken in the interval $(-\frac{\pi}{2}, \frac{\pi}{2})$.

The canonical transformation S_I (26) reduces the proof of the second of the equalities (26) to the first one (the difference in sign is the consequence of the fact that γ_1 and γ_3 have opposite directions). Thus, it is enough to prove the first of the equalities (109). To do this, we choose a special coordinate system $(\alpha_1, \ldots, \alpha_n)$, namely, the canonical coordinate system $(x_1, \ldots, x_n)$. In this system we obtain

$$J(0,\tau) = \det\{(E - iB)(E - i\tau B)^{-1}\} \equiv \det M(\tau)m, \tag{110}$$

where $B = \frac{\partial p}{\partial x}$. It was already discovered in the proof of Proposition 4 that B is a symmetric matrix with a non-negative (in fact zero in this section) imaginary part.

We assert that the eigenvalues $\lambda_m(\tau)$, $m = 1, \ldots, n$, of the matrix $M(\tau) = (E - iB)(E - i\tau B)^{-1}$ lie in the open right half-plane for all $\tau \geqslant 0$. Indeed, we have the representation

$$\lambda_m(\tau) = (1 - i\kappa_m)(1 - i\tau\kappa_m)^{-1}, \tag{111}$$

where κ_m are the eigenvalues of B, which lie in the upper half-plane by Lemma 2. We have thus

$$\begin{aligned} \operatorname{Re} \lambda_m(\tau) &= (1 + \tau^2|\kappa_m|^2)^{-1} \operatorname{Re}(1 - i\kappa_m)(1 + i\tau\bar{\kappa}_m) = \\ &= (1 + \tau^2|\kappa_m|^2)^{-1}(1 + \tau|\kappa_m|^2 + (1 + \tau)\operatorname{Im}\kappa_m) > 0. \end{aligned} \tag{112}$$

We have $M(1) = E$, $M(0) = \frac{\partial(q-ip)}{\partial x}$, thus $\lambda_m = \lambda_m(0)$. If we prescribe $\arg \lambda_m(1) = 0$ and define $\arg \lambda_m(\tau)$ for $\tau \geqslant 0$ by continuity, we obtain from the above that $-\frac{\pi}{2} < \arg \lambda_m(0) < \frac{\pi}{2}$. We have thus

$$\operatorname*{var}_{\gamma_1} \arg J(t,\tau) = \sum_{m=1}^{n} (\arg \lambda_m(1) - \arg \lambda_m(0)) = -\sum_{m=0}^{n} \arg \lambda_m, \tag{113}$$

where $\arg \lambda_m \in (-\frac{\pi}{2}, \frac{\pi}{2})$, $m = 1, \ldots, n$, and (109) is proved.

Now (104) follows from (105), (106), (109), and Definition (48) of the measure density arguments. Indeed, combining these identities, we obtain

$$\begin{aligned} \arg J(\tfrac{\pi}{2}, 0) &= \arg \sigma - \sum_{m=0}^{n} \arg \lambda_m - \frac{\pi}{2}|\bar{I}| + \\ &+ \sum_{m=0}^{n} \arg \mu_m = \sum_{m=0}^{n} \arg \mu_m - \frac{\pi}{2}|\bar{I}| = \arg \sigma_I. \end{aligned} \tag{114}$$

Next we show that

$$\arg J(\tfrac{\pi}{2}, 0) - \arg J(0,0) = -\arg \det \frac{\partial q}{\partial(x_I, \xi_{\bar{I}})} \tag{115}$$

(thus, the left-hand side of (95) is in fact precisely equal to zero). Since $\partial(q - ip)/\partial\alpha$ is a constant matrix,

$$\underset{\gamma}{\text{var}} \arg \mathcal{J}(t,\tau) = $$
$$= \underset{\gamma}{\text{var}} \arg \det\left[\frac{\partial(q_I - i\tau p_I, (q_{\bar{I}} - i\tau p_{\bar{I}})\cos t + (p_{\bar{I}} + i\tau q_{\bar{I}})\sin t)}{\partial(\alpha_1,\ldots,\alpha_n)} \right]. \tag{116}$$

Choosing the canonical coordinates $(x_1,\ldots,x_n)$ as $(\alpha_1,\ldots,\alpha_n)$, we obtain

$$\underset{\gamma}{\text{var}} \arg \mathcal{J}(t,\tau) = \underset{\gamma}{\text{var}} \arg \det(C(t,\tau) - iB(t,\tau)) = \sum_{j=1}^{m} \underset{\gamma}{\text{var}} \arg \lambda_j(t,\tau), \tag{117}$$

where $\lambda_j(t,\tau)$ are (continuously dependent on t,τ) eigenvalues of $C(t,\tau) - iB(t,\tau)$; $\arg \lambda_j(t,\tau)$ are continuously branches of their arguments,

$$C(t,\tau) = \frac{\partial(q_I, q_{\bar{I}}\cos t + p_{\bar{I}}\sin t)}{\partial x},$$
$$B(t,\tau) = \tau \frac{\partial(p_I, p_{\bar{I}}\cos t - q_{\bar{I}}\sin t)}{\partial x}. \tag{118}$$

Both C and B are symmetric matrices, and the imaginary part of $C - iB$ is negative semi-definite (recall that $t \in [0,\frac{\pi}{2}]$). By Lemma 2, the spectrum of $C - iB$ lies in the lower half-plane, as $(t,\tau) \in \gamma$; thus the branches $\arg \lambda_j(t,\tau)$ may be chosen satisfying the conditions

$$-\pi \leqslant \lambda_j(t,\tau) \leqslant 0. \tag{119}$$

Since

$$C(0,0) - iB(0,0) = E,$$
$$C(\tfrac{\pi}{2},0) - iB(\tfrac{\pi}{2},0) = \frac{\partial(q_I, p_{\bar{I}})}{\partial x} = \left(\frac{\partial q}{\partial(x_I, \xi_{\bar{I}})}\right)^{-1}, \tag{120}$$

we obtain immediately that (115) is valid.

Hence, we have proved that the assertion of Theorem 4 is valid under the choice of $C_{jk}^{(1)}$, $C_{jk}^{(2)}$ described in Theorem 5 (c). The uniqueness of $C_{jk}^{(1)}$, $C_{jk}^{(2)}$, V_{jk}^{s} and the property (a) of Theorem 5 are the consequences of the following:

Proposition 9. The elementary canonical operator $K_{e\ell}$ is asymptotically monomorphic; more precisely, this means that the following conditions are equivalent for any N:

(a) $\phi \in I_{\mathcal{U}}^{N}[h]$;

(b) $K_{e\ell}\phi = \hat{O}(h^N)$.

Also if $K_{e\ell}$ corresponds to a non-singular canonical chart, condition (a) is equivalent to:

(b') $K_{e\ell}\phi = O(h^N)$. See [52] for proof of Proposition 9.

In addition Theorem 5 (b) follows from Proposition 9; however, we give below its direct proof, which also provides the validity of Theorem 5 (d). We observe first that the formulas of Theorem 5 (c) give the correct definition of $C_{jk}^{(1)}(\omega)$, $C_{jk}^{(2)}$ for any non-empty intersection $\tilde{\mathcal{U}}_j \cap \tilde{\mathcal{U}}_k$. We should verify that

$$C_{jk}^{(i)} + C_{k\ell}^{(i)} + C_{\ell j}^{(i)} = 0, \quad i = 1,2, \tag{121}$$

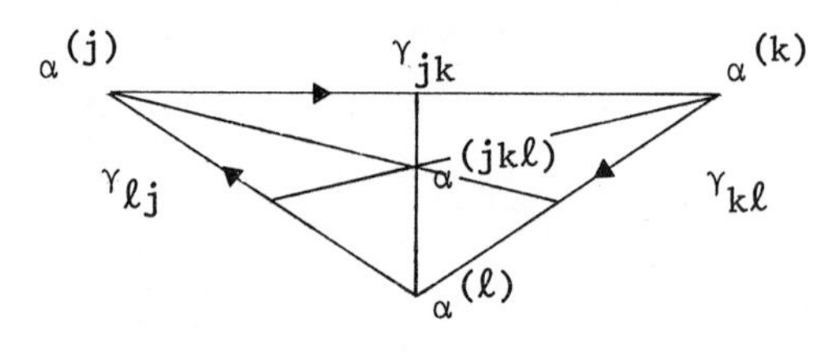

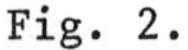

Fig. 2.

if $\tilde{U}_{jk\ell} = \tilde{U}_j \cap \tilde{U}_k \cap \tilde{U}_\ell \neq \emptyset$, and that

$$c_{jj}^{(i)} = 0, \quad i = 1,2, \quad j \in J. \tag{122}$$

The latter is evident since the path γ_{jj} is homotopic to a constant path $\tilde{\gamma}_{jj}(t) \equiv \alpha^{(j)}$, $t \in [0,1]$. To prove (121), consider the system of paths on L, shown in Figure 2.

Here $\alpha^{(jk\ell)}$ lies in $\tilde{U}_{jk\ell}$, each of the small "triangles" lies in a single chart and is therefore homotopic to zero. We have

$$c_{jk}^{(i)} + c_{k\ell}^{(i)} + c_{\ell j}^{(i)} = \int_\Gamma \omega^{(i)}, \quad i = 1,2, \tag{123}$$

where $\Gamma = \gamma_{jk} + \gamma_{k\ell} + \gamma_{\ell j}$ and $\omega^{(i)}$ is a closed 1-form,

$$\omega^{(1)} = i^*\omega^1, \quad \omega^{(2)} = \frac{1}{2}\, d \arg \frac{D\mu}{D\sigma}. \tag{124}$$

Thus (123) may be represented as the sum of integrals over small triangles, the latter being equal to zero, so that (121) is proved. It follows that $\{c_{jk}^{(i)}\}$ represent some cohomology classes of the covering $\{\tilde{U}_j\}$, hence cohomology classes of L, since the covering $\{\tilde{U}_j\}$ satisfies the conditions of Leray's theorem which asserts that cohomologies of the open covering are isomorphic to cohomologies of the manifold itself, provided that the sets forming the covering and all their finite intersections have trivial cohomologies. It is not hard to establish that these classes are exactly

$$c^{(1)} = [i^*\omega^1], \quad c^{(2)} = \frac{1}{2}\left[d \arg \frac{D\mu}{D\sigma}\right]. \tag{125}$$

Indeed,

$$c_{jk}^{(i)} = b_k^{(i)} - b_j^{(i)} + \tilde{c}_{jk}^{(i)} = (\partial b^{(i)})_{jk} + \tilde{c}_{jk}^{(i)}, \quad i = 1,2, \tag{126}$$

$$\tilde{c}_{jk}^{(1)} = \int_{\gamma_{jk}} i^*\omega^1, \quad \tilde{c}_{jk}^{(2)} = \frac{1}{2}\int_{\gamma_{jk}} d \arg \frac{D\mu}{D\sigma}, \tag{127}$$

i.e., $c_{jk}^{(i)}$ are cohomological to $\tilde{c}_{jk}^{(i)}$ given by (127) and hence represent the necessary classes. Thus, Theorems 4 and 5 are proved, and it remains only to prove Theorem 6.

(a) For fixed $\omega \in \Omega$, consider the subcovering $\{\tilde{U}_j\}_{j \in J_\omega}$ of the covering $\{\tilde{U}_j\}_{j \in J}$, consisting of all $\tilde{U}_j$ such that the projection of ${}^\omega U_j$ onto Ω contains ω. Clearly $\{\tilde{U}_j\}_{j \in J_\omega}$ is a covering of L, satisfying the conditions of Leray's theorem. Let now

$$\frac{i}{h}\, c_{jk}^{(1)}(\omega) + c_{jk}^{(2)} = 2\pi m, \quad m = m(j,k) \in \mathbb{Z} \tag{128}$$

for any $j,k \in J$, $\tilde{U}_{jk} \neq \emptyset$. Then all the more (128) is satisfied for $j,k \in \in J_\omega$, $U_{jk} \neq \emptyset$. If $\gamma : [0,1] \to L$ is any closed path on L, then it is obviously homological to a finite algebraic sum of the closed paths $\Gamma_{jk} = \gamma_j + \gamma_{jk} - \gamma_k$, $j,k \in J_\omega$, and therefore

$$< \Theta(\omega,h),[\gamma]> = \int_\gamma (\frac{i}{h} i^*\omega^1 + \frac{1}{2} d \arg \frac{D\mu}{D\sigma}) = 2\pi k(\gamma),\ k(\gamma) \in \mathbb{Z}. \quad (129)$$

Conversely, if the quantization condition is satisfied, then (128) is valid since it is a particular case of (129) for $\gamma = \Gamma_{jk}$. Note that we have proved that it is sufficient to require (128) only for $j,k \in J_\omega$. Next we show that it is also sufficient to require (128) only for j,k such that the projection of $U_j \cap U_k$ onto Ω contains ω. This means that, in a sense, the quantization condition is not only sufficient, but necessary as well for the existence of the canonical operator, mod $\hat{O}(h)$ independent of the choice of the unity partition and other auxiliary objects (we say "in a sense" since the structure $\tilde{\Omega}$ outside the subset $\{h \leqslant \varepsilon\}$, ε being any fixed positive, plays no role in the construction). Thus, assume that (128) is valid for that such j,k $U_j^\omega \cap U_k^\omega \neq \emptyset$ (here U^ω denotes the intersection of $U \subset L \times \Omega$ with $L \times \{\omega\}$).

Now let j,k be such that $\tilde{U}_j \cap \tilde{U}_k \neq \emptyset$. Since $\{U_j^\omega\}_{j \in J}$ is an open covering of L (and obviously $U_j^\omega \subset \tilde{U}_j$), we may cover the path γ_{jk} by a finite sequence $U_{j_1}^\omega, \ldots, U_{j_m}^\omega$ of open sets $j_1 = j$, $j_m = k$, and $U_{j_i}^\omega \cap U_{j_{i+1}}^\omega$, $i = 1,\ldots,m-1$. We obtain thus:

$$\frac{1}{h} C_{jk}^{(1)} + C_{jk}^{(2)} = \sum_{i=1}^{m-1} (\frac{1}{h} C_{j_i j_{i+1}}^{(1)} + C_{j_i j_{i+1}}^{(2)}) = 2\pi m(j,k),\ m(j,k) \in \mathbb{Z}. \quad (130)$$

(b) The independence of the canonical operator of the choice of a concordant family of cut-off functions is clear from Theorem 3 (b). As for its independence of the choice of central points $\alpha^{(j)}$ and the paths γ^j, this follows from the fact that if $\Theta(\omega,h)/2\pi$ is an integer cohomology class on L, we have

$$\frac{1}{h} S_j + \frac{1}{2} \arg \mu_{I_j} = (\frac{1}{h} W_j + \frac{1}{2} \arg_j \frac{D\mu}{D\sigma}) \circ \pi_{I_j}^{-1} \equiv \{\int_\gamma \frac{1}{h} i^*\omega^1 +$$
$$+ \int_\gamma \frac{1}{2} d \arg \frac{D\mu}{D\sigma} - p_{\overline{I}_j} q_{\overline{I}_j} \} \circ \pi_{I_j}^{-1} + \arg \frac{D\mu}{D\sigma} (\alpha^{(0)},\omega) \pmod{2\pi}, \quad (131)$$

where $\gamma = [0,1] \to L$ is any path with $\gamma(0) = \alpha^{(0)}$, $\gamma(1) = \alpha$, $(\alpha,\omega) \in U_j$, i.e., the left-hand side of (131) modulo 2π does not depend on the choice of $\alpha^{(j)}$ and $\gamma^{(j)}$.

(c) The independence of

$$K^{(0)} : A^0[h] \equiv A \to \hat{H}_{h,\ell oc}(\mathbb{R}^n \times \tilde{\Omega})/\text{mod } \hat{O}(h)$$

of the choice of the canonical atlas and the unity partition is also clear. Indeed, the union of canonical atlases (subdivided if necessary) is a canonical atlas itself; thus it is enough to prove the invariance under the choice of the unity partition. Let $\{e_j\}$, $\{e_j'\}$ be unity partitions, subordinate to the canonical covering; then since $V_{jk}^0 = 1$, we have the following sequence of identities in the quotient space $\hat{H}_{h,\ell oc}(\mathbb{R}^n \times \Omega)/$ /mod $\hat{O}(h)$:

$$\sum_j K_j e_j \phi \equiv \sum_{j,j'} K_j e_j e'_{j'} \phi \equiv \sum_{j,j'} K_{j'} e_j e'_{j'} \phi \equiv \sum_{j'} K_{j'} e'_{j'} \phi, \quad \phi \in A, \tag{132}$$

and the desired assertion is proved. The proof of Theorems 4, 5, and 6 is now complete.

C. Commutation of a Pseudo-Differential Operator and a Canonical Operator

In this item we establish the commutation formulas for λ-PDO and the canonical operator. Here $\lambda = 1/h$ is a large parameter; the proofs in the form given here were first proposed in [56], and we use here the material of [56] with slight shortenings.

Our task is to derive asymptotic solutions for $\lambda \to \infty$ of the equation

$$L(\overset{2}{x}, \lambda^{-1}\overset{1}{D_x}; (i\lambda)^{-1})u(x,\lambda) = 0, \quad (D_x = -i\partial/\partial x).$$

If $Lu = \sum_{k=0}^{m} a_k (\lambda^{-1}D_x)^k u$, $x \in R^1$, and the coefficients a_k are constant, then these solutions can be sought in the form $\exp(i\lambda S(x))$. Moreover, whenever the characteristic equation has simple roots, any solution will be a linear combination of the exponential solutions. Thus in accordance with this example, we shall look for a solution in the form of a formal series

$$\exp[i\lambda S(x)] \sum_{k=0}^{\infty} (i\lambda)^{-k} \phi_k(x),$$

or, more generally, in the form $K\phi$, where K is the canonical operator as described above. It is necessary to specify how a λ-PDO acts on a rapidly oscillating exponential, i.e., on a function of the form $\phi(x)\exp[i\lambda S(x)]$.

Let $x \in R^1$; then

$$\left(\frac{1}{i\lambda}\frac{d}{dx}\right)\exp[i\lambda S(x)] = S'(x)\exp[i\lambda S(x)];$$

$$\left(\frac{1}{i\lambda}\frac{d}{dx}\right)^2\exp[i\lambda S(x)] = [(S'(x))^2 + \frac{1}{i\lambda} S'(x)S''(x)]\exp[i\lambda S(x)];$$

$$\left(\frac{1}{i\lambda}\frac{d}{dx}\right)^m(\exp[i\lambda S(x)]) =$$

$$= \exp[i\lambda S(x)][(S'(x))^m + \frac{m(m-1)}{2i\lambda} (S'(x))^{m-1} S''(x) + O(\lambda^{-2})].$$

By using the Leibniz formula, we obtain

$$\left(\frac{1}{i\lambda}\frac{d}{dx}\right)^m[\phi(x)\exp(i\lambda S(x))] =$$

$$= \exp[i\lambda S(x)][(S'(x))^m\phi(x) + \frac{m}{i\lambda} (S''(x))^{m-1}\phi'(x) +$$

$$+ \frac{m(m-1)}{2i\lambda} (S'(x))^{m-2} S''(x)\phi(x) + \sum_{j=2}^{m} (i\lambda)^{-j} (R_j\phi)(x)],$$

where R_j is a differential operator of order j. Operator $((1/i\lambda)/(d/dx))^m$ has the symbol $L(p) = p^m$, and since $dL/dp = mp^{m-1}$, $d^2L/dp^2 = m(m-1)p^{m-2}$, we get for the operator in question the following expression:

$$L(\lambda^{-1}D_x)[\phi(x)\exp(i\lambda S(x))] = \exp(i\lambda S(x))[L(S'(x))\phi(x) +$$

$$+ \frac{1}{i\lambda} \frac{dL(S''(x))}{dp} \phi'(x) + \frac{1}{2i\lambda} \frac{d^2L(S'(x))}{dp^2} S''(x)\phi(x) + \ldots].$$

This formula is true also for differential operators $L(\overset{2}{x},\lambda^{-1}\overset{1}{D_x})$ with variable coefficients, since we differentiate first with respect to x, and then multiply the obtained expressions by functions of x. Finally, as the operators $\partial/\partial x_j$, $\partial/\partial x_k$ commute on smooth functions, we have

$$L(x,\lambda^{-1}D_x;(i\lambda)^{-1})[\phi(x)\exp(i\lambda S(x))] = \\ = \exp(i\lambda S(x)) \sum_{j=0}^{m} (i\lambda)^{-j} R_j(x,D_x)\phi(x), \tag{1}$$

where R_j are linear differential operators of order j. If in particular L is a differential operator of order m, the coefficients of which are polynomials in $(i\lambda)^{-1}$, then we obtain

$$(R_o\phi)(x) = L(x, \frac{\partial S(x)}{\partial x};0)\phi(x),$$

$$(R_1\phi)(x) = \left\langle \frac{\partial L(x,(\partial S/\partial x);0)}{\partial p}, \frac{\partial \phi(x)}{\partial x}\right\rangle + \tag{2}$$

$$+ [\frac{1}{2} \mathrm{Sp}(\frac{\partial^2 L(x,(\partial S(x)/\partial x);0)}{\partial p^2} \frac{\partial^2 S(x)}{\partial x^2}) + \frac{\partial}{\partial \varepsilon} L(x, \frac{\partial S(x)}{\partial x};\varepsilon)|_{\varepsilon=0}]\phi(x).$$

Theorem 1. Let symbol $L(x,p;(i\lambda)^{-1}) \in C^\infty([0,1],S^\infty(\mathbb{R}^{2n}))$ and function S(x) be real-valued. Then for $\lambda \geq 1$ and for an arbitrary integer $N \geq 0$, we have

$$L(\overset{2}{x},\lambda^{-1}\overset{1}{D_x},(i\lambda)^{-1})[\phi(x)\exp(i\lambda S(x))] = \\ = \exp(i\lambda S(x)) \sum_{j=0}^{N} (i\lambda)^{-j} R_j(x,D_x)\phi(x) + O_{-N-1}(x,\lambda). \tag{3}$$

Here $R_j(x,D_x)$ is a linear differential operator of order $\leq j$ with coefficients from the class $C^\infty(\mathbb{R}^n)$.

The estimate of the remainder is given by

$$|D_x^\alpha O_{-N-1}(x,\lambda)| \leq C_r \lambda^{-N-1+|\alpha|}(1+|x|)^{-r} \tag{4}$$

with arbitrary $r > 0$, $x \in R^n$. R_o and R_1 satisfy the formulas (2).

Proof. Let $u(x,\lambda)$ be the left-hand side of (3). Then

$$u(x,\lambda) = (\frac{\lambda}{2\pi})^n \int \exp[i\lambda\langle x,p\rangle] L(x,p;(i\lambda)^{-1}) I(p,\lambda)dp, \\ I(p,\lambda) = \int \phi(x)\exp[i\lambda S(x) - \langle x,p\rangle]dx. \tag{5}$$

Let $M = \{p : p = \partial S/\partial x, x \in \mathrm{supp}\ \phi\}$ and $G(p) \subset R_p^n$ be the exterior of a finite domain, $\overline{G(p)} \cap M = \emptyset$. We construct a C^∞-partition of the unity: $\eta_1(p) + \eta_2(p) = 1$, $p \in R^n$, where $\eta_1(p)$ has the compact support supp $\eta_2(p) \subset$ $\subset G(p)$, and we correspondingly set $u(x,\lambda) = u_1(x,\lambda) + u_2(x,\lambda)$. Further, we obtain

$$u_2(x,\lambda) = -\frac{1}{i\lambda|x|^2} \int \exp[i\lambda\langle x,p\rangle] \sum_{j=1}^{n} \frac{\partial}{\partial p_j}(I(p,\lambda)L(x,p))dp$$

for $x \neq 0$. Consequently, for $|x| \geq 1$ and arbitrary $N \geq 0$, we have

$$|u_2(x,\lambda)| \leq C\lambda^{-N}(1+|x|)^{m-1}.$$

Taking into account the above estimate for $|u_2|$, derive

$$|u_2(x,\lambda)| \leqq C_N\lambda^{-N}(1+|x|)^{-N}, \quad x \in R^n,$$

for arbitrary $N \geqq 0$, and find that the same estimates are true for all derivatives of u_2 with respect to x. Further,

$$u_1(x,\lambda) = \left(\frac{\lambda}{2\pi}\right)^n \int L(x,p;(i\lambda)^{-1})\phi(y)\eta_1(p)\exp[i\lambda\psi(x,y,p)]dy\,dp, \tag{6}$$
$$\psi(x,y,p) = \langle x-y,p\rangle + S(y),$$

where integration is performed over a finite domain in $R^n_y \times R^n_p$. Let $x \in K$. The function ψ (as a function of y and p) has a single stationary point $Q(x)$: $y = x$, $p = \partial S/\partial x$. Let H(x) be the matrix composed of the second derivatives with respect to y and p of the function ψ at the point Q(x), i.e., $H = \| \partial^2\psi/\partial y_j\partial p_k \|$, $1 \leqq j, k \leqq n$. Then $\det H(x) = (-1)^n$, the signature of H(x) is zero, and the eigenvalues of it are ± 1. Further,

$$\psi(x,Q(x)) = S(x).$$

If $|x| \leqq R$, then we obtain (3) and (4). If $|x| > R$, and $R > 0$ is large enough, then the integral for u_1 does not contain stationary points, and thus

$$|u_1(x,\lambda)| \leqq C_N\lambda^{-N}(1+|x|)^{-N}, \quad |x| \geqq R,$$

where $N \geqq 0$ is arbitrary; an analogous estimate holds for all derivatives of u with respect to x. Thus (3) is proved.

Corollary. The asymptotic expansion (3) can be differentiated with respect to x and λ any number of times.

The following theorem can be proved exactly in the same way as for Theorem 1:

Theorem 2. All statements of Theorem 1 remain valid for operator $L(\overset{1}{x},\lambda^{-1}\overset{2}{D_x};(i\lambda)^{-1})$ except that

$$(R_1,\phi)(x) = \left\langle\frac{\partial L(x,\partial S(x)/\partial x;0)}{\partial p}, \frac{\partial\phi(x)}{\partial x}\right\rangle +$$
$$+ \left[\frac{1}{2}\,\mathrm{Sp}\left(\frac{\partial^2 L(x,(\partial S(x)/\partial x);0)}{\partial x^2}\,\frac{\partial^2 S(x)}{\partial x^2}\right) + \right. \tag{7}$$
$$\left. + \frac{\partial}{\partial\varepsilon} L\left(x, \frac{\partial S(x)}{\partial y};\varepsilon\right)\Big|_{\varepsilon=0} + \mathrm{Sp}\,\frac{\partial^2 L(x,(\partial S/\partial x);0)}{\partial x\partial p}\right]\phi(x).$$

(Note that now

$$L(\overset{1}{x},-\frac{1}{\lambda}\overset{2}{D_x};(i\lambda)^{-1})[\phi(x)\exp(i\lambda S(x))] = \left(\frac{\lambda}{2\pi}\right)^n \iint L(y,p;(i\lambda)^{-1})\phi(y) \times \tag{8}$$
$$\times \exp[i\lambda(S(y)+\langle p,x-y\rangle)]dp\,dy.)$$

Now we arrive at establishing the commutation formula in general canonical coordinates. We decompose the set $(1,2,\ldots,n)$ into two disjoint subsets $(\alpha),(\beta)$: $(\alpha) = (\alpha_1,\ldots,\alpha_k)$, $(\beta) = (\beta_1,\ldots,\beta_\ell)$, where $k + \ell = n$, $\alpha_i \neq \beta_j$ for all i,j (one of the sets $(\alpha),(\beta)$ can be empty). We set $x = (x_{(\alpha)},x_{(\beta)})$, $x_{(\alpha)} = (x_{\alpha_1},\ldots,x_{\alpha_k})$, and analogously $p = (p_{(\alpha)},p_{(\beta)})$. We denote

$$\langle x_{(\alpha)},p_{(\alpha)}\rangle = \sum_{j=1}^{k} x_{\alpha_j}p_{\alpha_j}, \quad dx_{(\alpha)} = dx_{\alpha_1}\ldots dx_{\alpha_k},$$

and analogously

$$\langle x_{(\beta)}, p_{(\beta)} \rangle, \ dx_{(\beta)}, \ dp_{(\alpha)}, \ dp_{(\beta)}.$$

We introduce the λ-Fourier transformation over a part of the variables by

$$(F_{\lambda}, x_{(\alpha)} \to p_{(\alpha)} u(x))(p_{(\alpha)}, x_{(\beta)}) = \\ = \left(\frac{\lambda}{2\pi i}\right)^{k/2} \int \exp[-i\lambda \langle x_{(\alpha)}, p_{(\alpha)} \rangle] u(x) dx_{(\alpha)}, \tag{9}$$

where, as usual, $\sqrt{i} = e^{i\pi/4}$. Then well-known properties of the Fourier transformation yield

$$F_{\lambda}, x(\alpha) \to p_{(\alpha)} L(\overset{2}{x}, \lambda^{-1} \overset{1}{D}_x; (i\lambda)^{-1}) F^{-1}_{\lambda, p_{(\alpha)} \to x_{(\alpha)}} u(p_{(\alpha)}, x_{(\beta)}) = \\ = L(-\lambda^{-1} \overset{2}{D}_{p_{(\alpha)}}, \overset{2}{x}_{(\beta)}, \overset{1}{p}_{(\alpha)}, \lambda^{-1} \overset{1}{D}_{x_{(\beta)}}; (i\lambda)^{-1} u(p_{(\alpha)}, x_{(\beta)}) \tag{10}$$

for any λ-PDO, $\hat{L} = L(\overset{2}{x}, \lambda^{-1})$. The above formula allows us to construct a class of f.a. solutions of the form

$$u(x, \lambda) = F^{-1}_{\lambda, p_{(\alpha)} \to x_{(\alpha)}} (\phi(p_{(\alpha)}, x_{(\beta)}) \exp[i\lambda S(p_{(\alpha)}, x_{(\beta)})]). \tag{11}$$

<u>Theorem 3</u>. Let symbol $L(x, p; (i\lambda)^{-1}) \in C^{\infty}([0,1], S^{\infty}(\mathbb{R}^{2n}))$, function $S(p_{(\alpha)}, x_{(\beta)})$ be real-valued

$$\phi(p_{(\alpha)}, x_{(\beta)}) \in C_0^{\infty}(\mathbb{R}^k_{p_{(\alpha)}} \times \mathbb{R}^{n-k}_{x_{(\beta)}}),$$

$$S(p_{(\alpha)}, x_{(\beta)}) \in C^{\infty}(\mathbb{R}^k_{p_{(\alpha)}} \times \mathbb{R}^{n-k}_{x_{(\beta)}}).$$

Then

$$(L(\overset{2}{x}, \lambda^{-1} D_x; (-i\lambda)^{-1}) F^{-1}_{\lambda, p_{(\alpha)} \to x_{(\alpha)}} (\phi(p_{(\alpha)}, x_{(\beta)}) \exp[i\lambda S(p_{(\alpha)}, x_{(\beta)})]) = \\ = F^{-1}_{\lambda, p_{(\alpha)} \to x_{(\alpha)}} [\exp(i\lambda S(p_{(\alpha)}, x_{(\beta)}) \sum_{j=0}^{N-1} (i\lambda)^{-j} R_j \phi(p_{(\alpha)}, x_{(\beta)})] + O(\lambda^{-N}) \tag{12}$$

holds for $\lambda \geqq 1$ and for an arbitrary integer $N \geqq 0$. Here R_j is a linear differential operator of the order $\leqq j$ with C^{∞}-coefficients (depending on S).

<u>Proof</u>. Let function $V(p_{(\alpha)}, x_{(\beta)})$ be infinitely differentiable and with compact support with respect to all its arguments. Then because of (6) we have

$$L(\overset{2}{x}, \lambda^{-1} \overset{1}{D}_x; (i\lambda)^{-1}) F^{-1}_{\lambda, p_{(\alpha)} \to x_{(\alpha)}} V(p_{(\alpha)}, x_{(\beta)}) = \\ = F^{-1}_{\lambda, p_{(\alpha)} \to x_{(\alpha)}} \tilde{L} V(p_{(\alpha)}, x_{(\beta)}) = L(-\lambda^{-1} \overset{1}{D}_{p_{(\alpha)}}, \overset{2}{x}_{(\beta)}, \lambda^{-1} \overset{1}{D}_{x_{(\beta)}}, \overset{2}{p}_{(\alpha)}). \tag{13}$$

Let $V = e^{i\lambda S} \phi$. We obtain the expansions

$$\tilde{L} V = e^{i\lambda S} \sum_{j=0}^{N-1} (i\lambda)^{-j} R_j \phi + M_N(p_{(\alpha)}, x_{(\beta)}; (i\lambda)^{-1}).$$

The statement of the theorem concerning the remainder follows from (3). We write the first two terms of expansion (9):

$$R_o\phi = L^0\phi; \tag{14}$$

$$R_1\phi = \left\langle\frac{\partial L^0}{\partial x_{(\alpha)}}, \frac{\partial\phi}{\partial p_{(\alpha)}}\right\rangle - \left\langle\frac{\partial L^0}{\partial p_{(\beta)}}, \frac{\partial\phi}{\partial x_{(\beta)}}\right\rangle -$$

$$- \frac{1}{2}\Big[Sp\Big(\frac{\partial^2 L^0}{(\partial x_{(\alpha)})^2}\,\frac{\partial^2 S}{(\partial p_{(\alpha)})^2}\Big) + Sp\Big(\frac{\partial^2 L^0}{(\partial p_{(\beta)})^2}\,\frac{\partial^2 S}{(\partial x_{(\beta)})^2}\Big) - \tag{15}$$

$$- 2Sp\Big(\frac{\partial^2 L^0}{\partial x_{(\alpha)}\partial p_{(\beta)}}\,\frac{\partial^2 S}{\partial p_{(\alpha)}\partial x_{(\beta)}}\Big) - 2\sum_{j=1}^{k}\frac{\partial^2 L^0}{\partial x_{\alpha_j}\partial p_{\beta_j}}\Big]\phi + \Big(\frac{\partial L}{\partial(i\lambda)^{-1}}\Big)_o\phi.$$

Here

$$L^0 = L\Big(-\frac{\partial S(p_{(\alpha)},x_{(\beta)})}{\partial p_{(\alpha)}}, x_{(\beta)}, p_{(\alpha)}, \frac{\partial S(p_{(\alpha)},x_{(\beta)})}{\partial x_{(\beta)}}; 0\Big). \tag{16}$$

The derivative $(\partial L/\partial(i\lambda)^{-1})_o$ is taken for $(i\lambda)^{-1} = 0$ and for the same values of x,p as in L^0.

Next we pass to global commutation formulas. Let K_{Λ^n} be a canonical operator on a Lagrangian manifold Λ^n.

Theorem 4. The following commutation formula holds:

$$L(\overset{2}{x}, \lambda^{-1}\overset{1}{D_x}; (i\lambda)^{-1})K_{\Lambda^n}\phi = K_{\Lambda^n}(L(x,p;0)\phi) + O(\lambda^{-1}). \tag{17}$$

Proof. Let $\Omega \subset \Lambda^n$ be a canonical chart, $K_{\Lambda^n}(\Omega)$ the pre-canonical operator, and $\phi \in C_0^\infty(\Omega)$. Let Ω be a non-singular chart. Then $K_{\Lambda^n}\phi = \exp \times$ $\times\ (i\lambda S(x))|d\sigma^n/dx|^{1/2}\phi$. We have

$$\hat{L}K_{\Lambda^n}(\Omega)\phi = \exp(i\lambda S(x))\Big|\frac{d\sigma^n}{dx}\Big|^{1/2}L\Big(x, \frac{\partial S}{\partial x}; 0\Big)\phi + O(\lambda^{-1}) =$$
$$= K_{\Lambda^n}(\Omega)(L(x,p;0)\phi) + O(\lambda^{-1}), \tag{18}$$

since $p = \partial S(x)/\partial x$ on Λ^n. Let Ω be a singular chart with the focal coordinates $(p_{(\alpha)}, x_{(\beta)})$. Then

$$\hat{L}K_{\Lambda^n}(\Omega)\phi = F^{-1}_{\lambda,p_{(\alpha)}\to x_{(\alpha)}} L(-\lambda^{-1}D_{p_{(\alpha)}}, x_{(\beta)}, p_{(\alpha)}, D_{x_{(\beta)}}; (i\lambda)^{-1}) \times$$

$$\times \Big|\frac{d\sigma^n(r)}{\partial p_{(\alpha)}\partial x_{(\beta)}}\Big|^{1/2}\exp(i\lambda S(p_{(\alpha)},x_{(\beta)}))\phi(r),$$

where $r = r(p_{(\alpha)}, x_{(\beta)})$.

Function S has the form

$$S(P_{(\alpha)},x_{(\beta)}) = \int_{r^g}^{r}\langle p,dx\rangle - \langle p_{(\alpha)}, x_{(\alpha)}(p_{(\alpha)},x_{(\beta)})\rangle$$

so that

$$\frac{\partial S}{\partial p_{(\alpha)}} = -x_{(\alpha)}(p_{(\alpha)},x_{(\beta)}),\ \frac{\partial S}{\partial x_{(\alpha)}} = p_{(\alpha)}.$$

By applying Theorem 3, we obtain

$$\hat{L}K_{\Lambda^n}(\Omega)\phi = F^{-1}_{\lambda,p_{(\alpha)}\to x_{(\alpha)}}\left(\left|\frac{d\sigma^n(r)}{dp_{(\alpha)}dx_{(\beta)}}\right|^{1/2}\exp(i\lambda S(p_{(\alpha)},x_{(\beta)})\phi(r)\times\right.$$
$$\times L(x_{(\alpha)}(p_{(\alpha)},x_{(\beta)}),x_{(\beta)},p_{(\alpha)},p_{(\beta)}(p_{(\alpha)},x_{(\beta)});0)+O(\lambda^{-1}). \tag{19}$$

Consequently, formula (18) holds. Let $\phi \in C_0^\infty(\Lambda^n)$. Then (17) follows from (18) and (19). Thus, the theorem is proved.

The commutation formula in Theorem 4 was established under the assumption that the differential operator acts first and the operators of multiplication by the independent variables act second. Exactly the same formula is valid even if the mentioned operators act in the reverse order.

Next, using the terms of the first order in the expansions (1) and (12), we come, after simple but clumsy calculation, to:

Theorem 5. Let operator L and Lagrangian manifold Λ^n fulfill the assumptions of Theorem 4 and moreover the conditions:

(1) Function $L(x,p;0)$ is real-valued, and the equation

$$L(x,p;0) = 0 \tag{20}$$

determines a $(2n-1)$-dimensional C^∞-manifold $M^{2n-1}(L)$ in the phase space.

(2) $\Lambda^n \subset M^{2n-1}(L)$.

(3) Manifold Λ^n and volume element $d\sigma^n$ are invariant with respect to the dynamical system

$$\frac{dx}{dt} = \frac{\partial L(x,p;0)}{\partial p}, \quad \frac{dp}{dt} = -\frac{\partial L(x,p;0)}{\partial x}. \tag{21}$$

(4) There exists a solution of system (21) for all $t \in \mathbb{R}$; it is unique and infinitely differentiable for arbitrary initial data $(x,p) \in M^{2n-1}(L)$.

Then the commutation formula

$$L(\overset{2}{x},\lambda^{-1}\overset{1}{D}_x;(i\lambda)^{-1})(K_{\Lambda^n}\phi(r))(x) = \frac{1}{i\lambda}K_{\Lambda^n}\left(\frac{d}{dt}-\frac{1}{2}\,Sp\left(\frac{\partial^2 L(x,p;0)}{\partial x\partial p}\right)+\right.$$
$$\left.+\frac{\partial L(x,p;\varepsilon)}{\partial\varepsilon}\bigg|_{\varepsilon=0}\right)\phi + O(\ ^{-2}) \tag{22}$$

is true. Here d/dt is the derivative with respect to the Hamiltonian system (21), i.e.,

$$\frac{d\phi(r)}{dt} = -\left\langle\frac{\partial L(x,p;0)}{\partial x_{(\alpha)}},\frac{\partial\phi(r)}{\partial p_{(\alpha)}}\right\rangle + \left\langle\frac{\partial L(x,p;0)}{\partial p_{(\beta)}},\frac{\partial\phi(r)}{\partial x_{(\beta)}}\right\rangle, \tag{23}$$

provided that a neighborhood of the point $r \in \Lambda^n$ admits a diffeomorphic projection on the plane $(p_{(\alpha)},x_{(\beta)})$.

The generalization of results presented here in the general case, described in items A, B, is obvious, and we leave this reformulation to the reader.

2. THE CANONICAL OPERATOR ON A LAGRANGIAN SUBMANIFOLD OF A SYMPLECTIC MANIFOLD

In the previous section we gave the construction which enables us to solve h^{-1}-PDE's in R^n. Here we intend to generalize this construction, thus covering, by the way, the asymptotic solution of h^{-1}-PDE's on manifolds and, more generally, h^{-1}-PDE's in the space of sections of special sheaves on symplectic manifolds (at this stage, our construction appears to be somewhat like the one used in the orbit method and geometric quantization (see [46,42,43] and other papers)). Thus we define a special sheaf over a symplectic manifold M, and the canonical operator should take values in the space of sections of this sheaf. In the case $M = T^*N$ the sections of the sheaf reduce to functions on M. In order to make the presentation more smooth, we deal throughout the Section with the spaces of finite or rapidly decaying at infinity functions. Obvious modifications should be made to make it possible to consider the (quasi) homogeneous case.

A. 1/h-PDO and Wave Front Sets

Let $f(q,h)$ be a function of variables $q \in R^n$, $h \in (0,1)$. We denote for any $s \in R$

$$\|f\|_{H^s} = \sup_{h \in (0,1)} \|(1 + q^2 - h^2\Delta_q)^{s/2} f\|_{L_2(R^n)}, \tag{1}$$

where

$$\Delta_q = \frac{\partial^2}{\partial q_1^2} + \dots + \frac{\partial^2}{\partial q_n^2}$$

is the Laplace operator with respect to the variables $q_1,\dots,q_n$, and

$$\|u\|_{L_2(R^n)} = [\int |u(q)|^2 dq]^{1/2}$$

is the L_2-norm in R^n. Set

$$H(R^n) = \bigcap_s H^s(R^n).$$

By a Hamiltonian function we mean any function $H(q,p,h)$ of the variables $q \in R^n$, $p \in R^n$, $h \in [0,1)$ (the point $h = 0$ is not excluded), satisfying the estimates

$$\left| \frac{\partial^{|\alpha| + |\beta| + \gamma} H}{\partial q^\alpha \partial p^\beta \partial h^\gamma}(q,p,h) \right| \leqslant C_{\alpha\beta\gamma}(1 + |q| + |p|)^m \tag{2}$$

for some $m \in R$ independent of α, β, γ.

We denote by $S^m(R^n)$ the space of functions, satisfying (2), and by $S^\infty(R^n)$ the union

$$S^\infty(R^n) = \bigcup_m S^m(R^n).$$

<u>Proposition 1</u>. If $H(q,p,h) \in S^m$, then

$$(1 + q^2)^{-N}(1 + p^2)^{-N} H(q,p,h) \in S^{m-N}. \tag{3}$$

<u>Proof</u>. It is enough to use the Leibniz formula.

<u>Definition 1</u>. Let $H \in S^\infty$. A 1/h-PDO in $H(R^n)$ is a linear operator in $H(R^n)$ given by:

$$[\hat{H}u](q,h) \equiv H(q^2, -ih\frac{\partial^1}{\partial q}, h)u(q,h) = \tag{4}$$
$$= e^{i(\pi/4)}(2\pi h)^{-n/2}\int_{R^n} H(q,p,h)e^{(i/h)pq}\tilde{u}(p,h)dp.$$

Here $pq = p_1q_1 + \dots + p_nq_n$ and

$$\tilde{u}(p,h) = e^{-i(\pi/4)}(2\pi h)^{-n/2}\int_{R^n} u(q,h)e^{-(i/h)pq}dq = [F^{1/h}u](p,h). \tag{5}$$

Note that the inverse operator is given by

$$u(q,h) = e^{i(\pi/4)}(2\pi h)^{-n/2}\int_{R^n} \tilde{u}(p,h)e^{(i/h)pq}dp = [F^{-1/h}\tilde{u}](q,h). \tag{6}$$

Lemma 1. (a) Definition 1 is correct.

(b) 1/h-PDO form an algebra, namely, if $H_1, H_2 \in S^\infty$, then

$$\hat{H}_1 \circ \hat{H}_2 = \hat{H}_3,$$

where $H_3 \in S^\infty$ and possesses the asymptotic expansion

$$H_3(q,p,h) = \sum_{m=0}^{N} (-ih)^m \sum_{|\alpha|+k+\ell = m} \frac{\partial^{|\alpha|+k}H_1}{\partial p^\alpha \partial h^k}(q,p,0) \times$$
$$\times \frac{\partial^{|\alpha|+\ell}H_2}{\partial q^\alpha \partial h^\ell}(q,p,0) + (-ih)^{N+1}R_{N+1}(q,p,h) \tag{7}$$

for any natural N; the remainder in (7) belongs to S^∞. If at least one of the symbols H_1, H_2 lies in

$$S^{-\infty} = \bigcap S^m,$$

then so does the remainder in (7).

(c) If $\hat{H}$ is a 1/h-PDO with symbol $H \in S^\infty$ then $\hat{H}^*$ (formally adjoint operator in $L^2(R^n)$) is also a 1/h-PDO. Its symbol admits an expansion

$$\mathrm{Smb}(\hat{H}^*)(q,p,h) = \sum_{m=0}^{N} (-ih)^m \sum_{|\alpha|+k = m} \frac{\partial^{2|\alpha|+k}\bar{H}}{\partial q^\alpha \partial p^\alpha \partial h^k}(q,p,0) +$$
$$+ (-ih)^{N+1}R'_{N+1}(q,p,h) \tag{8}$$

(here the line over the symbol denotes the complex conjugation; the remainder R'_{N+1} lies in S^∞ (respectively, $S^{-\infty}$, if $H \in S^{-\infty}$)).

Proof. (a) Since $\tilde{u}$ belongs to a Schwartz space, while $H(q,p,h)$ grows not faster than a polynomial, the integral (4) converges. Prove that $\hat{H}u$ lies in $H(R^n)$. To do this make use of Proposition 1. Fix N large enough and set

$$(1+q^2)^{-N}(1+p^2)^{-N}H(q,p,h) = H_1(q,p,h). \tag{9}$$

We obtain

$$\hat{H}u = (1+q^2)^N\hat{H}_1(1-\Delta)^N u. \tag{10}$$

Since $1 + q^2$, $1 - \Delta$ are the operators which transform $H(R^n)$ into $H(R^n)$, it is enough to prove the required statement when $H \in S^{-m}(R^n)$ with m sufficiently large. In this case

$$[\hat{H}u](q,h) = e^{i(\pi/4)n}(2\pi h)^{-n/2}(2\pi h)^{-n}\int_{R^{3n}} e^{(i/h)q\xi} e^{i(qv+\xi\omega)} \overset{0}{H}(v,\omega,h) \times$$
$$\times \tilde{u}(\xi,h)d\xi dv d\omega = (2\pi)^{-n}\int_{R^{2n}} \overset{0}{H}(v,\omega,h)e^{iqv}u(q+h,\omega,h)dv d\omega, \tag{11}$$

where

$$\overset{0}{H}(v,\omega,h) = (2\pi)^{-n}\int_{R^{2n}} e^{-i(qv+\xi\omega)}H(q,\xi,h)dq d\xi \tag{12}$$

is the usual Fourier transform of H. By (12), $\hat{H}(v,\omega,h)$ is a continuous rapidly decaying function. Therefore it is easy to see that the function $e^{iqv}u(q + h,\omega,h)$ is, for any natural S, a continuous function of the parameters (v,ω) in the norm (1), and besides there is an estimate

$$\| e^{ixv}\phi(x+h,\omega,h)\|_k \leqslant C(1+|v|+|\omega|^k)\ \|\phi(x,h)\|_k. \tag{13}$$

It follows that (a) is valid and that for any natural S (cf. Chapter 2:4):

$$\|\hat{H}u\|_s \leqslant \text{const}\ \|u\|_s. \tag{14}$$

To prove (b) we make use of the composition formula (cf. Chapter 2:4):

$$\text{Smb}(\hat{H}_1 \circ \hat{H}_2) = H_1(\overset{2}{q}, p-ih\frac{\overset{1}{\partial}}{\partial q}, h)H_2(q,p,h). \tag{15}$$

Expand $H_1(q,p + h\xi,h)$, using the Taylor formula with the remainder:

$$H_1(q,p+h\xi,h) = \sum_{m=0}^{N} h^m \Big(\sum_{k+|\alpha|=m} \frac{\partial^{|\alpha|+k}H_1}{\partial p^\alpha \partial h^k}(q,p,0)\xi^\alpha\Big) + $$
$$+ h^N H^{(N)}(q,p,\xi,h). \tag{16}$$

It follows from Chapter 2:4 that the operator

$$H^{(N)}(\overset{2}{q}, p - i\frac{\overset{1}{\partial}}{\partial q}, h)$$

acts in $S^\infty \to S^\infty$. Thus (7) is an immediate consequence of (16).

The proof of (c) is quite analogous to that given above, and so we omit all details. The lemma is proved.

Definition 2. Let $f(q,h)$ be a function such that $h^{-r}f(q,h) \in H$ for some real r. Then we write $f = \hat{O}(h^r)$. The space of such functions will be denoted by I^r. From Lemma 1 it follows that $\hat{H}I^r \subset I^r$ for any 1/h-PDO $\hat{H}$ and any $r \in R$.

Let $H(p,q,h)$ be a Hamiltonian function. We say it is finite and write $H \in S_o$, if for any fixed h the support supp H lies in the ball $|q^2| + |p^2| \leqslant R$, where R does not depend on h. It is easy to see that $S_o \subset S^{-\infty}$.

Definition 3. Let $f = \hat{O}(h^r)$ for some real r. The wave front of order $k \in R \cup \{+\infty\}$ of the function f is the smallest of the closed subsets $K \subset R^{2n} = R^n_q \times R^n_p$, having the property: for any $H \in S_o$, such that $H = 0$ in the vicinity of K, the relation

$$\hat{H}f = \hat{O}(h^{k+1}) \tag{17}$$

(here $\infty + 1 = \infty$) is satisfied.

The wave front of order k of the function f will be denoted by $WF_h^{(k)}(f)$ ($WF_h(f)$ also, if $k = \infty$).

Lemma 2. (a) Definition 3 is correct.

(b) For any 1/h-PDO $\hat{H}$ and $f = O(h^r)$,

$$WF_h^{(k)}(\hat{H}f) \subset WF_h^{(k)}(f) \subset WF_h^{(k)}(\hat{H}f) \cup$$

$$\cup\{(q,p)/ H(q,p,0) = 0\}. \tag{18}$$

(c) If $f = \hat{O}(h^{k+1})$, then $WF_h^{(k)}(f) = \emptyset$.

(d) There is a relation

$$WF_h(f) = \bigcup_{k \in R} \overline{WF_h^{(k)}(f)}. \tag{19}$$

Proof. (a) We should show that the intersection K_o of sets K possessing the described property possesses this property as well. Let $H \in S_o$, $\operatorname{supp} H \cap K_o = \emptyset$. Since supp H is a compact set, one can find the sets $K_1, \ldots, K_r$ with the mentioned property, such that

$$\operatorname{supp} H \cap K_1 \cap \ldots \cap K_r = \emptyset. \tag{20}$$

By means of unity partition we represent H in the form of the sum $H = H_1 + \ldots + H_r$, where $\operatorname{supp} H_r \cap K_j = \emptyset$. We have

$$\hat{H}_j f = \hat{O}(h^{k+1}),\ j = 1,\ldots,r$$

and, consequently,

$$\hat{H}f = \hat{H}_1 f + \ldots + \hat{H}_r f = \hat{O}(h^{k+1}); \tag{21}$$

(a) is proved.

(b) The left inclusion (18) follows directly from the definition and from (7). To prove the right inclusion it is enough to mention that if $H(q_o,p_o,0) \neq 0$, then using (7) we easily construct the symbol $G(q,p,h)$ such that

$$\hat{H} \circ \hat{G} = F(\overset{2}{q}, -ih \overset{1}{\frac{\partial}{\partial q}}, h), \tag{22}$$

where $F(p,q,h) = 1$ in the vicinity of (q_o,p_o). The statements (c) and (d) follow directly from the definition. The lemma is proved.

The notion of a wave front set introduced here is quite analogous to Hörmander's one. We note that in [36] the wave front was defined in another way (namely, as a support of the limit in $D'(R^{2n})$ of the corresponding density function

$$\rho(q,p,h) = e^{(i/h)pq} f(q)\tilde{f}(p)(2\pi h)^{-n/2}).$$

Let $U \subset R^{2n}$ be some subset. We denote by $\tilde{H}^k(U)$ the subspace in $H(R^{2n})$, consisting of functions $f(q,h)$ such that

$$WF_h^{(k)}(f) \subset U,$$

and by $H^k(U)$ the factor-space

$$H^k(U) = \tilde{H}^k(R^n)/\tilde{H}^k(R^{2n} \setminus U). \tag{23}$$

Lemma 3. There are natural mappings and commutation diagrams:

(a)

$$\begin{array}{ccc} & H^{k_2}(\mathcal{U}) & \\ \nearrow & & \searrow \\ H^{k_1}(\mathcal{U}) & \longrightarrow & H^{k_3}(\mathcal{U}) \end{array} \tag{24}$$

for $k_1 \geq k_2 \geq k_3$;

(b)

$$\begin{array}{ccc} & H^{k}(\mathcal{U}_2) & \\ {}^{j\mathcal{U}_2\mathcal{U}_1}\nearrow & & \searrow^{j\mathcal{U}_3\mathcal{U}_2} \\ H^{k}(\mathcal{U}_1) & \xrightarrow{j\mathcal{U}_2\mathcal{U}_3} & H^{k}(\mathcal{U}_3) \end{array} \tag{25}$$

for $\mathcal{U}_1 \supset \mathcal{U}_2 \supset \mathcal{U}_3$;

(c)

$$\begin{array}{ccc} H^{k_1}(\mathcal{U}_1) & \longrightarrow & H^{k_1}(\mathcal{U}_2) \\ \downarrow & & \downarrow \\ H^{k_2}(\mathcal{U}_1) & \longrightarrow & H^{k_2}(\mathcal{U}_2) \end{array} \tag{26}$$

for $k_1 \geq k_2$, $\mathcal{U}_1 \supset \mathcal{U}_2$.

Proof. The existence of commutation diagrams (24) - (26) follows from the natural embeddings

$$\begin{aligned} \tilde{H}^k(\mathcal{U}) &\subset \tilde{H}^k(V), \quad \mathcal{U} \subset V; \\ \tilde{H}^k(\mathcal{U}) &\subset \tilde{H}^\ell(\mathcal{U}), \quad K \geqslant \ell, \end{aligned} \tag{27}$$

which, in their turn, immediately follow from the definition of the space $\tilde{H}^k(\mathcal{U})$.

Lemma 4. The space $\tilde{H}^k(\mathcal{U})$ is invariant with respect to 1/h-PDO (and thus 1/h-PDO are correctly defined in $H^k(\mathcal{U})$). If $H \in S^\infty$, $\mathcal{U} \subset V$, $K \geqslant \ell$, then the commutative diagram

$$\begin{array}{ccc} H^k(V) & \xrightarrow{\hat{H}} & H^k(V) \\ {\scriptstyle j\mathcal{U}V}\downarrow & & \downarrow{\scriptstyle j\mathcal{U}V} \\ H^k(\mathcal{U}) & \xrightarrow{\hat{H}} & H^k(\mathcal{U}) \\ \downarrow & & \downarrow \\ H^\ell(\mathcal{U}) & \xrightarrow{\hat{H}} & H^\ell(\mathcal{U}) \end{array} \tag{28}$$

exists. The 1/h-PDO in the space $H^k(\mathcal{U})$ is completely determined by restriction of its symbol on $\mathcal{U}$.

The proof is obvious. Thus, if for any domain $\mathcal{U} \subset R^{2n}$, we set

$$\begin{aligned} S^\infty(\mathcal{U}) &= \{H(q,p,h) \in C^\infty(\mathcal{U} \times [0,\ 1)) \,|, \\ &\exists H_1(q,p,h) \in S^\infty,\ H_1 = H|_{(q,p)\in\mathcal{U}}, \end{aligned} \tag{29}$$

then to any symbol $H \in S^\infty(\mathcal{U})$ there corresponds an operator

$$\hat{H} : H^k(\mathcal{U}) \to H^k(\mathcal{U}), \tag{30}$$

which we call the 1/h-PDO in $H^k(\mathcal{U})$ with the symbol $H \in S^\infty(\mathcal{U})$. By Lemma 4 the operators (30) are consistent with the restriction homomorphisms, introduced in Lemma 3.

B. Quantization of Canonical Transformations of Bounded Domains in R^{2n}

Let $\mathcal{U}$,V be bounded simply connected domains in R^{2n}, and let $g : \mathcal{U} \to V$ be a mapping, given by relations

$$(q',p') \to (q(q',p'),p(q',p')) = g(q',p'), \tag{31}$$

and which is a canonical transformation of $\mathcal{U}$ into V, i.e., a smooth mapping such that

$$g^*(dp \wedge dq) = dp' \wedge dq'. \tag{32}$$

Here

$$dp \wedge dq = dp_1 \wedge dq_1 + \dots + dp_n \wedge dq_n. \tag{33}$$

Assume that $g(\mathcal{U}) = V$; we will require that the mapping might be continued into some neighborhood $\mathcal{U}_o \supset \bar{\mathcal{U}}$ of the closure of $\mathcal{U}$.

Consider in $\mathbb{R}^{2n}_{(q',p')} \times \mathbb{R}^{2n}_{(q,p)}$ the graph $\Gamma(g)$ of the transformation g continued into $\mathcal{U}_o$:

$$\Gamma(g) = \{(q',p',q,p) \,|\, (q',p') \in \mathcal{U}_o, \ (q,p) = g(q',p')\}. \tag{34}$$

Lemma 1. The graph $\Gamma(g)$ is a Lagrangian sub-manifold in $\mathbb{R}^{4n}$ with respect to the 2-form

$$\Omega^2 = dp \wedge dq - dp' \wedge dq' \tag{35}$$

(i.e., the restriction of the form Ω^2 on $\Gamma(g)$ is equal to zero).

Proof. The statement of the lemma follows from the relation (32).

Let $\mathcal{K}_g$ be a canonical operator on the manifold $\Gamma(g)$, taking 1/h-oscillating functions of $(q,q') \in \mathbb{R}^{2n}$ into correspondence with smooth finite functions on $\Gamma(g)$, polynomially dependent on h. To define the canonical operator we must fix a non-degenerate measure μ on $\Gamma(g)$; we shall make use of

$$\mu = \underbrace{(dp \wedge dq) \wedge \dots \wedge (dp \wedge dq)}_{n \text{ times}}. \tag{36}$$

The unity partition, the initial point and the coincident choice of measure density arguments will be fixed in an arbitrary way.

Example 1. If the manifold $\Gamma(g)$ may be defined by virtue of the unique generating function ([1]) $S(q',q)$, i.e., may be described by the equations

$$p = \frac{\partial S}{\partial q}(q',q), \qquad p' = -\frac{\partial S}{\partial q'}(q',q), \tag{37}$$

then the result of action of the canonical operator K_g on the function $\phi \in C_o^\infty(\Gamma(g))$ may be given by the formula

$$[\mathcal{K}g\phi](q',q,h) = e^{(i/h)S(q',q)}\phi(q',-\frac{\partial S}{\partial q'}(q',q)) \times$$
$$\times \{\det \frac{\partial^2 S}{\partial q \partial q'}(q',q)\}^{-1/2} \tag{38}$$

(here and in the sequel we consider (q',p') as a coordinate system on $\Gamma(g)$; in particular, the function ϕ is written in terms of these coordinates).

In (38) we have fixed an arbitrary, continuous in $\mathcal{U}_o$ branch of the argument of the Jacobian

$$\det \frac{\partial^2 S}{\partial q \partial q'}(q',q(q',p')).$$

We construct now a linear operator

$$T(g) : H^k(\mathcal{U}) \to H^k(V). \tag{39}$$

In the following way: let $\phi \in C_o^\infty(\mathcal{U}_o)$, $\phi \equiv 1$ in $\bar{\mathcal{U}}$. We set

$$[T(g)f](q,h) = e^{-i(\pi/4)}(2\pi h)^{-n/2}\int_{R^n}[\mathcal{K}g\phi](q',q,h)f(q',h)dq'. \tag{40}$$

Lemma 2. (a) The operator $T(g) : H^k(\mathcal{U}) \to H^k(V)$ is defined by (40) correctly, i.e., it does not depend on the choice of function ϕ, satisfying the mentioned condtions, and maps

$$H^k(\mathcal{U}) \to H^k(V).$$

(b) The mapping $T(g)$ defined in (40) commutes with the restriction mappings, introduced in Lemma 7, i.e., the following diagram is commutative

$$\begin{array}{ccc} H^k(\mathcal{U}) & \xrightarrow{T(g)} & H^k(V) \\ {\scriptstyle i_{WV}}\downarrow & & \downarrow{\scriptstyle i_{g(W)V}} \\ H^k(W) & \xrightarrow{T(g)} & H^k(g(W)) \end{array} \tag{41}$$

Proof. Let $\hat{\mathcal{H}}$ be a 1/h-PDO with the symbol $\mathcal{H} \in S^{-\infty} \cap S(g(\mathcal{U}))$. The following commutation formula holds

$$\hat{\mathcal{H}} \circ T(g) = T(g) \circ \hat{\mathcal{H}}_1 + \hat{\mathcal{O}}(h^\infty), \tag{42}$$

where

$$\mathcal{H}_1(q',p',0) = (g^*\mathcal{H})(q',p',0) \stackrel{def}{=} \mathcal{H}(q(q',p'),0),$$
$$\operatorname{supp} \mathcal{H}_1 \subset g^{-1}(\operatorname{supp} \mathcal{H}). \tag{43}$$

The proof of (42) follows immediately from the stationary phase method.

It follows from (42) that the operator $T(g)$ maps $H^k(\mathcal{U})$ into $H^k(g(\mathcal{U}))$. The remaining assertions of Lemma 2 are proved quite analogously to the corresponding assertions of Lemma 4 in Section 2:A. The lemma is proved.

Note 1. It should be noted, however, that the operator $T(g)$ is defined not uniquely but only up to a factor of the form

$$\exp(iC^{(1)}/h + i\pi C^{(2)}),\ C^{(1)} \in \mathbb{R},\ C^{(2)} \in \mathbb{Z}, \tag{44}$$

dependent on the choice of action and the measure density argument on $\Gamma(g)$.

Study now the algebraic properties of the correspondence $g \to T(g)$. Let $\mathcal{U}, V$ be the bounded simply connected domains in $\mathbb{R}^{2n}$. Denote by $Sp(\mathcal{U},V)$ the set of canonical transformations $g : \mathcal{U} \to V$, $g(\mathcal{U}) = V$, which might be continued in some simply connected neighborhood of $\bar{\mathcal{U}}$.

If $\mathcal{U},V,W \subset \mathbb{R}^{2n}$ are bounded simply connected domains, the composition law is defined

$$Sp(\mathcal{U},V) \times Sp(V,W) \to Sp(\mathcal{U},W), \quad (g_1,g_2) \to g_2 \circ g_1. \tag{45}$$

Theorem 1. Let $\mathcal{U},V,W$ be bounded connected simply connected domains in $\mathbb{R}^{2n}$, $g_1 \in Sp(\mathcal{U},V)$, $g_2 \in Sp(V,W)$. Then

$$T(g_2)T(g_1)u = \exp\{\frac{i}{h} C^{(1)}_{g_1g_2} + i\pi C^{(2)}_{g_1g_2}\} T(g_2 \circ g_1) V_{g_1g_2} u \tag{46}$$

for any $u \in H^k(\mathcal{U})$. In (46) $C^{(1)}_{g_1g_2}$ is a real number and $C^{(2)}_{g_1g_2}$ is an integer, depending on the choice of action and the measure density arguments on the graphs $\Gamma(g_1)$, $\Gamma(g_2)$, $\Gamma(g_1g_2)$; $V_{g_1g_2}$ is a $1/h$-PDO of the form

$$V_{g_1g_2} = 1 - ihR_{g_1g_2}(\overset{2}{q}, -ih\frac{\partial}{\partial q}, h), \tag{47}$$

where $R_{g_1g_2} \in S^\infty(\mathcal{U})$.

Proof. We use the proof scheme used in [39] and [61].

Lemma 3. Let $g : \mathcal{U} \to \mathbb{R}^{2n}$ be a canonical transformation. There exists a smooth family $\{g^t\}$, $t \in [0,1]$, $g^t : \mathcal{U} \to \mathbb{R}^{2n}$ of canonical transformations, such that

$$g^0 = 1, \quad g^1 = g. \tag{48}$$

Proof. Set for $Z = (q,p) \in \mathcal{U}$

$$\tilde{g}(z,t) = \begin{cases} \psi_t(z) + 2tg^1(0), & t \in [0, \frac{1}{2}], \\ \frac{1}{t-1/2}[g^1((t-\frac{1}{2})z) - g^1(0)] + g^1(0), & t \in [\frac{1}{2}, \frac{3}{2}], \end{cases} \tag{49}$$

where ψ_t is a smooth family of linear canonical transformations,

$$\psi_o = id, \quad \psi_1 = dg^t|_{z=0}$$

(such a family exists since the group $Sp(n,\mathbb{R})$ is connected). Let the function $f[0,1] \to [0,3/2]$ be strictly monotone,

$$f(0) = 0, \quad f(1) = \frac{3}{2}, \quad f(\tfrac{1}{2}) = \frac{1}{2}, \quad f^{(k)}(\tfrac{1}{2}) = 0, \text{ for } k \geqslant 1. \tag{50}$$

Then $g^t(\cdot) \overset{def}{=} \tilde{g}(\cdot, f(t))$ is a smooth family of canonical transformations, satisfying the required conditions. The lemma is proved.

Consider now the vector field

$$V_t(z) = \frac{\partial g^t}{\partial t}((g^t)^{-1}z). \tag{51}$$

It is a Hamiltonian one; thus there exists a function $H_o(z,t)$ such that

$$V_t = \mathcal{T}\frac{\partial H_o}{\partial z}, \quad \mathcal{T} = \begin{pmatrix} 0 & E \\ -E & 0 \end{pmatrix}. \tag{52}$$

Thus $g^t = g^t_{H_o}$. Outside the bounded domain

$$\tilde{U} = \bigcap_{t\in[0,1]} g^t(U); \tag{53}$$

we continue the function $H_o(z,t)$ in such a way that the inclusion can be satisfied:

$$\partial^k H_o(z,t)/\partial t^k \in S^\infty(\mathbb{R}^{2n}), \; k = 0,1,2,\dots, \; t\in[0,1].$$

Set now

$$H(q,p,h,t) = H_o(q,p,t) - ihH_1(q,p,t), \tag{54}$$

where $H_1(q,p,t) \in S^\infty(\mathbb{R}^{2n})$,

$$H_1(q,p,t) = -\frac{1}{2}\,\mathrm{tr}\,\frac{\partial^2 H_o}{\partial q\partial p}(q,p,t). \tag{55}$$

Set $V_t = g^t(U)$. Consider the operator family

$$\begin{gathered} T(t) = e^{(i/h)a(t) + (i\pi/2)b(t)}T(g^t), \\ T(t) : H^k(U) \to H^k(V_t), \end{gathered} \tag{56}$$

where $a(t)$ and $b(t)$ are chosen in such a way that the operator (56) continuously depends on t.

Lemma 4. The operator (56) satisfies the equation

$$-ih\frac{\partial T(t)}{\partial t}u + \hat{H}(t)\circ T(t)u = \hat{O}(h^2), \quad u \in H(\mathbb{R}^n). \tag{57}$$

The proof is obvious.

Perform now the above construction for $g = g_2$. We obtain that the operators $T(g_2^t)\circ T(g_1)$ and $T(g_2^t g_1)$ both satisfy (57) and have coinciding initial data. Since $\hat{H}(t)$ is formally self-adjoint modulo $\hat{O}(h^2)$, the formula (46) is valid. The proof of the Theorem 1 is complete.

C. The Sheaves of Rapidly Oscillating Functions on a Symplectic Manifold

Let M be a smooth manifold of even dimension $2n$, supplied with the symplectic form ω^2. Let

$$\{U_\alpha, \gamma_\alpha : U_\alpha \to V_\alpha \subset \mathbb{R}^{2n}_{(q,p)}\}_{\alpha\in\mathcal{T}}$$

be a canonical atlas on M, i.e., an atlas such that

$$\gamma_\alpha(dp\wedge dq) = \omega^2 \tag{58}$$

for all $\alpha\in\mathcal{T}$. Let $\alpha,\beta,\dots \in \mathcal{T}$. We denote

$$\begin{aligned} U_{\alpha\beta\dots} &= U_\alpha \cap U_\beta \cap \dots, \\ V_{\alpha(\beta\dots)} &= \gamma_\alpha(U_{\alpha\beta\dots}) \subset V_\alpha, \\ H_{\alpha(\beta\dots)} &= H(U_{\alpha(\beta\dots)}), \\ H^0_{\alpha(\beta\dots)} &= H^0(U_{\alpha(\beta\dots)}). \end{aligned} \tag{59}$$

It is assumed that all the sets $U_\alpha, U_{\alpha\beta\ldots}$ (and, consequently, their coordinate images) are bounded, connected, and simply connected.

The canonical coordinate mappings are defined

$$g_{\alpha\beta} = \gamma_\alpha \circ \gamma_\beta^{-1} : V_{\beta(\alpha)} \to V_{\alpha(\beta)}$$

and, consequently, we may obtain the operators

$$T_{\alpha\beta} \equiv T(g_{\alpha\beta}) : H^0_{\beta(\alpha)} \to H^0_{\alpha(\beta)}. \tag{60}$$

From Theorem 1 of the previous item we obtain, that if $U_{\alpha\beta\gamma} \neq \emptyset$,

$$T_{\alpha\beta} \circ T_{\beta\gamma} = T_{\alpha\gamma} \cdot \exp\{i \frac{\varepsilon^{(1)}_{\alpha\beta\gamma}}{h} + i\pi\varepsilon^{(2)}_{\alpha\beta\gamma}\} \tag{61}$$

in the space $H^0_{\gamma(\beta\alpha)}$. In (61) $\varepsilon^{(1)}_{\alpha\beta\gamma} \in \mathbb{R}$, $\varepsilon^{(2)}_{\alpha\beta\gamma} \in \mathbb{Z}_2$. Thus, two-dimensional cochains $\varepsilon^{(1)}$ and $\varepsilon^{(2)}$ of the covering $\{U_\alpha\}$ with values in $\mathbb{R}$ and $\mathbb{Z}_2$, respectively, are defined. It turns out that the following lemma is valid.

Lemma 1. The cochains $\varepsilon^{(1)}$ and $\varepsilon^{(2)}$ are cocycles.

Proof. By Theorem 1 of previous item,

$$T_{\alpha\beta} \circ T_{\beta\alpha} = \lambda \cdot T(\mathrm{id}) = \lambda \cdot 1, \ \lambda \neq 0,$$

so that $T_{\alpha\beta}$ is an invertible operator. Let $U_{\alpha\beta\gamma\delta} \neq \emptyset$. We have

$$\begin{aligned} T_{\alpha\beta} \circ T_{\beta\gamma} \circ T_{\gamma\delta} &= T_{\alpha\gamma} \circ T_{\gamma\delta} \exp\{i \frac{\varepsilon^{(1)}_{\alpha\beta\gamma}}{h} + i\pi\varepsilon^{(2)}_{\alpha\beta\gamma}\} = \\ &= T_{\alpha\delta} \cdot \exp\{i \frac{\varepsilon^{(1)}_{\alpha\beta\gamma}}{h} + i \frac{\varepsilon^{(2)}_{\alpha\gamma\delta}}{h} + i\pi\varepsilon^{(1)}_{\alpha\beta\gamma} + i\pi\varepsilon^{(2)}_{\alpha\gamma\delta}\}; \end{aligned} \tag{62}$$

on the other hand,

$$\begin{aligned} T_{\alpha\beta} \circ T_{\beta\gamma} \circ T_{\gamma\delta} &= T_{\alpha\beta} \circ T_{\beta\delta} \exp\{i \frac{\varepsilon^{(1)}_{\beta\gamma\delta}}{h} + i\pi\varepsilon^{(2)}_{\beta\gamma\delta}\} = \\ &= T_{\alpha\delta} \cdot \exp\{i \frac{\varepsilon^{(0)}_{\beta\gamma\delta}}{h} + i \frac{\varepsilon^{(1)}_{\alpha\beta\delta}}{h} + i\pi\varepsilon^{(2)}_{\beta\gamma\delta} + i\pi\varepsilon^{(2)}_{\alpha\beta\delta}\}. \end{aligned} \tag{63}$$

From (62) - (63) we obtain, multiplying by

$$\begin{aligned} [d\varepsilon^{(1)}/h + \pi d\varepsilon^{(2)}]_{\alpha\beta\gamma\delta} &= (\varepsilon^{(1)}_{\alpha\beta\gamma} - \varepsilon^{(1)}_{\alpha\beta\delta} + \varepsilon^{(1)}_{\alpha\gamma\delta} - \\ &- \varepsilon^{(1)}_{\beta\gamma\delta})/h + \pi(\varepsilon^{(2)}_{\alpha\beta\gamma} - \varepsilon^{(2)}_{\alpha\beta\delta} + \varepsilon^{(2)}_{\alpha\gamma\delta} - \varepsilon^{(2)}_{\beta\gamma\delta}) \equiv 0 \pmod{2\pi} \end{aligned} \tag{64}$$

for every $h \in [0,1]$. It follows that

$$d\varepsilon^{(1)} = 0, \ d\varepsilon^{(2)} = 0 \pmod 2. \tag{65}$$

The lemma is proved. Thus, we have defined the cohomology classes

$$\varepsilon^{(1)} \in H^2(M,\mathbb{R}), \ \varepsilon^{(2)} \in H^2(M,\mathbb{Z}_2)$$

on the symplectic manifold M.

Note 1. It should be noted that the classes defined above do not depend on the choice of initial points, arguments of measure density, etc.

It may be shown that the class $\varepsilon^{(1)}$ is simply the class of symplectic form ω^2 on M. We derive the quantization conditions.

From now on assume that M depends smoothly on parameters $\omega \in \Omega$, where Ω is a manifold.

Definition 1. The pair (ω, h) satisfies the quantization condition (two-dimensional) if for the given values of ω and h, we have

$$\frac{1}{\pi h}\varepsilon^{(1)} + \varepsilon^{(2)} \equiv 0 \pmod 2, \tag{66}$$

i.e., $\varepsilon^{(1)}/\pi h + \varepsilon^{(2)}$ defines a zero cohomology class in $H^2(M, \mathbb{R}_2)$.

Let the quantization conditions be satisfied. Then there exists a cochain $\mu = \{\mu_{\alpha\beta}\}$ of the covering $\{\mathcal{U}_\alpha\}$ with values in $\mathbb{R}_2$ such that for $\mathcal{U}_{\alpha\beta\gamma} \neq \emptyset$:

$$(d\mu)_{\alpha\beta\gamma} \equiv \mu_{\alpha\beta} - \mu_{\alpha\gamma} + \mu_{\beta\gamma} = \frac{1}{\pi h}\varepsilon^{(1)}_{\alpha\beta\gamma} + \varepsilon^{(2)}_{\alpha\beta\gamma} \pmod 2. \tag{67}$$

Set

$$\begin{aligned} \tilde{T}_{\alpha\beta} &= T_{\alpha\beta}\exp(-i\pi\mu_{\alpha\beta}), \\ \tilde{T}_{\alpha\beta} &= T_{\alpha\beta}\exp(-\pi\mu_{\alpha\beta}). \end{aligned} \tag{68}$$

From (67) it follows that the operators $\tilde{T}_{\alpha\beta}$ satisfy the agreement conditions

$$\tilde{T}_{\alpha\beta} \circ \tilde{T}_{\beta\gamma} = \tilde{T}_{\alpha\gamma} \tag{69}$$

and, consequently, define some sheaf $\mathcal{F}^0$ of linear spaces on M.

Definition 2. The sheaf $\mathcal{F}^0$ is called the sheaf of rapidly oscillating functions on M. Give an explicit description of the sheaf $\mathcal{F}^0$. Recall, first of all, the definition of the sheaf (cf. [20]).

The sheaf of linear spaces over manifold M is a collection of linear spaces $\mathcal{F}(\mathcal{U})$ where $\mathcal{U}$ varies in the set of all open subsets of M and linear operators

$$i_{\mathcal{U}\mathcal{U}'} : \mathcal{F}(\mathcal{U}') \to \mathcal{F}(\mathcal{U}),$$

defined for $\mathcal{U} \subset \mathcal{U}'$ such that the following conditions are satisfied:

(a) For $\mathcal{U} \subset \mathcal{U}' \subset \mathcal{U}''$, we have

$$i_{\mathcal{U}\mathcal{U}'} \circ i_{\mathcal{U}'\mathcal{U}''} = i_{\mathcal{U}\mathcal{U}''}. \tag{70}$$

(b) If $\mathcal{U} = \bigcup_{j \in \mathcal{T}} \mathcal{U}_j$ and $f \in \mathcal{F}(u)$ is such that

$$i_{\mathcal{U}_j\mathcal{U}} f = 0 \tag{71}$$

for all j, then f = 0.

(c) If $\mathcal{U} = \bigcup_{j \in \mathcal{T}} \mathcal{U}_j$ and the tuple of elements $f_j \in \mathcal{U}$, $j \in \mathcal{T}$ satisfies the agreement conditions

$$i_{\mathcal{U}_j \cap \mathcal{U}_k, \mathcal{U}_j} f_j = i_{\mathcal{U}_j \cap \mathcal{U}_k, \mathcal{U}_k} f_k \tag{72}$$

for all $j,k \in T$, then there exists a unique element such that

$$i_{U_j U} f = f_j \tag{73}$$

for all $j,k \in T$.

$F(U)$ is called the space of sections over U, while the mappings are called the restriction homomorphisms. The sheaf F^0 is constructed in the following way. For each open subset $U \subset M$ denote by $F^0(U)$ the subspace in the direct product

$$\prod_{\alpha \in T} H^0(\gamma_\alpha(U \cap U_\alpha)),$$

consisting of tuples $\{\psi_\alpha\}_{\alpha \in T}$ such that

$$\tilde{T}_{\alpha\beta}\psi_\beta = \psi_\alpha \tag{74}$$

in the space $H^0(\gamma_\alpha U \cap U_{\alpha\beta})$ for all pairs (α,β) such that $U_{\alpha\beta} \neq \emptyset$. If $U \subset U'$, then by

$$i_{UU'} : \mathcal{F}^0(U') \to \mathcal{F}^0(U)$$

we denote the mapping induced by natural projections

$$H^0(\gamma_\alpha(U' \cap U_\alpha)) \to H^0(\gamma_\alpha(U \cap U_\alpha)).$$

It is not difficult to verify the correctness of the given construction, while the validity of sheaf axioms (a) - (c) follows readily from the definitions.

We see that the construction of the sheaf F^0 depends on the choice of the one-dimensional cochain μ of the covering $\{U_\alpha\}$ such that

$$d\mu = \varepsilon^{(1)}\pi h + \varepsilon^{(2)} \pmod 2. \tag{75}$$

Let μ,μ' be the two cochains, satisfying (75); then $d(\mu - \mu') = 0$ and thus $\mu - \mu'$ is a one-dimensional cocycle. Let

$$T'_{\alpha\beta} = T_{\alpha\beta}\exp(-i\pi\mu_{\alpha\beta}), \tag{76}$$

and let $\mathcal{F}^0$ be a sheaf constructed by the means described above, using the transition operators $\tilde{T}'_{\alpha\beta}$. Assume that $\mu - \mu'$ is a coboundary modulo 2. Then

$$\mu_{\alpha\beta} - \mu_{\alpha\beta} = \lambda_\alpha - \lambda_\beta \pmod 2 \tag{77}$$

for $U_{\alpha\beta} \neq \emptyset$; and the mapping

$$\{\psi_\alpha\}_{\alpha \in T} \to \{\exp(i\pi\lambda_\alpha)\psi_\alpha\}_{\alpha \in T} \tag{78}$$

induces the sheaf isomorphism $F \to F^{0\prime}$. Thus we have proved the following:

Theorem 1. The sheaf of rapidly oscillating functions on a symplectic manifold M exists if and only if the cohomology class

$$\frac{\varepsilon^{(1)}}{\pi h} + \varepsilon^{(2)} \in H^2(M,\mathbb{R}_2)$$

equals zero. If this class vanishes, the sheaf of rapidly oscillating functions is fixed (up to an isomorphism) by the choice of some (also dependent on parameters) cohomology class $\mu \in H^1(\mathbb{R}_2)$.

Now let $U \subset M$ be an open subset, $\psi \in F^0(U)$.

Definition 3. The wave front of ψ is the following subset of M:

$$WF_h(\psi) = \bigcup_\alpha \gamma_\alpha^{-1}(WF_h(\psi_\alpha)) \subset \bar{U}. \tag{79}$$

It is easy to see that the set (79) depends only on ψ, not on a canonical atlas and the representatives ψ_α.

Let H be a smooth function on M, growing slowly in all the charts of some canonical atlas.

Definition 4. The operator $\hat{H}$ in $F^0(U)$ given by

$$\hat{H}\{\psi_\alpha\} = \{(H^0\gamma_\alpha^{-1})(\overset{2}{q}, -ih\overset{1}{\frac{\partial}{\partial q}})\psi_\alpha\} \tag{80}$$

is called a 1/h-PDO in the space of sections of the sheaf F^0.

Lemma 2. The operator (80) is a sheaf homomorphism $F^0 \to F^0$. The statement of Lemma 2 of item A (k = 0 only) remains valid for 1/h-PDO in the space of sections of the sheaf of rapidly oscillating functions. The proof is obvious.

D. Quantized Lagrangian Sub-Manifolds and the Canonical Operator

Now let M be a quantized symplectic manifold with the form ω^2, and Λ be a connected Lagrangian sub-manifold in M (i.e., the restriction $\omega^2|_{T \in \Lambda}$ is equal to zero for all $z \in \Lambda$). Let also $d\sigma$ be a non-degenerative measure on Λ. We construct here an operator

$$K : C_0^\infty(\Lambda) \to \mathcal{F}^0(U) \tag{81}$$

for any open subset $U \subset M$, which will be called a canonical operator.

Let again

$$\{\gamma_\alpha : U_\alpha \to V_\alpha \subset \mathbb{R}^{2n}_{(q,p)}\}_{\alpha \in T}$$

be a canonical atlas on M. For any α

$$\Lambda_\alpha = \gamma_\alpha(\Lambda \cap U_\alpha) \tag{82}$$

is a Lagrangian sub-manifold in $\mathbb{R}^{2n} \supset V_\alpha$; we assume that the covering $\{U_\alpha\}$ is such that Λ_α is connected and simply connected. Under these conditions on each Λ_α there exists a canonical operator, amd we may consider it acting in the spaces

$$K_\alpha : C_0^\infty(\Lambda \cap U_\alpha) \to C_0^\infty(\Lambda_\alpha) \to H_\alpha^0. \tag{83}$$

Lemma 1. Let the intersection $\Lambda \cap U_\alpha \cap U_\beta$ be non-empty. Then for any function $\phi \in C_0^\infty(\Lambda \cup U_\alpha \cup U_\beta)$ there is an equality in the space $H^0_{\alpha(\beta)}$:

$$K_\alpha\phi = K_\beta\phi \exp(i\pi b_{\alpha\beta}), \tag{84}$$

where $b_{\alpha\beta}$ is the cocycle on Λ with the values in the group of real numbers modulo 2.

Proof. This is an obvious consequence of the stationary phase method.

Let now the cohomology class

$$b \in H_1(\Lambda, \mathbb{R}_2) \tag{85}$$

be trivial. We say in this case that the quantization condition on Λ is satisfied. Multiplying K_α by the factor ξ_α, $|\xi_\alpha| = 1$, we can reach the identity $K_\alpha\phi = K_\beta\phi$ for all α,β such that $\Lambda \cap U_\alpha \cap U_\beta$ is non-empty and $\phi \in C_0^\infty(\Lambda \cap U_\alpha \cap U_\beta)$. It follows that the global operator

$$K : C_0^\infty(\Lambda) \to F(u) \tag{86}$$

is defined for any open subset $U \subset M$, coinciding with K_α on $C_0^\infty(\Lambda \cap U_\alpha)$.

Definition 1. The operator (86) is called a canonical operator on the quantized Lagrangian submanifold Λ of quantized symplectic manifold.

3. THE CANONICAL SHEAF ON A SYMPLECTIC $\mathbb{R}_+$-MANIFOLD

A. Some Function Spaces

From now on we consider the R_+-homogeneous version of asymptotic constructions which, in particular, enables us to prove the main quasi-invertibility theorem (this is performed in Chapter 4). This section presents a necessary preliminary to the subsequent considerations. We list a number of function spaces to serve further as local models of the Poisson algebras, spaces of canonical sheaf sections over a R_+-homogeneous symplectic manifold, etc. We define also pseudo-differential operators in the introduced function spaces and give some supplementary information and notations.

Let R^n be an arithmetic n-dimensional space with elements denoted by $x = (x_1,\ldots,x_n)$ and let $\lambda = (\lambda_1,\ldots,\lambda_n)$ be a fixed n-tuple of non-negative real numbers. Set

$$\begin{aligned} I_o &= \{j \in [n] \,|\, \lambda_j = 0\}, \\ I_+ &= [n] \setminus I_o = \bar{I}_o. \end{aligned} \tag{1}$$

Here $[n]$ is the set of integers from 1 to n. We assume I_+ to be non-empty. The equation

$$\tau x = (\tau^{\lambda_1}x_1,\ldots,\tau^{\lambda_n}x_n), \quad \tau \in R_+ \tag{2}$$

defines free smooth action of the group R_+ (multiplicative group of positive real numbers) on the space

$$R^{|I_0|} \times (R^{|I_+|} \setminus \{0\}) \stackrel{\text{def}}{=} R^n_*. \tag{3}$$

For any $x \in R^n_*$, set

$$\tilde{\Lambda}(x) = \tau^{-1}, \tag{4}$$

where $\tau = \tau(x) \in R_+$ satisfies the property

$$\sum_{j\in I_+} (\tau x)_j^2 = \sum_{j\in I_+} \tau^{2\lambda_j} x_j^2 = 1. \tag{5}$$

It is evident that $\tilde{\Lambda}(x)$ is a smooth non-vanishing function on R^n_* satisfying the properties:

$$\tilde{\Lambda}(\tau x) = \tau\tilde{\Lambda}(x), \tag{6}$$

$$c\tilde{\Lambda}(x) \leqslant \sum_{j\in I_+} |x_j|^{1/\lambda_j} \leqslant C\tilde{\Lambda}(x) \tag{7}$$

Table 1. Functional Spaces

Notation of the functional space	Conditions for f(x) to belong to this space
$S^m_{(\lambda_1,\dots,\lambda_n)}(\mathcal{U})$	$\lvert f^{(\alpha)}(x_1,\dots,x_n)\rvert \leqslant C_\alpha \Lambda(x)^{m-\Sigma_j\lambda_j\alpha_j}$
$S^m_{(\lambda_1,\dots,\lambda_n)}(\mathcal{U})$	$\lvert f^{(\alpha)}(x_1,\dots,x_n)\rvert \leqslant C_\alpha \Lambda(x)^{m}$
$L^m_{(\lambda_1,\dots,\lambda_n)}(\mathcal{U})$	$\lvert f^{(\alpha)}(x_1,\dots,x_n)\rvert \leqslant C_\alpha \Lambda(x)^{m+\Sigma_j(1-\lambda_j)\alpha_j}$
$O^m_{(\lambda_1,\dots,\lambda_n)}(\mathcal{U})$	same as for $S^m_{(\lambda_1,\dots,\lambda_n)}(\mathcal{U})$; in addition, $f(\tau x) = \tau^m f(x)$ for $\tau \geqslant 1$, $\Lambda(x) \geqslant R$ where R depends on f
$P^m_{(\lambda_1,\dots,\lambda_n)}(\mathcal{U})$	same as for $S^m_{(\lambda_1,\dots,\lambda_n)}(\mathcal{U})$; in addition, for any $N > 0$, $f(x) = \sum_{j=0}^{N_1(N)} f_j(x) + R_N(x)$, where $f_j(x) \in O^{m_j}_{(\lambda_1,\dots,\lambda_n)}(\mathcal{U})$, $m_j \leqslant m$; $j = 0$; $N_1(N)$; $R_N(x) \in S^{m-N}_{(\lambda_1,\dots,\lambda_n)}(\mathcal{U})$

with some positive c and C. We introduce the real-valued function $\Lambda(x)$ on R^n, setting $\Lambda(x) = \tilde{\Lambda}(x)$ for $\tilde{\Lambda}(x) \geqslant 1$ and regarding that Λ be a smooth function depending on x and $\geqslant 1/2$ everywhere. For $I \cap I_+ \neq \emptyset$, we set

$$\Lambda_I(x) \equiv \Lambda_I(x_I) = \Lambda(x)\big|_{y_{I_+\setminus I}=0}. \tag{8}$$

For example, if $n = 2$, $\lambda_1 = \lambda_2 = 1$ and therefore $\Lambda(x) = (x_1^2 + x_2^2)^{1/2}$ for large x, then $\Lambda_{\{1\}}(x_1,x_2) = \Lambda(x_1,0) = |x_1|$ for large x_1. Obviously, the following inequalities are valid

$$\Lambda_I(x) \leqslant \Lambda(x), \tag{9}$$

$$c\Lambda(x) \leqslant 1 + \sum_{j\in I_+} |x_j|^{1/\lambda_j} \leqslant C\Lambda(x) \tag{10}$$

with positive constants c and C in (10) different, in general, from those in (7).

We have gathered the functional spaces most frequently used in the sequel in Table 1 given above. Some elucidation is necessary. $\mathcal{U}$ denotes an arbitrary domain in R^n; the case $\mathcal{U} = R^n$ is also included. The elements of a space are smooth functions in $\mathcal{U}$ up to the boundary of $\mathcal{U}$. In the second column of our table conditions for f(x) to belong to a specified functional space are presented; they must be satisfied by in $\mathcal{U}$. The lowest possible values of constants in the estimates form the tuple of seminorms, defining the topology in the described functional space. m denotes any real number. The notation of the functional space may include an additional subscript 0 (which makes sense for $\mathcal{U} \neq R^n$ and indicates that supp $f \subset \mathcal{U}$ for any f from the space) or c (which means that the elements of the space have compact supports with respect to x_{I_0}).

The elements of the space $\mathcal{O}^m_{(\lambda_1,\ldots,\lambda_n)}(\mathcal{U})$ are called $(\lambda_1,\ldots,\lambda_n)$-quasi-homogeneous functions of degree m for large x_{I_+}; those of the space $\mathcal{P}^m_{(\lambda_1,\ldots,\lambda_n)}(\mathcal{U})$ are called asymptotically $(\lambda_1,\ldots,\lambda_n)$-quasi-homogeneous functions of degree $\leqslant m$. Except for $\mathcal{O}^m_{(\lambda_1,\ldots,\lambda_n)}(\mathcal{U})$, we allow m to take values $+\infty$ and $-\infty$, denoting, respectively, the union and the intersection of the corresponding spaces for all finite m.

Obviously, the inclusions between these spaces are valid:

(a) Except for $\mathcal{O}^m_{(\lambda_1,\ldots,\lambda_n)}(\mathcal{U})$ the space with the lower value of index m is contained in the space with the greater one;

(b) $\mathcal{O}^m_{(\lambda_1,\ldots,\lambda_n)}(u) \subset \mathcal{P}^m_{(\lambda_1,\ldots,\lambda_n)}(u) \subset S^m_{(\lambda_1,\ldots,\lambda_n)}(u) \subset \tilde{S}^m_{(\lambda_1,\ldots,\lambda_n)} \times (u) \subset \mathcal{L}^m_{(\lambda_1,\ldots,\lambda_n)}(u)$.

Also it is evident that $\Lambda(x) \in \mathcal{O}^1_{(\lambda_1,\ldots,\lambda_n)}(\mathcal{U})$. We present the useful property of quasi-homogeneous functions:

<u>Proposition 1 (the Euler identity)</u>. If $f \in \mathcal{O}^m_{(\lambda_1,\ldots,\lambda_n)}(\mathcal{U})$, then for $x \in u$, $\Lambda(x)$ large enough, we have

$$mf(x) = \sum_{j=1}^{n} \lambda_j \partial f(x)/\partial x_j. \tag{11}$$

<u>Proof</u>. Differentiate $\tau^m f(x) = f(\tau x)$ with respect to τ and set $\tau = 1$. The space $S^m_{(\lambda_1,\ldots,\lambda_n)}(\mathcal{U})$ will be a local model for the space of the Poisson algebra, and the space $\mathcal{L}^m_{(\lambda_1,\ldots,\lambda_n)}(\mathcal{U})$ will be a local model for the space of canonical sheaf sections. We need some considerations of pseudo-differential operators in the latter space.

Let $\phi \in \mathcal{L}^\infty_{(\lambda_1,\ldots,\lambda_n)}(R^n)$. We intend to define the action of pseudo-differential operators by means of the common formula (cf. Chapter 2):

$$f(\overset{2}{x}, -i\overset{1}{\frac{\partial}{\partial x}})\phi(x) = (2\pi)^{-n}\iint f(x,p)e^{i(x-y)p}\phi(y)dydp. \tag{12}$$

The integral on the right in (12) is, generally speaking, divergent. However, under certain conditions it may be regularized and give rise to a continuous operator in $\mathcal{L}^\infty_{(\lambda_1,\ldots,\lambda_n)}(R^n)$.

<u>Definition 1</u>. We denote by $T^m_{(\lambda_1,\ldots,\lambda_n)}(R^{2n})$ the space of functions $f(x,p)$, $x,p \in R^n$ satisfying the following conditions:

$$\left|\frac{\partial^{|\alpha|+|\beta|} f(x,p)}{\partial x^\alpha \partial p^\beta}\right| \leqslant C_{\alpha\beta}\Lambda(x)^{m-\Sigma_j\lambda_j\alpha_j+\Sigma_j(\lambda_j-1)\beta_j} \times (1+\Sigma_j|p_j\Lambda(x)^{\lambda_j-1}|)^{m_1+|\alpha|}, \quad |\alpha|+|\beta| = 0,1,2,\ldots, \tag{13}$$

where m_1 does not depend on α,β (but depends on f). Set also

$$T^{\infty}_{(\lambda_1,\ldots,\lambda_n)}(R^{2n}) = \bigcup_m T^m_{(\lambda_1,\ldots,\lambda_n)}(R^{2n});$$

$$T^{-\infty}_{(\lambda_1,\ldots,\lambda_n)}(R^{2n}) = \bigcap_m T^m_{(\lambda_1,\ldots,\lambda_n)}(R^{2n}).$$

Theorem 1. For $f \in T^{\infty}_{(\lambda_1,\ldots,\lambda_n)}(R^{2n})$ the integral (12) admits a regularization (described later in the proof of this theorem); once f belongs to $T^m_{(\lambda_1,\ldots,\lambda_n)}(R^{2n})$, the operator (12) is continuous in the spaces

$$f(\overset{2}{x},-i\frac{\overset{1}{\partial}}{\partial x}) : L^s_{(\lambda_1,\ldots,\lambda_n)}(R^{2n}) \to L^{s+m+\delta}_{(\lambda_1,\ldots,\lambda_m)}(R^{2n}) \tag{14}$$

for any $s \in R$ and any $\delta > 0$.

Proof. Denote the integral on the right-hand side of (12) by $(2\pi)^{-n}I(x)$. We perform some variable changes and transformations over I(x) (our calculations to this end being purely formal). We introduce the new variables of integration ξ_j, η_j, $j = 1,\ldots,n$ by the formulas:

$$y_j = x_j + \xi_j\Lambda(x)^{\lambda_j};\ p_j = \eta_j\Lambda(x)^{1-\lambda_j}. \tag{15}$$

Substitution of (15) into I(x) yields

$$I(x) = [\iint e^{i\Lambda(x)\xi\eta} f(x,\eta\Lambda(x)^{1-\lambda})\phi(x+\xi\Lambda(x)^{\lambda})d\xi d\eta]\Lambda(x)^{\eta}. \tag{16}$$

In (16) and subsequent formulas $\eta\Lambda(x)^{1-\lambda}$ stands (for short) for $(\eta_1\Lambda(x)^{1-\lambda_1}, \ldots,\eta_n\Lambda(x)^{1-\lambda_n})$ etc. Then (16) may be written in the form:

$$I(x) = \Lambda(x)^n\iint e^{i\Lambda(x)\xi\eta}F(x,\eta)\Phi(x,\xi)d\xi d\eta, \tag{17}$$

where $F(x,\eta) = f(x,\eta\Lambda(x)^{1-\lambda})$, $\Phi(x,\xi) = \phi(x + \xi\Lambda(x)^{\lambda})$.

The conditions that $f(x,p) \in T^m_{(\lambda_1,\ldots,\lambda_n)}(R^{2n})$ and $\phi \in L^s_{(\lambda_1,\ldots,\lambda_n)}(R^n)$ turn into the following estimates on F and Φ are:

$$\left|\frac{\partial^{|\alpha|+|\beta|}F(x,\eta)}{\partial x^{\alpha}\partial\eta^{\beta}}\right| \leqslant C_{\alpha\beta}\Lambda(x)^{m-\Sigma_j\lambda_j\alpha_j}(1+|\eta|)^{m_1+|\alpha|}, \quad |\alpha|+|\beta| = 0,1,2,\ldots; \tag{18}$$

$$\left|\frac{\partial^{\gamma}\Phi(x,\xi)}{\partial\xi^{\gamma}}\right| \leqslant C_{\gamma}\Lambda(x+\xi\Lambda(x)^{\lambda})^{s+\Sigma_j(1-\lambda_j)\gamma_j}\Lambda(x)^{\Sigma_j\lambda_j\gamma_j}, \quad |\gamma| = 0,1,2,\ldots. \tag{19}$$

We consider first the case when m_1 is negative and sufficiently large, namely, $m_1 < -n$ and then we point out how the general case may be reduced to this one.

Let $e_1(z)$ be a smooth finite function in R^n equal to 1 in the vicinity of the origin. For any $\varepsilon > 0$, we have the partition of the unity

$$1 = e_1(\Lambda(x)^{1-\varepsilon}\xi) + (1 - e_1(\Lambda(x)^{1-\varepsilon}\xi)) \equiv$$

$$\equiv e_1(\Lambda(x)^{1-\varepsilon}\xi) + e_2(\Lambda(x)^{1-\varepsilon}\xi)$$

and, respectively, $I(x) = I_1(x) + I_2(x)$, where $I_i(x)$ is obtained from $I(x)$ via the multiplication of the integrand by $e_i(\Lambda(x)^{1-\varepsilon}\,\xi)$, $i = 1,2$.

By (18), taking into account that $m_1 < -n$, $I_1(x)$ converges. The regularization of $I_2(x)$ goes as follows: consider the differential operator

$$L = \frac{1}{i\Lambda(x)\xi^2} \sum_{j=1}^{n} \xi_j \frac{\partial}{\partial \eta_j}. \tag{20}$$

Obviously, the coefficients of L are smooth on supp e_2 and

$$L e^{i\Lambda(x)\xi\eta} = e^{i\Lambda(x)\xi\eta}. \tag{21}$$

Therefore we obtain by formally integrating by parts N times:

$$I_2(x) = \frac{1}{i^N} \Lambda(x)^{n-N} \iint e^{i\Lambda(x)\xi\eta} e_2(\Lambda(x)^{1-\varepsilon}\xi)\Phi(x,\xi) \times \times \frac{1}{(\xi^2)^N}\left[\left(\sum_{j=1}^{n} \xi_j \frac{\partial}{\partial\eta_j}\right)^N F(x,\eta)\right] d\xi d\eta. \tag{22}$$

Estimate the integrand in (22). Denoting it by $J_N(x,\xi,\eta)$, we have

$$|J_N(x,\xi,\eta)| \leqslant \text{const}\cdot\Lambda(x)^m (1+|\eta|)^{m_1}|\xi|^{-N} \times \times \Lambda(x+\xi\Lambda(x)^{\lambda})^s \leqslant \text{const}\cdot\Lambda(x)^{m+s}(1+|\eta|^{m_1}|\xi|^{-N}(1+\sum_{j\in I_+}|\xi_j|^{1/\lambda_j})^{|s|}. \tag{23}$$

We see that for N large enough the integral on the right-hand side of (22) is absolutely convergent (note that $|\xi| \geq C\Lambda(x)^{\varepsilon-1} > 0$ on supp J_N).

We take the formula (22) as the definition of the regularization of the integral $I_2(x)$ and therefore we have defined the regularization for $I(x)$ as well, since $I_1(x)$ needs no regularization. It is obvious that the regularization does not depend on the numbers N,ε and on the choice of $e_1(\xi)$. We also see from (22) that

$$|I_2(x)| \leqslant \text{const}\cdot\Lambda(x)^{-N_1} \tag{24}$$

for any N_1.

Indeed, since $|\xi| \geqslant C\Lambda(x)^{\varepsilon-1}$, we have

$$(1+|\xi|)C\Lambda(x)^{\varepsilon-1} \leqslant C\Lambda(x)^{\varepsilon-1} + c_0|\xi| \leqslant (1+c_0)|\xi|,$$

where $c_0 = \sup_{\alpha} C\Lambda(x)^{\varepsilon-1} < \infty$. Therefore

$$\frac{1}{|\xi|^N} \leqslant \left(\frac{1+c_0}{c}\right)^N \Lambda(x)^{N-N\varepsilon} \frac{1}{(1+|\xi|)^N}$$

on supp $e_2(|\xi|\Lambda(x)^{1-\varepsilon})$. Substituting this into (22) and (23), we obtain

$$|I_2(x)| \leqslant \text{const}\cdot\Lambda(x)^{m+n+s-N\varepsilon}$$

which immediately yields (24).

Consider now the case $m_1 \geqslant -n$. We introduce the C^∞ partition of unity $\{\rho_j(z)\}_{j=0}^{n}$ in the space R_z^n with the properties:

i) supp $\rho_0 \subset \{|z| < 1\}$, supp $\rho_j \subset \{|z| > \frac{1}{2}, |z_j| \geqslant \frac{1}{2}|z_k|, k = 1,\ldots,n\}$;

ii) $|\rho_j^{(\alpha)}(z)| \leqslant C_\alpha$, $|\alpha| = 0,1,2,\ldots$, $j = 0,\ldots,n$

and represent I(x) in the form:

$$I(x) = \sum_{j=0}^{n} I^{(j)}(x) \equiv \sum_{j=0}^{n} \iint f(x,p)\rho_j(p\Lambda(x)^{\lambda-1})e^{i(x-y)p}\phi(y)dydp \equiv$$
$$\equiv \sum_{j=0}^{n} \iint f_{(j)}(x,p)e^{i(x-y)p}\phi(y)dydp \tag{25}$$

(note that $f_{(j)}(x,p) \in T^m_{(\lambda_1,\ldots,\lambda_n)}(R^n)$, $j = 0,\ldots,n$; the reader can easily verify it).

Each of the integrals $I^{(j)}(x)$ may be considered separately. As for $I^{(0)}(x)$, nothing has to be proved, for the resulting integral in (17) is taken over $\{|\eta| \leqslant 1\}$. In the expression (25) for $I^{(j)}(x)$ we perform (formal) integration by parts over y_j, which yields the formula (12) with $\phi(y)$ replaced by $\partial^M\phi(y)/\partial y_j^M$ and $f(y,p)$ by $f_{(i)}(y,p)(ip_j)^{-M}$. After that we go along the familiar ways and obtain

$$I^{(j)}(x) = \Lambda(x)\iint e^{i\Lambda(x)\xi\eta} f_{(j)}(x,\eta\Lambda(x)^{1-\lambda}) \times$$
$$\times (i\eta_j\Lambda(x)^{1-\lambda_j})^{-M} \frac{\partial^M\phi}{\partial y_j^M}(x+\xi\Lambda(x)^\lambda)d\xi d\eta = \tag{26}$$
$$= \Lambda(x)^{-M(1-\lambda_j)}\iint e^{i\Lambda(x)\xi\eta}F_{jM}(x,\eta)\Phi_M(x,\xi)d\xi d\eta,$$

where

$$\left|\frac{\partial^{|\alpha|+|\beta|}F_{jM}(x,\eta)}{\partial x^\alpha \partial\eta^\beta}\right| \leqslant C_{\alpha\beta}\Lambda(x)^{m-\Sigma_k\lambda_k\alpha_k}(1+|\eta|)^{m_1-M+|\alpha|} \tag{27}$$

(we have taken into account that $|\eta| \leqslant C|\eta_i|$ on supp $\rho_i(\eta)$) and

$$\left|\frac{\partial^\gamma\Phi_M(x,\xi)}{\partial\xi^\gamma}\right| \leqslant C_\alpha\Lambda(x+\xi\Lambda(x)^\lambda)^{s+M(1-\lambda_j)+\Sigma_k(1-\lambda_k)\gamma_k}\Lambda(x)^{\Sigma\lambda_k\gamma_k}. \tag{28}$$

Choosing M so that $m_1 - M < -n$, we can perform the regularization just as above. The correctness of this procedure can be easily verified. Again we obtain $|I_2^{(j)}(x)| \leqslant C_{N_1}\Lambda(x)^{-N_1}$ for any N_1. Estimate now $I_1(x)$ (resp. $I_1^{(j)}(x)$), which according to (24) is the "essential" part of I(x). We have

$$I_1(x) = \Lambda(x)^n\iint e^{i\Lambda(x)\xi\eta}F(x,\eta)e_1(\Lambda(x)^{1-\varepsilon}\xi)\Phi(x,\xi)d\xi d\eta, \tag{29}$$

where $F(x,\eta)$ satisfies (18), while (19) on supp $e_1(\Lambda(x)^{1-\varepsilon}\xi)$ reduces to

$$\left|\frac{\partial^\gamma(e_1(\xi)\Phi(x,\xi))}{\partial\xi^\gamma}\right| \leqslant C_\gamma\Lambda(x)^{s+|\gamma|}. \tag{30}$$

Combining these estimates, we obtain, taking into consideration that the volume of integration over ξ in (29) does not exceed const$\cdot\Lambda(x)^{n(\varepsilon-1)}$:

$$|I_1(x)| \leqslant C\Lambda(x)^{m+s+n\varepsilon}, \tag{31}$$

and exactly the same is valid for $I_1^{(j)}(x)$. Thus we have estimated $I(x)$ itself and we need to estimate the derivatives $\frac{\partial^{|\alpha|}}{\partial x^\alpha} I(x)$, $|\alpha| = 0,1,2,\ldots$, as well. It is an easy computational exercise to verify that the derivatives may be obtained as follows: one should apply the derivatives under the integral sign in (12) and then follow the regularization procedure described above. The calculation omitted here leads to the desired result:

$$\left| \frac{\partial^{|\alpha|} I(x)}{\partial x^\alpha} \right| \leqslant C_\alpha \Lambda(x)^{m+s+\Sigma\alpha_j(1-\lambda_j)+n\varepsilon}, \quad |\alpha| = 0,1,2,\ldots, \tag{32}$$

where the constants C_α depend linearly on the constants in the estimates for $\phi(x)$. It remains to set $\varepsilon = \delta/n$. Theorem 1 is thereby proved.

We need also to establish an (approximate) composition law for pseudo-differential operators. This is the matter of the following theorem. Set $\hat{T}^m_{(\lambda_1,\ldots,\lambda_n)} = \{f(\overset{2}{x}, -i \overset{1}{\frac{\partial}{\partial x}}) | f \in T_{(\lambda_1,\ldots,\lambda_n)}(R^{2n})\}$; $\hat{T}^\infty_{(\lambda_1,\ldots,\lambda_n)} = \bigcup_m \hat{T}^m_{(\lambda_1,\ldots,\lambda_n)}$.

Theorem 2. The set $\hat{T}^\infty_{(\lambda_1,\ldots,\lambda_n)}$ is an operator algebra in the space $L^\infty_{(x_1,\ldots,x_n)}(R^n)$. More precisely, the operator multiplication maps $\hat{T}^m_{(\lambda_1,\ldots,\lambda_n)} \cdot \hat{T}^\ell_{(\lambda_1,\ldots,\lambda_n)}$ into $\hat{T}^{m+\ell}_{(\lambda_1,\ldots,\lambda_n)}$ and for given $f \in T^m_{(\lambda_1,\ldots,\lambda_n)} \times (R^{2n})$ and $g \in T^\ell_{(\lambda_1,\ldots,\lambda_n)}(R^{2n})$, the symbol $h \in T^{m+\ell}_{(\lambda_1,\ldots,\lambda_n)}(R^{2n})$ of the product $[f(\overset{2}{x}, -i \overset{1}{\frac{\partial}{\partial x}})][g(\overset{2}{x}, -i \overset{1}{\frac{\partial}{\partial x}})]$ has an asymptotic expansion:

$$h_{(x,p)} \simeq \sum_{|\alpha|=0}^{\infty} \frac{(-i)^{|\alpha|}}{\alpha!} \frac{\partial^{|\alpha|} f}{\partial p^\alpha}(x,p) \frac{\partial^{|\alpha|} g}{\partial x^\alpha}(x,p). \tag{33}$$

The equality (33) means, by definition, that the difference $h_{(x,p)} - \sum_{|\alpha|=0}^{R} \frac{(-i)^{|\alpha|}}{\alpha!} \frac{\partial^\alpha f}{\partial p^\alpha} \frac{\partial^{|\alpha|} g}{\partial x^\alpha}$ belongs to $T^{m+\ell-R-1}_{(\lambda_1,\ldots,\lambda_n)}(R^{2n})$ for any natural R.

Proof. We sketch the critical points of the argument. Let f and g be as in the formulation of the theorem. Using (12) twice and performing the simple variable change, we come to the following statement: for any $\phi \in L^\infty_{(\lambda_1,\ldots,\lambda_n)}(R^n)$

$$(\hat{f} \circ \hat{g})\phi = \hat{h}\phi, \tag{34}$$

where*

$$h(x,p) = (2\pi)^{-n} \iint e^{i(x-y)\tilde{p}} f(x,\tilde{p}+p) g(y,p) dy d\tilde{p} \tag{35}$$

and the integral (35) is defined via regularization as in Theorem 1. Now estimate the integral (35). Perform the change of variables (15) and set also

$$\tilde{p}_j = \tilde{\eta}_j \Lambda(x)^{1-\lambda_j}. \tag{36}$$

---

* We denote $\Sigma_j(x_j - y_j)\tilde{p}_j = (x - y)\tilde{p}$ for the sake of brevity.

We obtain:

$$h(x,\eta\Lambda(x)^{1-\lambda}) = (2\pi)^{-n}\Lambda(x)^n\iint e^{i\Lambda(x)\xi\tilde{\eta}} f(x,(\tilde{\eta}+\eta)\Lambda^{1-\lambda}) \times$$
$$\times\, g(x+\xi\Lambda(x)^{\lambda},\eta\Lambda(x)^{1-\lambda})d\xi d\tilde{\eta} = \tag{37}$$
$$= (\frac{\Lambda(x)}{2\pi})^n\iint e^{i\Lambda(x)\xi\tilde{\eta}} F(x,\tilde{\eta}+\eta)G(x,\xi,\eta)d\xi d\tilde{\eta},$$

where the functions $F(x,z) = f(x,z\Lambda(x)^{1-\lambda})$ and $G(x,\xi,\eta) = g(x + \xi\Lambda(x)^{\lambda}, \eta\Lambda(x)^{1-\lambda})$ satisfy the estimates

$$\left| \frac{\partial^{|\alpha|+|\beta|}F(x,z)}{\partial x^{\alpha}\partial z^{\beta}} \right| \leqslant C_{\alpha\beta}\Lambda(x)^{m-\Sigma\lambda_i\alpha_i}(1+|z|)^{m_1+|\alpha|}; \tag{38}$$

$$|G(x,\xi,\eta)| \leqslant C\Lambda(x+\xi\Lambda(x)^{\lambda})^{\ell}(1+|\eta|(\frac{\Lambda(x)}{\Lambda(x+\xi\Lambda(x)^{\lambda})})^{1-\lambda})^{m_2}. \tag{39}$$

Let $m_1 < -n$. By the Peetre inequality [45], we have

$$(1+|\eta+\tilde{\eta}|)^{m_1} \leqslant (1+|\tilde{\eta}|)^{m_1}(1+|\eta|)^{|m_1|}. \tag{40}$$

Following the lines of the previous proof, we obtain

$$|h(x,\eta\Lambda(x)^{1-\lambda})| \leqslant C\Lambda(x)^{m+\ell+\delta}(1+|\eta|)^{m_2+|m_1|} \tag{41}$$

(the technical details are left to the reader), or

$$h(x,p) \leqslant C\Lambda(x)^{m+\ell+\delta}(1+|p\Lambda(x)^{\lambda-1}|)^{m_2+|m_1|}. \tag{42}$$

Assume now that $m_1 \geqslant -n$. We use again the partition of unity $\rho(z)$ introduced in the proof of the previous theorem and represent the j-th term of the sum in the form

$$h_{(j)}(x,p) = (2\pi)^{-n}\iint e^{i(x-y)\tilde{p}} \frac{f_{(j)}(x,\tilde{p}+p)}{(\tilde{p}_j+p_j)^M}(\tilde{p}_j+p_j)^M g(y,p)dyd\tilde{p} =$$
$$= (2\pi)^{-n}\iint e^{i(x-y)\tilde{p}} \frac{f_{(j)}(x,\tilde{p}+p)}{(\tilde{p}_j+p_j)^M}[(p_j - i\frac{\partial}{\partial y_j})^M g(y,p)]dyd\tilde{p}. \tag{43}$$

The symbol $g_{jM}(y,p) = (p_j - i\frac{\partial}{\partial y_i})^M g(y,p)$ satisfies the estimates:

$$\left| \frac{\partial^{|\alpha|+|\beta|}g_{jM}(y,p)}{\partial y^{\alpha}\partial p^{\beta}} \right| \leqslant C_{\alpha\beta}\Lambda(y)^{\ell+M(1-\lambda_j)-\Sigma\lambda_j\alpha_j+\Sigma(\lambda_j-1)\beta_j} \times$$
$$\times\,(1+|p\Lambda(y)^{\lambda-1}|)^{m_2+M+|\alpha|}; \tag{44}$$

after changing the variables as in (26), we obtain

$$h_{(j)}(x,p) = (2\pi)^{-n}\Lambda(x)^{n-M(1-\lambda_j)}\iint e^{i\Lambda(x)\xi\tilde{\eta}} F_{jM}(x,\eta+\tilde{\eta})G_{jM}(x,\xi\eta)d\xi d\tilde{\eta}, \tag{45}$$

where F_{jM} satisfies the estimates (27). Choosing $M > n + m_1$, we obtain

$$|h(x,p)| \leqslant C\Lambda(x)^{m+\ell+\delta}(1+|p\Lambda^{\lambda-1}|)^{m_2+M+|m_1-M|}. \tag{46}$$

In any case we may take $M = m_1 + n + 1$, thus obtaining

$$|h(x,p)| \leqslant C\Lambda(x)^{m+\ell+\delta}(1 + |p\Lambda^{\lambda-1}|)^{m_2+m_1+2(n+1)}. \tag{47}$$

Next we should estimate the derivatives of h(x,p). Performing the derivation in (35) and using the elementary properties of Fourier transformation, we obtain immediately:

$$\frac{\partial^{|\alpha+\beta|}h(x,p)}{\partial x^\alpha \partial p^\beta} = \sum_{\substack{0\leqslant\gamma\leqslant\alpha \\ 0\leqslant\theta\leqslant\beta}} (2\pi)^{-n}\iint e^{i(x-y)\tilde{p}} \frac{\partial^{|\gamma|+|\theta|}f}{\partial x^\gamma \partial p^\theta}(x,p+\tilde{p}) \times \frac{\partial^{|\alpha-\gamma|+|\beta-\theta|}g}{\partial x^{\alpha-\gamma}\partial p^{\beta-\theta}}(y,p)dyd\tilde{p}. \tag{48}$$

Each term in the sum may be estimated exactly in the same way as h(x,p) itself. We obtain finally

$$\left|\frac{\partial^{|\alpha+\beta|}h(x,p)}{\partial x^\alpha \partial p^\beta}\right| \leqslant C_{\alpha\beta}\Lambda(x)^{m+\ell+\delta-\Sigma_j\lambda_j\alpha_j+\Sigma_j(\lambda_j-1)\beta_j} \times (1+|p\Lambda(x)^{\lambda-1}|)^{m_1+m_2+|\alpha|+2n+1}. \tag{49}$$

Thus we have proved that $h(x,p) \in T^{m+\ell+\delta}_{(\lambda_1,\ldots,\lambda_n)}(R^{2n})$ for any $\delta > 0$. It remains to prove the expansion (33) which shows, in particular, that we may set $\delta = 0$ in the above inclusion.

To perform this we express $f(x,\tilde{p} + p)$ in (35) in the form of the Taylor's expansion with the remainder. To avoid extended calculations we derive the principal term of the expansion (33) only; we hope that the reader has already grasped the idea of our estimates and can proceed with the general case without assistance.

We have

$$f(x,\tilde{p}+p) = f(x,p) + \int_o^1 \sum_{j=1}^n \tilde{p}_j \frac{\partial g}{\partial p_j}(x,p+\theta\tilde{p})d\theta. \tag{50}$$

Substituting (50) into (35) yields

$$h(x,p) = f(x,p)g(x,p) - i\sum_{j=1}^n \iint e^{i(x-y)\tilde{p}}\left\{\int_o^1 \frac{\partial f}{\partial p_j}(x,p+\theta\tilde{p})d\theta\right\}\frac{\partial g}{\partial x_j}(y,p)dyd\tilde{p}. \tag{51}$$

Denote the expression in parentheses by $f_j(x,p,\tilde{p})$. It is obvious that the following estimates are valid for $f_j(x,p,\tilde{p})$ and $\frac{\partial y}{\partial x_j}(x,p)$:

$$\left|\frac{\partial^{|\alpha|+|\beta|+|\gamma|}f_j(x,p,\tilde{p})}{\partial x^\alpha \partial p^\beta \partial \tilde{p}^\gamma}\right| \leqslant C_{\alpha\beta\gamma}\Lambda(x)^{m+\lambda_j-1-\Sigma\lambda_k\alpha_k+\Sigma(\lambda_k-1)(\beta_k+\gamma_k)} \times (1+|(p+\tilde{p})\Lambda(x)^{\lambda-1}|)^{\max(m_1,0)+|\alpha|}; \tag{52}$$

$$\left|\frac{\partial^{|\alpha|+|\beta|}}{\partial x^\alpha \partial p^\beta}\frac{\partial g}{\partial x_j}\right| \leqslant C_{\alpha\beta}\Lambda(x)^{\ell-\lambda_j-\Sigma\lambda_k\alpha_k+\Sigma(\lambda_k-1)\beta_k} \times (1+|p\Lambda(x)^{\lambda-1}|)^{m_2+|\alpha|}. \tag{53}$$

Therefore, applying the estimating techniques developed previously, we obtain

$$h(x,p) - f(x,p)g(x,p) \in T^{m+\ell-1+\delta}_{(\lambda_1,\ldots,\lambda_n)}(R^{2n}) \tag{54}$$

for any $\delta > 0$. In particular, $h(x,p) \in T^{m+\ell}_{(\lambda_1,\ldots,\lambda_n)}(R^{2n})$. Thus, the principal term of the expansion (33) is obtained; all the subsequent terms are derived in a completely analogous way. Theorem 2 is proved.

To end this item, we introduce the notion of asymptotic expansion in the homogeneous case, i.e., in the space $L^{\infty}_{(\lambda_1,\ldots,\lambda_n)}(R^n)$.

Definition 2. Let $\phi(x)$ be a given element of the space $L^{\infty}_{(\lambda_1,\ldots,\lambda_n)}(R^n)$. We say that the formal series $\sum_{j=0}^{\infty} \phi_j(x)$ gives an asymptotic expansion of ϕ and write $\phi \sim \sum_{j=0}^{\infty} \phi_j(x)$ if $\phi_j(x) \in L^{m_j}_{(\lambda_1,\ldots,\lambda_n)}(R^n)$ and $\lim_{j\to\infty} m_j = -\infty$ and if for any N there exists n_o such that

$$\phi - \sum_{j=0}^{\ell} \phi_j(x) \in L^{-N}_{(\lambda_1,\ldots,\lambda_n)}(R^n), \tag{55}$$

once $\ell \geq n_0$.

In terms of Definition 2 one may say that Theorem 2 gives an asymptotic expansion of $(\hat{f} \circ \hat{g})\phi$ in the space $L^{\infty}_{(\lambda_1,\ldots,\lambda_n)}(R^n)$.

B. Stationary Phase Method, the Canonical Operator, and Wave Fronts in the Quasi-Homogeneous Case

Among the elements of the space $L^{\infty}_{(\lambda_1,\ldots,\lambda_n)}(R^n)$ there is a class of special interest for us - namely the class of canonically representable functions. This class will be described and investigated here in some detail as well as some geometric objects connected with these functions.

The simplest example of the canonically representable function (CRF in the sequel) is

$$\psi(x) = e^{iS(x)}\phi(x), \tag{1}$$

where $S(x) \in O^{1}_{(\lambda_1,\ldots,\lambda_n)}(R^n)$, $\phi(x) \in S^{m}_{(\lambda_1,\ldots,\lambda_n)}(R^n)$, and $\mathrm{Im}S(x) \geq 0$. It is not hard to establish that under these conditions $\psi(x) \in L^{m}_{(\lambda_1,\ldots,\lambda_n)}(R^n)$. In the present item we confine ourselves to only real-valued phase functions S(x). The experience in 1/h-theory suggests that the consideration of the Lagrangian manifold associated with the phase function S(x) of CRF would be useful. This Lagrangian manifold L is given by the equation

$$p_j = p_j(x) \equiv \frac{\partial S}{\partial x_j}(x), \quad j = 1,\ldots,n, \tag{2}$$

and possesses the properties which are formalized in the following definition (here and in the sequel we assume the numbers $\lambda_1,\ldots,\lambda_n$ to be given and fixed).

Definition 1. The Lagrangian manifold L is called proper if:

(a) the inequalities

$$|p_j| \leqslant C\Lambda(x)^{1-\lambda_j}, \ j = 1,\dots,n \tag{3}$$

are valid on L with some constant $C = C_L$;

(b) for $\Lambda(x)$ sufficiently large L is invariant under the action of the group R_+ on R^{2n}, defined as follows:

$$\tau(x,p) = (\tau^{\lambda}x, \tau^{1-\lambda}p) \equiv (\tau^{\lambda_1}x_1,\dots,\tau^{\lambda_n}x_n, \tau^{1-\lambda_1}p_1,\dots,\tau^{1-\lambda_n}p_n),$$
$$(x,p) \in R^{2n}, \ \tau \in R_+. \tag{4}$$

In other words, (b) means that if $(x,p) \in L$ and $\Lambda(x) > R = R_L$, then for $\tau \geqslant 1$ the point $\tau(x,p)$ necessarily belongs to L. However, in what follows we use a somewhat different version of condition (b), more convenient for our needs: L is supposed to be everywhere invariant under the action of R_+, but the equation (2) (or similar equations in the mixed coordinate-momenta representation) defines L only for $\Lambda(x)$ large enough.

The equation (2) defines not a general proper Lagrangian manifold, but one which diffeomorphically projects onto the x-plane. To deal with the general case one should be concerned with the mixed coordinate-momenta representation. To motivate our considerations thoroughly, we first study the partial Fourier transform of the function (1) under some additional assumptions.

Fix some subset $I \subset [n]$ and suppose that the inequalities

$$|x_j| \leqslant C\Lambda_I(x)^{\lambda_j}, \ j \in \bar{I} \tag{5}$$

holds on the support supp ϕ. Adding to ψ a suitable function bounded with its derivatives and finite with respect to $x_{\bar{I} \cup I_+}$, therefore belonging to $L^{-\infty}_{(\lambda_1,\dots,\lambda_n)}(R^n)$, we may assume that $\phi(x) = 0$ for $\Lambda(x) < R$ and therefore $S(x)$ is quasi-homogeneous on supp ϕ. Consider the partial Fourier transform

$$I(x_I, p_{\bar{I}}) = \left(\frac{-i}{2\pi}\right)^{|\bar{I}|/2} \int e^{i[S(x) - p_{\bar{I}}x_{\bar{I}}]}\phi(x)dx_{\bar{I}}. \tag{6}$$

The integral (6) converges since, due to (5), it is taken over the finite domain. The stationary point of its phase is given by

$$p_{\bar{I}} = p_{\bar{I}}(x_I, x_{\bar{I}}) \equiv \frac{\partial S}{\partial x_{\bar{I}}}(x_I, x_{\bar{I}}). \tag{7}$$

To calculate the asymptotic expansion of the integral (6) perform the change of variables

$$x_j = \xi_j \Lambda_I(x)^{\lambda_j},$$
$$p_j = \eta_j \Lambda_I(x)^{1-\lambda_j}, \ j \in \bar{I}. \tag{8}$$

In the new variables we have

$$I(x_I, p_{\bar{I}}) = \left(\frac{-i}{2\pi}\right)^{|\bar{I}|/2} \Lambda(x)^{m+|\lambda_{\bar{I}}|} \int e^{i\Lambda_I(x)[S(\xi_I,\xi_{\bar{I}}) - \eta_{\bar{I}}\xi_{\bar{I}}]} \times$$
$$\times \Phi(\Lambda_I(x), \xi_I, \xi_{\bar{I}})d\xi_{\bar{I}}, \tag{9}$$

where $|\lambda_{\bar{I}}| = \sum_{j \in \bar{I}} \lambda_j$

$$\Phi(\Lambda,\xi) = \phi(\xi\Lambda^{\lambda})/\Lambda^m. \tag{10}$$

The integral in (9) is taken over a compact set $\{|\xi_j| \leqslant C, j \in \bar{I}\}$. Estimates for derivatives of $\Phi(\Lambda,\xi)$ with respect to $\xi_{\bar{I}}$ are

$$\left| \frac{\partial^\alpha}{\partial \xi_{\bar{I}}^\alpha} \Phi(\Lambda,\xi) \right| = \left| \Lambda^{\Sigma\lambda_j\alpha_j - m} \phi^{(\alpha)}(\xi\Lambda^\lambda) \right| \leqslant C_\alpha \Lambda^{m-\Sigma\lambda_j\alpha_j+\Sigma\lambda_j\alpha_j-m} = C_\alpha. \quad (11)$$

Thus, all the $\xi_{\bar{I}}$-derivatives of $\Phi(\Lambda,\xi)$ are bounded uniformly with respect to Λ and we may apply the stationary phase method to the integral (9), considering $\Lambda_I(x)$ as a large parameter provided that the stationary point or the phase in (9) is unique and non-degenerate on supp Φ. The equations of the stationary point coincide up to notation with (7). Let $\xi_{\bar{I}} = \xi_{\bar{I}}(\xi_I,\eta_{\bar{I}})$ be the (resolved) equation of the stationary point and assume that $\det \frac{\partial^2 S}{\partial\xi_{\bar{I}}\partial\xi_{\bar{I}}}(\xi_I,\xi_{\bar{I}}(\xi_I,\eta_{\bar{I}})) \neq 0$, provided that $(\xi_I,\xi_{\bar{I}}(\xi_I,\eta_{\bar{I}})) \in$ supp Φ.

The usual stationary phase method (see, for example, [56]) gives now

$$I(x_I,p_{\bar{I}}) = \left(\frac{-i}{2\pi}\right)^{|\bar{I}|/2} \Lambda_I(x)^{m+|\lambda_{\bar{I}}|-|\bar{I}|/2} \times$$

$$\times \frac{e^{i\Lambda_I(x)[S(\xi_I),\xi_{\bar{I}}(\xi_I,\eta_{\bar{I}}),-\eta_I - \xi_{\bar{I}}(\xi_I,\eta_{\bar{I}})]}}{\left(\det \frac{\partial^2 S}{\partial\xi_{\bar{I}}\partial\xi_{\bar{I}}}(\xi_I,\xi_{\bar{I}}(\xi_I,\eta_{\bar{I}}))\right)^{1/2}} \{\Phi(\Lambda_I(x),\xi_I,\xi_{\bar{I}}(\xi_I,\eta_{\bar{I}})) + \quad (12)$$

$$+ \Lambda_I(x)^{-1}\Phi_1(\Lambda_I(x),\xi_I,\eta_{\bar{I}}) + \ldots\} + R(\Lambda_I(x),\xi_I,\eta_{\bar{I}}),$$

where $\arg\det \frac{\partial^2 S}{\partial\xi_{\bar{I}}\partial\xi_{\bar{I}}}$ is chosen in a special way, and all the derivatives of the remainder R have the estimates

$$\left| \frac{\partial^{k+|\alpha|} R}{\partial z^k \partial\xi_I^{\alpha_I} \partial\eta_{\bar{I}}^{\alpha_{\bar{I}}}}(z,\xi_I,\eta_{\bar{I}}) \right| \leqslant C_{k,\alpha,N} z^{-N}(1+|\eta_{\bar{I}}|)^{-N},$$

$$k,|\alpha|,N = 0,1,2\ldots. \quad (13)$$

Returning back to initial variables, we obtain

$$I(x_I,p_{\bar{I}}) = e^{i\bar{S}(x_I,p_{\bar{I}})} \tilde{\phi}(x_I,p_{\bar{I}}) + \tilde{R}(x_I,p_{\bar{I}}), \quad (14)$$

where $\bar{S}$ is equal to the phase value in the stationary point (7) and the following estimates are valid with $m_1 = m + |\lambda_{\bar{I}}| - |\bar{I}|/2$:

$$|p_j| \leqslant C\Lambda_I(x)^{1-\lambda_j}, \ j \in \bar{I}, \text{ on supp } \tilde{\phi}; \quad (15)$$

$$\left| \frac{\partial^{|\alpha|}\tilde{\phi}(x_I,p_{\bar{I}})}{\partial x_I^{\alpha_I}\partial p_{\bar{I}}^{\alpha_{\bar{I}}}} \right| \leqslant C_\alpha \Lambda_I(x)^{m_1 - \sum_{j\in I}\alpha_j\lambda_j - \sum_{j\in\bar{I}}\alpha_j(1-\lambda_j)}, \quad |\alpha| = 0,1,2,\ldots; \quad (16)$$

$$\left| \frac{\partial^{|\alpha|}\tilde{S}(x_I,p_{\bar{I}})}{\partial x_I^{\alpha_I}\partial p_{\bar{I}}^{\alpha_{\bar{I}}}} \right| \leqslant \tilde{C}_\alpha \Lambda_I(x)^{1-\sum_{j\in I}\alpha_j\lambda_j - \sum_{j\in I}\alpha_j(1-\lambda_j)}, \quad (17)$$

$$\text{on supp } \tilde{\phi}, \ |\alpha| = 0,1,2,\ldots;$$

$$\left| \frac{\partial^{|\alpha|} R(x_I, p_{\bar I})}{\partial x_I^{\alpha_I} \partial p_{\bar I}^{\alpha_{\bar I}}} \right| \leqslant \tilde{\tilde{C}}_\alpha \Lambda_I(x)^{-N} (1 + \sum_{j \in \bar I} |p_j \Lambda_I^{\lambda_j - 1}|)^{-N}, \qquad (18)$$

$$|\alpha|, N = 0, 1, 2, \ldots .$$

The estimates (15) - (18) are just those corresponding to the essence of the matter, as follows from:

Lemma 1. (a) If the estimates (18) are valid then*

$$\bar{F}_{p_{\bar I} \to x_{\bar I}} (\tilde{R}) \in L^{-\infty}_{(\lambda_1, \ldots, \lambda_n)} (R^n). \qquad (19)$$

(b) If the estimates (15) - (17) are valid and $\mathrm{Im}\, \tilde{S}' \geq 0$, then

$$\bar{F}_{p_{\bar I} \to x_{\bar I}} (e^{i\tilde S} \tilde\phi) \in L^{m_1 + |\bar I| - |\lambda_{\bar I}|}_{(\lambda_1, \ldots, \lambda_n)} (R^n). \qquad (20)$$

Proof. (a) In the integral

$$I(x) = (\bar{F}_{p_{\bar I} \to x_{\bar I}} \tilde{R})(x) = \left(\frac{1}{2\pi}\right)^{|\bar I|/2} \int e^{i x_{\bar I} p_{\bar I}} \tilde{R}(x_I, p_{\bar I}) dp_{\bar I}, \qquad (21)$$

perform the change of variables (8) and obtain

$$I(x) = \left(\frac{i}{2\pi}\right)^{|\bar I|/2} \Lambda_I(x)^{|\bar I| - |\lambda_{\bar I}|} \int e^{i \Lambda_I(x) \xi_{\bar I} \eta_{\bar I}} R(\Lambda_I(x), \xi_I, \eta_{\bar I}) d\eta_{\bar I}, \qquad (22)$$

where $R(z, \xi_I, \eta_{\bar I})$ satisfies the estimates (13). Integrating by parts then yields the estimates:

$$I(x) \leqslant C_N \Lambda_I(x)^{-N} (1 + |\xi_{\bar I}|)^{-N_1}, \quad N, N_1 = 0, 1, 2, \ldots, \qquad (23)$$

and the same estimates are valid for derivatives of I. It remains to note that

$$\Lambda_I(x)^{-N} (1 + |\xi_{\bar I}|)^{-N_1} \leq \Lambda(x)^{-N}, \text{ for large } N_1. \qquad (24)$$

(b) Perform the change of variables (8) in the integral (20). We have

$$I(x) \equiv \bar{F}_{p_{\bar I} \to x_{\bar I}} (e^{i\tilde S} \tilde\phi) = \left(\frac{i}{2\pi}\right)^{|\bar I|/2} \int e^{i(\tilde S(x_I, p_{\bar I}) + x_{\bar I} p_{\bar I})} \tilde\phi(x_I, p_{\bar I}) dp_I =$$

$$= \left(\frac{i}{2\pi}\right)^{|\bar I|/2} \Lambda_I(x)^{|\bar I| - |\lambda_{\bar I}|} \int e^{i \Lambda_I(x) (\tilde S(\xi_I \Lambda_I(x)^\lambda, \eta_{\bar I} \Lambda_{\bar I}(x)^{1-\lambda}) / \Lambda_I(x) +}$$

$$^{+ \xi_{\bar I} \eta_{\bar I})} \tilde\phi(\xi_I \Lambda_I(x)^\lambda, \eta_{\bar I} \Lambda_{\bar I}(x)^{1-\lambda}) d\eta_{\bar I} = \left(\frac{i}{2\pi}\right)^{|\bar I|/2} \Lambda(x)^{m_1 + |\bar I| - |\lambda_{\bar I}|} \times$$

$$\times \int e^{i \Lambda_I(x) [S(\Lambda_I(x), \xi_I, \eta_{\bar I}) + \xi_{\bar I} \eta_{\bar I}]} \phi(\Lambda_I(x), \xi_I, \eta_{\bar I}) d\eta_{\bar I}, \qquad (25)$$

where the functions

---

* $\bar{F}_{p_{\bar I} \to x_{\bar I}}$ denotes the inverse Fourier transformation.

$$S(z,\xi_I,\eta_{\bar{I}}) = \tilde{S}(\xi_I z^{\lambda},\eta_{\bar{I}} z^{1-\lambda})/z, \tag{26}$$

$$\phi(z,\xi_I,\eta_{\bar{I}}) = \tilde{\phi}(\xi_I z^{\lambda},\eta_{\bar{I}} z^{1-\lambda})/z^{m_1} \tag{27}$$

satisfy the estimates

$$\left| \frac{\partial^{|\alpha|}\phi}{\partial\xi_I^{\alpha_I}\partial\eta_{\bar{I}}^{\alpha_{\bar{I}}}} (z,\xi_I,\eta_{\bar{I}})\right| \leqslant C_\alpha, \ |\alpha| = 0,1,2,\ldots, \tag{28}$$

$$\left| \frac{\partial^{|\alpha|}S}{\partial\xi_I^{\alpha_I}\partial\eta_{\bar{I}}^{\alpha_{\bar{I}}}} (z,\xi_I,\eta_{\bar{I}})\right| \leqslant C_\alpha, \ |\alpha| = 0,1,2,\ldots, \text{ on supp } \phi. \tag{29}$$

Set $M = \max\limits_j \sup\limits_{\text{supp }\phi} \left|\frac{\partial S}{\partial\eta_j}\right|$. For $|\xi_{\bar{I}}| > M$ one may integrate by parts in (25) obtaining the factor $|\xi_{\bar{I}}|^{-N}$ for any N; thus we derive that

$$|I(x)| \leqslant \text{const}\cdot\Lambda_I(x)^{m_1+|\bar{I}|-|\lambda_{\bar{I}}|}(1+|\xi_{\bar{I}}|)^{-N} \leqslant \text{const}\cdot\Lambda(x)^{m_1+|\bar{I}|-|\lambda_{\bar{I}}|} \tag{30}$$

if we choose N sufficiently large. The derivatives of I(x) may be estimated in a similar way. Lemma 1 is thereby proved. (20) will be called a CRF in the mixed coordinate-momenta representation.

Now we are almost in a position to define the canonical operator since there are enough observations to be summarized. Let $L \subset R^{2n}$ be a proper Lagrangian manifold. It is well known that there exists a canonical atlas on L (see, for instance, Section 2 of the present chapter). However, an arbitrary canonical atlas is not sufficient for our needs; roughly speaking, the validity of the conditions of Lemma 1 (b) should be guaranteed. Thus we introduce the following:

<u>Definition 2</u>. Let $\{(u_j,\gamma_j : u_j \to R^n_{(x_{I(j)},p_{\bar{I}(j)})})\}_{j\in J}$, $\bigcup\limits_j u_j = L$ be a canonical atlas of a proper Lagrangian manifold L. This atlas is called <u>proper</u> if the following conditions are satisfied:

(a) $\{u_j\}$ is a locally finite R_+-invariant covering of L.

(b) Let $(u,\gamma : u \to R^n_{(x_I,p_{\bar{I}})})$ be any chart of the atlas. In $V = \gamma(u)$ the inequality holds

$$|p_j| \leqslant C\Lambda_I(x)^{1-\lambda_j}, \ j\in\bar{I}. \tag{31}$$

Also the function

$$S_I(x_I,p_{\bar{I}}) = \sum_{j\in I}\lambda_j p_j(x_I,p_{\bar{I}})x_j + \sum_{j\in\bar{I}}(\lambda_j - 1)p_j x_j(x_I,p_{\bar{I}}), \tag{32}$$

where

$$\begin{cases} p_I = p_I(x_I,p_{\bar{I}}) \\ x_{\bar{I}} = x_{\bar{I}}(x_I,p_{\bar{I}}) \end{cases} \tag{33}$$

are the equations of L in the local coordinates $(x_I,p_{\bar{I}})$, satisfies the estimates (17) in u for $\Lambda_I(x) \geqslant R_o = R_o(u)$.

(c) All the intersections $u_j \cap u_k$ are R_+-precompact sets (that is, the sets $u_j \cap u_k/R_+$ are precompact; it is equivalent to the assertion that the set $u_j \cap u_k \cap \{\Lambda(x) = 1\}$ is bounded).

(d) There exists a partition of unity $\{e_j\}_{j\in J}$ on L, homogeneous of degree zero, such that for any $j \in J$ the function $(e_j \circ \gamma_j^{-1})(x_{I(j)}, p_{\bar{I}(j)})$ satisfies for $\Lambda_{I(j)}(x) \geqslant R_o = R_o(u_j)$ the estimates (16) with $m_1 = 0$.

Lemma 2. The proper canonical atlas always exists on a proper Lagrangian manifold $L \subset R^{2n}$.

Proof. Let $\alpha_o \in L$ be any point. Let $j \in I_+$ be chosen from the condition $|x_j(\alpha_o)| = \max_{k\in I_+} |x_k(\alpha_o)|$. Then, obviously, $\Lambda(x) \leq c\cdot|x_j|^{1/\lambda_j}$ in some R_+-invariant neighborhood of α_o. Since $x_j(\alpha_o) \neq 0$, $\lambda_j \neq 0$, we have $dx_j|_{\alpha_o} \neq 0$ on L (use Proposition 1 of item A and homogeneous local coordinates to prove this fact). By the lemma on local coordinates [2] in some neighborhood u of α_o, a canonical system of coordinates $(x_I, p_{\bar{I}})$ may be chosen with $j \in I$; since L is R_+-invariant this neighborhood may also be chosen R_+-invariant. Since L is proper, (31) is satisfied. For each $\alpha_o \in L$ take a R_+-precompact coordinate neighborhood of the described type and select a locally finite subcovering of L; it is not hard to see that this canonical covering is a proper one. The lemma is proved.

Remark 1. The proper canonical covering constructed in the proof of Lemma 2 consists of R_+-precompact elements; however, the case of particular interest is the one where the charts are not R_+-precompact (not like their intersections). It is difficult to formulate general existence theorems concerned with this matter; nevertheless practical problems supply a lot of examples with non-R_+-compact charts.

Let a proper Lagrangian manifold L with a proper canonical covering $\{u_j, \nu_j : u_j \to V_j \subset R^{2n}_{(x_{I(j)}, p_{\bar{I}(j)})}\}_{j\in J}$ be given.

Definition 3. The space $\mathcal{D}^m(L) \equiv \mathcal{D}^m(L, \{u_j\})$ is the space of functions f on L, satisfying the following conditions:

(a) The support supp f intersects at most a finite number of u_j-s.

(b) For any $k \in J$ the function

$$f_k(x_I, p_{\bar{I}}) = (f \circ \nu_k^{-1})(x_I, p_{\bar{I}}), \quad I = I(k) \tag{34}$$

satisfies in V_k the estimates

$$\left| \frac{\partial^{|\alpha|} f_k(x_I, p_{\bar{I}})}{\partial x_I^{\alpha_I} \partial p_{\bar{I}}^{\alpha_{\bar{I}}}} \right| \leqslant C_\alpha \Lambda_I(x)^{m - \Sigma_{j\in I} \alpha_j \lambda_j - \Sigma_{j\in \bar{I}} (1-\lambda_j)\alpha_j}. \tag{35}$$

We set $\mathcal{D}^\infty(L) = \bigcup_m \mathcal{D}^m(L)$, $\mathcal{D}^{-\infty}(L) = \bigcap_m \mathcal{D}^m(L)$. The canonical operator we intend to define acts on functions from $\mathcal{D}^\infty(L)$, taking them into $L^\infty_{(\lambda_1,\ldots,\lambda_n)} \times (R^n)$. To define it we need to discuss the quantization conditions in the quasi-homogeneous case. We do not repeat the discussion of these conditions performed in Section 1 of the present chapter, but merely indicate some essential modifications engendered by the additional R_+-structure.

First of all, we note that the "first quantization condition" connected with the cohomology class of the form $pdx|_L$ disappears in the quasi-homogeneous case, namely, the form $pdx|_L$ is an exact one.

To prove this, consider a point $(x,p) \in L$ and the trajectory of the group R_+ starting at this point:

$$(x(t),p(t)) = (t^{\lambda_1}x_1,\ldots,t^{\lambda_n}x_n,t^{1-\lambda_1}p_1,\ldots,t^{1-\lambda_n}p_n). \tag{36}$$

The tangent vector of the trajectory (36) at (x,p) has the form

$$V = \sum_{j=1}^{n} (\lambda_j x_j \frac{\partial}{\partial x_j} + (1-\lambda_j)p_j \frac{\partial}{\partial p_j}). \tag{37}$$

V is tangent to L and therefore L being Lagrangian $V \lrcorner\, dp \wedge dx$ vanishes on the tangent space of L:

$$\begin{aligned} 0 = V \lrcorner\, dp \wedge dx|_L &= \sum_{j=1}^{n} [p_j dx_j - \lambda_j(p_j dx_j + x_j dp_j)]|_L = \\ &= (pdx - d(\Sigma\lambda_j p_j x_j))|_L. \end{aligned} \tag{38}$$

Thus we have the equality

$$pdx = d(\Sigma\lambda_j p_j x_j) \overset{\text{def}}{=\!=} dS \tag{39}$$

on L, which proves our assertion.

It follows that the function $S_I(x_I,p_{\bar I})$ given by (32) is a generating function of L in the sense that

$$\begin{aligned} p_I &= \frac{\partial S_I}{\partial x_I}(x_I,p_{\bar I}), \\ x_{\bar I} &= -\frac{\partial S_I}{\partial p_{\bar I}}(x_I,p_{\bar I}) \end{aligned} \tag{40}$$

are the equations of L on the local canonical coordinates $(x_I,p_{\bar I})$. Indeed, we have (omitting everywhere the sign $|_L$ of restriction onto L):

$$\begin{aligned} S_I &= S - p_{\bar I}x_{\bar I}, \\ dS_I &= pdx - p_{\bar I}dx_{\bar I} - x_{\bar I}dp_{\bar I} = p_I dx_I - x_{\bar I}dp_{\bar I}, \end{aligned} \tag{41}$$

which immediately proves (40). Also we note that in the intersection $u_j \cap u_k$ with $I = I(j)$, $\mathcal{T} = J = I(k)$, we have

$$S_I = S_J + p_{\bar J}x_{\bar J} - p_{\bar I}x_{\bar I}; \tag{42}$$

that is (running ahead), the collection of phases $\{S_{I(j)}\}_{j\in\mathcal{T}}$ on L is concordant with respect to the stationary phase method. Thus only the second quantization condition remains in the quasi-homogeneous case.

Let μ be a smooth non-degenerate measure on L, homogeneous of degree r with respect to the action of the group R_+. In canonical local coordinates*

$$\mu = \mu_I(y_I,p_{\bar I})dx_I \wedge dp_{\bar I}, \tag{43}$$

where the function $\mu_I(x_I,p_{\bar I})$, which is the density of the measure μ in local coordinates $(x_I,p_{\bar I})$, is quasi-homogeneous of the degree $r - \sum_{j\in I}\lambda_j - \sum_{j\in\bar I} \times$ $\times\ (1-\lambda_j)$:

$$\mu_I(\tau^\lambda x_I,\tau^{1-\lambda}p_{\bar I}) = \tau^{r-\sum_{j\in I}\lambda_j - \sum_{j\in\bar I}(1-\lambda_j)}\mu_I(x_I,p_{\bar I}). \tag{44}$$

We require that (a) the pair (L,μ) satisfy the second quantization condition (i.e., the class of $d\ln(\mu/\sigma)$ be trivial modulo 4π, see Section 1 of the present chapter); (b) for any chart of the given proper canonical atlas the measure density $\mu_I(x_I,p_{\bar I})$ satisfies the estimates

$$\left|\frac{\partial^\alpha\mu_I(x_I,p_{\bar I})}{\partial x_I^{\alpha_I}\partial p_{\bar I}^{\alpha_{\bar I}}}\right| \leqslant C_\alpha\Lambda_I(x)^{r-\sum_{j\in I}(1+\alpha_j)\lambda_j - \sum_{j\in\bar I}(1+\alpha_j)(1-\lambda_j)} \tag{45}$$

for $\Lambda_I(x)$ large enough. We assume that the concordant branches of $\arg\mu_I$ in canonical charts are chosen and fixed.

Definition 4. Let all the above requirements be satisfied. Then the canonical operator is the mapping

$$K = K_{(L,\mu)} : \mathcal{D}^\infty(L) \to L^\infty_{(\lambda_1,\ldots,\lambda_n)}(R^n) \tag{46}$$

defined by

$$K\phi = \Sigma K_j(e_j\phi), \tag{47}$$

where the "elementary canonical operator" K_j is given by the equality (here $\phi\in\mathcal{D}^\infty(L)$ supp $\phi\subset u_j$):

$$[K_j\phi](x) = F_{p_{\bar I}\to x_{\bar I}}\{e^{iS_I(x_I,p_{\bar I})}\xi(\Lambda_I(x))(\mu_I(x_I,p_{\bar I}))^{1/2}(\phi\circ\nu_j^{-1})(x_I,p_{\bar I})\}, \tag{48}$$

where $I = I(j)$, $1-\xi(z)\in C_0^\infty(R^1)$, $\xi(z) = 0$ for $z \le R_0(u_j)$. The functions of the form [47] are called canonically representable functions (CRF).

The correctness of Definition 4 is verified easily. Since $\phi\in\mathcal{D}^\infty(L)$, only a finite number of non-zero terms occur in the sum (47); we use Lemma 1 to estimate the elementary canonical operator and obtain the following:

Proposition 1. The elementary canonical operator acts in the spaces**

$$K_j : \mathcal{D}^m(u_j) \to L^{m-\frac12\Sigma_{j=1}^n\lambda_j+\frac12(r+|\bar I|)}_{(\lambda_1,\ldots,\lambda_n)}(R^n). \tag{49}$$

We need several properties of the canonical operator which are established in the subsequent theorems.

---

* In (43) $dx_I \wedge dp_{\bar I}$ denotes the exterior product of all dx_j, $j\in I$ and dp_j, $j\in\bar I$ rearranged in the order of increasing index.

** Here and below we denote $\mathcal{D}^m(u) = \{\phi\in\mathcal{D}^m(L)\,|\,\text{supp}\,\phi\subset u\}$ for $u\subset L$.

Theorem 1 (The cocyclicity theorem). Let $u_j \cap u_k \neq \emptyset$. There exists a sequence of differential operators $V_{jk\ell}$, $\ell = 0,1,2,\dots$, acting in the spaces

$$V_{jk\ell} : \mathcal{D}^m(u_j \cap u_k) \to \mathcal{D}^{m-\ell}(u_j \cap u_k) \tag{50}$$

(m is arbitrary) with the properties:

(a) $V_{jko} = 1$,

(b) $V_{jk\ell}$ is the operator of order $\leqslant 2\ell$ with real, smooth coefficients.

If we denote

$$V_{jk}^{(N)} = \sum_{\ell=0}^{N-1} (-i)^{\ell} V_{jk\ell}, \quad N = 0,1,2,\dots, \tag{51}$$

then for any $\phi \in \mathcal{D}^m(u_j \cap u_k)$:

$$K_k\phi - K_j V_{jk}^{(N)}\phi \in L_{(\lambda_1,\dots,\lambda_n)}^{m-\frac{1}{2}\Sigma_{j=1}^n \lambda_j + \frac{1}{2}(r+|\bar{I}(j)|)-N}(R^n). \tag{52}$$

Thus, K_k and K_j coincide "in the principal term."

Theorem 2 (The commutation theorem). Let $K = K_{(L,\mu)}$ be a canonical operator described in Definition 4. Let also $H(x,p) \in T^s_{(\lambda_1,\dots,\lambda_n)}(R^{2n})$. There exist operators P_ℓ, $\ell = 0,1,2,\dots$, acting in the spaces

$$P_\ell : \mathcal{D}^m(L) \to \mathcal{D}^{s+m-\ell}(L) \tag{53}$$

(m is arbitrary) with the properties:

(a) P_ℓ is a differential operator of order $\leq 2\ell$. Its coefficients are linear combinations (with smooth real coefficients) of restrictions on L of $H(x,p)$ and its derivatives up to the order 2ℓ.

(b) The operator P_o is an operator of multiplication by $H(x,p)|_L$.

(c) If we denote

$$P^{(N)} = \sum_{\ell=0}^{N-1} (-i)^{\ell} P_\ell, \tag{54}$$

then for any $\phi \in \mathcal{D}^m(L)$:

$$H(\overset{2}{x}, -i\overset{1}{\frac{\partial}{\partial x}})K\phi - KP^{(N)}\phi \in L_{(\lambda_1,\dots,\lambda_n)}^{m+s-\frac{1}{2}\Sigma_{j=1}^n\lambda_j + \frac{1}{2}(r+n-1)-N}(R^n). \tag{55}$$

Also, if $H(x,p)$ is associated with (L,μ) in the sense that

$$H(x,p)|_L = 0, \quad V(H)|_L\mu = 0, \tag{56}$$

where

$$V(H) = \frac{\partial H}{\partial p}\frac{\partial}{\partial x} - \frac{\partial H}{\partial x}\frac{\partial}{\partial p} \tag{57}$$

is a Hamiltonian vector field generated by H, then

$$P_o = 0, \quad P_1 = \{V(H) - \frac{1}{2}\sum_{j=1}^n \frac{\partial^2 H}{\partial x_j \partial p_j}\}|_L. \tag{58}$$

Proof of Theorems 1 and 2. As for the purely computational aspects, no new ideas in comparison with the 1/h-case are involved. The main problem is to show that the performed expansions are valid in the considered situation, it being proved that the formulas for the terms of asymptotic expansion merely coincide (with obvious renotations) with that for the 1/h-case. We intend to prove the expansions to be valid and thus complete the proof.

We begin with Theorem 1. Let $\phi \in \mathcal{D}_m(u_j \cap u_k)$. Denote $I = I(j)$, and $\mathcal{T} = I(k)$,

$$\phi_I(x_I, p_{\bar{I}}) = \xi(\Lambda_I(x))(\mu_I(x_I, p_{\bar{I}}))^{1/2}(\phi \circ \nu_j^{-1})(x_I, p_{\bar{I}}), \tag{59}$$

and analogously for $\phi_{\mathcal{T}}(x_{\mathcal{T}}, p_{\bar{\mathcal{T}}})$. We have

$$(K_j\phi)(x) = \left(\frac{i}{2\pi}\right)^{|\bar{I}|/2} \int e^{i[S_I(x_I, p_{\bar{I}}) + x_{\bar{I}}p_{\bar{I}}]} \phi_I(x_I, p_{\bar{I}})dp_{\bar{I}}, \tag{60}$$

$$(K_k\phi)(x) = \left(\frac{i}{2\pi}\right)^{|\bar{\mathcal{T}}|/2} \int e^{i[S_{\mathcal{T}}(x_{\mathcal{T}}, p_{\bar{\mathcal{T}}}) + x_{\bar{\mathcal{T}}}p_{\bar{\mathcal{T}}}]} \phi_{\mathcal{T}}(x_{\mathcal{T}}, p_{\bar{\mathcal{T}}})dp_{\bar{\mathcal{T}}}. \tag{61}$$

Let $\chi(z_{\bar{I}}) \in C_o^\infty(R^{|\bar{I}|})$, $\chi(z_{\bar{I}}) = 1$ in the vicinity of the origin. We claim that for $\varepsilon > 0$ small enough

$$(1 - \chi(\varepsilon\xi_{\bar{I}}(x)))(K_k\phi)(x) \in L^{-\infty}_{(\lambda_1,\dots,\lambda_n)}(R^n), \tag{62}$$

where

$$\xi_j(x) = x_j/\Lambda_I(x)^{\lambda_j}. \tag{63}$$

Indeed, consider the set $M = \pi(u_j \cap u_k) \subset R^n_x$, where $\pi : L \to R^n_x$, $(x,p) \to x$ is a natural projection. Since $u_j \cap u_k$ is R_+-precompact, so is M and, henceforth,

$$|x_j| \leqslant C\Lambda(x)^{\lambda_j}, \; j = 1,\dots,n \tag{64}$$

with some constant C for $x \in M$. Since $M \subset \pi(u_j)$, $\Lambda(x) \leqslant C_I\Lambda_{\bar{I}}(x)$ in M, and finally

$$|\xi_j(x)| \leqslant C_o, \; j = 1,\dots,n \tag{65}$$

for $x \in M$, where $\xi_j(x)$ are given by the equality (63). In terms of function $S_{\mathcal{T}}(x_{\mathcal{T}}, p_{\bar{\mathcal{T}}})$, M may be described in the following way:

$$M = \{x \in R^n | \exists p_{\bar{\mathcal{T}}} : (x_{\mathcal{T}}, p_{\bar{\mathcal{T}}}) \in \nu_k(u_j \cap u_k) \text{ and } x_{\bar{\mathcal{T}}} + \frac{\partial S}{\partial p_{\bar{\mathcal{T}}}}(x_{\mathcal{T}}, p_{\bar{\mathcal{T}}}) = 0\}. \tag{66}$$

Thus, necessarily, $x_j + \frac{\partial S}{\partial p_j}(x_{\mathcal{T}}, p_{\mathcal{T}}) \neq 0$ for some $j \in \bar{\mathcal{T}}$ if x does not satisfy (65). In (61) perform the change of variables

$$x_j = \Lambda(x)^{\lambda_j} y_j, \; j = 1,\dots,n;$$
$$p_j = \Lambda(x)^{1-\lambda_j}\eta_j, \; j \in \bar{\mathcal{T}}. \tag{67}$$

The function (61) then has the form

$$(K_k\phi)(x) = \left(\frac{i}{2\pi}\right)^{|\bar{\mathcal{T}}|/2} \Lambda(x)^{|\bar{\mathcal{T}}| - |\lambda_{\bar{\mathcal{T}}}|} \times$$
$$\times \int e^{i\Lambda(x)[S_{\mathcal{T}}(y_{\mathcal{T}}, \eta_{\bar{\mathcal{T}}}) + y_{\bar{\mathcal{T}}}\eta_{\bar{\mathcal{T}}}]} \phi_{\mathcal{T}}(\Lambda(x)^\lambda y_{\mathcal{T}}, \Lambda(x)^{1-\lambda}\eta_{\bar{\mathcal{T}}})d\eta_{\bar{\mathcal{T}}}. \tag{68}$$

It should be emphasized that in (68) integration is over a compact set independent of x and, supp ϕ being R_+-precompact, y_T varies also in a compact set. Consider the set

$$M_1 = \{x \in R^n \mid |\xi_j| \geqslant Co + 1 \text{ for some } j \in [n]\}. \tag{69}$$

The above considerations make it evident that

$$\inf_{\substack{x \in M_1 \\ (x_T, p_{\bar{T}}) \in \operatorname{supp} \phi_T}} |\operatorname{grad}_{\eta_{\bar{T}}}[S_T(y_T, \eta_{\bar{T}}) + y_{\bar{T}}]| > 0. \tag{70}$$

Thus we may set ε so small that $\chi(\varepsilon\xi_{\bar{I}}(x)) = 1$ for $|\xi_j(x)| \leqslant Co + 1$, $j \in \bar{I}$, and integration by parts in (68) yields (62). The reader may easily recover technical details of this proof.

Now we represent $K_k\phi = \chi(\varepsilon\xi_{\bar{I}}(x))K_k\phi + (1 - \chi(\varepsilon\xi_{\bar{I}}(x)))K_k\phi$. The second term is shown to be inessential; applying the Fourier transformation of $F_{x_{\bar{I}} \to p_{\bar{I}}}$ to the first term, we obtain the integral

$$T(x_I, p_{\bar{I}}) = \left(\frac{i}{2\pi}\right)^{(|\bar{T}|-|\bar{I}|)/2} \iint e^{i[S_T(x_T, q_{\bar{T}}) + x_{\bar{T}}q_{\bar{T}} - x_{\bar{I}}p_{\bar{I}}]} \times \tag{71}$$
$$\times \chi(\varepsilon\xi_{\bar{I}}(x))\phi_T(x_T, q_{\bar{J}})\, dq_{\bar{T}}\, dx_{\bar{I}}.$$

It is not hard to see that integral (71) is taken (for fixed $(x_I, p_{\bar{I}})$) over a compact set. We perform the change of variables

$$x_j = \Lambda_I(x)^{\lambda_j}\xi_j,\ j = 1,\dots,n,$$
$$q_j = \Lambda_I(x)^{1-\lambda_j}\eta_j,\ j \in \bar{T}, \tag{72}$$
$$p_j = \Lambda_I(x)^{1-\lambda_j}\theta_j,\ j \in \bar{I}.$$

The integral (71) takes the form

$$I(x_I, p_{\bar{I}}) = \left(\frac{i}{2\pi}\right)^{(|\bar{T}|-|\bar{I}|)/2} \Lambda_I(x)^{|\bar{T}| - |\lambda_{\bar{T}}| + |\lambda_{\bar{I}}|} \times \tag{73}$$
$$\times \iint e^{i\Lambda(x)[S_T(\xi_T, \eta_{\bar{T}}) + \xi_{\bar{T}}\eta_{\bar{T}} - \xi_{\bar{I}}\theta_{\bar{I}}]} \chi(\varepsilon\xi_{\bar{I}})\phi_T(\xi_{\bar{T}}, \eta_{\bar{T}})\, d\eta_{\bar{T}}\, d\xi_{\bar{I}},$$

where the integral is taken over a <u>fixed</u> compact set. The further treatment of this integral is completely analogous to that of (9), the usual phase method thus being applicable to this integral. We obtain (60) as the first term of expansion, and all the subsequent terms, thus proving Theorem 1.

Next we give a sketch of the proof of Theorem 2. First of all, it suffices to prove the theorem when the elementary canonical operator K_j is substituted for K; the general case then follows via successive use of the unity partition and transition operators $V_{jk}^{(N)}$.

Let $H(x,p) \in T^s_{(\lambda_1,\dots,\lambda_n)}(R^{2n})$, $\phi \in \mathcal{D}^m(u_j)$, and assume, to shorten the the calculations, that $\bar{I}(j) = \emptyset$. We have, by definition:

$$H(\overset{2}{x}, -i\overset{1}{\frac{\partial}{\partial x}})K_j\phi =$$

$$= (2\pi)^{-n}\iint H(x,p)e^{i[(x-y)p+S(y)]}\xi(\Lambda(y))(\mu(y))^{1/2}(\phi \circ \nu_j^{-1})(y)dydp = \tag{74}$$

$$= (2\pi)^{-n}\iint H(x,p)e^{i[(x-y)p+S(y)]}\Phi(y)dydp.$$

Perform in (74) the change of variables

$$\begin{aligned} x_j &= \Lambda(x)^{\lambda_j}\theta_j, \\ y_j &= (\theta_j + \xi_j)\Lambda(x)^{\lambda_j}, \\ p_j &= \eta_j\Lambda(x)^{1-\lambda_i}, \end{aligned} \tag{75}$$

and we obtain the integral

$$\begin{aligned} I(x) = \Lambda(x)^n\iint e^{i\Lambda(x)[\xi\eta+S(\theta+\xi)]} \times \\ \times H(\theta\Lambda(x)^{\lambda}, \eta\Lambda(x)^{1-\lambda})\ \Phi((\theta+\xi)\Lambda(x)^{\lambda})d\xi d\eta. \end{aligned} \tag{76}$$

This integral is regularized as in the proof of Theorem 1 of item A. Applying the stationary phase method with $\Lambda(x)$ as a large parameter, we come to Theorem 2. The reader can easily recover the technical details omitted here.

To finish with this item, we introduce the notion of the wave front for elements of the space $L^{\infty}_{(\lambda_1,\ldots,\lambda_n)}(R^n)$.

<u>Definition 5</u>. Let $\psi \in L^{\infty}_{(\lambda_1,\ldots,\lambda_n)}(R^n)$. The wave front $WF(\psi)$ is the minimal of all closed R_+-invariant subsets $K \subset R^{2n}$, satisfying the property: if the support of the function $H(x,p) \in T^{\infty}_{(\lambda_1,\ldots,\lambda_n)}(R^{2n})$ lies in a closed R_+-invariant R_+-compact set K_1 and $K_1 \cap K = \emptyset$, then

$$H(\overset{2}{x}, -i\overset{1}{\frac{\partial}{\partial x}})\psi \in L^{-\infty}_{(\lambda_1,\ldots,\lambda_n)}(R^n). \tag{77}$$

It is obvious that Definition 5 is correct and $WF(\psi)$ may be defined as the intersection of all closed R_+-invariant subsets $K \subset R^{2n}$, satisfying the mentioned property.

Theorem 2 of item A and Theorem 2 of the present item, being combined with Definition 5, show that the following obvious assertion is valid:

<u>Theorem 3</u>. (1) For any $H \in T^{\infty}_{(\lambda_1,\ldots,\lambda_n)}(R^{2n})$ and $\psi \in L^{\infty}_{(\lambda_1,\ldots,\lambda_n)}(R^n)$:

$$WF(\hat{H}\psi) \subset WF(\psi) \subset WF(\hat{H}\psi) \cup \{(x,p) \mid \lim_{\tau\to\infty} H(\tau^{\lambda}x, \tau^{1-\lambda}p) = 0\}. \tag{78}$$

(2) For any canonically representable function ψ the wave front $WF(\psi)$ is contained in the corresponding Lagrangian manifold.

C. Quantization of Quasi-Homogeneous Canonical Transformations

In the previous item we constructed a class of canonically representable functions (CRF's) associated with a given proper R_+-invariant Lagrangian manifold $L \subset R^{2n}$. Here we establish the correspondence between a R_+-homogeneous canonical transformation g satisfying several additional conditions and linear operators $T(g)$ with the main property: if ψ is a CRF associated with the Lagrangian manifold L, then $T(g)\psi$ is also a CRF, associated with $g(L)$. We establish also the composition formulas and the commutation formulas with pseudo-differential operators for the operators $T(g)$. We proceed to precise definitions and formulations.

First of all, we introduce a notion somewhat different to that of the wave front (defined in the previous item) but much more convenient to our needs. The notion of the wave front is well adapted to the case when one considers pseudo-differential operators with R_+-finite symbols and/or R_+-compact wave fronts, etc. On the other hand, removing the R_+-finiteness condition for symbols of "test" operators in Definition 5 of item B would yield severe difficulties when proving that $WF(\psi)$ exists for general ψ. Fortunately, to develop the theory of quantization of canonical transformations there is no need to use such precise information as $WF(\psi)$ should give; it suffices to employ the following:

Definition 1. Let $\psi \in L^{\infty}_{(\lambda_1,\dots,\lambda_n)}(R^n)$. For a given closed R_+-invariant subset $K \subset R^{2n}$ we say that K is an essential subset for ψ and write that $K \in \mathrm{Ess}(\psi)$, if for any operator $\hat{f} = f(\overset{2}{x}, -i\overset{1}{\frac{\partial}{\partial x}}) \in \hat{T}^{\infty}_{(\lambda_1,\dots,\lambda_n)}(R^{2n})$ such that $\mathrm{supp}\, f \cap K = \phi$, we have $\hat{f}\psi \in L^{-\infty}_{(\lambda_1,\dots,\lambda_n)}(R^n)$. The main properties of $\mathrm{Ess}(\psi)$ are collected in the following proposition:

Lemma 1. (a) If $K \in \mathrm{Ess}(\psi)$ and $K_1 \supset K$ is a closed R_+-invariant subset, then $K_1 \in \mathrm{Ess}(\psi)$.

(b) If $K \in \mathrm{Ess}(\psi)$, then $WF(\psi) \subset K$ provided that $WF(\psi)$ exists.

(c) If ψ is a CRF associated with the Lagrangian manifold L, then $L \in \mathrm{Ess}(\psi)$.

(d) For any $\hat{H} = H(\overset{2}{x}, -\overset{1}{\frac{i\partial}{\partial x}}) \in \hat{T}^{\infty}_{(\lambda_1,\dots,\lambda_n)}(R^n)$ the implication is valid:

$$K \in \mathrm{Ess}(\psi) \Rightarrow K \cap \mathrm{supp}_+ H \in \mathrm{Ess}(\hat{H}\psi).$$

where $\mathrm{supp}_+ H$ is the R_+-invariant envelope of $\mathrm{supp}\, H$.

(e) $K_i \in \mathrm{Ess}(\psi_i)$, $i = 1,2 \Rightarrow K_1 \cup K_2 \in \mathrm{Ess}(\psi_1 + \psi_2)$.

(f) $\emptyset \in \mathrm{Ess}(\psi)$ if and only if $\psi \in L^{-\infty}_{(\lambda_1,\dots,\lambda_n)}(R^{2n})$.

Proof. The properties (a), (e), and (f) are evident. (b) If $a \in WF(\psi) \setminus K$ then by definition of $WF(\psi)$ there is a symbol $f \in T^{\infty}_{(\lambda_1,\dots,\lambda_n)}(R^n)$ with the support in a small enough R_+-invariant neighborhood of a (so that $\mathrm{supp}\, f \cap K = \emptyset$) and such that $\hat{f}\psi \notin L^{-\infty}_{(\lambda_1,\dots,\lambda_n)}(R^n)$, which contradicts the definition of K.

(c) This immediately follows from Theorem 2 of item B, and (d) also immediately follows from Theorem 2 of item A. Lemma 1 is proved.

The set $U \subset R^{2n}$ will be called <u>inessential</u> for ψ if $R^{2n} \setminus U \in \mathrm{Ess}(\psi)$.

We denote by $\overset{0}{L}{}^{\infty}_{(\lambda_1,\ldots,\lambda_n)}(R^n) \subset L^{\infty}_{(\lambda_1,\ldots,\lambda_n)}(R^n)$ the subspace of functions ψ such that $\mathrm{Ess}(\psi)$ contains the set $\{(x,p) \mid |p_j| \le C\Lambda(x)^{1-\lambda_j},\ j = 1,\ldots,n\}$ for some $C = C_\psi$. One should note that all CRF's belong to $\overset{0}{L}{}^{\infty}_{(\lambda_1,\ldots,\lambda_n)}(R^n)$ (see Definition 1 of item B).

The operators T(g), which we are going to define, act on $\overset{0}{L}{}^{\infty}_{(\lambda_1,\ldots,\lambda_n)}(R^n)$ rather than on $L^{\infty}_{(\lambda_1,\ldots,\lambda_n)}(R^n)$. To begin with, describe the class of admissible canonical transformations g.

Let $(\lambda_1,\ldots,\lambda_n)$ and $(\mu_1,\ldots,\mu_n)$ be two non-zero n-tuples of non-negative numbers. Each of these tuples defines an action of the group R_+ on the space R^n and, consequently, on the cotangent space $T^*R^n = R^{2n}$:

$$\tau x = (\tau^{\lambda_1}x_1,\ldots,\tau^{\lambda_n}x_n),\ x \in R^n,$$
$$\tau(x,p) = (\tau^{\lambda_1}x_1,\ldots,\tau^{\lambda_n}x_n,\tau^{1-\lambda_1}p_1,\ldots,\tau^{1-\lambda_n}p_n),\ (x,p) \in T^*R^n = R^{2n}, \tag{1}$$

and

$$\tau y = (\tau^{\mu_1}y_1,\ldots,\tau^{\mu_n}y_n),\ y \in R^n,$$
$$\tau(y,q) = (\tau^{\mu_1}y_1,\ldots,\tau^{\mu_n}y_n,\tau^{1-\mu_1}q_1,\ldots,\tau^{1-\mu_n}q_n),\ (y,q) \in T^*R^n = R^{2n}. \tag{2}$$

Here and below we adopt the convention: if the coordinates are denoted by (x,p), then the action (1) is assumed, while for coordinates (y,q) we employ (2). Although this notation is not completely rigorous, confusion should not occur.

<u>Definition 2</u>. The mapping

$$g = T^*R^n \to T^*R^n,\quad (y,q) \to (x,p) = (x(y,q),p(y,q)) \tag{3}$$

is a proper R_+-invariant canonical transformation if the following conditions are satisfied:

(a) g is smooth outside the set $\{\Lambda(y) = 0\}$.

(b) g is a canonical transformation, i.e., preserves the symplectic 2-form:

$$g^*(dp \wedge dx) = dq \wedge dy. \tag{4}$$

(c) g commutes with the action of the group R_+:

$$\tau g(x,p) = g(\tau(x,p)),\quad \text{or}$$
$$x(\tau^{\mu}y,\tau^{1-\mu}q) = \tau^{\lambda}x(y,q), \tag{5}$$
$$p(\tau^{\mu}y,\tau^{1-\mu}q) = \tau^{1-\lambda}p(y,q).$$

(d) g "preserves the equivalence class of Λ" in the following sense: there exist positive constants c and C such that

$$c\Lambda(x(y,q)) \leqslant \Lambda(y) \leqslant C\Lambda(x(y,q)) \tag{6}$$

for all (y,q).

(e) Functions $p_j(y,q)$ and all the derivatives of the functions $x_j(y,q)$, $p_j(y,q)$, $j = 1,\ldots,n$, are bounded on any set of the form:

$$K = \{(y,q) \mid m_1 \leqslant \Lambda(y) \leqslant m_2, |q| \leqslant M\}, \tag{7}$$

where m_1, m_2, and M are positive numbers.

(f) The projection $\pi(\Gamma(g))$ of the graph $\Gamma(g) \subset T^*R^n \times T^*R^n$ of transformation g on the space $R^n_x \times R^n_y$ is a uniformly proper set in the following sense: if we set

$$I_o = \{j \mid \lambda_j = 0\},\ J_o = \{j \mid \mu_j = 0\},\ I_+ = [n] \setminus I_o,\ J_+ = [n] \setminus J_o,$$

then for any compact set $K_1 \in R^{|I_o|}_{x_{I_o}}$ the set $K_2 = \{y_{J_0} \mid (x, y_{J_o}, y_{J_+}) \in \pi(\Gamma(g))$ for some y_{J_+} and x with $x_{I_o} \in K_1\}$ is bounded and the diameter of K_2 depends only on the diameter of K_1, and the same property holds if we also exchange x and y.

We should also be concerned with canonical transformations defined on some R_+-invariant connected simply connected open subset $\mathcal{U} \subset T^*R^n$. Definition 2 applies to this case as well except that, instead of (3), we have

$$g : \mathcal{U} \to T^*R^n. \tag{3'}$$

<u>Proposition 1</u>. If g is a proper R_+-invariant canonical transformation* then:

(a) g takes proper Lagrangian manifolds into proper ones.

(b) The graph $\Gamma(g) \subset T^*R^n \times T^*R^n$ of the transformation g is a Lagrangian manifold with respect to the form

$$\Omega^2 = dp \wedge dx - dq \wedge dy. \tag{8}$$

<u>Proof</u>. (a) Let L be a proper Lagrangian manifold. To prove that g(L) is proper we need only to establish the inequalities $|p_j| < C\Lambda(x)^{1-\lambda_j}$ on g(L) or, equivalently, that $|p|$ is bounded on g(L) for $\Lambda(x) = 1$. From (6) the property $\Lambda(x) = 1$ yields $m_1 \leqslant \Lambda(y) \leqslant m_2$ with positive m_1, m_2, and thus $|q| \leqslant M$ since L is proper. By item (c) of Definition 2 $|p|$ is then bounded. Q.E.D. (b) is obvious.

The operator T(g) will be defined in terms of its kernel, $T_g(x,y)$, on the space $R^n \times R^n$, this kernel being given by the canonical operator on the graph $\Gamma(g)$. Thus we need to define the corresponding auxiliary objects.

The 2n-tuple $(\lambda_1,\ldots,\lambda_n,\mu_1,\ldots,\mu_n)$ defines the action of R_+ on $R^{2n}_{(x,y)} = R^n_x \times R^n_y$ and the corresponding function $\Lambda(x,y)$. The Lagrangian manifold $\Gamma(g) \subset T^*R^{2n} \simeq T^*R^n \times T^*R^n$ is <u>not</u> proper since the basic inequalities

$$\begin{gathered} |p_j| \leqslant C\Lambda(x,y)^{1-\lambda_j}, \\ |q_j| \leqslant C\Lambda(x,y)^{1-\mu_j},\ j = 1, \ldots, n \end{gathered} \tag{9}$$

---

* We drop the words "R_+-invariant" in the sequel since non-R_+-invariant objects are not the matter of our consideration.

do not hold on $\Gamma(g)$. However, for any constant C the intersection $\Gamma_C(g)$ of $\Gamma(g)$ with the set defined by the inequalities (9) is a proper Lagrangian manifold, and therefore the canonical operator may be defined on it (provided that the measure is chosen).

There is a special distinguished measure homogeneous of degree n defined on $\Gamma(g)$, namely, the pullback of the Euclidean volume measure on T^*R^n:

$$\mu = (\pi_1|_{\Gamma(g)})^*(dq_1 \wedge \ldots \wedge dq_n \wedge dy_1 \wedge \ldots \wedge dy_n). \tag{10}$$

Here

$$\pi_i = T^*R^n \times T^*R^n \to T^*R^n, \ i = 1,2 \tag{11}$$

is the natural projection on the i-th factor. Since being canonical g preserves the Euclidean volume as well, and μ can also be defined as

$$\mu = (\pi_2|_{\Gamma(g)})^*(dp_1 \frown \ldots \frown dp_n \wedge dx_1 \frown \ldots \frown dx_n). \tag{12}$$

For the sake of convenience we make use of a special canonical atlas on $\Gamma_C(g)$.

Proposition 2. There exists a proper canonical atlas on $\Gamma_C(g)$ such that the canonical coordinates in any chart of this atlas have the form $(x, y_I, q_{\bar{I}})$ for some $I \subset [n]$.

Proof. For any fixed x the set $\{(y,q) | x(y,q) = x\}$ is a Lagrangian manifold in $R^{2n}_{(y,q)}$. Thus, by the lemma on local coordinates, we may choose canonical coordinates of the form $(x, y_I, q_{\bar{I}})$ in R_+-invariant neighborhood of any point of $\Gamma_C(g)$. We need only to prove that, in obvious notations, the inequalities

$$|q_j| \leqslant \text{const}\cdot\Lambda_{[n],I}(x,y_I)^{1-\mu_j}, \ j \in \bar{J}, \tag{13}$$

hold in such neighborhoods; other properties of the proper canonical atlas may be satisfied by suitable choice of the covering. But (13) immediately follows from (9) and (6) since

$$\Lambda(x,y) \leqslant \text{const}\cdot\Lambda(x) \leqslant \text{const}\cdot\Lambda_{[n],I}(x,y_I). \tag{14}$$

Thus Proposition 2 is proved.

We assume in addition that there is a finite number of fixed canonical charts $\{\mathcal{U}_j\}_{j=1}^N$ on $\Gamma(g)$ so that a proper canonical atlas on any $\Gamma_C(g)$ may be given by $\{\mathcal{U}_j \cap \Gamma_C(g)\}_{j=1}^N$, and that the density of the measure μ in these charts satisfies the conditions imposed in item B.

We note that $\Gamma(g)$ is a simply connected manifold; therefore the quantization conditions are necessarily satisfied on $\Gamma(g)$. Denote by $\mathcal{D}^m(g)$ the union of spaces $\mathcal{D}^m(\Gamma_C(g))$ for all $C > 0$, $\mathcal{D}^\infty(g) = \bigcup \mathcal{D}^n(g)$, $\mathcal{D}^{-\infty}(g) = \bigcap \mathcal{D}^n(g)$. The above considerations yield the validity of the following:

Proposition 3. The canonical operator K_g on the graph $\Gamma(g)$ is defined and acts in the spaces:

$$K_g : \mathcal{D}^m(g) \to L^{m+n-\frac{1}{2}(|\lambda|+|\mu|)}_{(\lambda_1,\dots,\lambda_n,\mu_1,\dots,\mu_n)}(R^n_x \times R^n_y) \tag{15}$$

for any m.*

Proof. The above statement immediately follows from Proposition 1 of item B.

Let $K_o = \pi_{I_o,J_o}(\Gamma(g)) \subset R^n_{o\,y_{J_o}} \times R^n_{x_{I_o}}$ be the projection of the set $\Gamma(g) \subset R^{2n}_{(y,q)} \times R^{2n}_{(x,p)}$ on $R^n_{y_{J_o}} \times R^n_{x_{I_o}}$, and define

$$\tilde{K}_o = \{(y_{J_o},x_{I_o}) \mid \mathrm{dist}((y_{J_o},x_{I_o}),K_o) \leqslant 2\} \tag{16}$$

(here $\mathrm{dist}(z,K_o)$ is the distance between the point z and the set K_o in the usual Euclidean metrics), and set

$$\chi_1(x_{I_o},y_{J_o}) = \int\tilde{\chi}_1(x_{I_o}-\xi_{I_o},y_{J_o}-\eta_{J_o})d\xi_{I_o}d\eta_{J_o},\quad (\xi_{I_o},\eta_{J_o}) \in K_o, \tag{17}$$

where $\chi_1(z_{I_o},\mathcal{U}_{J_o}) \in C^\infty_o(R^{|I_o|+|J_o|})$, $\chi(z_{I_o},\mathcal{U}_{J_o}) = 0$ for $|z_{I_o}|^2+|\mathcal{U}_{J_o}|^2 \geqslant 1$ and $\int\chi_1(z_{I_o},\mathcal{U}_{J_o})dz_{I_o}d\mathcal{U}_{T_o} = 1$. It is obvious that $\chi_1(x_{I_o},y_{J_o})$ is a smooth function equal to one in the neighborhood of the set K_o, bounded with all the derivatives. The requirement (e) of Definition 2 guarantees that the diameter of the set

$$\{y_{J_o} \mid \chi(x_{I_o},y_{J_o}) \neq 0\}$$

is bounded uniformly with resepct to x_{I_o}. Then set also

$$\chi_2(x_{I_+},y_{J_+}) = \tilde{\chi}_2(\Lambda(y)/\Lambda(x)), \tag{18}$$

where $\tilde{\chi}_2(z) \in C^\infty_o(R^1)$, $\tilde{\chi}_2(z) = 0$ for $z > 2C$ or $z < c/2$, and $\tilde{\chi}_2(z) = 1$ for $\frac{2c}{3} \leqslant z \leqslant \frac{3}{2}C$ (here the constants c and C are the same as in (6)) and define finally:

$$\chi(x,y) = \tilde{\chi}_1(x_{I_o},y_{I_o})\chi_2(x_{I_+},y_{J_+}). \tag{19}$$

Clearly $\chi(x,y)$ equals 1 in the neighborhood of $\Gamma(g)$ and belongs also to $S^0_{(\lambda_1,\dots,\lambda_n,\mu_1,\dots,\mu_n)}(R^n_x \times R^n_y)$, therefore to $T^m_{(\lambda_1,\dots,\lambda_n,\mu_1,\dots,\mu_n)}(R^{4n})$. Now it follows from Definition 1 and Lemma 1, that for any $\phi \in \mathcal{D}^\infty(g)$

$$\chi(x,y)[K_g\phi](x,y) - [K_g\phi](x,y) \in L^{-\infty}_{(\lambda_1,\dots,\lambda_n,\mu_1,\dots,\mu_n)}(R^n_x \times R^n_y), \tag{20}$$

and consequently modulo $L^{-\infty}_{(\lambda_1,\dots,\lambda_n,\mu_1,\dots,\mu_n)}(R^n_x \times R^n_y)$ $\chi[K_g\phi]$ does not depend on the freedom of the choice of $\chi(x,y)$.

Definition 3. By $T(g,\phi)$ we denote the operator (as usual, $\arg i = \pi/4$)

$$[T(g,\phi)f](x) = \left(\frac{i}{2\pi}\right)^{n/2}\int\chi(x,y)[K_g\phi](x,y)f(y)dy. \tag{21}$$

---

* The reader should take into account that the phase functions, etc. in canonical charts for K_y are constructed in concordance with the choice of signs in (8).

Lemma 2. If $\phi \in \mathcal{D}^m(g)$ then the operator $T(g,\phi)$ acts in the spaces

$$T(g,\phi) = L^S_{(\mu_1,\ldots,\mu_n)}(R^n_y) \to L^{S+m+n+\frac{1}{2}(|\mu|-|\lambda|)}_{(\lambda_1,\ldots,\lambda_n)}(R^n_x) \tag{22}$$

for any $S \in R$. Modulo $L^{-\infty}_{(\lambda_1,\ldots,\lambda_n)}(R^n_x)$, the result of action of $T(g,\phi)$, does not depend on the choice of the cut-off function $\chi(x,y)$.

Proof. By Proposition 3, the kernel $K(x,y)$ of the integral operator (21) belongs to $L^{m+n-\frac{1}{2}(|\lambda|+|\mu|)}_{(\lambda_1,\ldots,\lambda_n,\mu_1,\ldots,\mu_n)}(R^n_x \times R^n_y)$. In the expression

$$[T(g,\phi)f](x) = \int K(x,y)f(y)dy \tag{23}$$

we perform the change of variables

$$y_j = \Lambda(x)^{\mu_j}\eta_j, \quad j = 1,\ldots,n. \tag{24}$$

Then we obtain

$$\int K(x,y)f(y)dy = \Lambda(x)^{|\mu|}\int K(x,\eta\Lambda(x)^{\mu})f(\eta\Lambda(x)^{\mu})d\eta , \tag{25}$$

where the volume of integration is bounded uniformly with respect to x by construction of the cut-off function $\chi(x,y)$. Estimating the integral on the right-hand side of (25), we obtain that it does not exceed

$$\text{const}\cdot\Lambda(x)^{|\mu|+S+m+n-\frac{1}{2}(|\lambda|+|\mu|)} = \text{const}\cdot\Lambda(x)^{S+m+n+\frac{1}{2}(|\mu|-|\lambda|)}.$$

The x-derivatives are estimated in an analogous way and thus we obtain the proof of Lemma 2 (the second assertion of the lemma is self-evident).

Our next aim is to establish the composition formulas for operators $T(g,\phi)$. We begin with formulas of composition with pseudo-differential operators.

Remark 1. In establishing and writing the composition formulas, it is convenient to slightly modify our notations in the following way. Thus consider the projection

$$\begin{aligned} \pi_1 \quad & T^*R^n \times T^*R^n \to T^*R^n \\ & (y,q,x,p) \to (y,q) \end{aligned} \tag{26}$$

(cf. (11)) and its restriction on $\Gamma(g)$, which will be denoted by the same letter. We consider now functions on $T^*R^n_y$ rather than on $\Gamma(g)$ (or, in other words, consider (y,q) as standard coordinates on $\Gamma(g)$) and denote by $T(g,\phi)$ the operator, which in the notations adopted above should read $T(g,\phi \circ \pi_1)$.

In these new notations our results take the simple and readable form:

Theorem 1. (a) Let $H(x,p) \in T^S_{(\lambda_1,\ldots,\lambda_n)}(R^{2n})$. There exists differential operators P_j on $\Gamma(g)$ with the properties: (i) P_j is a differential operator of order $\leqslant j$ with coefficients which are linear combinations of the derivatives of the function $(g^*H)(y,q)$ of order $\leqslant 2j$ with smooth real coefficients. The operator P_j acts in the spaces

$$P_j : \mathcal{D}^m(g) \to \mathcal{D}^{m+S-j}(g), \quad j = 0,1,2,\ldots \tag{27}$$

for any m. (ii) In particular,

$$\mathcal{P}_0 = (g^*H)(y,q). \tag{28}$$

(iii) For any N,

$$H(\overset{2}{x}, -i\overset{1}{\frac{\partial}{\partial x}})T(g,\phi) = T(g, \sum_{j=0}^{N-1} (-i)^j \mathcal{P}_j\phi), \tag{29}$$

modulo operators acting from $L^s_{(\mu_1,\ldots,\mu_n)}(R^n)$ to $L^{S+m+n-N+\frac{1}{2}(|\mu|-|\lambda|)}_{(\lambda_1,\ldots,\lambda_n)}(R^n)$ for any S.

(b) Let $H(y,q) \in T^S_{(\mu_1,\ldots,\mu_n)}(R^{2n})$. Then there exist differential operators $\mathcal{P}_j$ on $\Gamma(g)$ such that (i) and (ii) of item (a) are valid with H substituted for g^*H and

$$T(g,\phi)H(\overset{2}{y}, -i\overset{1}{\frac{\partial}{\partial y}}) = T(g, \sum_{j=0}^{N-1} (-i)^j \mathcal{P}_j\phi) \tag{30}$$

to within the same modulus as in (29).

Proof. (a) Theorem 2 of item B would give the desired result but it cannot be applied directly since $H(x,p) \notin T^\infty_{(\lambda_1,\ldots,\lambda_n,\mu_1,\ldots,\mu_n)}(R^{4n})$. In order to avoid this difficulty, we represent the kernel of the operator (29) in the form

$$\begin{aligned} K_1(x,y) &= \tilde{\chi}_2(\Lambda(y)/\Lambda(x))H(\overset{2}{x}, -i\overset{1}{\frac{\partial}{\partial x}})K(x,y) + (1-\tilde{\chi}_2(\Lambda(y)/\Lambda(x))) \times \\ &\times H(\overset{2}{x}, -i\overset{1}{\frac{\partial}{\partial x}})K(x,y) \equiv H_1(\overset{2}{x},y, -i\overset{1}{\frac{\partial}{\partial x}})K(x,y) + H_2(\overset{2}{x},y, -i\overset{1}{\frac{\partial}{\partial x}})K(x,y). \end{aligned} \tag{31}$$

Here K(x,y) is the kernel of the operator $T(g,\phi)$ and the function $\tilde{\chi}_2$ is the same as in (18). The function $H_1(x,y,p)$ belongs to $T^\infty_{(\lambda_1,\ldots,\lambda_n,\mu_1,\ldots,\mu_n)}(R^{4n})$ and it is easy to see that application of Theorem 2 of item B yields the result (computational details are left to the reader) once we prove that the second term in (31) is inessential. To do this we write it in the integral form:

$$\begin{aligned} &H_2(\overset{2}{x},y, -i\overset{1}{\frac{\partial}{\partial x}})K(x,y) = \\ &= (2\pi)^{-n}\int e^{ip(x-\xi)}H(x,p)[1-\tilde{\chi}_2(\Lambda(y)/\Lambda(x))]K(\xi,y)\,d\xi dp. \end{aligned} \tag{32}$$

Perform the variable change

$$\begin{aligned} x_j &= \Lambda(x)^{\lambda_j}\omega_j, \quad \xi_j = \Lambda(x)^{\lambda_j}\eta_j, \\ y_j &= \Lambda(x)^{\mu_j}\xi_j, \quad p_j = \Lambda(x)^{1-\lambda_j}\theta_j, \quad j = 1,\ldots,n; \end{aligned} \tag{33}$$

then (32) takes the form

$$H_2(\overset{2}{x},y, -i\overset{1}{\frac{\partial}{\partial x}})K(x,y) = \left(\frac{\Lambda(x)}{2\pi}\right)^n\int e^{i\Lambda(x)\theta(\omega-\eta)}H[1-\tilde{\chi}_2]K\,d\eta d\theta, \tag{34}$$

where $\omega - \eta \neq 0$ on the support of the integrand. Integrating by parts over θ, we obtain the factor $(\Lambda(x))^{-N}$ with N arbitrarily large; the case when the integral over p diverges is considered in just the same way as in items A and B. We do not go into further detail.

(b) We have

$$[T(g,\phi)H(\overset{2}{y},-i\overset{1}{\frac{\partial}{\partial y}})\psi](x) =$$
$$= (2\pi)^{-n}\iiint K(x,y)H(y,\xi)e^{i\xi(y-\eta)}\psi(\eta)d\eta d\xi dy = \int K_1(x,y)\psi(y)dy, \tag{35}$$

where

$$K_1(x,y) = (2\pi)^{-n}\iint K(x,\theta)H(\theta,\xi)e^{i\xi(\theta-y)}d\theta d\xi = H(\overset{1}{y}, i\overset{2}{\frac{\partial}{\partial y}})K(x,y). \tag{36}$$

Although we did not consider such operators in previous items, their theory is just parallel to that of operators of the form $H(\overset{2}{y},-i\overset{1}{\frac{\partial}{\partial y}})$. Considerations analogous to those performed in the proof of (a) lead to the identity (30). We have no space to present a more comprehensive study of the question at this point. The proof of the theorem is now complete.

Our next theorem establishes the relations between the composition of canonical transformations and composition of the correspondent operators $T(g,\phi)$. In addition to previous considerations, fix a non-zero tuple $(\theta_1,\ldots,\theta_n)$ of non-negative numbers and the corresponding action of the group R_+ on the space R^n_v and the cotangent space $T^*R^n_v \simeq R^n_v \oplus R^n_\eta$.

Consider the proper canonical transformations

$$\begin{aligned} g_1 &: T^*R^n_x \to T^*R^n_y \\ g_2 &: T^*R^n_y \to T^*R^n_v \end{aligned} \tag{37}$$

and their composition

$$g_2 \cdot g_1 = g_3 : T^*R^n_x \to T^*R^n_v. \tag{38}$$

Obviously, g_3 is a proper canonical transformation. We assume that the canonical transformations g_j, $j = 1,2,3$, satisfy all the additional conditions imposed in this item.

Under these assumptions and some further restrictions on canonical atlases, the following theorem is valid:

<u>Theorem 2</u>. Let $\phi_j \in \mathcal{D}^{m_j}(g_j)$, $j = 1,2$. There exist functions $\Phi_k \in \mathcal{D}^{m_1+m_2-k}(g_3)$, $k = 0,1,2,\ldots$, such that the following properties are satisfied:

(i) The function Φ_k is a bilinear form of the values of functions ϕ_1 and $g_1{}^*\phi_2$ and their derivatives up to the order 2k. The coefficients of these forms are smooth real functions.

(ii) The function Φ_0 has the form

$$\Phi_0 = \phi_1 \cdot g_1{}^*\phi_2. \tag{39}$$

(iii) For any natural N the equality

$$T(g_2,\phi_2)T(g_1,\phi_1) = \pm\, T(g_3, \sum_{j=0}^{N-1}(-i)\Phi_j) \tag{40}$$

holds modulo operators, acting from $L^{S}_{(\lambda_1,\ldots,\lambda_n)}(R^n)$ to $L^{S-N+m_1+m_2+n+\frac{1}{2}\times(|\lambda|-|\theta|)}(R^n)$, for any S.

The sign + or - in (40) depends on the choice of the branch of the argument of the density of canonical measure on graphs $\Gamma(g_1)$, $\Gamma(g_2)$, and $\Gamma(g_3)$ (recall that for a connected simply connected Lagrangian manifold there exist exactly two ways of choosing the concordant family of arguments of the measure of density).

Remark 2. Before proving the theorem, we wish to concretize the further restrictions on canonical atlases mentioned above. The point is that without additional assumptions even the function (39) may not belong to $t\circ\mathcal{D}^{m_1+m_2}(g_3)$. However, such a situation is rather pathological; in practically interesting cases the space $\mathcal{D}^m(g)$ merely coincides* with the set $\tilde{T}^m_{(\lambda_1,\ldots,\lambda_n)}(R^{2n})$ of functions $f\in T^m_{(\lambda_1,\ldots,\lambda_n)}(R^{2n})$ such that $|p\Lambda^{\lambda-1}|$ is bounded on supp f (here we assume in our notation that g acts on elements of $T^*R^n_x$, where $\tau x=(\tau^{\lambda_1}x_1,\ldots,\tau^{\lambda_n}x_n)$. Since conditions of Definition 2 imply that g takes $T^m_{(\lambda_1,\ldots,\lambda_n)}(R^{2n})$ into $T^k_{(\lambda_1,\ldots,\lambda_n)}(R^{2n})$, we see that additional conditions mentioned are necessarily satisfied in these cases. To ensure that $\mathcal{D}^m(g)$ coincides with $\tilde{T}^m_{(\lambda_1,\ldots,\lambda_n)}(R^{2n})$ it suffices to require that:

(1) As before, the number of charts in a proper canonical atlas is finite; (2) in any chart the Jacobian of canonical coordinates with respect to "standard" coordinates (y,g) is greater than some positive constant on any set of the form (7). Thus, (2) is an example of additional conditions under which Theorem 2 is necessarily true.

Proof of Theorem 2. Using the partition of unity and transition operators $V_{jk\ell}$ (see Theorem 1 of item B), we reduce the theorem to the case when only one canonical chart is employed on each of manifolds $\Gamma(g_i)$, i = 1, 2,3. The proof is based on the stationary phase method; as before, the critical point is to verify the applicability of this method, since the formula (40) itself is not new and unknown from the purely computational viewpoint. In fact it was established by Hörmander [29] (in other notations) for Fourier integral operators; in the 1/h-case the method of the Cauchy problem was used in [39] to prove this result - see also Section 2 of the present chapter.

Thus, taking into account that there is lack of space to repeat the known arguments, we restrict ourselves only to demonstrating the applicability of the stationary phase method. To establish (40), we need to calculate the asymptotic expansion of the kernel of the product of operators on the left-hand side of (40). According to previous considerations, we assume that the kernel of $T(g_1,\phi_1)$ has, in obvious notations, the form

$$K_1(y,x) = \chi_1(y,x)\int e^{i[S_1(y,x_I,p_{\bar{I}})-x_{\bar{I}}p_{\bar{I}}]}a_1(x,y_I,p_{\bar{I}})dp_{\bar{I}}, \qquad (41)$$

and the kernel of $T(g_2,\phi_2)$ has the form

$$K_2(v,y) = \chi_2(v,y)\int e^{i[S_2(v,y_T,q_{\bar{T}})-y_{\bar{T}}q_{\bar{T}}]}a_2(g,v,y_Tq_{\bar{T}})dq_{\bar{T}}. \qquad (42)$$

* See Remark 1.

Here a_1, a_2 are the total amplitude functions (including square roots of the measure density), and χ_1, χ_2 are the cut-off functions. Thus, the kernel of the product $T(g_2,\phi_2)T(g_1,\phi_1)$ has the form:

$$K_3(v,x) = \int K_2(v,y)K_1(y,x)dy =$$

$$= \iiint e^{i[S_1(y,x_I,p_{\bar{I}})-x_{\bar{I}}p_{\bar{I}}+S_2(v,y_T,q_{\bar{T}})-y_{\bar{T}}q_{\bar{T}}]}\chi_1(y,x) \times \tag{43}$$

$$\times \chi_2(v,y)a_1(x,y_I,p_{\bar{I}})a_2(v,y_T,q_{\bar{T}})dp_{\bar{I}}dq_{\bar{T}}dy$$

(note that (43) is an integral over a finite domain due to our construction of the cut-off functions). To establish (40) we need to consider the partial Fourier transformation

$$F_{x_{\bar{k}}\to p_{\bar{k}}}(K_3(v,x)) =$$

$$= \iiiint e^{i[S_1(y,x_I,\eta_{\bar{I}})+S_2(v,y_T,q_{\bar{J}})-x_{\bar{I}}\eta_{\bar{I}}-y_{\bar{J}}q_{\bar{J}}+x_{\bar{k}}p_{\bar{k}}]}\chi_1(y,x) \times \tag{44}$$

$$\times \chi_2(v,y)a_1(x,y_I,\eta_{\bar{I}})a_2(v,y_T,q_{\bar{J}})dp_{\bar{I}}dq_{\bar{J}}dydx_{\bar{k}},$$

where as above the integral is taken over the finite domain. Perform in (44) the variable change:

$$\begin{aligned}
v_j &= \Lambda(v)^{\theta_j}z_j^{(1)}, & j &= 1,\dots,n,\\
x_j &= \Lambda(v)^{\lambda_j}z_j^{(2)}, & j &\in K,\\
y_j &= \Lambda(v)^{\mu_j}\xi_j^{(1)}, & j &= 1,\dots,n,\\
p_j &= \Lambda(v)^{1-\lambda_j}z_j^{(2)}, & j &\in \bar{K},\\
\eta_j &= \Lambda(v)^{1-\lambda_j}z_j^{(3)}, & &\\
q_j &= \Lambda(v)^{1-\mu_j}\xi_j^{(4)}, & &\\
x_j &= \Lambda(v)^{\lambda_j}\xi_j^{(5)}, & j &\in \bar{K}.
\end{aligned} \tag{45}$$

After obvious transformations, we obtain

$$\bar{F}_{x_k\to p_k}(K_3(v,x)) =$$

$$= \Lambda(v)^{|\mu_T|+|\bar{T}|+|\lambda_k|+|\bar{I}|-|\lambda_{\bar{I}}|}\int e^{i\Lambda(v)\Phi(z,\xi)}A(z,\xi,\Lambda(v))d\xi, \tag{46}$$

where $z = (z^{(1)},z^{(2)})$, $\xi = (\xi^{(1)},\dots,\xi^{(5)})$; the functions $\Phi(z,\xi)$ and $A(z,\xi,\Lambda)$ are smooth and bounded uniformly with respect to Λ with all the (z,ξ)-derivatives. Also Φ is real-valued and the integration in (46) is performed over a bounded domain whose size does not depend on z and Λ. Application of the stationary phase method to (46) is therefore valid and after prolonged computations, we come eventually to (40).

Then, as an easy consequence of Theorem 1, we obtain the following:

Proposition 4. The operators $T(g,\phi)$ satisfy the property: if $K \in \mathrm{Ess}(\psi)$, then

$$g(K \cap \mathrm{supp}\ \phi) \in \mathrm{Ess}(T(g,\phi)\psi). \tag{47}$$

Proposition 5. This proposition is somewhat more complicated. If ψ is a CRF associated with a proper Lagrangian manifold L, then $T(g,\phi)\psi$ is a CRF associated with the proper Lagrangian manifold g(L). The proof is based on routine application of the stationary phase method.

Now we may define the operator T(g) corresponding to a proper R_+-invariant canonical transformation satisfying the additional conditions described previously.

Definition 4. The operator

$$T(g) : \overset{0}{L}{}^{\infty}_{(\mu_1,\ldots,\mu_n)}(R^n_y) \to \overset{0}{L}{}^{\infty}_{(\lambda_1,\ldots,\lambda_n)}(R^n_x) \tag{48}$$

is defined as follows: for any $\psi \in \overset{0}{L}{}^{\infty}_{(\mu_1,\ldots,\mu_n)}(R^n_y)$ choose the function $\phi(y,q) \in \tilde{T}^0_{(\mu_1,\ldots,\mu_n)}(R^{2n})$ such that $\phi(y,q) = 1$ in the neighborhood of some set $K \subset \mathrm{Ess}(\psi)$ of the form

$$K = \{(y,q) \,|\, |q_j| \leqslant C\Lambda(y)^{1-\mu_j},\ j = 1,\ldots,n\}. \tag{49}$$

ϕ may be interpreted as an element of $\mathcal{D}^0(g)$ (see Remark 1). Set

$$T(g)\psi \overset{\mathrm{def}}{=} T(g,\phi)\cdot\psi . \tag{50}$$

Proposition 6. Modulo $L^{-\infty}_{(\lambda_1,\ldots,\lambda_n)}(R^n_x)$, the operator T(g) is a linear one and the result of action T(g) on ψ in (49) does not depend on the choice of the function ϕ, satisfying the conditions of Definition 4.

Proof. Let ϕ_1,ϕ_2 be two functions satisfying these conditions. Then we have

$$\psi = \phi_3(\overset{2}{y},-i\overset{1}{\frac{\partial}{\partial y}})\psi (\mathrm{mod}\ L^{-\infty}_{(\mu_1,\ldots,\mu_n)}(R^1_y)) \tag{51}$$

for suitable $\phi_3 \in \tilde{T}^0_{(\mu_1,\ldots,\mu_n)}$ with $\phi_j = 1$ on supp ϕ_3, $j = 1,2$. Using Theorem 1, we obtain immediately

$$T(g,\phi_1)\psi \overset{\mathrm{def}}{\equiv} T(g,\phi)\cdot\hat{\phi}_3\psi \equiv T(g,\phi_2)\hat{\phi}_3\psi \equiv T(g,\phi_2)\psi \tag{52}$$

(all the equalities are valid modulo $L^{-\infty}_{(\lambda_1,\ldots,\lambda_n)}(R^n)$). Thus Proposition 6 is proved.

Thus we have defined the operator T(g), which is the "quantization" of the canonical transformation g. It should be emphasized that T(g) depends essentially (except for the "principal term") on the choice of the partition of unity on the graph $\Gamma(g)$. In the sequel we sometimes work with the families of canonical transformations, families of functions ψ to which the operators T(g) should be applied, etc. In all these cases we assume that the representatives of our operators may be chosen and that they are chosen in such a way that the function ϕ in (51) and the elements of the partition of unity depend smoothly on the parameters involved.

Before formulating the theorem on composition of "quantized" canonical transformations, we enlarge the class of admissible symbols of pseudo-differential operators acting in the space $\overset{0}{L}{}^{\infty}_{(\lambda_1,\dots,\lambda_n)}(R^n)$. Denote by $\overset{0}{T}{}^{m}_{(\lambda_1,\dots,\lambda_n)}(R^{2n})$ the space of smooth functions $f(x,p)$ satisfying estimates:

$$\left| \frac{\partial^{|\alpha+\beta|} f(x,p)}{\partial x^{\alpha}\partial p^{\beta}} \right| \leqslant C_{\alpha\beta M}\Lambda(x)^{m-\Sigma_j\lambda_j\alpha_j+\Sigma_j(\lambda_j-1)\beta_j} \tag{53}$$

for

$$|p_j\Lambda(x)^{\lambda_j-1}| \leqslant M; \quad |\alpha|,|\beta| = 0,1,2,\dots . \tag{54}$$

Thus we do not impose any restrictions on the rate of growth of f as $|p_j\Lambda(x)^{\lambda_j-1}| \to \infty$. We set $\overset{0}{T}{}^{\infty}_{(\lambda_1,\dots,\lambda_n)}(R^{2n}) = \bigcup_m \overset{0}{T}{}^{m}_{(\lambda_1,\dots,\lambda_n)}(R^{2n})$. Let $\psi \in \overset{0}{L}{}^{\infty}_{(\lambda_1,\dots,\lambda_n)}(R^{2n})$. For some $C > 0$, the set

$$K = \{(x,p) \,|\, |p_j| \leqslant C\Lambda(x)^{1-\lambda_j},\ j = 1,\dots,n\} \tag{55}$$

belongs to Ess ψ. If $f \in \overset{0}{T}{}^{\infty}_{(\lambda_1,\dots,\lambda_n)}(R^{2n})$, we clearly may choose the function $f \in \tilde{T}^{\infty}_{(\lambda_1,\dots,\lambda_n)}(R^{2n})$ such that $f - f_o = 0$ on K (make use of the cut-off function of the form $\chi(p_1\Lambda(x)^{\lambda_1-1},\dots,p_n\Lambda(x)^{\lambda_n-1})$). We set

$$\hat{f}\psi(x) \equiv f(\overset{2}{x},-i\overset{1}{\frac{\partial}{\partial x}})\psi(x) \overset{\text{def}}{=} f_o(\overset{2}{x},-i\overset{1}{\frac{\partial}{\partial x}})\psi(x). \tag{56}$$

Clearly $\hat{f}\psi(x)$ modulo $L^{-\infty}_{(\lambda_1,\dots,\lambda_n)}(R^{2n})$ does not depend on the choice of function f_o. It is also clear that the composition theorem, such as Theorem 2 of item B and Theorem 1 of the present item, remain valid for this extended set of pseudo-differential operators if we consider all the operators to act only on $\psi \in \overset{0}{L}{}^{\infty}_{(\lambda_1,\dots,\lambda_n)}(R^{2n})$.

The symbol space $\overset{0}{T}{}^{\infty}_{(\lambda_1,\dots,\lambda_n)}(R^{2n})$ possesses the useful property: it admits "asymptotic summation." More precisely, the following proposition is valid:

<u>Lemma 3</u>. Let $f_j(x,p) \in \overset{0}{T}{}^{m_j}_{(\lambda_1,\dots,\lambda_n)}(R^{2n})$, $j = 0,1,2,\dots$, where $m_j \to -\infty$ as $j \to \infty$. There exist $f(x,p) \in \overset{0}{T}{}^{m}_{(\lambda_1,\dots,\lambda_n)}(R^{2n})$, where $m = \max_j m_j$, such that

$$f(x,p) - \sum_{j=0}^{N} f_j(x,p) \in \overset{0}{T}{}^{\max_{j>N} m_j}_{(\lambda_1,\dots,\lambda_n)}(R^{2n}) \tag{57}$$

for any natural N.

<u>Proof</u>. This is a variant of the famous Borel lemma, proved for the classical PDO in [30]. The function $f(x,p)$ may be taken in the form

$$f(x,p) = \sum_{j=0}^{\infty} (1-\chi(\varepsilon_j\Lambda(x))f_j(x,p), \tag{58}$$

where $\chi \in C_o^{\infty}(R^1)$ equals 1 in the vicinity of zero, and the positive numbers ε_j tend to zero rapidly enough to ensure the validity of estimates

(53) and inclusions (57) (note that the sum (58) is finite for any fixed x). The details of the choice of ε_j are completely analogous to those in [30] and are omitted.

The property established in Lemma 3 enables us to obtain the asymptotic composition formulas to within operators whose image lies in the space $L^{-\infty}_{(\lambda_1,\dots,\lambda_n)}(R^{2n})$ rather than in $L^{-N}_{(\lambda_1,\dots,\lambda_n)}(R^{2n})$ for N arbitrary, but fixed previously. For example, the following theorem holds:

Theorem 3. Let $g_1 : T^*R^n_x \to T^*R^n_y$ and $g_2 : T^*R^n_y \to T^*R^n_v$ be canonical transformations satisfying all the above conditions together with their composition $g_2 \circ g_1$. Then

$$T(g_2)\circ T(g_1) = \pm(1+\hat{R})T(g_2\circ g_1), \tag{59}$$

modulo operators acting from $\overset{0}{L}{}^{\infty}_{(\lambda_1,\dots,\lambda_n)}(R^n_x)$ to $L^{-\infty}_{(\theta_1,\dots,\theta_n)}(R^n_v)$, where the sign + or - in (59) depends on the choice of the branch of the square root of the measure density on Lagrangian manifolds $\Gamma(g_1)$, $\Gamma(g_2)$, and $\Gamma(g_1g_2)$; and $\hat{R} = R(\overset{2}{v},-i\overset{1}{\frac{\partial}{\partial v}})$ is a pseudo-differential operator with the symbol $R(v,\zeta) \in \overset{0}{T}{}^{-1}_{(\theta_1,\dots,\theta_n)}(R^{2n})$ which has the following asymptotic expansion:

$$R(v,\zeta) \simeq \sum_{j=1}^{\infty} (-i)^j R_j(v,\zeta); \tag{60}$$

$R_j^{(j)} \in \overset{0}{T}{}^{-j}_{(\theta_1,\dots,\theta_n)}(R^{2n})$ is a real symbol satisfying

$$R_j(\tau^{\theta}v,\tau^{1-\theta}\zeta) = \tau^{-j}R_j(v,\zeta) \text{ for large } \Lambda(v), \tag{61}$$

and the sign $\simeq$ in (60) is used to denote that the difference between R and the sum of N terms of the series belongs to $\overset{0}{T}{}^{-N-1}_{(\theta_1,\dots,\theta_n)}(R^{2n})$.

Proof. Using Theorem 2, we come to the problem of solving the equation (for any $\psi \in \overset{0}{L}{}^{\infty}_{(\lambda_1,\dots,\lambda_n)}(R^n)$)

$$(1+\hat{R})T(g_2\circ g_1)\psi = T(g_2\circ g_1,\phi\chi)\psi \tag{62}$$

with respect to $\hat{R}$, where ϕ is a certain element of $\overset{0}{T}{}^{0}_{(\lambda_1,\dots,\lambda_n)}(R^{2n})$ independent of ψ, and χ is a cut-off function with $\{\chi = 1\} \in \mathrm{Ess}(\psi)$. Furthermore, ϕ is asymptotically quasi-homogeneous, $\phi \simeq \sum_{j=0}^{\infty}\phi_j$, where $\phi_j \in \overset{0}{T}{}^{-j}_{(\lambda_1,\dots,\lambda_n)}(R^{2n})$, $\phi_j(\tau^{\lambda}x,\tau^{1-\lambda}p) = \tau^{-j}\phi_j(x,p)$ for large $\Lambda(x)$, and $\phi_o = 1$. Set $\hat{H} = 1+\hat{R}$. Using Theorem 1, we obtain the system of equations on R_j of the form

$$\phi_o(g_2\circ g_1)^*(R_j) = F_j[\phi,R_1,\dots,R_{j-1}], \tag{63}$$

where F_j is a given differential expression. This system may be solved recurrently since $\phi_o = 1$. Applying Lemma 3, we obtain (62) and henceforth (59). Theorem 3 is proved.

We intend to finish this item with consideration of the case when the canonical transformation is defined on some R_+-invariant open subset $\mathcal{U} \subset \subset T^*R^n_x$. Let

$$g : \mathcal{U} \to T^*R^n_y \tag{64}$$

be such a transformation.

We assume that g is defined on (or may be extended to) some large open set $\tilde{\mathcal{U}} \supset \mathcal{U}$ such that there exists a function $\phi_{\mathcal{U}\tilde{\mathcal{U}}}$ with the properties*:

$$\phi_{\mathcal{U}\tilde{\mathcal{U}}} \in \overset{0}{T}{}^0_{(\lambda_1,\ldots,\lambda_n)}(R^{2n}); \; \phi_{\mathcal{U}\tilde{\mathcal{U}}} = 1 \text{ in } \mathcal{U}; \; \operatorname{supp} \phi_{\mathcal{U}\tilde{\mathcal{U}}} \subset \tilde{\mathcal{U}}. \tag{65}$$

The formula (51) is then modified and reads

$$T(g)\psi \overset{\text{def}}{=} T(g, \phi \circ \phi_{\mathcal{U}\tilde{\mathcal{U}}})\psi, \tag{66}$$

where ϕ is chosen as in Definition 4. The formula (66) makes sense for any $\psi \in \overset{0}{L}{}^\infty_{(\lambda_1,\ldots,\lambda_n)}(R^{2n})$; however, the result is independent of the choice of $\phi_{\mathcal{U}\tilde{\mathcal{U}}}$ only for $\psi \in \overset{0}{L}{}^\infty_{(\lambda_1,\ldots,\lambda_n)}(\mathcal{U})$ where

$$\overset{0}{L}{}^m_{(\lambda_1,\ldots,\lambda_n)}(\mathcal{U}) = \{\psi \in \overset{0}{L}{}^m_{(\lambda_1,\ldots,\lambda_n)}(R^n) \,|\, \exists K \subset \mathcal{U} : K \in \operatorname{Ess}(\psi)\} \tag{67}$$

(here $m \in R \cup \{\infty\}$). Let now $g_1 = \mathcal{U}_1 \to T^*R^n_y$, $\mathcal{U}_1 \subset T^*R^n_x$ and $g_2 : \mathcal{U}_2 \to T^*R^n_v$, $\mathcal{U}_2 \subset T^*R^n_y$ be the canonical transformations satisfying the above conditions together with

$$g_2 \circ g_1 : \mathcal{U}_1 \cap g_1^{-1}(\mathcal{U}_2) \to T^*R^n_v. \tag{68}$$

The statement of Theorem 3 remains valid if both sides of (59) are considered as operators acting on elements of the space $\overset{0}{L}{}^\infty_{(\lambda_1,\ldots,\lambda_n)}(\mathcal{U}_1 \cap g_1^{-1} \times \times (\mathcal{U}_2))$.

D. The Canonical Sheaf F_+

In Section 2 of the present chapter we defined the sheaf of rapidly oscillating functions on the symplectic manifold M in terms of factor-spaces, their elements being the equivalence classes, etc. It seems to be the most convenient way in view to clarify the exposition as far as possible.

Here we choose another approach. We work in terms of local representatives rather than their equivalence classes and construct only the space of sections of the sheaf over the whole symplectic manifold M. This is because we have in mind our general aim, to apply the described results to the theory of operator equations. There is no other way (known to the authors) to do this except working with representatives, since we shall have to substitute in the latter the tuples of non-commuting operators instead of their arguments.

On the other hand, the experience induced by the material of Section 2 enables the reader to think over the homogeneous situation from a different viewpoint which, however, appears to be ill-adapted to the mentioned applications. We begin with the notion of a proper symplectic R_+-manifold.

---

* In this case $(\mathcal{U},\tilde{\mathcal{U}})$ will be called a proper pair.

Definition 1. M is a proper symplectic R_+-manifold if and only if the following conditions are satisfied:

(a) The free infinitely differentiable action of the group R_+ on M is given such that the set of orbits M/R_+ admits the structure of a smooth manifold for which the projection $M \to M/R_+$ is a smooth mapping.

(b) The closed non-degenerate differential 2-form ω^2 on M is given, homogeneous of degree one:

$$\tau^* \omega^2 = \tau\omega^2, \quad \tau \in R_+. \tag{1}$$

(c) The smooth non-vanishing function Λ_M, homogeneous of degree one, is given:

$$\Lambda_M(\tau z) = \tau\Lambda_M(z), \quad \tau \in R_+, \quad z \in M. \tag{2}$$

Λ_M will be called the weight function on M.

(d) The finite atlas $M = \bigcup_{\alpha=1}^{N_o} \tilde{u}_a$ of connected shrinkable coordinate charts $\{\tilde{u}_\alpha, \mu_\alpha : \tilde{u}_\alpha \to \tilde{\mathcal{D}}_\alpha \subset R^{2n}\}_{\alpha=1}^{M_o}$ is given* with the properties:

(d_1) For each α the non-zero n-tuple $(\lambda_1,\dots,\lambda_n) = (\lambda_1(\alpha),\dots,\lambda_n(\alpha))$ of non-negative numbers is given and μ_α is a homogeneous canonical transformation with respect to action of R_+ on R^{2n}, defined by this tuple:

$$\mu_\alpha^*(dp^{(\alpha)} \wedge dx^{(\alpha)}) = \omega^2, \tag{3}$$

$$\begin{aligned} x_j^{(\alpha)}(\tau z) &= \tau^{\lambda_j(\alpha)} x_j^{(\alpha)}(z), \\ p_j^{(\alpha)}(\tau z) &= \tau^{1-\lambda_j(\alpha)} p_j^{(\alpha)}(z), \quad j = 1,\dots,n, \quad \tau \in R_+, \quad z \in \tilde{u}_\alpha. \end{aligned} \tag{4}$$

Here $(x_1^{(\alpha)},\dots,x_n^{(\alpha)},p_1^{(\alpha)},\dots,p_n^{(\alpha)})$ are the coordinates in $R^{2n} \supset \tilde{\mathcal{D}}_\alpha$, $(x_1^{(\alpha)}(z), \dots, x_n^{(\alpha)}(z), p_1^{(\alpha)}(z),\dots,p_n^{(\alpha)}(z))$ are the components of the mapping μ_α.

(d_2) For each α the function $\Lambda_M(z)$ and $\mu_\alpha^*\Lambda_\alpha(z)$, where $\Lambda_\alpha(x^{(\alpha)})$ is constructed in correpondence with the tuple, are equivalent on $\tilde{u}_\alpha$ in the sense that there exist positive constants c and C such that for $f_o\Lambda_M(z) \geqslant 1$

$$c\Lambda_M(z) \leqslant \mu_\alpha^*\Lambda_\alpha(z) \leqslant C\Lambda_M(z), \quad z \in \tilde{u}_\alpha. \tag{5}$$

(d_3) All non-empty intersections of charts $\tilde{u}_\alpha$ are shrinkable sets (in other words, the covering $\tilde{u}_\alpha$ satisfies the conditions of Leray's theorem), and if the intersection $\tilde{u}_\alpha \cap \tilde{u}_\beta$ is non-empty then the mapping

$$\gamma_{\beta\alpha} = \mu_\beta\mu_\alpha^{-1} : \mu_\alpha(\tilde{u}_\alpha \cap \tilde{u}_\beta) \to \mu_\beta(\tilde{u}_\alpha \cap \tilde{u}_\beta) \tag{6}$$

is a proper R_+-invariant canonical transformation satisfying the conditions of item C.

(d_4) There exists an inscribed covering $M = \bigcup_\alpha \mathcal{U}_\alpha$, $\mathcal{U}_\alpha \subset \tilde{\mathcal{U}}_\alpha$ satisfying all the above conditions such that for any α $(\mathcal{D}_\alpha, \tilde{\mathcal{D}}_\alpha)$, where $\mathcal{D}_\alpha = \mu_\alpha(\mathcal{U}_\alpha)$, is a proper pair (see last footnote in item C).

---

* Our methods can be extended to consider locally finite atlases.

(d_5) There exists a partition of unity $\{\theta_\alpha\}_{\alpha=1}^{M_0}$, subordinate to the covering u_α, such that θ_α is a homogeneous function of degree zero for large $\Lambda_M(z)$ and $(\mu_\alpha^{-1})^*\theta_\alpha \in T^0_{(\lambda_1(\alpha),\dots,\lambda_n(\alpha))}(R^{2n})$. Here saying that $\{\theta_\alpha\}$ is a partition of unity we mean that $\sum_\alpha \theta_\alpha = 1$ for $\Lambda_M(z) \geqslant$ const. This is because we intend to avoid (where possible) explicit use of the numerous cut-off functions.)

<u>Example 1</u>. Let an n-tuple $(\lambda_1,\dots,\lambda_n)$ be given. The space R^{2n} with the action of R_+ given by $\tau(x,p) = (\tau^{\lambda_1}x_1,\dots,\tau^{\lambda_n}x_n,\tau^{1-\lambda_1}p_1,\dots,\tau^{1-\lambda_n}p_n)$, the form $\omega^2 = dp \wedge dx$, and the atlas, consisting of exactly one chart, do not satisfy the conditions of Definition 1. However, if we delete the set $\{x_{I_+} = 0\}$, the space $R^{2n} \setminus \{(x,p) | x_{I_+} = 0\}$ gives a simple example illustrating the definition.

Next we should define transition operators. For the sake of convenience and in order to shorten the formulas, we introduce the following notations:

$$\overset{0}{L}{}^m_\alpha = \overset{0}{L}{}^m_{(\lambda_1(\alpha),\dots,\lambda_n(\alpha))}(\mathcal{D}_\alpha),$$
$$\overset{0}{L}{}^m_{\alpha\beta} = \overset{0}{L}{}^m_{(\lambda_1(\alpha),\dots,\lambda_n(\alpha))}(\mu_\alpha(u_\alpha \cap u_\beta)), \tag{7}$$
$$\overset{0}{L}{}^m_{\alpha\beta\delta} = \overset{0}{L}{}^m_{(\lambda_1(\alpha),\dots,\lambda_n(\alpha))}(\mu_\alpha(u_\alpha \cap u_\beta \cap u_\delta)),$$

etc., where $m \in R \cup \{\infty\}$. Denote also

$$L^{-\infty}_\alpha \equiv L^{-\infty}_{(\lambda_1(\alpha),\dots,\lambda_n(\alpha))}(R^n_x(\alpha)). \tag{8}$$

We assume that the partitions of unity subordinate to canonical coverings on graphs $\Gamma(\gamma_{\beta\alpha})$ are chosen and fixed for all $\alpha_1\beta$ such that $u_\alpha \cap u_p \neq \emptyset$. Set $T_{\alpha\alpha} = 1$ and for $\beta \neq \alpha$ for $u_\alpha \cap u_\beta \neq \emptyset$

$$\tilde{T}_{\beta\alpha} = T(\gamma_{\beta\alpha}) : \overset{0}{L}{}^\infty_{\alpha\beta} \to \overset{0}{L}{}^\infty_{\beta\alpha}, \tag{9}$$

where $T(\gamma_{\beta\alpha})$ is given by Definition 4 of item C.

<u>Lemma 1</u>. The operators $\tilde{T}_{\beta\alpha}$ of (9) satisfy the conditions: for any non-empty intersection $u_\alpha \cap u_\beta \cap u_\gamma$ there exists an asymptotically homogeneous symbol

$$R_{\alpha\beta\gamma}(x^{(\alpha)},p^{(\alpha)}) \in \overset{0}{T}{}^{-1}_{(\lambda_1(\alpha),\dots,\lambda_n(\alpha)}(R^{2n}), \tag{10}$$

having expansion of the form ((60) of item C), and integer number $\varepsilon_{\alpha\beta\gamma}$ such that the equality holds

$$\tilde{T}_{\alpha\beta} \circ \tilde{T}_{\beta\gamma} = (1 + \hat{R}_{\alpha\beta\gamma}) \circ \tilde{T}_{\alpha\gamma}\exp(i\pi\varepsilon_{\alpha\beta\gamma}), \tag{11}$$

modulo operators acting from $\overset{0}{L}{}^\infty_{\gamma\alpha\beta}$ to $L^{-\infty}_\alpha$. $\{\varepsilon_{\alpha\beta\gamma}\}$ is a (Čech) 2-cocycle on M.

<u>Proof</u>. We readily obtain (11), applying Theorem 3 of item C. The cochain $\varepsilon_{\alpha\beta\gamma}$ is none other than the cochain $\varepsilon^{(2)}$ from Section 2; it is a cocycle according to Lemma 1 of Section 2:C; we denote the corresponding cohomology class by

$$[\varepsilon] \in H^2(M,Z). \tag{12}$$

Definition 2. A proper symplectic R_+-manifold M is quantized or satisfies the quantization condition if the cohomology class $[\varepsilon]$ is trivial modulo 2.

We assume that M is quantized. Then there exists an integer 1-cochain $\{\mu_{\alpha\beta}\}$ of the covering $\{u_\alpha\}$ such that $\varepsilon_{\alpha\beta\gamma}$ is a coboundary of $\mu_{\alpha\beta}$,

$$\varepsilon_{\alpha\beta\gamma} = \mu_{\alpha\beta} - \mu_{\alpha\gamma} + \mu_{\beta\gamma} \pmod 2. \tag{13}$$

Replacing $\tilde{T}_{\alpha\beta}$ by $\tilde{T}_{\alpha\beta}\exp(-i\pi\mu_{\alpha\beta})$, that is, making a proper choice of arguments of measure density on graphs $\Gamma(\gamma_{\alpha\beta})$, we cancel out the factor of $\exp(i\pi\varepsilon_{\alpha\beta\gamma})$ in the equality (11). From now on we assume that M is quantized and the arguments of the measure are chosen in such a way that (11) holds with $\varepsilon_{\alpha\beta\gamma} = 0$. It appears that the factor $(1 + \hat{R}_{\alpha\beta\gamma})$ also can be eliminated by means of an appropriate adjustment of the operators $\tilde{T}_{\alpha\beta}$.

Theorem 1. There exist symbols $Q_{\alpha\beta}(x^{(\beta)},p^{(\beta)}) \in \overset{0}{T}{}^{-1}_{(\lambda_1(\beta),\ldots,\lambda_n(\beta))} \times (R^{2n})$, having the expansion

$$Q_{\alpha\beta} \simeq \sum_{j=1}^{\infty} Q_{\alpha\beta j}(x^{(\beta)},p^{(\beta)}) \tag{14}$$

where $Q_{\alpha\beta j} \in \overset{0}{T}{}^{-j}_{(\lambda_1(\beta),\ldots,\lambda_n(\beta))}(R^{2n})$ is a real-valued symbol satisfying

$$Q_{\alpha\beta j}(\tau^{\lambda(\beta)}x^{(\beta)},\tau^{1-\lambda(\beta)}p^{(\beta)}) = \tau^{-j}Q_{\alpha\beta j}(x^{(\beta)},p^{(\beta)}), \tag{15}$$

such that the "adjusted" operators

$$T_{\alpha\beta} = \tilde{T}_{\alpha\beta}(1 + \hat{Q}_{\alpha\beta}) \tag{16}$$

satisfy the cocyclicity conditions

$$T_{\alpha\beta} \circ T_{\beta\gamma} = T_{\alpha\gamma} \text{ for } U_\alpha \cap U_\beta \cap U_\gamma \neq \emptyset, \tag{17}$$

modulo operators acting from $\overset{0}{L}{}^{\infty}_{\gamma\alpha\beta}$ to $L^{-\infty}_{\alpha}$.

Proof. (a) Preliminary stage. We first construct the operators

$$T^{(0)}_{\alpha\beta} = \tilde{T}_{\alpha\beta}(1 + \hat{Q}^{(0)}_{\alpha\beta}),\ Q^{(0)}_{\alpha\beta} \in \overset{0}{T}{}^{-1}_{(\lambda_1(\beta),\ldots,\lambda_n(\beta))}(R^{2n}), \tag{18}$$

satisfying the properties:

$$T^{(0)}_{\alpha\beta}T^{(0)}_{\beta\gamma} = (1 + \hat{R}^{(0)}_{\alpha\beta\gamma})T^{(0)}_{\alpha\gamma},\ R^{(0)}_{\alpha\beta\gamma} \in \overset{0}{T}{}^{-1}_{(\lambda_1(\alpha),\ldots,\lambda_n(\alpha))}(R^{2n}), \tag{19}$$

modulo operators acting from $\overset{0}{L}{}^{\infty}_{\gamma\alpha\beta}$ to $L^{-\infty}_{\alpha}$;

$$T^{(0)}_{\alpha\beta}T^{(0)}_{\beta\alpha} = 1, \tag{20}$$

modulo operators acting from $\overset{0}{L}_{\alpha}$ to $\overset{0}{L}{}^{-\infty}_{\alpha}$. To do this, set

$$Q^{(0)}_{\alpha\beta} = 0 \text{ for } \beta < \alpha, \tag{21}$$

and for $\beta > \alpha$ consider the equation on $\hat{Q}^{(0)}_{\alpha\beta}$

$$\tilde{T}_{\beta\alpha} T_{\alpha\beta}(1 + \hat{Q}^{(0)}_{\alpha\beta}) = 1, \tag{22}$$

or

$$(1 + \hat{R}_{\beta\alpha\beta})(1 + \hat{Q}^{(0)}_{\alpha\beta}) = 1. \tag{23}$$

(Here and below we do not write the moduli to within which the equations are valid since they are obvious from above considerations.) We write formally the Neumann series

$$Q^{(0)}_{\alpha\beta} = 1 - \hat{R}_{\beta\alpha\beta} + (\hat{R}_{\beta\alpha\beta})^2 - (\hat{R}_{\beta\alpha\beta})^3 \dots \tag{24}$$

Using Theorem 2 of item A and Lemma 3 of item C, we sum this series asymptotically and come to (20). It follows from composition formulas that (19) is also valid.

(b) Induction process. Assume now that for some N we have constructed the operators

$$T^{(N)}_{\alpha\beta} = \tilde{T}_{\alpha\beta}(1 + \hat{Q}^{(N)}_{\alpha\beta}), \quad \hat{Q}^{(N)}_{\alpha\beta} \in \overset{0}{T}{}^{-1}_{(\lambda_1(\beta),\dots,\lambda_n(\beta))}(R^{2n}), \tag{25}$$

so that

$$T^{(N)}_{\alpha\beta} T^{(N)}_{\beta\alpha} = 1 \tag{26}$$

and

$$T^{(N)}_{\alpha\beta} T^{(N)}_{\beta\gamma} = (1 + \hat{R}^{(N)}_{\alpha\beta\gamma}) T^{(N)}_{\alpha\gamma}, \quad R^{(N)}_{\alpha\beta\gamma} \in \overset{0}{T}{}^{-N-1}_{(\lambda_1(\alpha),\dots,\lambda_n(\alpha))}(R^{2n}). \tag{27}$$

The problem is to pass from N to N + 1. We seek $T^{(N+1)}_{\alpha\beta}$ in the form

$$T^{(N+1)}_{\alpha\beta} = T^{(N)}_{\alpha\beta}(1 + \Delta\hat{Q}^{(N+1)}_{\alpha\beta}), \quad \Delta Q^{(N+1)}_{\alpha\beta} \in \overset{0}{T}{}^{-N-1}_{(\lambda_1(\beta),\dots,\lambda_n(\beta))}(R^{2n}) \tag{28}$$

so that

$$Q^{(N+1)}_{\alpha\beta} - Q^{(N)}_{\alpha\beta} \in \overset{0}{T}{}^{-N-1}_{(\lambda_1(\beta),\dots,\lambda_n(\beta))}(R^{2n}). \tag{29}$$

Recall that all the symbols involved are asymptotically homogeneous, i.e., admit expansions of the type (60) of item C.

Let $r^{(N)}_{\alpha\beta\gamma}$ denote the principal homogeneous term of $R^{(N)}_{\alpha\beta\gamma}$, and $\Delta q^{(N+1)}_{\alpha\beta}$ denote that of $\Delta Q^{(N+1)}_{\alpha\beta}$; set also

$$\begin{aligned} \rho^{(N)}_{\alpha\beta\gamma} &= \mu^*_\alpha(r^{(N)}_{\alpha\beta\gamma}), \\ \sigma^{(N+1)}_{\alpha\beta} &= \mu^*_\beta(\Delta q^{(N+1)}_{\alpha\beta}). \end{aligned} \tag{30}$$

Calculate the composition $T^{(N+1)}_{\alpha\beta} \circ T^{(N+1)}_{\beta\alpha}$. We denote by z with various subscripts the terms of lower order in our expansions. We have, using the composition formulas,

$$T_{\alpha\beta}^{(N+1)} \circ T_{\beta\gamma}^{(N+1)} = T_{\alpha\beta}^{(N)}(1 + \Delta\hat{Q}_{\alpha\beta}^{(N+1)})T_{\beta\gamma}^{(N)}(1 + \Delta\hat{Q}_{\beta\gamma}^{(N+1)}) =$$

$$= (1 + (\mu_\alpha^{-1})^*(\rho_{\alpha\beta\gamma}^{(N)} + \sigma_{\alpha\beta}^{(N+1)} + \sigma_{\beta\gamma}^{(N+1)}) + \hat{z}_1)\hat{T}_{\alpha\gamma}^{(N)} = \tag{31}$$

$$= (1 + (\mu_\alpha^{-1})^*(\rho_{\alpha\beta\gamma}^{(N)} + \sigma_{\alpha\beta}^{(N+1)} + \sigma_{\beta\gamma}^{(N+1)} - \sigma_{\alpha\gamma}^{(N+1)}) + \hat{z}_2)\hat{T}_{\alpha\gamma}^{(N+1)}$$

(we omit the detailed calculations). Thus, we need to solve the system of equations

$$\sigma_{\alpha\beta}^{(N+1)} - \sigma_{\alpha\gamma}^{(N+1)} + \sigma_{\beta\gamma}^{(N+1)} = \rho_{\alpha\beta\gamma}^{(N)} \text{ in } \mathcal{U}_\alpha \cap \mathcal{U}_\beta \cap \mathcal{U}_\gamma \tag{32}$$

on M.

Lemma 2. $\rho_{\alpha\beta\gamma}^{(N)}$ is a cocycle on M, i.e., it is antisymmetric with respect to its indices and for $\mathcal{U}_\alpha \cap \mathcal{U}_\beta \cap \mathcal{U}_\gamma \cap \mathcal{U}_\delta$

$$\rho_{\alpha\beta\gamma}^{(N)} - \rho_{\alpha\beta\delta}^{(N)} + \rho_{\alpha\gamma\delta}^{(N)} - \rho_{\beta\gamma\delta}^{(N)} = 0 \text{ in } \mathcal{U}_\alpha \cap \mathcal{U}_\beta \cap \mathcal{U}_\gamma \cap \mathcal{U}_\delta. \tag{33}$$

Proof. For any non-empty intersection $u_{\alpha\beta\gamma\delta} = u_\alpha \cap u_\beta \cap u_\gamma \cap u_\delta$, consider the product

$$T_{\alpha\beta\gamma\delta} = T_{\alpha\beta}^{(N)} \circ T_{\beta\gamma}^{(N)} \circ T_{\gamma\delta}^{(N)}. \tag{34}$$

We may calculate this product in two different ways:

$$\begin{aligned} T_{\alpha\beta\gamma\delta} &= (1 - i\hat{r}_{\alpha\beta\gamma}^{(N)} + \hat{z}_3)T_{\alpha\beta}^{(N)} \circ T_{\gamma\delta}^{(N)} = \\ &= (1 - i\hat{r}_{\alpha\beta\gamma}^{(N)} + \hat{z}_3)(1 - i\hat{r}_{\alpha\gamma\delta}^{(N)} + \hat{z}_4)T_{\alpha\delta}^{(N)} = (1 - i\hat{r}_{\alpha\beta\gamma}^{(N)} - i\hat{r}_{\alpha\gamma\delta}^{(N)} + \hat{z}_5)T_{\alpha\delta}^{(N)}; \end{aligned} \tag{35}$$

on the other hand,

$$\begin{aligned} T_{\alpha\beta\gamma\delta} &= T_{\alpha\beta}^{(N)}(1 - i\hat{r}_{\beta\gamma\delta}^{(N)} + \hat{z}_6)T_{\beta\delta}^{(N)} = \\ &= [1 - i(\overset{*}{\gamma}_{\beta\alpha} r_{\beta\gamma\delta}^{(N)}) + \hat{z}_7]T_{\alpha\beta}^{(N)} \circ T_{\beta\delta}^{(N)} = [1 - i(\overset{*}{\gamma}_{\beta\alpha} r_{\beta\gamma\delta}^{(N)}) - i\hat{r}_{\alpha\beta\delta}^{(N)} + \hat{z}_8]T_{\alpha\delta}^{(N)}. \end{aligned} \tag{36}$$

Multiplying by $T_{\delta\alpha}^{(N)}$ from the right, we obtain

$$\hat{r}_{\alpha\beta\gamma}^{(N)} + \hat{r}_{\alpha\gamma\delta}^{(N)} - \hat{r}_{\alpha\beta\delta}^{(N)} - (\overset{*}{\gamma}_{\beta\alpha} r_{\beta\gamma\delta}^{(N)}) + \hat{z}_8 = 0. \tag{37}$$

We claim that (37) yields

$$r_{\alpha\beta\gamma}^{(N)} + r_{\alpha\gamma\delta}^{(N)} - r_{\alpha\beta\delta}^{(N)} - \overset{*}{\gamma}_{\beta\alpha} r_{\beta\gamma\delta}^{(N)} = 0 \tag{38}$$

in $\mu_\alpha(u_\alpha \cap u_\beta \cap u_\gamma \cap u_\delta)$.

Indeed, denote the left-hand side of (38) by $F(x^{(\alpha)}, p^{(\alpha)})$. $F(x^{(\alpha)}, p^{(\alpha)}) \in \overset{0}{\in} T^{-1}_{(\lambda_1(\alpha),\ldots,\lambda_n(\alpha))}(R^{2n})$ and is quasi-homogeneous of degree 1 for large $\Lambda_\alpha(x)$. Then if $F(x_o^{(\alpha)}, p_o^{(\alpha)}) \neq 0$ for some $(x_o^{(\alpha)}, p_o^{(\alpha)}) \in \mu_\alpha(u_\alpha \cap u_\beta \cap u_\gamma \cap u_\delta)$ consider the function

$$\psi(x) = \exp(iS(x^{(\alpha)}))\phi(x), \tag{39}$$

where $S \in \mathcal{O}^1_{(\lambda_1(\alpha),\ldots,\lambda_n(\alpha))}(R^n)$ is a real-valued function satisfying

$$\frac{\partial S}{\partial x}(x_o^{(\alpha)}) = p_o^{(\alpha)}, \tag{40}$$

and $\phi(x) \in S^0_{(\lambda_1(\alpha),\ldots,\lambda_n(\alpha))}(R^n)$ is a cut-off function in the small R_+-invariant neighborhood of $x_o^{(\alpha)}$. Substituting ψ into (37) yields a contradiction in view of Theorem 1 of item A, Theorem 2 of item B, and Lemma 1 (c) of item C. Passing to functions on M, (38) gives (33). The antisymmetry follows from (26). Lemma 2 is proved.

It is well known from the theory of sheaves that equations (32) are solvable on M. One may set

$$\sigma_{\alpha\beta}^{(N+1)} = \sum_\delta \rho_{\alpha\beta\delta}^{(N)} \theta_\delta, \tag{41}$$

where θ_δ is the partition of unity described in Definition 1. It is easy to establish that the function $(\mu_\beta^*)^{-1}(\sigma_{\alpha\beta}^{(N+1)})$ may be prolonged to a function from $\overset{0}{T}{}^{-N-1}_{(\lambda_1(\beta),\ldots,\lambda_n(\beta))}$. For arbitrary choice of lower-order terms in $\Delta Q_{\alpha\beta}^{(N+1)}$ the operators (28) satisfy

$$T_{\alpha\beta}^{(N+1)} \circ T_{\beta\gamma}^{(N+1)} = (1 + \hat{R}_{\alpha\beta\gamma}^{(N+1)}) T_{\alpha\gamma}^{(N+1)} \tag{42}$$

with $R_{\alpha\beta\gamma}^{(N+1)} \in \overset{0}{T}{}^{-N-2}_{(\lambda_1(\alpha),\ldots,\lambda_n(\alpha))}(R^{2n})$. We have, in particular,

$$T_{\alpha\beta}^{(N+1)} T_{\beta\alpha}^{(N+1)} = 1 + \hat{R}_{\alpha\beta\alpha}^{(N+1)}. \tag{43}$$

Adjusting $T_{\beta\alpha}^{(N+1)}$ for $\alpha < \beta$ with the help of the Neumann series analogous to (24), we obtain

$$T_{\alpha\beta}^{(N+1)} T_{\beta\alpha}^{(N+1)} = 1, \tag{44}$$

while (42) remains valid. Thus the induction process is performed. It remains to set

$$Q_{\alpha\beta} \simeq Q_{\alpha\beta}^{(0)} + \sum_{N=1}^{\infty} (Q_{\alpha\beta}^{(N)} - Q_{\alpha\beta}^{(N-1)}) \tag{45}$$

(the sum of asymptotic series is obtained via Lemma 3 of item C). Theorem 1 is proved.

We come to the definition of the sheaf $\mathcal{F}_+$; more precisely, the space $\mathcal{F}_+(M)$ of sections of $\mathcal{F}_+$ over M.

Now consider the set of tuples $\{\psi_\alpha\}_{\alpha=1}^{N_o}$, where

$$\psi_\alpha \in \overset{0}{L}{}^\infty_\alpha. \tag{46}$$

We say that the tuple $\{\psi_\alpha\}$ is equivalent to zero, if for any α the sum

$$F_\alpha = \sum_\gamma T_{\alpha\gamma}\psi_\gamma \in \overset{0}{L}{}^\infty_{(\lambda_1(\alpha),\ldots,\lambda_n(\alpha))}(R^n_{x(\alpha)}) \tag{47}$$

satisfies the property: there exists a set $K_\alpha \subset R^n_{x(\alpha)} \setminus \mathcal{D}_\alpha$ such that

$$K_\alpha \in \mathrm{Ess}(F_\alpha). \tag{48}$$

Definition 3. $F_+(M)$ is a factor-space

$$F_+(M) = \prod_\alpha \overset{0}{L}{}^{\infty}_\alpha / \{\{\psi_\alpha\} \text{ equivalent to } 0\}. \tag{49}$$

Also by $F^m_+(M)$ we denote the space

$$F^m_+(M) = \prod_\alpha \overset{0}{L}{}^{m}_\alpha / \{\{\psi_\alpha\} \text{ equivalent to } 0\}. \tag{50}$$

4. PSEUDO-DIFFERENTIAL OPERATORS AND THE CAUCHY PROBLEM IN THE SPACE $F_+(M)$

In this section, the notions of pseudo-differential equations and the Cauchy problem in the space of sections of the sheaf, constructed in the previous sections of a canonical sheaf on the basis of this discussion, the theorem on sufficient conditions for the asymptotic solvability of the Cauchy problem for a pseudo-differential equation of the first order is established.

A. Pseudo-Differential Operators in the Space $F_+(M)$

Let M be a proper quantized symplectic R_+-manifold, and let $F_+(M)$ be the space of sections of a canonical sheaf on M, constructed in the previous section. To define pseudo-differential operators in $F_+(M)$, first of all turn to local representation. Let $\hat{H} = \{\hat{H}_\alpha\}$ be a tuple of pseudo-differential operators, $\hat{H}_\alpha$ being an operator in $\overset{0}{L}{}^{\infty}_{(\lambda_1(\alpha),\ldots,\lambda_n(\alpha))}(R^n_{x}(\alpha))$ with the symbol

$$H_\alpha(x^{(\alpha)}, p^{(\alpha)}) \in \overset{0}{T}{}^{\infty}_{(\lambda_1(\alpha),\ldots,\lambda_n(\alpha))}(R^{2n}). \tag{1}$$

Let $\psi \in F_+(M)$; $\{\Psi_\alpha\} \subset \prod_\alpha \overset{0}{L}{}^{\infty}_\alpha$ being a representative of ψ. Consider the tuple

$$\hat{H}\{\Psi_\alpha\} \overset{\text{def}}{=} \{\hat{H}_\alpha \Psi_\alpha\}. \tag{2}$$

$\hat{H}$ correctly defines an operator in the space $F_+(M)$ if and only if the equivalence class of the tuple (2) depends only on ψ or, in other words, if $\{\Psi_\alpha\} \sim 0$ implies $\{\hat{H}_\alpha \Psi_\alpha\} \sim 0$.

We assume that the operators $\hat{H}_\alpha$ satisfy the compatibility condition: for any α,β with $U_\alpha \cap U_\beta \neq 0$

$$\hat{H}_\alpha T_{\alpha\beta} = T_{\alpha\beta} \hat{H}_\beta, \tag{3}$$

modulo operators acting from $\overset{0}{L}{}^{\infty}_{\beta\alpha}$ to $L^{-\infty}_\alpha$. Let $\{\Psi\}$ define a zero class in $F_+(M)$. We have

$$\sum_\alpha T_{\gamma\alpha} \hat{H}_\alpha \Psi_\alpha = \hat{H}_{\gamma} \sum_\alpha T_{\gamma\alpha} \psi_\alpha + \sum_\alpha (T_{\gamma\alpha} \hat{H}_\alpha - \hat{H}_\gamma T_{\gamma\alpha}) \psi_\alpha \equiv \Phi_1 + \Phi_2. \tag{4}$$

As for the first summand in (4), there exists $K \subset R^{2n} \setminus D_\gamma$ such that $K \in$ $\in \text{Ess } \Phi_1$. Indeed, this is valid for $\sum_\alpha T_{\gamma\alpha} \psi_\alpha$ and the pseudo-differential operator does not enlarge essential subsets (see Lemma 1 (d) of item C). As for the second summand, the condition (3) yields that the operator $T_{\gamma\alpha} \hat{H}_\alpha - \hat{H}_\gamma T_{\gamma\alpha}$ may be represented in the form

$$T_{\gamma\alpha} \hat{H}_\alpha - \hat{H}_\gamma T_{\gamma\alpha} = T(\gamma_{\gamma\alpha}, \phi), \tag{5}$$

where $\phi \in \overset{0}{T}{}^{\infty}_{(\lambda_1(\alpha),\ldots,\lambda_n(\alpha))}(R^{2n})$ and

$$\text{supp } \phi \cap \overline{\mu_\alpha(U_\alpha \cap U_\gamma)} = \emptyset \tag{6}$$

(we omit in (5) the cut-off factor depending on the function to which $T(\gamma_{\gamma\alpha},\phi)$ is applied).

From the statements on the behavior of the essential sets (see Proposition 4 of item C) it follows that there is $K_\alpha \in \mathrm{Ess}((T_{\gamma\alpha}\hat{H}_\alpha - \hat{H}_\gamma T_{\gamma\alpha})\psi_\alpha)$ such that $K_\alpha \subset R^{2n}\setminus \mathcal{D}_\gamma$. Since the sum (4) is finite, we have $K_0 \equiv K \cup \cup(\bigcup_\alpha K_\alpha) \in \mathrm{Ess}(\Phi_1 + \Phi_2)$, $K_0 \subset R^{2n}\setminus\mathcal{D}_\gamma$.

Thus, the tuple $\{\hat{H}_\alpha \Psi_\alpha\}$ is equivalent to zero, and the mapping (2) gives rise to the mapping of equivalence classes, i.e., of elements of $F_+(M)$. We come to the following natural definition:

Definition 1. The pseudo-differential operator in $F_+(M)$ is a linear mapping

$$\hat{H} : F_+(M) \to F_+(M) \tag{7}$$

induced by the mapping (2) of the representatives, where the tuple $\{\hat{H}_\alpha\}$ satisfies the compatibility condition (3). The operator (7) and the tuple $\{\hat{H}_\alpha\}$ are denoted by the same letter, but there should be no confusion since everything is clear from the context.

We denote by $T^m(M)$ the space of pseudo-differential operators $\hat{H}$ for which $\hat{H}_\alpha \in \hat{T}^m_{(\lambda_1(\alpha),\dots,\lambda_n(\alpha))}(R^{2n})$. It is obvious that if $\hat{H} \in \hat{T}^m(M)$, then $\hat{H}$ acts in the spaces

$$\hat{H} : F^s_+(M) \to F^{s+m+\delta}_+(M) \tag{8}$$

for any $s \in R$ and any $\delta > 0$.

Next we intend to define several global objects associated with pseudo-differential operators. To perform this, we restrict ourselves to the consideration of "classical" pseudo-differential operators. By saying that the pseudo-differential operator $H(\overset{2}{x},-i\overset{1}{\frac{\partial}{\partial x}})$ is classical of order m, we mean here that its symbol $H(x,p) \in \overset{0}{T}{}^m_{(\lambda_1,\dots,\lambda_n)}(R^{2n})$ has an asymptotic expansion of the form:

$$H(x,p) \simeq \sum_{j=0}^{\infty} H_j(x,p), \tag{9}$$

where $H_j(x,p) \in \overset{0}{T}{}^{m-j}_{(\lambda_1,\dots,\lambda_n)}(R^{2n})$ and

$$H_j(\tau^\lambda x, \tau^{1-\lambda}p) = \tau^{m-j}H_j(x,p) \tag{10}$$

for $\tau \geqslant 1$, $\Lambda(x)$ large enough. The operator $\hat{H} \in \hat{T}^m(M)$ is classical if each $\hat{H}_\alpha$ is a classical pseudo-differential operator.

Introduce some function spaces on M. The space of functions satisfying (9) and (10) will be denoted by $\overset{0}{P}{}^m_{(\lambda_1,\dots,\lambda_n)}(R^{2n})$; set also $\overset{0}{P}{}^\infty_{(\lambda_1,\dots,\lambda_n)} \times (R^{2n}) = \bigcup_m \overset{0}{P}{}^n_{(\lambda_1,\dots,\lambda_n)}(R^{2n})$, $\overset{0}{P}{}^{-\infty}_{(\lambda_1,\dots,\lambda_n)}(R^{2n}) = \bigcap_m \overset{0}{P}{}^m_{(\lambda_1,\dots,\lambda_n)}(R^{2n})$.

We denote by $\overset{0}{P}{}^m(M)$ the space of functions such that for any α the function $(\mu^*_\alpha)^{-1}(f)$ defined in $\tilde{\mathcal{D}}_\alpha$ may be prolonged to a function belonging to $\overset{0}{P}{}^{m}_{(\lambda_1(\alpha),\ldots,\lambda_n(\alpha))}(R^{2n})$ (here $m \in R \cup \{-\infty,+\infty\}$). Then by $\overset{0}{P}{}^{m}(u)$, $u \subset M$ we denote the space of functions on u, which are restrictions of elements of $\overset{0}{P}{}^{m}(M)$. Properties of the transition homomorphisms $\gamma_{\beta\alpha}$ (see Definition 1 of Section 3:D) make it evident that

$$\overset{0}{P}{}^{m}(u_\alpha) = \mu_\alpha^*(\overset{0}{P}{}^{m}_{(\lambda_1(\alpha),\ldots,\lambda_n(\alpha))}(R^{2n}).$$

Let $\hat{H} : F_+(M) \to F_+(M)$ be a classical pseudo-differential operator of order m. Set

$$h_\alpha = \mu_\alpha^*(H_\alpha) \in \overset{0}{P}{}^{m}(u_\alpha). \tag{11}$$

The compatibility condition (3) yields, via composition theorems (Section 3:C), certain conditions on the functions h_α. These conditions read

$$h_\alpha \simeq t_{\alpha\beta}h_\beta \text{ in } U_\alpha \cap U_\beta; \tag{12}$$

here $t_{\alpha\beta}$ is an asymptotic differential operator:

$$t_{\alpha\beta} \simeq \sum_{j=0}^{\infty} (-i)^j t_{\alpha\beta j}, \tag{13}$$

where

$$t_{\alpha\beta j} : \overset{0}{P}{}^{s}(u_\alpha \cap u_\beta) \to P^{s-j}(u_\alpha \cap u_\beta) \text{ for any s.} \tag{14}$$

$t_{\alpha\beta j}$ is a homogeneous of degree $-j$ differential operator of order $\leqslant r_j$ with real coefficients in $u_\alpha \cap u_\beta$, $t_{\alpha\beta o} = 1$. Consider the asymptotic expansion with respect to degree of homogeneity

$$h_\alpha \simeq \sum_{j=0}^{\infty} (-i)^j h_{\alpha j}, \quad h_{\alpha j} \in \overset{0}{P}{}^{m-j}(u_\alpha); \tag{15}$$

$h_{\alpha j}$ is quasi-homogeneous of degree $m - j$ for large Λ_M, and consider also the analogous expansion for h_β. Collecting in (12) the terms with equal degree of homogeneity, we obtain the infinite system of equations:

$$h_{\alpha o} = h_{\beta o}; \quad h_{\alpha 1} = h_{\beta 1} + t_{\alpha\beta 1}h_{\beta o}; \quad \ldots\ldots\ldots\ldots . \tag{16}$$

The first of equations (16) leads to the following:

Proposition 1. The principal term $h_{\alpha o}$ of $\mu_\alpha^*(H_\alpha)$ is a globally defined function on M. This function will be denoted by $h_o \in \overset{0}{P}{}^{m}(M)$ and called the principal symbol of the operator $\hat{H}$.

As for lower-order invariants, the situation is rather more complicated. It is probably impossible to define the "subprincipal symbol" of the operator invariant to any canonical coordinate changes. However, a substitute may be defined in our case; this substitute is valid only in the canonical charts of the atlas. It follows from the cocyclicity conditions for $T_{\alpha\beta}$ that the operators $t_{\alpha\beta}$ satisfy the (asymptotic) cocyclicity conditions:

$$t_{\alpha\beta}t_{\beta\gamma} \simeq t_{\alpha\gamma} \text{ in } U_\alpha \cap U_\beta \cap U_\gamma, \tag{17}$$

$$t_{\alpha\beta}t_{\beta\alpha} \simeq 1 \text{ in } U_\alpha \cap U_\beta. \tag{18}$$

Since the principal term of the operator $t_{\alpha\beta}$ equals the identiy operator the conditions (17) and (18), in particular, yield the following conditions for the operators $t_{\alpha\beta 1}$:

$$t_{\alpha\beta 1} = -t_{\beta\alpha 1} \text{ in } U_\alpha \cap U_\beta, \tag{19}$$

$$t_{\alpha\beta 1} - t_{\alpha\gamma 1} + t_{\beta\gamma 1} = 0 \text{ in } U_\alpha \cap U_\beta \cap U_\gamma; \tag{20}$$

that is, $t_{\alpha\beta 1}$ is an operator-valued 1-cocycle on M.

Again it is not difficult to find the operator-valued 0-cochain σ_α on M such that $t_{\alpha\beta 1}$ is a coboundary of it:

$$\sigma_\alpha - \sigma_\beta = t_{\alpha\beta 1} \text{ in } U_\alpha \cap U_\beta. \tag{21}$$

We make use of the partition of the unity $\{\theta_\alpha\}$ and set

$$\sigma_\alpha = \sum_\gamma \theta_\gamma t_{\alpha\gamma 1} \text{ in } U_\alpha. \tag{22}$$

Clearly

$$\sigma_\alpha - \sigma_\beta = \sum_\gamma \theta_\gamma (t_{\alpha\gamma 1} - t_{\beta\gamma 1}) = \sum_\gamma \theta_\gamma t_{\alpha\beta 1} = t_{\alpha\beta 1} \text{ in } U_\alpha \cap U_\beta; \tag{23}$$

here we made use of the cocyclicity condition (20).

Proposition 2. The functions $h_{sub,\alpha}$ defined by

$$h_{sub,\alpha} = h_{\alpha 1} - \sigma_\alpha h_o \tag{24}$$

coincide on the intersections $U_\alpha \cap U_\beta$. Indeed, we have

$$\begin{aligned} h_{sub,\alpha} - h_{sub,\beta} &= h_{\alpha 1} - h_{\beta 1} + \sigma_\beta h_o - \sigma_\alpha h_o = \\ &= h_{\beta 1} + t_{\alpha\beta 1} h_o - h_{\beta 1} - t_{\alpha\beta 1} h_o = 0 \text{ in } U_\alpha \cap U_\beta. \end{aligned} \tag{25}$$

Thus, there is a globally defined function h_{sub} which coincides in U_α with the function (24). The function h_{sub} will be called a subprincipal symbol of the operator h.

The knowledge of the principal and subprincipal symbol allows the reconstruction of the first two terms of the expansion of the symbol $H_\alpha(x^{(\alpha)}, p^{(\alpha)})$ in each canonical coordinate system $(U_\alpha, \mu_\alpha : U \to D_\alpha))$. Further "invariants" (we put this word in quotes since these quantities depend, however, on the choice of the partition of unity) may be constructed in a similar way with only technical complications, but we do not go into detail.

Also a question arises: the functions $h_o \in \overset{0}{P}{}^m(M)$ and $h_{sub} \in \overset{0}{P}{}^{m-1}(M)$ being given, can we construct a pseudo-differential operator $\hat{H} \in \hat{T}^m(M)$ having these functions as its principal and subprincipal symbol, respectively? The answer to this question is affirmative, as the following proposition shows:

Proposition 3. There exists an operator $\hat{H} \in \hat{T}^m(M)$ such that its principal symbol is equal to h_o and the subprincipal symbol is equal to h_{sub}.

Proof. For any α, set

$$h_\alpha \simeq \sum_\gamma t_{\alpha\gamma} \theta_\gamma \tilde{h}, \tag{26}$$

where

$$\tilde{h} = (1 - \sum_\delta [t_{\alpha\delta 1}, \theta_\delta])h_0 + h_{sub} \text{ in } \mathcal{U}_\alpha . \quad (27)$$

Here the square brackets denote the commutator of operators. The function $\tilde{h}$ does not depend on the choice of α; indeed,

$$\sum_\delta [t_{\alpha\delta 1}, \theta_\delta] - \sum_\delta [t_{\beta\delta 1}, \theta_\delta] = \sum_\delta [t_{\alpha\delta 1} - t_{\beta\delta 1}, \theta_\delta] =$$
$$= \sum_\delta [t_{\alpha\beta 1}, \theta_\delta] = [t_{\alpha\beta 1}, 1] = 0 \text{ in } \mathcal{U}_\alpha \cap \mathcal{U}_\beta . \quad (28)$$

Thus,

$$t_{\beta\alpha} h_\alpha \simeq \sum_\gamma t_{\beta\alpha} t_{\alpha\gamma} \theta_\gamma \tilde{h} \simeq \sum_\gamma t_{\beta\gamma} \theta_\gamma \tilde{h} = h_\beta , \quad (29)$$

i.e., the compatibility conditions are satisfied. Next, the first two terms of the asymptotic expansion of h_α have the form:

$$h_{\alpha o} = h_o , \quad (30)$$

$$h_{\alpha 1} = \sum_\gamma t_{\alpha\gamma 1} \theta_\gamma h_o + h_{sub} - \sum_\delta [t_{\alpha\delta 1}, \theta_\delta] h_o = h_{sub} + \sum_\gamma \theta_\gamma t_{\alpha\gamma 1} h_o . \quad (31)$$

Thus the principal symbol of the constructed operator equals h_0, and comparison of (31) with (22) and (24) yields that its subprincipal symbol equals h_{sub}. Proposition 3 is thereby proved.

B. Pseudo-Differential Equations and Statement of the Cauchy Problem

Let $\hat{H}$ be a pseudo-differential operator in $F_+(M)$, $\hat{H} \in \hat{T}^m(M)$. The equation of the form

$$\hat{H}\psi = v, \quad (1)$$

where v is known and ψ unknown elements of $F_+(M)$, is called a pseudo-differential equation in the space $F_+(M)$. Our aim in this book does not include the solution of general pseudo-differential equations in the space $F_+(M)$. We consider only one special case which is a necessary stage in the procedure of solving operator equations, considered in Chapter 4. We mean the Cauchy problem.

Let M be a proper quantized symplectic R_+-manifold. Consider the line R^1 with the coordinate $x_o \equiv 1$ and the trivial action of R_+. The direct product $\tilde{M} = M \times T^*R^1$ is obviously a proper quantized symplectic R_+-manifold. Indeed, the canonical charts on $M \times T^*R^1$ may be obtained as the direct products of those on M and of T^*R^1:

$$u_\alpha = \mathcal{U}_\alpha \times T^*R^1, \quad (2)$$

and all the conditions of Definition 1 of Section 3:D are easily verified.

Let $\psi \in F_+(\tilde{M})$. For each fixed $t \in R^1$, the mapping

$$i_+^* : F_+(\tilde{M}) \to F_+(M) \quad (3)$$

is defined taking the equivalence class of $\{\Psi_\alpha\}$ to the equivalence class of $\{\Psi|_{t=t_o}\}$. The mapping acts in the spaces

$$i_+^* : F_+^m(\tilde{M}) \to F_+^m(M) \quad (4)$$

for any m. We denote $i_+^*(\Psi) = \Psi(t)$.

Let $\hat{H} \in \hat{T}^1(M)$ be a pseudo-differential operator in the space $F_+(M)$. Then $\hat{\mathrm{H}} \overset{\text{def}}{=} \hat{p}_0 + \hat{H} \in \hat{T}^1(\tilde{M})$.

Definition 1. The Cauchy problem in $F_+(M)$ is a problem of solving the equation

$$(-i \frac{\partial}{\partial t} + \hat{H})\psi \equiv \hat{\mathrm{H}}\psi = 0 \text{ in } F_+(\tilde{M}) \tag{5}$$

with the initial data

$$\psi_0 \in F_+(M). \tag{6}$$

We also consider the non-homogeneous Cauchy problem, which differs from (5) - (6) in that on the right-hand side of (5) we have some given element of $F_+(M)$. Also the case may be considered when the operator H depends on t, $\hat{H} - \hat{H}(t)$. Finding the solution of the problem (5) - (6) usually fails since the estimates which one manages to obtain are not uniform with respect to $t \in R^1$.

Hence we may consider some fixed segment $K \subset R^1$, $K \ni 0$; for example, $K = [0,T]$, then set $\tilde{M} = M \times T^*K$ and repeat the above arguments. Thus we obtain the definition of the Cauchy problem on the segment K. The initial data may be imposed for any fixed $t_0 \in K$, not necessary for $t_0 = 0$.

Proposition 4. (Duhamel's principle). Consider the non-homogeneous Cauchy problem

$$\begin{cases} (-i \frac{\partial}{\partial t} + \hat{H}(t))\psi = v \\ \psi(0) = 0 \end{cases} \tag{7}$$

on the segment $t \in [0,T]$. If the solution of the auxiliary homogeneous Cauchy problem

$$\begin{cases} (-i \frac{\partial}{\partial t} + \hat{H}(t))\chi_t = 0 \\ \chi_{t_0}(t_0) = v(t_0) \end{cases} \tag{8}$$

on the segment $[t_0,T]$ exists for any $t_0 \in [0,T]$ and depends continuously on t_0 (in the sense that there exists a family of representatives $\{\chi_{t_0\alpha}\}$, the elements of which depend on t_0 continuously), then the problem (7) has a solution of the form:

$$\psi(t) = i\int_0^t \chi_{t_0}(t)dt_0, \tag{9}$$

where the integral with respect to the parameter is defined via integrals of the representatives.

Proof. Passing to representatives, we have

$$-i \frac{\partial \chi_{t_0\alpha}}{\partial t} + \hat{H}_\alpha \chi_{t_0\alpha} = \Delta_{t,\alpha}, \tag{10}$$

where $\mathrm{Ess}(\Sigma T_{\alpha\gamma}\Delta_{t_0\gamma}) \ni K_\alpha : K_\alpha \cap \mathcal{D}_\alpha = \emptyset$. Set

$$\psi_\alpha(t) = i\int_0^t \chi_{t_0\alpha}(t)dt_0. \tag{11}$$

We have then

$$-i\frac{\partial}{\partial t}(\psi_\alpha(t)) = \int_0^t(-i\frac{\partial}{\partial t}\chi_{t_o\alpha}(t))dt_o + \chi_{t_o\alpha}(t_o) =$$

$$= v(t_o) + i\int_0^t(\Delta_{t_o\alpha}(t) - \hat{H}(t)\chi_{t_o\alpha}(t))dt_o = v(t_o) - \hat{H}(t)\psi_\alpha(t) + \Sigma_\alpha, \tag{12}$$

where

$$\Sigma_\alpha = \int_o^t {}_{t_o\alpha}(t)dt_o, \tag{13}$$

and therefore Σ_α satisfies the same condition as $\Delta_{t_o\alpha}$. Proposition 4 is proved.

To obtain solutions of homogeneous Cauchy problems for classical pseudo-differential operators in $F_+(M)$ we need to construct the canonical operator acting into the space $F_+(M)$. This is performed in our next item.

C. Canonical Operator on a Lagrangian Submanifold of a Proper Quantized Symplectic R_+-Manifold M

Define first of all essential subsets for elements of the space $F_+(M)$. Let $\hat{H} \in \hat{T}^\infty(M)$ be a classical pseudo-differential operator. Consider any point $z \in u_\alpha$. The symbol H_α has an asymptotic expansion:

$$H_\alpha(x^{(\alpha)},p^{(\alpha)}) \simeq \Sigma(-i)^j H_{\alpha j}(x^{(\alpha)},p^{(\alpha)}) \tag{1}$$

(cf. (9) of item A). We say that $z \notin \text{ess supp}(\hat{H})$ if and only if there is a neighborhood of the point $\mu_\alpha(z)$ such that for $(x^{(\alpha)},p^{(\alpha)})$ belonging to this neighborhood the function $H_{\alpha j}(\tau^{\lambda(\alpha)}x^{(\alpha)},\tau^{1-\lambda(\alpha)}p^{(\alpha)})$ vanishes for τ large enough for each $j = 0,1,\ldots$. If $z \in u_\alpha \cap u_\beta$, this condition on z does not depend on the choice of the chart in view of the compatibility conditions (12) of item A. It is clear that ess supp $\hat{H}$ is a closed R-invariant subset of M.

Definition 1. Let $\psi \in F_+(M)$. The closed R_+-invariant subset $K \subset M$ is called an essential subset for ψ (we write $K \subset \text{Ess}(\psi)$), if we have $\hat{H}\psi = 0$ for any classical pseudo-differential operator H satisfying the condition:

$$K \cap \text{ess supp}(\hat{H}) = \emptyset. \tag{2}$$

Proposition 1. Let $\psi \in F_+(M)$. There exists a representative $\{\psi_\alpha\}$ of ψ such that $K \subset \text{Ess}(\psi)$ if and only if the closure $\mu_\alpha(K) \cap \mathcal{D}_\alpha$ is an essential subset for ψ_α, $\alpha = 1,\ldots,M_o$.

Proof. First of all we construct the family of operators $\hat{\rho}_{(\alpha)} \in \hat{T}^\infty(M)$ such that (i) ess supp$(\hat{\rho}_{(\alpha)}) \subset u_\alpha$ and (ii) $\Sigma_\alpha\hat{\rho}_{(\alpha)} = 1$. Each operator $\hat{\rho}_{(\alpha)}$ is given by a collection of symbols, whose image on M we denote by $\rho_{(\alpha)\beta}$, $\beta = 1,\ldots,M_o$. Note that by virtue of (i) the definition defines the operator $\hat{\rho}_{(\alpha)}$ completely. We denote $\rho_{(\alpha)\alpha}$ by ρ_α for short. The condition (ii) reads

$$\Sigma_\beta t_{\alpha\beta}\rho_\beta \sim 1 \text{ in } U_\alpha \text{ for all } \alpha. \tag{3}$$

Note that

$$t_{\alpha\beta}1 = 1 \tag{4}$$

and, consequently,

$$t_{\alpha\beta j}1 = 0,\ j = 1,2,\dots. \tag{5}$$

(It follows from the fact that $1\cdot T_{\alpha\beta} = T_{\alpha\beta}\cdot 1$.) We seek ρ_β in the form

$$\rho_\beta = \sum_{j=0}^{\infty}\rho_{\beta j},\ \rho_{\beta j}\in \overset{0}{P}{}^{-j}(M). \tag{6}$$

It suffices to satisfy the system of equations

$$\begin{aligned}\sum_\beta \rho_{\beta o} &= 1,\\ \sum_\beta \rho_{\beta 1} &= -\sum_\gamma t_{\alpha\gamma 1}\rho_{\gamma o} \text{ in } \mathcal{U}_\alpha \text{ for any } \alpha,\\ \sum_\beta \rho_{\beta 2} &= -\sum_\gamma t_{\alpha\gamma 1}\rho_{\gamma 1} - \sum_\gamma t_{\alpha\gamma 2}\rho_{\gamma o} \text{ in } \mathcal{U}_\alpha \text{ for any } \alpha,\end{aligned} \tag{7}$$

in such a way that $\operatorname{supp}\rho_{\gamma j}\subset \mathcal{U}_\gamma$. Set

$$\rho_{\beta o} = \theta_\beta, \tag{8}$$

where θ_β is a partition of unity (see Definition 1, Section 3:D). The second equation in (7) reads now

$$\sum_\beta \rho_{\beta 1} = -\sum_\gamma t_{\alpha\gamma 1}\theta_\gamma \text{ in } \mathcal{U}_\alpha. \tag{9}$$

The right-hand side of (9) does not depend on α. Indeed, in view of the cocyclicity conditions ((20), item A) it may be rewritten in the form (in $\mathcal{U}_\alpha\cap\mathcal{U}_\beta$):

$$\sum_\gamma t_{\alpha\gamma 1}\theta_\gamma = \sum_\gamma t_{\delta\gamma 1}\theta_\gamma + \sum_\gamma t_{\alpha\delta 1}\theta_\gamma = \sum_\gamma t_{\delta\gamma 1}\theta_\gamma + t_{\alpha\delta 1}\cdot 1 = \sum_\gamma t_{\delta\gamma 1}\theta_\gamma. \tag{10}$$

Thus we may set

$$\rho_{\beta 1} = \theta_\beta\Big(\sum_\gamma t_{\alpha\gamma 1}\theta_\gamma\Big) \text{ in } \mathcal{U}_\beta\cap\mathcal{U}_\alpha. \tag{11}$$

Repeating the process (the next stages are, however, somewhat more complicated), we construct the desired operators $\hat\rho_{(\alpha)}$. We obviously have

$$\psi = \sum_\alpha \hat\rho_{(\alpha)}\psi \overset{\text{def}}{=} \sum_\alpha \psi_{(\alpha)}. \tag{12}$$

For each $\psi_{(\alpha)}$ we may choose a representative $\{\psi_{(\alpha)\beta}\}_{\beta=1}^{M_o}$ such that only $\psi_{(\alpha)\alpha}$ is different from zero. Indeed, it suffices to set

$$\psi_{(\alpha)\alpha} = \sum_\beta T_{\alpha\beta}(\hat\rho_{(\alpha)}\psi)_\beta. \tag{13}$$

It is clear that $\{\psi_{(\alpha)\alpha}\}_{\alpha=1}^{M_o}$ is a representative of ψ. Now let $K\subset M$, $\mu_\alpha^*(K)\cap\mathcal{D}_\alpha\in \operatorname{Ess}(\psi_{(\alpha)\alpha})$ for any α. Then if $K\cap \operatorname{ess\,supp}(\hat{H}) = 0$ for any α, we have α ess $\operatorname{supp}(\hat H_\alpha)\cap\mu_\alpha^*(K)\cap\mathcal{D}_\alpha = \phi$, and it is easy to establish that $\hat H_\alpha\psi_{(\alpha)\alpha} = 0$.

Conversely, let $K\subset \operatorname{Ess}(\psi)$, and let $\hat H_\alpha$ be an operator in $\overset{0}{L}{}^{\infty}_{(\lambda_1(\alpha),\dots,\lambda_n(\alpha))}(R^n_{x(\alpha)})$ such that $\operatorname{supp}\hat H_\alpha\cap\mu_\alpha^*(K)\cap\mathcal{D}_\alpha = \emptyset$. Consider the element $\tilde\psi_{(\alpha)}\in F_+(M)$ induced by the representative

$$\{\psi_{(\alpha)\beta}\} = \begin{cases}0, & \beta\neq\alpha\\ \hat H_\alpha\psi_{(\alpha)\alpha}, & \beta = \alpha.\end{cases} \tag{14}$$

We have obviously

$$\tilde{\psi}_{(\alpha)} = \hat{H}_{(\alpha)},\hat{\rho}_{(\alpha)}\psi, \tag{15}$$

where $\hat{H}_{(\alpha)}$ is any element of $\hat{T}^{\infty}(M)$, prolonging the operator $\hat{H}_{\alpha}$, and therefore $\tilde{\psi}_{\alpha} = 0$ since ess supp $(\hat{H}_{(\alpha)},\hat{\rho}_{(\alpha)}) \cap K = \emptyset$. Proposition 1 is proved. We now come to the construction of the canonical operator.

Definition 2. Let $L \subset M$ be a Lagrangian manifold. L is called a proper Lagrangian manifold if for any α the manifold

$$\tilde{L}_{\alpha} = \mu_{\alpha}(L \cap \mathcal{U}_{\alpha}) \subset R^{2n}_{(x^{(\alpha)},p^{(\alpha)})} \tag{16}$$

is a proper Lagrangian manifold in the sense of Definition 1 of Section 3:B.

Definition 3. Let $L \subset M$ be a proper Lagrangian manifold. The proper atlas on L is a locally finite covering $\bigcup_{\sigma} V_{\sigma} = L$ of L by open sets together with the coordinate mapping, defined on these sets, such that the following conditions are satisfied:

i) For any σ the corresponding $\alpha = \alpha(\sigma)$ is given such that $V_{\sigma} \subset L \cap \mathcal{U}_{\alpha}$.

ii) The coordinate mappings have the form

$$V_{\sigma} \ni z \to (x_I^{(\alpha)},p_{\bar{I}}^{(\alpha)}) = (x_I^{(\alpha)}(z),p_{\bar{I}}^{(\alpha)}(z)), \tag{17}$$

where $\alpha = \alpha(\sigma)$ and $I = I(\sigma)$.

iii) All the intersections $V_{\sigma} \cap V_{\sigma'}$ of the canonical charts on L are also R_+-precompact.

iv) For any α the charts $\{V_{\sigma}\}_{\alpha(\sigma)=\alpha}$ form on the Lagrangian manifold

$$L_{\alpha} = \mu_{\alpha}(\bigcup_{\alpha(\sigma)=\alpha} V_{\sigma}), \tag{18}$$

a proper atlas in the sense of Definition 2 of Section 3:B.

We denote by $D^m(L)$ the space of functions ϕ on L such that ϕ may be represented in the form

$$\phi = \sum_{\alpha}\phi_{\alpha}, \tag{19}$$

where $(\mu_{\alpha}^{*})^{-1}\phi_{\alpha} \in \mathcal{D}^m(L_{\alpha})$, and by $\mathcal{D}^m(V_{\sigma})$ the space $\mathcal{D}^m(V_{\sigma}) = \{\phi \in \mathcal{D}^m(L) | \text{supp} \times V_{\sigma}\}$.

Let μ be a given measure on L, homogeneous of degree r, and such that in any canonical chart V_{σ} the density $\mu_I(x_I^{(\alpha)},p_{\bar{I}}^{(\alpha)})$ (here $\alpha = \alpha(\sigma)$, $I = I(\sigma)$) satisfies the estimates

$$|\partial^{|\gamma|}\mu_I(x_I^{(\alpha)},p_{\bar{I}}^{(\alpha)})| \leqslant \leqslant C_{\gamma}\Lambda_{\alpha,I}(x_I^{(\alpha)})^{r-|\lambda_I|-|\bar{I}|+|\lambda_{\bar{I}}| - \Sigma_{j\in I}\gamma_j\lambda_j - \Sigma_{j\in\bar{I}}(1-\lambda_j)\gamma_j}. \tag{20}$$

We choose arbitrarily and fix the argument arg $\mu_I(x_I^{(\alpha)},p_{\bar{I}}^{(\alpha)})$ for any chart of the canonical atlas and define the elementary canonical operator,

$$K_{\sigma} : \mathcal{D}^{\infty}(V_{\sigma}) \to F_+(M) \tag{21}$$

by means of the formula

$$K_\sigma \phi = j_\sigma \{ \bar{F}_{p_{\bar{I}}^{(\alpha)} \to x_{\bar{I}}^{(\alpha)}} \int e^{iS_I(x_I^{(\alpha)}, p_{\bar{I}}^{(\alpha)})} \zeta(\Lambda_{\alpha I}(x_I^{(\alpha)})) \times$$

$$\times \sqrt{\mu_I(x_I^{(\alpha)}, p_{\bar{I}}^{(\alpha)})} (\phi \circ \nu_\sigma^{-1})(x_I^{(\alpha)}, p_{\bar{I}}^{(\alpha)}) \}, \quad \phi \in \mathcal{D}^\infty(V_\sigma). \tag{22}$$

Here ν_σ is the coordinate mapping in the canonical chart V_σ; the expression in outer braces is a usual elementary canonical operator (see Definition 4 of Section 3:B),

$$j_\sigma : \overset{0}{L}{}^\infty_\alpha \to F_+(M), \tag{23}$$

where $\alpha = \alpha(\sigma)$ is a natural mapping, taking $\psi \in \overset{0}{L}{}^\infty_\alpha$ into the equivalence class of $\{\psi_\beta\}$, where $\psi_\beta = 0$ for $\beta \neq \alpha$, $\psi_\alpha = \psi$. The application of the stationary phase method immediately yields the following:

Theorem 1. For any non-empty intersection $V_\sigma \cap V_{\sigma'}$, there exists an integer $H_{\sigma\sigma'}$ and a sequence of differential operators $v_{\sigma\sigma' j}$, $j = 0,1,2,$ in $V_\sigma \cap V_{\sigma'}$ with the following properties:

i) The operator $v_{\sigma\sigma' j}$ is a differential operator of order $\leqslant 2j$ with the real coefficients acting in the spaces

$$v_{\sigma\sigma' j} : \mathcal{D}^m(V_\sigma \cap V_{\sigma'}) \to \mathcal{D}^{m-j}(V_\sigma \cap V_{\sigma'}) \tag{24}$$

for any m and homogeneous of degree $-j$ (i.e., decreasing the degree of homogeneity by j).

ii) If we denote by $v_{\sigma\sigma'}$ the asymptotic sum*

$$v_{\sigma\sigma'} \simeq \sum_{j=0}^{\infty} (-i)^j v_{\sigma\sigma' j}, \tag{25}$$

then for any $\phi \in \mathcal{D}^\infty(V_\sigma \cap V_{\sigma'})$

$$K_{\sigma'} \phi = \exp(i\pi H_{\sigma\sigma'}) K_\sigma (v_{\sigma\sigma'} \phi). \tag{26}$$

iii) The operators $v_{\sigma\sigma'}$ satisfy the cocyclicity condition

$$v_{\sigma\sigma'} v_{\sigma'\sigma''} \simeq v_{\sigma\sigma''} \text{ in } V_\sigma \cap V_{\sigma'} \cap V_{\sigma''}$$
$$\text{if } V_\sigma \cap V_{\sigma'} \cap V_{\sigma''} \neq \emptyset, \tag{27}$$

and for the principal term $v_{\sigma\sigma' o}$ we have

$$v_{\sigma\sigma' o} = 0 \text{ in } V_\sigma \cap V_{\sigma'} \tag{28}$$

for any σ, σ' with $V_\sigma \cap V_{\sigma'} \neq \emptyset$.

iv) $H_{\sigma\sigma'}$ is a cocycle modulo 2 on L.

Proof of Theorem 1. This is clear from our previous discussions (see Theorem 1 of Section 3:B; Proposition 5 of Section 3:C).

---

* The asymptotic summation follows along the lines of Lemma 3 of Section 3:C.

Definition 4. L is called quantized if $H_{\sigma\sigma'}$ is a coboundary modulo 2 on L (and henceforth the branches of arg $\mu_I(x_I^{(\alpha)}, p_{\bar{I}}^{(\alpha)})$ in different charts may be chosen concordant so that factor $(i\pi H_{\sigma\sigma'})$ in (25) is eliminated).

Remark 1. Whether L will be quantized or not depends on two factors: first, on the choice of the measure μ on L (more precisely on the equivalence class of μ with respect to the equivalence relation: $\mu_1 \sim \mu_2$ if and only if $\mu_1/\mu_2 : L \to \mathbb{C}\setminus\{0\}$ is a mapping, homotopic to a constant one), and second, on the choice of concordant branches of the argument when defining the transition operators $T_{\alpha\beta}$ of the canonical sheaf $F_+(M)$. We do not go into further detail here; see Section 2 and also [38], [55].

Consider a quantized proper Lagrangian manifold $L \subset M$ (we assume, consequently, that the atlas $\{V_\sigma\}$ and the measure μ are given and fixed). We assume that the concordant branches of arg μ_I are chosen ($H_{\sigma\sigma'} = 0$) and that the partition of unity $\{e_\sigma\}$ is chosen subordinate to the covering $\{V_\sigma\}$ such that $e_\sigma \in \mathcal{D}^0(L)$ for all σ.

We define the canonical operator on L,

$$K : \mathcal{D}^\infty(L) \to F_+(M) \tag{29}$$

by the equality

$$K\phi \overset{\text{def}}{=} \sum_\sigma K_\sigma(e_\sigma\phi), \quad \phi \in \mathcal{D}^\infty(L). \tag{30}$$

Theorem 2. For any "classical" pseudo-differential operator $\hat{H} \in \hat{T}^s(M)$ the commutation formula is valid

$$\hat{H}K\phi = KP\phi, \tag{31}$$

where the operator P is the asymptotic sum

$$P \simeq \sum_{j=0}^{\infty} (-i)^j P_j, \tag{32}$$

and the operators P_j possess the following properties:

i) P_j is a differential operator of order $\leqslant 2j$ on L, acting in the spaces

$$P_j : \mathcal{D}^m(L) \to \mathcal{D}^{m+s-j}(L) \tag{33}$$

for any S, homogeneous of order m - j.

ii) The coefficients of the operator P_j are linear forms with the smooth real coefficients of terms of the asymptotic expansion of the symbol H_α and its derivatives up to order 2j .*

iii) The operator P_o coincides with the restriction of the principal symbol of the operator $\hat{H}$ on L.

iv) If the pair (L,μ) is associated with $\hat{H}$ in the sense that

$$h_o|_L = 0 \tag{34}$$

and

$$L_{V(h_o)}\mu = 0, \tag{35}$$

* This is valid for any coordinate chart U_α, containing the considered point of L.

where $L_{V(h_o)}$ is a Lie derivative along the trajectory of the Hamiltonian vector field*

$$V(h_o) \lrcorner\, \omega = -dh_o, \tag{36}$$

then $P_o = 0$ and

$$P_1 = V(h_o) + F[h_o, h_{sub}], \tag{37}$$

where $F[h_o, h_{sub}]$ is a linear form with the smooth real coefficients of h_{sub} and the derivatives of h_o of order $\leqslant 2$, restricted on L.

Proof. Employing the unity partition $\{e_\sigma\}$ and the operators $v_{\sigma\sigma}$, reduces the problem by virtue of our remarks about essential subsets to the "local" case considered in Theorem 2 of Section 3:B. The only novelty here is the asymptotic sum (32). This (purely technical) point of the argument is treated similarly to Lemma 3 of Section 3:C.

Corollary. If $\psi = K\phi$, where K is a canonical operator on a proper quantized Lagrangian manifold $L \subset M$, then $L \subset \mathrm{Ess}(\psi)$.

The elements $\psi \in F_+(M)$ of the form $\psi = K\phi$, where K is a canonical operator on L, will be called the canonically representable functions (CRF's), associated with the proper quantized Lagrangian manifold L.

D. Solution of the Cauchy Problem for a Pseudo-Differential Equation in $F_+(M)$

In $F_+(M)$ we consider the Cauchy problem

$$\left(-i\frac{\partial}{\partial t} + \hat{H}(t)\right)\psi = 0, \tag{1}$$

$$\psi(0) = \psi_o \in F_+(M) \tag{2}$$

on the segment $[0,T]$, where $\hat{H}(t) \in \hat{T}^1(M)$ is a "classical" pseudo-differential operator in $F_+(M)$, $\psi_o \in F_+(M)$ is a CRF associated with a proper quantized Lagrangian manifold L. Let $h_o(z,t)$, where z is a point of M, be the principal symbol of $\hat{H}$ (homogeneous of degree 1) and $h_{sub}(z,t)$ be the subprincipal symbol of $\hat{H}$ (homogeneous of degree 0). Consider first the case when $h_o(z,t)$ is real-valued.

Then the procedure of finding a solution to (1) - (2) is very familiar (from experience in various other situations). We seek the solution to (1) - (2) in the form

$$\psi = K\phi, \tag{3}$$

where K is a canonical operator on the Lagrangian manifold $L \subset \tilde{M} = M \times T^*R^1_t$, which is constructed in the following way. Let

$$\psi_o = K_o\phi_o, \tag{4}$$

where K_o is a canonical operator on the Lagrangian manifold $L_o \subset M$. Now consider the Hamiltonian system

$$\begin{cases} \dot{t} = 1 \\ \dot{E} = (\partial h_o(z,t)/\partial t) \\ \dot{z} = V(h_o)(z,t) \end{cases} \tag{5}$$

* (35) makes sense since under the condition (34) the field is tangent to L.

(here E is the momentum dual to t) on $\tilde{M}$ with the initial data

$$(t_o,E_o,z_o) = (0,h_o(z_o,0),z_o), \; z_o \in L_o. \tag{6}$$

The set of functions of system (5) with the initial data (6) for various $z_o \in L_o$ forms a Lagrangian manifold $L \subset \tilde{M}$. We define the measure on L, setting

$$\mu = \mu_o \wedge dt \tag{7}$$

in coordinates (z_o,t) on L, where μ_o is the given measure on L_o. It is easy to verify that

$$(h_o + E)|_L = 0, \tag{8}$$

$$\mathcal{L}_{V(h_o+E)}\mu = 0; \tag{9}$$

that is, L is associated with the operator $\hat{E} + \hat{H} = -i \frac{\partial}{\partial t} + \hat{H}$.

We impose the requirement that L be a proper Lagrangian manifold (generally this may not be the case since the Hamiltonian system (5) does not necessarily preserve the inequality (in local coordinates))

$$|p_j| \leqslant C\Lambda(x)^{1-\lambda_j}. \tag{10}$$

Since L is shrinkable to L_o, L is necessarily quantized if L_o is. Thus the form (3) may be employed, and we may choose in such a way that

$$(K\phi)(0) = K_o\phi(0) \tag{11}$$

for any $\phi \in \mathcal{D}^\infty(L)$.

Substituting (3) into (1) and using Theorem 2 of item C, we obtain the following asymptotic problem for ϕ:

$$\mathcal{P}\phi = 0, \tag{12}$$

$$\phi(0) = \phi_o. \tag{13}$$

This system is easily solved since the principal term of the operator $\mathcal{P}$ has the form (in coordinates (z_o,t) on L):

$$\mathcal{P}_1 = \frac{\partial}{\partial t} + F(z_o,t), \tag{14}$$

where $F(z_o,t)$ is a given function.

We require that the solution of (12) - (13) belong to $\mathcal{D}^m(L)$ (again this may not be the case since in pathological cases for some values of t, infinitely many canonical charts on L, intersecting with the trajectories of the Hamiltonian vector field $V(h_o + E)$ coming from supp ϕ_o, may occur). If our requirements are satisfied, we readily obtain the solution to the problem (1) - (2) on the segment [0,T].

Consider now the case when the principal symbol is essentially complex-valued. The solution then would be given by a canonical operator on a Lagrangian manifold with the complex germ. In this book we have no space to present its construction in R_+-quasi-homogeneous case. This construction is a result of complicated synthesis of the ideas, used here in the case of a real Lagrangian manifold and the theory of complex germ [52] (see also [54,59,61]). Fortunately, the existence theorem for this case, which will be used in Chapter 4, may be formulated in terms of real geometrical objects and a few additional notions should be introduced to give the formulation.

We assume as before that the initial data have the form (4), i.e., ψ_o is a CRF associated with the real Lagrangian manifold $L_o \subset M$. (However, this restriction is not essential.) Denote

$$H(z,t) = \mathrm{Re}\,h_o(z,t),$$
$$\tilde{H}(z,t) = \mathrm{Im}\,h_o(z,t); \tag{15}$$

$H(z,t)$ and $\tilde{H}(z,t)$ are homogeneous functions of degree 1.

For any $K \subset M$ and $\varepsilon > 0$, denote by $\mathcal{U}^{\varepsilon}(K)$ the subset in M consisting of all points z such that there is $z_o = z_o(z) \in K$ so that: $z, z_o \in \mathcal{U}_{\alpha}$ for some $\alpha = \alpha(z)$, and the points $(x^{(\alpha)}, p^{(\alpha)}) = \mu_{\alpha}(z)$ and $(x_o^{(\alpha)}, p_o^{(\alpha)}) = \mu_{\alpha}(z_o)$ satisfy

$$|x_j^{(\alpha)} - x_{oj}^{(\alpha)}| \leqslant \varepsilon\Lambda_{\alpha}(x_o^{(\alpha)})^{\lambda_j(\alpha)},$$
$$|p_j^{(\alpha)} - p_{oj}^{(\alpha)}| \leqslant \varepsilon\Lambda_{\alpha}(x_o^{(\alpha)})^{1-\lambda_j(\alpha)}, \quad j = 1,\dots,n. \tag{16}$$

Condition 1. For some $\varepsilon > 0$ the inequality is valid:

$$\tilde{H}(z,t) \leqslant 0 \text{ for } z \in \mathcal{U}^{\varepsilon}(L_o), \quad t \in [0,\varepsilon]. \tag{17}$$

Condition 2. Set

$$\Omega_{\varepsilon} = \mathcal{U}^{\varepsilon}(\{z \in M \mid z \in L_o, \tilde{H}(z,0) = 0\}). \tag{18}$$

Trajectories $z(z_o,t)$ of a (non-autonomous) Hamiltonian system

$$\dot{z}(z_o,t) = V(H(z(z_o,t),t)) \tag{19}$$

exist for $t \in [0,T]$, $z_o \in \Omega_{\varepsilon}$. There is a constant C such that

$$|p_j^{(\alpha)}(z(z_o,t))| \leqslant C\Lambda_{\alpha}(x^{(\alpha)}(z(z_o,t)))^{1-\lambda_j(\alpha)}, \quad j = 1,\dots,n \tag{20}$$

on these trajectories for any α, such that $z(z_o,t) \in \mathcal{U}_{\alpha}$. Also

$$\tilde{H}(z(z_o,t),t) \leqslant 0 \tag{21}$$

on the trajectories.

Condition 3. For any $t \in [0,T]$

$$\phi_t \in \mathcal{D}^{\infty}(L(t)), \tag{22}$$

where ϕ_t is the solution of the transport equation, corresponding to $H(z,t) = \mathrm{Re}\,h_o(z,t)$, and $L(t)$ is the shift of L_o along the trajectories (19) during the time t.

Condition 4. For $t = T$, we have

$$\tilde{H}(z(z_o,T),T) < -\varepsilon \tag{23}$$

for $z_o \in \Omega_{\varepsilon}$.

Theorem 1. Let Conditions 1 - 3 be satisfied. Then there exists a solution ψ of the problem (1) - (2). If in addition Condition 4 is satisfied, this solution satisfies the condition

$$\psi(T) = 0. \tag{24}$$

Remark 1. The solution is given by a canonical operator on a Lagrangian manifold with a complex germ, obtained by the construction somewhat similar to that in the case of a real-valued symbol. See [52] for the theory of the canonical operator on a Lagrangian manifold with a complex germ.

Remark 2. Conditions 1 - 4 are called the absorption conditions (for the initial manifold L_0 and given $\varepsilon > 0$).

IV
Quasi-inversion theorem for functions of a tuple of non-commuting operators

1. EQUATIONS WITH COEFFICIENTS GROWING AT INFINITY

We consider in this section the quasi-inversion theorem (presented in general form in the subsequent sections) in relation to a particular problem, namely, to partial differential equations in $\mathbb{R}^n$, whose coefficients may have polynomial growth as $|x| \to \infty$. Numerous papers were devoted to the "elliptic" case, when the principal symbol of the operator in question is a non-vanishing function of (x,p), e.g., homogeneous in (x,p) of some degree. However, the non-elliptic case remained uninvestigated. The quasi-inversion theorem is the very tool that enables us to consider it and to construct asymptotic solutions of the differential equation such that the consequent terms of asymptotics (and, respectively, error terms on the right-hand side of the equation) become more and more smooth and decay more rapidly at infinity.

A single example is presented in item A, while the general equation with coefficients growing at infinity is considered in item B.

A. Model Example

Consider the following example. Let an equation be given in $\mathbb{R}^2 \ni (x,t)$ of the form:

$$[Lu](x,t) \equiv \frac{\partial^2 u}{\partial t^2} - \frac{\partial^2 u}{\partial x^2} + c(x,t)x^{2m}u(x,t) = 0, \tag{1}$$

with initial data

$$u|_{t=0} = u_o(x), \quad \partial u/\partial t|_{t=0} = u_1(x), \tag{2}$$

where m is a positive integer, c(x,t) is a smooth function, bounded with all its derivatives and satisfying $c(x,t) \geqslant \varepsilon > 0$, and the initial data $u_o(x)$ and $u_1(x)$ are tempered distributions in the space $\mathbb{R}^1$.

We seek the solution of (1) - (2), asymptotic in the following sense:

Definition 1. The functional sequence $u_N(x,t)$, N = 1,2,..., is an asymptotic solution of (1) - (2) on the interval [0,T] if, for any N, (2) is satisfied and

$$\sup_{t\in[0,T]} \sum_{k+r\leqslant N} \| x^k(-i\frac{\partial}{\partial x})^r L u_N \|_{L^2(\mathbb{R}^1)} < \infty. \tag{3}$$

More precisely, in the situation described we shall speak of $\{u_N\}$ as an asymptotic solution with respect to powers of operators x and $(-i\frac{\partial}{\partial x})$.

For m = 1, the asymptotics in the above sense were constructed for (1) - (2) in [52] by reduction to a system of first order.

In what follows, we denote

$$c_o = \inf_{(x,t)\in\mathbb{R}\times[0,T]} c(x,t). \tag{4}$$

By assumption c_o is strictly positive.

To solve (1) - (2) asymptotically, we set

$$u_N(x,t) = G_{N,o}(\overset{1}{A_1},\overset{2}{A_2},\overset{3}{B},t)u_o(x) + G_{N,1}(\overset{1}{A_1},\overset{2}{A_2},\overset{3}{B},t)u_1(x), \tag{5}$$

where A_i, i = 1,2, and B are the self-adjoint operators in $L_2(\mathbb{R}^1)$:

$$A_1 = -i(\partial/\partial x),\ A_2 = x,\ B = x, \tag{6}$$

satisfying the commutation relations

$$[A_1,A_2] = [A_1,B] = -i;\ [A_2,B] = 0, \tag{7}$$

and $G_{N,i}(y_1,y_2,y_3,t)$, i = 0,1, are symbols to be determined. The equation in question may be written in the form

$$-(-i\frac{\partial}{\partial t})^2 u + \{A_1^2 + c(B,t)A_2^{2m}\}u = 0. \tag{8}$$

The major sense of introducing the operator $B = A_2$ is that we can explicitly separate the polynomial growth at infinity (powers of A_2), while the remaining coefficients are required to be bounded functions of the operator B. Then it is sufficient to construct an asymptotics with respect to powers of A_1 and A_2, uniform with respect to B. The left regular representation for the tuple $(\overset{1}{A_1},\overset{2}{A_2},\overset{3}{B})$ has the form

$$L_1 \equiv L_{A_1} = y_1 - i\frac{\partial}{\partial y_2} - i\frac{\partial}{\partial y_3},\ L_2 \equiv L_{A_2} = y_2,\ L_3 \equiv L_B = y_3 \tag{9}$$

(see Chapter 2, Section 4 for the definition and the method of evaluation of regular representations. In our particular case the calculation is rather simple and therefore left to the reader.)

Theorem 1. For any natural N there exists a natural $N_1 = N_1(N)$ such that the estimates

$$\left| \frac{\partial^\alpha T(y)}{\partial y^\alpha} \right| \leqslant C_\alpha (1 + |y_1| + |y_2|)^{-N_1+|\alpha|},\ |\alpha| = 0,1,\ldots,N_1, \tag{10}$$

valid for the symbol $T(y_1,y_2,y_3)$, imply the operator norm estimates

$$\sum_{k+r\leqslant N} \| x^k(-i\frac{\partial}{\partial x})^r T(\overset{1}{A_1},\overset{2}{A_2},\overset{3}{B}) \|_{L^2\to L^2} < \infty. \tag{11}$$

Proof. N being fixed, let N_1 be large enough. Then, once $k + r \leqslant N$, the function

$$f_{kr}(y_1,y_2,y_3) = y_2^k(y_1 - i \frac{\partial}{\partial y_2})^r T(y_1,y_2,y_3) \tag{12}$$

satisfies

$$\left| \frac{\partial^{\alpha_1+\alpha_2} f_{kr}}{\partial y_1^{\alpha_1} \partial y_2^{\alpha_2}} (y_1,y_2,y_3) \right| \leqslant C(1 + |y_1| + |y_2|)^{-3} \tag{13}$$

for $\alpha_1 + \alpha_2 \leqslant 3$. By Theorem 1 of Chapter 2, Section 4:F, the operator

$$f_{kr}(\overset{1}{A}_1,\overset{2}{A}_2,\overset{3}{B}) = (L_2^k L_1^r T)(\overset{1}{A}_1,\overset{2}{A}_2,\overset{3}{B}) = x^k(-i \frac{\partial}{\partial x})^r T(\overset{1}{A}_1,\overset{2}{A}_2,\overset{3}{B}) \tag{14}$$

is bounded in $L^2(\mathbb{R})$, and the theorem is proved.

Next we derive the equations to determine symbols $G_{N,i}(y,t)$. We obtain through substituting (5) into (8) and using the left regular representation operators (9):

$$-(-i \frac{\partial}{\partial t})^2 G_{N,i} + \{(y_1 - i \frac{\partial}{\partial y_2} - i \frac{\partial}{\partial y_3})^2 + c(y_3,t)y_2^{2m}\}G_{N,i} = R_{N,i}(y,t), \tag{15}$$

with initial conditions

$$G_{N,i}(y,0) = \delta_{io}, \quad \frac{\partial G_{N,i}}{\partial t}(y,0) = \delta_{il} \tag{16}$$

(here δ_{ij} is the Kronecker delta). Symbols $R_{N,i}(y,t)$ in (15) are the arbitrary ones, satisfying the estimate (11) uniformly with respect to $t \in [0,T]$.

Next we perform the change of variables in (15), depending on the parameter $\lambda > 0$. Namely,

$$y_1 = \lambda x_1, \; y_2 = \lambda^{1/m} x_2, \; y_3 = x_3. \tag{17}$$

Dividing by λ^2, we obtain

$$\begin{aligned} -(-i\lambda^{-1} \frac{\partial}{\partial t})^2 g_{N,i}(x,\lambda,t) + \{(x_1 + \lambda^{-1/m}(-i\lambda^{-1} \frac{\partial}{\partial x_2}) - \\ - i\lambda^{-1} \frac{\partial}{\partial x_3})^2 + c_3(x,t)x_2^{2m}\}g_{N,i}(x,\lambda,t) = \lambda^{-2} r_{N,i}(x,\lambda,t), \end{aligned} \tag{18}$$

where small letters denote expressions of functions, denoted by capitals, in the variables (x,λ,t).

We construct below an asymptotic solution for $\lambda \to +\infty$ of (18) with initial data (16) and next show that this asymptotic leads to functions $G_{N,i}(y,t)$ satisfying the desired estimates. Thus, consider the equation

$$\begin{aligned} -(-i\lambda^{-1} \frac{\partial}{\partial t})^2 \psi(x,\lambda,t) + \{(x_1 + \lambda^{-1/m}(-i\lambda^{-1} \frac{\partial}{\partial x_2}) - \\ - i\lambda^{-1} \frac{\partial}{\partial x_3})^2 + c(x_3,t)x_2^{2m}\}\psi(x,\lambda,t) = \hat{O}(\lambda^{-s}) \end{aligned} \tag{19}$$

(we write $\phi(x,\lambda) = \hat{O}(\lambda^{-s})$ iff for any compact set $K \subset \mathbb{R}^2_{(x_1,x_2)} \setminus \{0\}$ and any multi-index $\alpha = (\alpha_1,\alpha_2,\alpha_3)$ there exists a constant $C_{K,\alpha}$, independent of x_3, such that

$$\left| \frac{\partial^{|\alpha|}\phi(x,\lambda)}{\partial x_1^{\alpha_1}\partial x_2^{\alpha_2}\partial x_3^{\alpha_3}} \right| \leqslant C_{K,\alpha}\lambda^{-s+|\alpha|} \tag{20}$$

for $(x_1,x_2) \in K$, $\lambda \geqslant 1$).

The solution of (19) is sought in the form of a linear combination of functions

$$\psi(x,\lambda,t) = e^{i\lambda S(x,\lambda^{-1/m},t)} \cdot \sum_{k=0}^{s-1} (-i\lambda)^{-k}\phi_k(x,\lambda^{-1/m},t), \tag{21}$$

where $S(x,\varepsilon,t)$ and $\phi_k(x,\varepsilon,t)$ are smooth functions in all their arguments, and S is a real function. Substituting the function (21) into equation (19), we obtain

$$\begin{gathered}\{-[\frac{\partial S}{\partial t} - i\lambda^{-1}\frac{\partial}{\partial t}]^2 + (x_1 + \varepsilon\frac{\partial S}{\partial x_2} + \frac{\partial S}{\partial x_3} - i\varepsilon\lambda^{-1}\frac{\partial}{\partial x_2} - i\lambda^{-1}\frac{\partial}{\partial x_3})^2 + \\ + c(x_3,t)x_2^{2m}\}^{s-1} \cdot \sum_{k=0}^{s-1} (-i\lambda)^{-k}\phi_k(x,\varepsilon,t) = O(\lambda^{-s}),\end{gathered} \tag{22}$$

where $\varepsilon = \lambda^{-1/m}$, or after opening brackets and collecting the terms with equal powers of λ and ε,

$$\begin{gathered}\{\{[-(\frac{\partial S}{\partial t})^2 + (x_1 + \frac{\partial S}{\partial x_3})^2 + c(x_3,t)x_2^{2m}] + 2\varepsilon\frac{\partial S}{\partial x_2}(x_1 + \frac{\partial S}{\partial x_3}) + \\ + \varepsilon^2(\frac{\partial S}{\partial x_2})^2\} - i\lambda^{-1}\{[-2\frac{\partial S}{\partial t}\frac{\partial}{\partial t} + 2(x_1 + \frac{\partial S}{\partial x_3})\frac{\partial}{\partial x_3} - \frac{\partial^2 S}{\partial t^2} + \frac{\partial^2 S}{\partial x_3^2}] + \\ + 2\varepsilon[(x_1 + \frac{\partial S}{\partial x_3})\frac{\partial}{\partial x_2} + \frac{\partial S}{\partial x_2}\frac{\partial}{\partial x_3} + \frac{\partial^2 S}{\partial x_2 \partial x_3}] + \varepsilon^2[2\frac{\partial S}{\partial x_2}\frac{\partial}{\partial x_2} + \\ + \frac{\partial^2 S}{\partial x_2^2}]\} - \lambda^2\{[-\frac{\partial^2}{\partial t^2} + \frac{\partial^2}{\partial x_3^2}] + 2\varepsilon\frac{\partial^2}{\partial x_2\partial x_3} + \varepsilon^2\frac{\partial^2}{\partial x_2^2}\}\} \times \\ \times \sum_{k=0}^{s-1} (-i\lambda)^{-k}\phi_k(x,\varepsilon,t) = O(\lambda^{-s}).\end{gathered} \tag{23}$$

The notation $f = O(\lambda^{-s})$ means that the function f is locally uniformly decreasing as $\lambda \to \infty$ together with all its derivatives as λ^{-s}.

The equations for S and ϕ_k follow from (23). These equations are somewhat different in cases $m = 1, m = 2$, and $m \geqslant 3$, and we present all these cases below:

(a) $m = 1$ (i.e., $\varepsilon = \lambda^{-1}$); then we have

$$(\frac{\partial S}{\partial t})^2 = (x_1 + \frac{\partial S}{\partial x_3})^2 + c(x_3,t)x_2^2 \tag{24}$$

(Hamilton-Jacobi equation);

$$\begin{gathered}\{-2\frac{\partial S}{\partial t}\frac{\partial}{\partial t} + 2(x_1 + \frac{\partial S}{\partial x_3})\frac{\partial}{\partial x_3} - \frac{\partial^2 S}{\partial t^2} + \frac{\partial^2 S}{\partial x_3^2} + 2i\frac{\partial S}{\partial x_2}(x_1 + \frac{\partial S}{\partial x_3})\} \times \\ \times \phi_k(x,t) + \{-\frac{\partial^2}{\partial t^2} + \frac{\partial^2}{\partial x_3^2} + 2i[(x_1 + \frac{\partial S}{\partial x_3})\frac{\partial}{\partial x_2} + \frac{\partial S}{\partial x_2}\frac{\partial}{\partial x_3} +\end{gathered}$$

$$+\frac{\partial^2 S}{\partial x_2 \partial x_3}]-(\frac{\partial S}{\partial x_2})^2\}\phi_{k-1}(x,t)+\{-[2\frac{\partial S}{\partial x_2}\frac{\partial}{\partial x_2}+\frac{\partial^2 S}{\partial x_2^2}]+ \tag{25}$$

$$+2i\frac{\partial^2}{\partial x_2 \partial x_3}\}\phi_{k-2}(x,t)-\frac{\partial^2\phi_{k-3}}{\partial x_2^2}=0,\ k=0,1,\ldots,s-1$$

(transport equation; for the sake of convenience the notation $\phi_\ell(x,t) \equiv 0$ is used for $\ell < 0$).

(b) $m = 2$ (i.e., $\varepsilon = \lambda^{-1/2}$), then we have

$$-(\frac{\partial S}{\partial t})^2+(x_1+\frac{\partial S}{\partial x_3})^2+c(x_3,t)x_2^4+2\varepsilon\frac{\partial S}{\partial x_2}(x_1+\frac{\partial S}{\partial x_3})=\varepsilon^2 A(x,\varepsilon,t) \tag{26}$$

(Hamilton-Jacobi equation);

$$\{[-2\frac{\partial S}{\partial t}\frac{\partial}{\partial t}+2(x_1+\frac{\partial S}{\partial x_3})\frac{\partial}{\partial x_3}-\frac{\partial^2 S}{\partial t^2}+\frac{\partial^2 S}{\partial x_3^2}+iA(x,\varepsilon,t)-$$

$$-i(\frac{\partial S}{\partial x_2})^2]+2\varepsilon[(x_1+\frac{\partial S}{\partial x_3})\frac{\partial}{\partial x_2}+\frac{\partial S}{\partial x_2}\frac{\partial}{\partial x_3}]\}\phi_k(x,\varepsilon,t)+$$

$$+\{[-\frac{\partial^2}{\partial t^2}+\frac{\partial^2}{\partial x_3^2}-2i\frac{\partial S}{\partial x_2}\frac{\partial}{\partial x_2}-i\frac{\partial^2 S}{\partial x_2^2}]+2\varepsilon\frac{\partial^2}{\partial x_2\partial x_3}\}\phi_{k-1}(x,\varepsilon,t)- \tag{27}$$

$$-i\varepsilon^2\frac{\partial^2\phi_{k-2}}{\partial x_2^2}(x,\varepsilon,t)+ia_{k-1}(x,\varepsilon,t)=\varepsilon^2 a_k(x,\varepsilon,t),\ k=0,1,\ldots,s-1$$

(transport equation).

(c) $m \geqslant 3$, then we have

$$-(\frac{\partial S}{\partial t})^2+(x_1+\frac{\partial S}{\partial x_3})^2+c(x_3,t)x_2^{2m}+2\varepsilon\frac{\partial S}{\partial x_2}(x_1+\frac{\partial S}{\partial x_3})+$$

$$+\varepsilon^2(\frac{\partial S}{\partial x_2})^2=\varepsilon^m A(x,\varepsilon,t) \tag{28}$$

(Hamilton-Jacobi equation);

$$\{[-2\frac{\partial S}{\partial t}\frac{\partial}{\partial t}+2(x_1+\frac{\partial S}{\partial x_3})\frac{\partial}{\partial x_3}-\frac{\partial^2 S}{\partial t^2}+\frac{\partial^2 S}{\partial x_3^2}+iA(x,\varepsilon,t)]+$$

$$+2\varepsilon[(x_1+\frac{\partial S}{\partial x_3})\frac{\partial}{\partial x_2}+\frac{\partial S}{\partial x_2}\frac{\partial}{\partial x_3}+\frac{\partial^2 S}{\partial x_2\partial x_3}]+\varepsilon^2[2\frac{\partial S}{\partial x_2}\frac{\partial}{\partial x_2}+\frac{\partial^2 S}{\partial x_2^2}]\}\phi_k(x,\varepsilon,t)+ \tag{29}$$

$$+\{[-\frac{\partial^2}{\partial t^2}+\frac{\partial^2}{\partial x_3^2}]+2\varepsilon\frac{\partial^2}{\partial x_2\partial x_3}+\varepsilon^2\frac{\partial^2}{\partial x_2^2}\}\phi_{k-1}(x,\varepsilon,t)+ia_{k-1}(x,\varepsilon,t)=$$

$$=\varepsilon^m a_k(x,\varepsilon,t),\ k=0,\ldots,s-1$$

(transport equation). In (26) - (29) $A(x,\varepsilon,t)$ and $a_k(x,\varepsilon,t)$ are arbitrary smooth functions, $a_{-1}(x,\varepsilon,t) \overset{def}{\equiv} 0$.

First we construct the solution in case (a). Equation (24) splits into two equations

$$\frac{\partial S}{\partial t}+\sqrt{(x_1+(\partial S/\partial x_3))^2+c(x_3,t)x_2^2}=0 \tag{30}$$

and

$$\frac{\partial S}{\partial t} - \sqrt{(x_1 + (\partial S/\partial x_3))^2 + c(x_3,t)x_2^2} = 0. \qquad (31)$$

We construct the solution S_+ of equation (30) and the solution S_- of equation (31), satisfying the following initial conditions

$$S_+|_{t=0} = S_-|_{t=0} = 0. \qquad (32)$$

(As it will be shown below, these initial conditions agree with initial conditions (16) for functions $g_{N,i}$.) Equations (30) and (31) are Hamilton-Jacobi equations with the Hamiltonian function

$$H_\pm(x,p,t) = \pm\sqrt{(x_1 + p_3)^2 + c(x_3,t)x_2^2}. \qquad (33)$$

Their solutions have the form [2]:

$$S_\pm(x,t) = \{\int_0^t [p_\pm(x_o,\tau)H_{\pm p}(x_\pm(x_o,\tau),p_\pm(x_o,\tau'),\tau') - H_\pm(x_\pm(x_o,\tau),p_\pm(x_o,\tau),\tau)d\tau\}|_{x_o = x_{o\pm}(x,\tau)}. \qquad (34)$$

In formula (34) $x_\pm(x_o,\tau), p_\pm(x_o,\tau)$ are solutions of the Hamiltonian system, corresponding to the Hamiltonian function $H_\pm$ with initial conditions $x|_{\tau=0} = x_o$, $p|_{\tau=0} = 0$ and $x_o = x_{o\pm}(x,t)$ is the solution of equation $x = x_\pm(x_o,\tau)$. (We assume that the segment [0,T] is sufficiently small, so that then the Jacobian $\det(\partial x_\pm/\partial x_o)$ does not vanish, and this solution exists; the fact that such $T > 0$ may be chosen is the consequence of homogeneity of functions $H_\pm$ in (x_1,x_2) and of uniform boundedness of derivatives of $c(x_3,\tau)$ in x_3.)

Write the Hamiltonian system corresponding to the Hamiltonian function $H_\pm$. The variables x_1 and x_2 are parameters in equations (30) - (31), hence we have non-trivial equations only for x_3 and p_3. (Here and below we drop indices $\pm$ and arguments x_1 and x_2 of x_3 and p_3.) These equations have the form

$$\dot{x}_3 = \frac{x_1 + p_3}{H_\pm(x,p,t)}, \quad \dot{p}_3 = -\frac{x_2^2(\partial c/\partial x_3)(x_3,\tau)}{2H_\pm(x,p,\tau)}, \qquad (35)$$
$$x_3(0) = x_{3o}, \quad p_3(0) = 0.$$

It is easily seen that the solution (x_3,p_3) of system (35) is a pair of homogeneous functions of degree 0 and 1 respectively in variables x_1 and x_2. Hence $S_\pm$ is a homogeneous function of degree 1 in the same variables.

Now we turn to transport equation (25). It can be rewritten using equation (35) and the fact that $H_\pm \neq 0$ for $x_1^2 + x_2^2 > 0$ as follows:

$$2H_\pm(x,p,t)\left(\left[\frac{\partial}{\partial t} + \dot{x}_3\frac{\partial}{\partial x_3}\right] + f_\pm(x,t)\right)\phi_{k\pm}(x,t) + R_{1\pm}\phi_{k-1\pm} + R_{2\pm}\phi_{k-2\pm} + R_{3\pm}\phi_{k-3\pm} = 0, \quad k = 0,1,\ldots, \qquad (36)$$

where $f_\pm(x,t)$, $R_{i\pm}$, $i = 1,2,3$, are functions and differential operators, which can easily be calculated once the function $S_\pm$ is obtained. Let

$$A_\pm(x,t) = \exp(-\int_o^t f_\pm d\tau) \tag{37}$$

(the integral in (37) and below is taken along the trajectory of system (35) which meets the point x at time t). Then the recurrent formulas

$$\phi_{k\pm}(x,t) = A_\pm(x,t)(\phi_{ko\pm}(x_o(x,t)) - \frac{1}{2}\int_o^t A_\pm^{-1}H_\pm^{-1}\sum_{j=1}^{3} R_{j\pm}\phi_{k-j\pm}d\tau) \tag{38}$$

define the solution of transport equation (25) (here $\phi_{ko\pm}(x)$ are initial data for functions $\phi_{k\pm}(x,t)$).

Constructing the functions $S_\pm$, $\phi_{k\pm}$, we see that the functions

$$\psi_\pm(x,\lambda,t) = e^{i\lambda S_\pm(x,t)} \cdot \sum_{k=0}^{s-1} (-i\lambda)^{-k}\phi_{k\pm}(x,t) \tag{39}$$

are smooth for $x_1^2 + x_2^2 \neq 0$ and satisfy equation (19). We set

$$\phi^{(1)}_{oo\pm}(x) = \pm i(2\sqrt{x_1^2 + c(x_3,0)x_2^2})^{-1}; \tag{40}$$

$$\phi^{(1)}_{ko\pm}(x) = 0, \; k = 1,2,\ldots,s-1,$$

$$\psi^{(1)}(x,\lambda,t) = \psi^{(1)}_+(x,\lambda,t) + \psi^{(1)}_-(x,\lambda,t). \tag{41}$$

The upper index in brackets means that the functions, which we construct, correspond to $g_{N,1}$. Then

$$\psi^{(1)}|_{t=0} = 0, \; \frac{\partial\psi^{(1)}}{\partial t}|_{t=0} = \lambda. \tag{42}$$

We set (the number $s = s(N)$ will be chosen below)

$$g_{N,1}(x,\lambda,t) = (\psi^{(1)}(x,\lambda,t)/\lambda)\chi(x_1\lambda,x_2\lambda), \tag{43}$$

where $\chi(x_1,x_2) = 0$ in a neighborhood of zero in $\mathbb{R}^2$,

$$1 - \chi \in C_o^\infty(\mathbb{R}^2).$$

By setting

$$\phi^{(0)}_{oo\pm}(x) = \frac{1}{2}, \; \phi^{(0)}_{ko\pm}(x) = 0, \; k = 1,2,\ldots,s-1, \tag{44}$$

we obtain for $\psi^{(0)} = \psi^{(0)}_+ + \psi^{(0)}_-$,

$$\psi^{(0)}|_{t=0} = 1, \; \frac{\partial\psi^{(0)}}{\partial t}|_{t=0} = 0. \tag{45}$$

We set

$$g_{N,o}(x,\lambda,t) = \psi^{(0)}(x,\lambda,t)\chi(x_1\lambda,x_2\lambda). \tag{46}$$

It is easily seen that equations (24) - (25) are homogeneous with respect to (x_1,x_2), as well as the initial conditions (32), (40), (41), and (44) so that the obtained solutions can be written in the form

$$g_{N,i}(x,\lambda,t) = G_{N,i}(x_1\lambda,x_2\lambda,x_3,t), \; i = 0,1, \tag{47}$$

where

$$G_{N,i}(y,t) = g_{N,i}(y,1,t). \tag{48}$$

The functions $G_{N,i}(y,t)$ satisfy initial conditions (16) up to smooth functions finite with respect to (y_1,y_2) and independent of y_3. If we subtract these functions from $G_{N,i}$, we obtain on the right-hand side of (15) and additional term which is finite with respect to (y_1,y_2) and, consequently, satisfies the estimate (11). Thus we may assume the initial conditions to be satisfied precisely. Substituting functions $G_{N,i}$ into equation (15), we obtain

$$R_{N,i} \equiv -(-i \frac{\partial}{\partial t})^2 G_{N,i} + \{(y_1 - i \frac{\partial}{\partial y_2} - i \frac{\partial}{\partial y_3})^2 + c(y_3,t)y_2^2\} G_{N,i} \equiv$$

$$\equiv [-(-i \frac{\partial}{\partial t})^2 + (y_1 - i \frac{\partial}{\partial y_2} - i \frac{\partial}{\partial y_3})^2 + c(y_3,t)y_2^2, \chi(y_1,y_2)]\psi^{(i)}(y,1,t) + \tag{49}$$

$$+ \chi(y_1,y_2)\{(-(-i \frac{\partial}{\partial t})^2 + (y_1 - i \frac{\partial}{\partial y_2} - i \frac{\partial}{\partial y_3})^2 + c(y_3,t)y_2^2)\psi^{(i)}(y,1,t)\}$$

(in formula (49) the square brackets denote a commutator). The properties of function χ yield that the first term on the right-hand side of (49) is a smooth function finite with respect to (y_1,y_2). Using (47), we rewrite the expression in curly brackets on the right-hand side of (49) as follows:

$$F(y) \stackrel{\text{def}}{=} \{\lambda^2(-(-i\lambda^{-1} \frac{\partial}{\partial t})^2 + (x_1 - i\lambda^{-2} \frac{\partial}{\partial x_2} - i\lambda^{-1} \frac{\partial}{\partial x_3})^2 + c(x_3,t)x_2^2) \times$$

$$\times \psi^{(i)}(x,\lambda,t)\}\Big|_{\substack{x_1 = y_1/\lambda \\ x_2 = y_2/\lambda \\ x_3 = y_3}} = \lambda^2\hat{O}(\lambda^{-s})\Big|_{\substack{x_1 = y_1/\lambda \\ x_2 = y_2/\lambda \\ x_3 = y_3}} = \hat{O}(\lambda^{-s+2})\Big|_{\substack{x_1 = y_1/\lambda \\ x_2 = y_2/\lambda \\ x_3 = y_3}} \tag{50}$$

since the functions $\psi^{(i)}(x,\lambda,t)$, $i = 1,2$, which have been constructed, satisfy the equation (19). For $(y_1,y_2) \in \operatorname{supp} \chi$, $y_1^2 + y_2^2 \geqslant r^2 > 0$ holds, so setting $\lambda = \frac{1}{r}\sqrt{(y_1^2 + y_2^2)} \geqslant 1$, $x_i = y_i/\lambda$, $i = 1,2$, $x_3 = y_3$, we obtain by (20) (choosing $K = \{(x_1,x_2) | x_1^2 + x_2^2 = r\} \subset \mathbb{R}^2 \setminus \{0\}$) that

$$\left| \frac{\partial^2 F(y)}{\partial y_1^{\alpha_1} \partial y_2^{\alpha_2} \partial y_3^{\alpha_3}} \right| \leqslant C_{k,\alpha}(y_1^2 + y_2^2)^{-s+\alpha_3+2}, \quad (y_1^2 + y_2^2 \geqslant 1). \tag{51}$$

Hence, $R_{N,i}$ also satisfies the estimate (51), probably with other constants $C_{k,\alpha}$.

Choosing $s = N_1(N) + 2$, we obtain by Theorem 1 that the function (5) with symbols $G_{N,o}(y,t)$, $G_{N,1}(y,t)$ constructed above solves the problem (1) - (2) in case (a).

Now consider the case (b) (the case (c) is quite analogous). The Hamilton-Jacobi equation in case (b) has the form

$$\frac{\partial S}{\partial t} + H_{\pm}(x,p,\varepsilon,t) = O(\varepsilon^m) \tag{52}$$

with the Hamiltonian

$$H_{\pm}(x,p,\varepsilon,t) = \pm\sqrt{(x_1 + p_3)^2 + c(x_3,t)x_2^{2m} + 2\varepsilon p_2(x_1 + p_3) + \varepsilon^2 p_2^2},$$

$$H_{\pm}(x,p,0,t) \equiv H_{o\pm} = \pm\sqrt{(x_1 + p_3)^2 + c(x_3,t)x_2^{2m}}. \tag{53}$$

Equation (52) can be solved by the successive approximations method. Let

$$S_\pm = S_\pm(x,t,\varepsilon); \quad S_\pm^{(k)}(x,t) = \left(\frac{\partial^k}{\partial \varepsilon^k} S_\pm(x,t,\varepsilon)\right)\Big|_{\varepsilon=0};$$

then we obtain the following system of equations, which allows us to find the functions $S_\pm^{(k)}(x,t)$:

$$\frac{\partial S_\pm^{(0)}}{\partial t} + H_{o\pm}\left(x, \frac{\partial S_\pm^{(0)}}{\partial x}, t\right) = 0,$$

$$\frac{\partial S_\pm^{(1)}}{\partial t} + \frac{\partial H_{o\pm}}{\partial p}\left(x, \frac{\partial S_\pm^{(0)}}{\partial x}, t\right)\frac{\partial S_\pm^{(1)}}{\partial x} + \frac{\partial H_\pm}{\partial \varepsilon}\left(x, \frac{\partial S_\pm^{(0)}}{\partial x}, 0, t\right) = 0, \quad (54)$$

$$\cdots\cdots\cdots\cdots\cdots\cdots\cdots\cdots\cdots\cdots$$

$$\frac{\partial S_\pm^{(k)}}{\partial t} + \frac{\partial H_{o\pm}}{\partial p}\left(x, \frac{\partial S_\pm^{(0)}}{\partial x}, t\right)\frac{\partial S_\pm^{(k)}}{\partial x} + F_k\left[x, t, \frac{\partial S_\pm^{(0)}}{\partial x}, \ldots, \frac{\partial S_\pm^{(k-1)}}{\partial x}\right] = 0,$$

where the functions F_k can be easily obtained in a recurrent way. The solution of the first equation in (54) coincides with the above-constructed solution of equations (30) - (31) (one should only replace x_2^2 by x_2^{2m}).

All subsequent equations in (54) are ordinary linear differential equations along the trajectories of the Hamiltonian system, corresponding to the Hamiltonian function $H_{o\pm}$; hence they can be solved by ordinary integration. Quite analogously the transport equations can be solved by the methods of perturbation theory with respect to the small parameter ε.

Consider now some special cases. Let $m = 1$. If the function $c(x_3,t) \equiv c(x_3)$, i.e., it is independent of t, equations (35) can be integrated in a more explicit way. Namely, the Hamiltonian function $H_\pm$ is independent of t in this case and therefore it is constant along the trajectories of system (35) [2]. Differentiating the first equation with respect to τ, we obtain

$$\ddot{x}_3 = -\frac{x_2^2 c'(x_3)}{2(x_1^2 + c(x_{30})x_2^2)}; \quad x_3(0) = x_{30}; \quad \dot{x}_3(0) = \frac{\pm x_1}{\sqrt{(x_1^2 + c(x_{30})x_2^2)}}; \quad (55)$$

hence

$$(\dot{x}_3)^2 = -\int_{x_{30}}^{x_3} \frac{x_2^2 c'(x_3)dx_3}{x_1^2 + c(x_{30})x_2^2} + \frac{x_1^2}{x_1^2 + c(x_{30})x_2^2} = 1 - \frac{x_2^2 c(x_3)}{x_1^2 + c(x_{30})x_2^2}, \quad x_3(0) = x_{30}. \quad (56)$$

The sign of $\dot{x}_3$ coincides with the sign of x_1 when τ belongs to a neighborhood of zero, which depends on (x_1,x_2), but as $x_1 = 0$ the sign of $\dot{x}_3$ is opposite to the sign of $c'(x_{30})$.

Equation (56) is an equation with separable variables and thus can be integrated:

$$d\tau = \pm \frac{dx_3}{\sqrt{\left(1 - \frac{x_2^2 c(x_3)}{x_1^2 + c(x_{30})x_2^2}\right)}} \quad (57)$$

(the signs ± in (57) are, generally speaking, alternative on different segments of trajectories and, to be more precise, one should know the concrete form of function $c(x_3)$; an example will be studied below).

It is easily seen that

$$(pH_{\pm p} - H_{\pm})(x,p,t) = x_1\dot{x}_3 \mp \frac{c(x_3)x_2^2}{\sqrt{(x_1^2 + c(x_{30})x_2^2)}} ; \tag{58}$$

hence

$$S_{\pm}(x,t) = x_1(x_3 - x_{30}(x,t)) \mp \\ \pm \frac{x_2^2}{\sqrt{(x_1^2 + c(x_{30}(x,t))x_2^2)}} \int_0^t c(x_3(x_{30}(x,t),\tau))d\tau. \tag{59}$$

In particular, if $c(x_3) = \cos x_3 + 2$, then x_3 satisfies the equation of pendulum oscillations:

$$\ddot{x}_3 + \frac{x_2^2}{2(x_1^2 + x_2^2(\cos x_{30} + 2))} \sin x_3 = 0, \\ x_3(0) = x_{30}, \quad \dot{x}_3(0) = \pm x_1/\sqrt{x_1^2 + x_2^2(2 + \cos x_{30})} \tag{60}$$

with the parameter depending on x_1, x_2, and initial conditions.

Now consider the case $c(x_3) \equiv 1$. In this case we shall compare the constructed asymptotics with the usual asymptotics with respect to smoothness. Note that the equation obtained in this case can be solved precisely by other methods and is used here only as an example.

The Hamilton-Jacobi equation has the form

$$\frac{\partial S_{\pm}}{\partial t} \pm \sqrt{((x_1 + \frac{\partial S}{\partial x_3})^2 + x_2)} = 0, \quad S_{\pm}|_{t=0} = 0. \tag{61}$$

It has the solution

$$S_{\pm}(x,t) = \mp t\sqrt{x_1^2 + x_2^2}. \tag{62}$$

The transport equation (the solution of which is evidently independent of x_3, so we omit the terms including derivatives in x_3) has the form:

$$\pm 2\sqrt{x_1^2 + x_2^2}\frac{\partial \phi_k}{\partial t} \mp 2i \frac{x_1 x_2 t}{\sqrt{(x_1^2 + x_2^2)}} \phi_k + \{-\frac{\partial^2}{\partial t^2} + 2ix_1 \frac{\partial}{\partial x_2} - t^2 \frac{x_2^2}{x_1^2 + x_2^2}\}\phi_{k-1} \pm \\ \pm \{2\frac{tx_2}{\sqrt{(x_1^2 + x_2^2)}} \frac{\partial}{\partial x_2} + \frac{tx_1^2}{(x_1^2 + x_2^2)^{3/2}}\}\phi_{k-2} - \frac{\partial^2 \phi_{k-3}}{\partial x_2^2} = 0, \quad k = 0,1,\ldots, \tag{63}$$

or

$$\frac{\partial \phi_o}{\partial t} - i\frac{x_1 x_2 t}{x_1^2 + x_2^2}\phi_o = 0, \\ \frac{\partial \phi_1}{\partial t} - i\frac{x_1 x_2 t}{x_1^2 + x_2^2}\phi_1 = \mp \frac{1}{2\sqrt{(x_1^2 + x_2^2)}} \{-\frac{\partial^2}{\partial t^2} + 2ix_1 \frac{\partial}{\partial x_2} - t^2 \frac{x_2^2}{x_1^2 + x_2^2}\}\phi_o, \tag{64}$$

and so on. Hence

$$\phi_o(x,t) = \phi_o(x,0)\exp\{\frac{i}{2} t^2 \frac{x_1x_2}{x_1^2+x_2^2}\},$$

$$\phi_1(x,t) = \mp \int_o^t (\frac{1}{2\sqrt{(x_1^2+x_2^2)}} [-\frac{\partial^2}{\partial t^2} + 2ix_1 \frac{\partial}{\partial x_2} - \tau^2 \frac{x_2^2}{x_1^2+x_2^2}] \times \tag{65}$$

$$\times \phi_o(x,\tau))\exp\{\frac{i}{2}\frac{x_1x_2}{x_1^2+x_2^2}(t^2-\tau^2)\}d\tau,$$

and so on (here we used the fact that $\phi_1(x,0) = 0$). So taking initial conditions into consideration, we have

$$\phi^{(1)}_{o\pm}(x,t) = \pm \frac{i}{2\sqrt{(x_1^2+x_2^2)}} \exp\{\frac{i}{2} t^2 \frac{x_1x_2}{x_1^2+x_2^2}\},$$

$$\phi^{(0)}_{o\pm}(x,t) = \frac{1}{2}\exp\{\frac{i}{2} t^2 \frac{x_1x_2}{x_1^2+x_2^2}\}; \tag{66}$$

$$G_{N,1}(y,t) = \chi(y_1,y_2)\frac{i}{2\sqrt{(y_1^2+y_2^2)}}\exp\{\frac{i}{2} t^2 \frac{y_1y_2}{y_1^2+y_2^2}\} \times$$

$$\times [e^{it\sqrt{y_1^2+y_2^2}} - e^{-it\sqrt{y_1^2+y_2^2}}] = \frac{\chi(y_1,y_2)}{\sqrt{(y_1^2+y_2^2)}}\exp\{\frac{it^2y_1y_2}{2(y_1^2+y_2^2)}\} \times \tag{67}$$

$$\times \sin(t\sqrt{y_1^2+y_2^2}) \text{ (+ lower-order terms);}$$

$$G_{N,o}(y,t) = \chi(y_1,y_2)\exp\{\frac{it^2y_1y_2}{2(y_1^2+y_2^2)}\}\cos(t\sqrt{y_1^2+y_2^2}) \text{ (+ lower-order terms).} \tag{68}$$

The usual asymptotics with respect to smoothness of the problem (1) - (2) can be constructed as follows ($m = 1$, $c(x,t) \equiv 1$). We shall seek the asymptotic solution in the following form:

$$u(x,t) = \bar{G}_o(\overset{1}{-i\frac{\partial}{\partial x}}, \overset{2}{x}, t)u_o(x) + \bar{G}_1(\overset{1}{-i\frac{\partial}{\partial x}}, \overset{2}{x}, t)u_1(x). \tag{69}$$

The function $\bar{G}_i(y_1,y_2,t)$ satisfies the equation

$$-(-i\frac{\partial}{\partial t})^2\bar{G}_i + \{(y_1 - i\frac{\partial}{\partial y_2})^2 + y_2^2\}\bar{G}_i = \bar{R}_i, \tag{70}$$

where $\bar{R}_i$ is the symbol of the smoothing operator, or after the change of variables

$$y_1 = \lambda x_1, \quad y_2 = x_2, \tag{71}$$

$$-(-i\lambda^{-1}\frac{\partial}{\partial t})^2\bar{g}_i + \{(x_1 - i\lambda^{-1}\frac{\partial}{\partial x_2})^2 + \lambda^{-2}x_2^2\}\bar{g}_i = \hat{O}(\lambda^{-s}). \tag{72}$$

Seeking $\bar{g}_i$ in the form of a linear combination of the functions

$$\bar{\psi}(x,\lambda,t) = e^{i\lambda\bar{S}(x,t)}\cdot\sum_{k=0}^{s-1}(-i\lambda)^{-k}\cdot\bar{\phi}_k(x,t), \tag{73}$$

we obtain the Hamilton-Jacobi equation for $\bar{S}$:

$$\left(\frac{\partial\bar{S}}{\partial t}\right)^2 = \left(x_1 + \frac{\partial\bar{S}}{\partial x_2}\right)^2, \tag{74}$$

and the transport equation for $\bar{\phi}_k(x,t)$:

$$\begin{aligned}&\left[-2\frac{\partial\bar{S}}{\partial t}\frac{\partial}{\partial t} + 2\left(x_1 + \frac{\partial\bar{S}}{\partial x_2}\right)\frac{\partial}{\partial x_2} - \frac{\partial^2\bar{S}}{\partial t^2} + \frac{\partial^2\bar{S}}{\partial x_2^2}\right]\bar{\phi}_k(x,t) + \\ &+ \left[-\frac{\partial^2}{\partial t^2} + \frac{\partial^2}{\partial x_2^2} - x_2^2\right]\bar{\phi}_{k-1}(x,t) = 0, \quad k = 0,1,2,\ldots\end{aligned} \tag{75}$$

(recall that $\bar{\phi}_k = 0$ for $k = -1$ as it was assumed before). By solving equation (74) with zero initial condition, we obtain

$$\bar{S}_{\pm}(x,t) = \pm t x_1. \tag{76}$$

Then the transport equation becomes

$$\mp\frac{\partial\bar{\phi}_k}{\partial t} + \frac{\partial\bar{\phi}_k}{\partial x_2} + \frac{1}{x_1}\left[-\frac{\partial^2\bar{\phi}_{k-1}}{\partial t^2} + \frac{\partial^2\bar{\phi}_{k-1}}{\partial x_2^2} + x_2^2\bar{\phi}_{k-1}\right] = 0 \tag{77}$$

or

$$\frac{\partial\bar{\phi}_o}{\partial t} \mp \frac{\partial\bar{\phi}_o}{\partial x_2} = 0, \tag{78}$$

$$\frac{\partial\bar{\phi}_k}{\partial t} \mp \frac{\partial\bar{\phi}_k}{\partial x_2} = \pm\frac{1}{x_1}\left[-\frac{\partial^2\bar{\phi}_{k-1}}{\partial t^2} + \frac{\partial^2\bar{\phi}_{k-1}}{\partial x_2^2} + x_2^2\bar{\phi}_{k-1}\right]. \tag{79}$$

By setting $\bar{\phi}^{(1)}_{oo\pm}(x) = \mp i/2x_1$, $\bar{\phi}^{(0)}_{oo\pm}(x) = \frac{1}{2}$, we obtain

$$\bar{\phi}^{(1)}_{o\pm}(x,t) = \mp\frac{i}{2x_1}\ ;\ \bar{\phi}^{(0)}_{o\pm}(x,t) = \frac{1}{2}. \tag{80}$$

Introducing new variables $\xi = (x_2 + t)/2$, $\eta = (x_2 - t)/2$, we rewrite equation (79) (for the upper sign) as follows:

$$\frac{\partial\bar{\phi}_{k+}}{\partial\eta} = -\frac{1}{x_1}\left[(\xi+\eta)^2 + \frac{\partial^2}{\partial\xi\partial\eta}\right]\bar{\phi}_{k-1+} \tag{81}$$

with the initial conditions

$$\bar{\phi}_{k+}|_{\xi=\eta} = 0, \quad k = 1,2,\ldots\ .$$

Hence

$$\begin{aligned}\bar{\phi}_{k+}(\xi,\eta) &= -\frac{1}{x_1}\int_{\xi}^{\eta}\left[(\xi+\eta)^2 - \frac{\partial^2}{\partial\xi\partial\eta}\right]\bar{\phi}_{k-1+}(\xi,\eta)d\eta, \quad k = 1,2,\ldots, \\ \bar{\phi}^{(1)}_{1+} &= \frac{i}{2x_1^2}\left[\frac{x_2^3}{3} - \frac{(x_2+t)^3}{3}\right],\end{aligned} \tag{82}$$

and so on. Analogously,

$$\begin{aligned}\bar{\phi}^{(1)}_{1-} &= -\frac{i}{6x_1^2}\left[x_2^3 - (x_2+t)^3\right], \\ \bar{\phi}^{(0)}_{1\pm} &= -\frac{1}{2x_1}\left[\frac{x_2^3}{3} - \frac{(x_2+t)^3}{3}\right].\end{aligned} \tag{83}$$

Finally, we obtain

$$\bar{G}_1(y_1,y_2,t) = \frac{i}{2y_1}(e^{-ity_1} - e^{ity_1}) + \frac{i}{6y_1^2}[y_2^3 - (y_2+t)^3](e^{ity_1} - e^{-ity_1})\ldots,$$

$$\bar{G}_o(y_1,y_2,t) = \frac{1}{2}(e^{ity_1} - e^{-ity_1}) - \frac{1}{6y_1}[y_2^3 - (y_2+t)^3](e^{ity_1} - e^{-ity_1})\ldots, \tag{84}$$

$$\bar{G}_1 = \sin(ty_1)/y_1 + (\text{lower-order terms}),$$

$$\bar{G}_o = \cos(ty_1) + (\text{lower-order terms}).$$

If concrete initial conditions u_o, u_1 are given, then the asymptotics with respect to smoothness and the mixed asymptotics with respect to smoothness and growth at infinity of the solution of problem (1) - (2) are given by formulas (69) and (5) respectively. These formulas may be rewritten more explicitly:

$$u_{sm}(x,t) = \frac{1}{\sqrt{(2\pi)}}\int e^{ipx}\{\bar{G}_o(p,x,t)\tilde{u}_o(p) + \bar{G}_1(p,x,t)\tilde{u}_1(p)\}dp \tag{85}$$

and

$$u_{mix}(x,t) = \frac{1}{\sqrt{(2\pi)}}\int e^{ipx}\{G_o(p,x,x,t)\tilde{u}_o(p) + G_1(p,x,x,t)\tilde{u}_1(p)\}dp, \tag{86}$$

where $\tilde{u}_o(p)$ and $\tilde{u}_1(p)$ are Fourier transforms of initial data

$$\tilde{u}_i(p) = \frac{1}{\sqrt{(2\pi)}}\int e^{-ipx}u_i(x)dx,\ i = 0,1, \tag{87}$$

and the functions G_i, $\bar{G}_i$ are given by formulas (67), (68), and (84).

Formulas (82) and (83) show that the asymptotics $u_{sm}(x,t)$ (see (85)) is not uniform in x. One can show that the more terms of asymptotics taken, the stronger this non-uniformity becomes. Since the functions $\phi_{k\pm}$ behave as positive powers of x_2, the mixed asymptotics $u_{mix}(x,t)$ gives not only uniform smoothness of the remainder on the whole axis, but also its decrease as $|x| \to \infty$.

If we restrict our consideration to bounded domain $\mathcal{D} \in \mathbb{R}^1$ only, then for $x \subset \mathcal{D}$ the asymptotics u_{mix} yields the asymptotics u_{sm}. Here we show this only for the principal term of asymptotics. Consider the behavior of function (67) for $y_2 \in \mathcal{D}$ and $|y_1| \to \infty$. We have

$$\frac{1}{\sqrt{(y_1^2+y_2^2)}}\exp\{\frac{it}{2}\frac{y_1y_2}{y_1^2+y_2^2}\}\sin(t\sqrt{y_1^2+y_2^2}) =$$

$$= \frac{\sin(ty_1\sqrt{1+\alpha^2})}{y_1\sqrt{(1+\alpha^2)}}\exp\{\frac{it}{2}\frac{\alpha}{1+\alpha^2}\} + \frac{\sin(ty_1)}{y_1} + O(\alpha) \tag{88}$$

as $\alpha \to 0$ (here $\alpha = y_2/y_1 \to 0$ as $|y_1| \to \infty$ uniformly in $y_2 \in \mathcal{D}$). Thus $G_{N,1}$ differs from $\bar{G}_1$ by a symbol of the smoothing operator, which is uniformly smoothing in any bounded domain. Analogously, one can show that $G_{N,o}$ turns into $\bar{G}_o$ in any bounded domain.

In conclusion we present the principal term of asymptotics for special initial data. Let initial data have the form:

$$u_o(x) = 0,\quad u_1(x) = e^{ix/h}, \tag{89}$$

where $h \to 0$ is a small parameter. Then the principal term of asymptotics at infinity, which is uniformly smooth with respect to h, has the form

$$u_{mix}(x,t) = \frac{\chi(h^{-1},x)h}{\sqrt{(1+x^2h^2)}} e^{ix/h} e^{(it^2/2)(xh/1+x^2h^2)} \sin(\frac{t}{h}\sqrt{1+x^2h^2}). \quad (90)$$

Note. The asymptotics constructed above in the form given here are valid only on segment $t \in [0,T]$ such that there are no focal points when the solution of the Hamilton-Jacobi equation is being constructed. In the general case the solution is given by means of a canonical operator (see Chapter 3), but we do not give these formulas to avoid cumbersome analytic constructions.

B. Theorem on Asymptotic Solutions

In this item we formulate in general form the theorem on quasi-invertibility of (pseudo) differential operators with growing coefficients and give the scheme of its proof. The details of the proof are omitted since this theorem is a special case of the general theorem on quasi-invertibility which will be formulated and proved in subsequent sections. First of all we give necessary definitions.

Definition 1. The function $f(z) = f(z_1,\ldots,z_n)$ is called $(\pi_1,\ldots,\pi_n)$-quasi-homogeneous of degree r (here $\pi_1,\ldots,\pi_n,r$ are real numbers, $\pi_j > 0$, $j = 1,\ldots,n$), if for any $\lambda > 0$ the following equality holds:

$$f(\lambda^{\pi_1}z_1,\ldots,\lambda^{\pi_n}z_n) = \lambda^r f(z_1,\ldots,z_n). \quad (1)$$

Definition 2. The function $\phi(z_1,\ldots,z_n)$ is called $(\pi_1,\ldots,\pi_n)$-small of order s, if the estimates

$$\left| \frac{\partial^\alpha \phi(z)}{\partial z^2} \right| \leqslant C_\alpha [\sqrt{\Sigma |z_i|^{2/\pi_i}}]^{-s-|\alpha|}, \quad |\alpha| = 0,1,2,\ldots \quad (2)$$

hold for $\Sigma |z_i|^{2/\pi_i} \geqslant 1$.

Definition 3. Smooth function $f(z_1,\ldots,z_n)$ is called asymptotically $(\pi_1,\ldots,\pi_n)$-quasi-homogeneous of degree r, if for arbitrarily large s the following representation is valid:

$$f(z_1,\ldots,z_n) = \sum_{k=0}^{N(s)} f_k(z_1,\ldots,z_n) + \phi_s(z_1,\ldots,z_n), \quad (3)$$

where $\phi_s(z_1,\ldots,z_n)$ is a $(\pi_1,\ldots,\pi_n)$-small function of the order s, and $f_k(z_1,\ldots,z_n)$ is a $(\pi_1,\ldots,\pi_n)$-quasi-homogeneous function of the degree r_k, $r = r_o > r_1 > r_2 > \ldots$.

If f also depends on additional variables ω, then it is called asymptotically $(\pi_1,\ldots,\pi_n)$-quasi-homogeneous of degree r with respect to $(z_1, \ldots,z_n)$ uniformly in ω, if the following conditions hold:

(a) The functions f_k from (3) are bounded together with all their derivatives on the quasi-sphere $\Sigma |z_i|^{2/\pi_i} = 1$ uniformly in ω.

(b) The constants from (2) in the estimate of the "remainder" ϕ_s in (3) may be chosen independent of ω.

Theorem 1. Let $F(x_1,\ldots,x_n,z_1,\ldots,z_n,\xi_1,\ldots,\xi_n)$ be an asymptotically $(\rho_1,\ldots,\rho_{2n})$-quasi-homogeneous function of degree r with respect to $(z_1,\ldots,z_n,\xi_1,\ldots,\xi_n)$ uniformly in $x_1,\ldots,x_n$ and $\min\{\rho_{n+1},\ldots,\rho_{2n}\} = 1$. Let the Hamiltonian function

$$H(\omega,h,q,p) \equiv H(\omega_1,\ldots,\omega_n,h,q_1,\ldots,q_{2n},p_1,\ldots,p_{2n}) =$$

$$\stackrel{\text{def}}{=} h^r F(q_1,\ldots,q_n,h^{-\rho_1}q_{n+1},\ldots,h^{-\rho_n}q_{2n},h^{-\rho_{n+1}}(\omega_1 + h^{\rho_{n+1}-1}p_1 + \tag{4}$$

$$+ h^{\rho_1+\rho_{n+1}-1}p_{n+1},\ldots,h^{-\rho_{2n}}(\omega_n + h^{\rho_{2n}-1}p_n + h^{\rho_n+\rho_{2n}-1}p_{2n}))$$

satisfy absorption conditions (see below). Then the operator

$$\hat{F} = F(\overset{2}{x}_1,\ldots,\overset{2}{x}_n,\overset{2}{x}_1,\ldots,\overset{2}{x}_n,\overset{1}{-i\frac{\partial}{\partial x_1}},\ldots,\overset{1}{-i\frac{\partial}{\partial x_n}}) \tag{5}$$

in $L^2(\mathbb{R}^n)$ has the right quasi-inverse in the following sense: there exists the sequence $\{\hat{G}_N\}$ of operators in $L^2(\mathbb{R}^n)$ such that

$$\hat{F} \circ \hat{G}_N = 1 + \hat{R}_N, \tag{6}$$

where for $\mu + \nu \leqslant N$ the operator

$$T_{\mu\nu N} = x^\mu \left(-i\frac{\partial}{\partial x}\right)^\nu R_N \tag{7}$$

is bounded in $L^2(\mathbb{R}^n)$.

Formulate now the absorption conditions. The quasi-homogeneity of the function F and formula (4) imply that, for arbitrary N, function $H(\omega,h,q,p)$ can be expanded in a series in terms of (fractional) powers of h modulo h^N,

$$H(\omega,h,q,p) \sim \sum_{j=0}^{\infty} h^{\varepsilon_j} H_j(\omega,q,p); \tag{8}$$

here $0 = \varepsilon_o < \varepsilon_1 < \ldots$, $\lim_{j\to\infty} \varepsilon_j = +\infty$. We set

$$H_{ess}(\omega,h,q,p) = \sum_{j\,:\,\varepsilon_j<1} h^{\varepsilon_j} H_j(\omega,q,p) \tag{9}$$

and call function (9) the essential Hamiltonian of the operator (5) (see [52]).

Thus the essential Hamiltonian contains together with the zero term of the expansion of function $H(\omega,h,q,p)$ in powers of h, all the subsequent terms of this expansion with $\varepsilon_j < 1$. Denote by Ω the set of points (ω,q,p) such that

$$p = 0,\ \sum_{i=1}^{n} (|\omega_i|^{2/\rho_{n+i}} + |q_{n+i}|^{2/\rho_i}) = 1,\ H_o(\omega,q,p) = 0 \tag{10}$$

and by Ω_ε, $\varepsilon > 0$, the domain

$$\Omega_\varepsilon = \{(\omega,q,p)\,|\,\mathrm{dist}((\omega,q,p),\Omega) < \varepsilon\}.$$

Definition 4. (cf. [52]). We shall say that the considered problem satisfies the absorption conditions, if there exist $\varepsilon > 0$, $T > 0$, and a continuous function

$$\tau = \tau(\omega,q,p),\ 0 \leqslant \tau(\omega,q,p) \leqslant T,\ (\omega,q,p) \in \Omega_\varepsilon, \tag{11}$$

such that:

(a) The trajectories $(q(q_o,p_o,\omega,t),p(q_o,p_o,\omega,t))$ of the Hamiltonian system

$$\dot{q} = \frac{\partial \operatorname{Re} H_o}{\partial p}(\omega,q,p),$$

$$\dot{p} = -\frac{\partial \operatorname{Re} H_o}{\partial q}(\omega,q,p), \tag{12}$$

$$q|_{t=0} = q_o,\ p|_{t=0} = p_o,\ (\omega,q_o,p_o) \in \Omega_\varepsilon$$

are defined for $0 \leqslant t \leqslant \tau(\omega,q_o,p_o)$ and the mapping

$$[0,T] \times \Omega_\varepsilon \to [0,T] \times \mathbb{R}^{2n}_q \times \mathbb{R}^n_\omega,\ (t,\omega,q_o,p_o) \to (t,q(q_o,p_o,\omega,t),\omega)$$

is a proper one.

(b) The inequality

$$\operatorname{Im} H_{ess} \leqslant 0 \tag{13}$$

holds on these trajectories; besides

$$\operatorname{Im} H_{ess}(\omega,h,q(q_o,p_o,\omega,\tau(\omega,q_o,p_o)),p(q_o,p_o,\omega,\tau(\omega,q_o,p_o))) \leqslant -\varepsilon. \tag{14}$$

<u>Proof of the theorem.</u> We shall seek the operator $\hat{G}_N$ in the form

$$\hat{G}_N = G_N(\overset{1}{-i\frac{\partial}{\partial x}},\overset{2}{x},\overset{2}{x}); \tag{15}$$

then we obtain the following equation for the symbol $G_N(\xi,z,x)$:

$$F(\overset{2}{x_1},\ldots,\overset{2}{x_n},\overset{2}{z_1},\ldots,\overset{2}{z_n},\xi_1 - i\overset{1}{\frac{\partial}{\partial x_1}} - i\overset{1}{\frac{\partial}{\partial z_n}},\ldots,\xi_n - i\overset{1}{\frac{\partial}{\partial x_n}} - i\overset{1}{\frac{\partial}{\partial z_n}}) \times$$
$$\times\ G_N(\xi,z,x) = 1 + R_N(\xi,z,x), \tag{16}$$

where one should take G_N in a form such that R_N satisfy the conditions of Theorem 1 of item A (more precisely, the multi-dimensional version of this theorem is considered).

We change the variables:

$$\xi_i = h^{-\rho_{n+i}}\omega_i,\ x_i = q_i,\ z_i = h^{-\rho_i} q_{n+i},\ i = 1,\ldots,n. \tag{17}$$

Taking into account Theorem 1 of item A, one may rewrite equation (16) in new variables in the form (here $g_N(q,\omega,h) = G_N(h^{-\rho_{n+1}}\omega_1,\ldots,h^{-\rho_{2n}}\omega_n,h^{-\rho_1} \times$ $\times\ q_{n+1},\ldots,h^{-\rho_n}q_{2n},q_1,\ldots,q_n))$:

$$h^{-r}H(\omega,h,\overset{2}{q},-ih\overset{1}{\frac{\partial}{\partial q}})(g_N(q,\omega,h)) \equiv h^{-r}\hat{H}g_N(q,\omega,h) = 1 + \hat{O}(h^{N_1}). \tag{18}$$

Thus we come to the h^{-1}-pseudo-differential equation on the function of $g_N(q,\omega,h)$.

Using the partition of unity, we can rewrite the right-hand side of (18) in the form:

$$1 = \rho_1(q,\omega) + \rho_2(q,\omega), \tag{19}$$

where

$$\text{supp } \rho_1(q,\omega) \subset \{(q,\omega) \,|\, (\omega_1/R^{\rho_{n+1}/2},\ldots,\omega_n/R^{\rho_{2n}/2}, q_1,\ldots,q_n, \times$$
$$\times\, q_{n+1}/R^{\rho_1/2},\ldots,q_{2n}/R^{\rho_n/2}, 0) \subset \Omega_\varepsilon\}. \tag{20}$$

Here

$$R = \sum_j (|\omega_j|^{2/\rho_{n+j}} + |q_{n+j}|^{2/\rho_j});$$

$$H_o(\omega,q,0) \neq 0$$

for

$$(q,\omega) \in \text{supp } \rho_2.$$

We shall seek g_N in the form

$$g_N = h^r(g_{N1} + g_{N2}), \tag{21}$$

where g_{Ni} satisfy the following equations:

$$\hat{H} g_{Ni} = \rho_i(q,\omega) + \hat{O}(h^{N_1}), \quad i = 1,2. \tag{22}$$

First we solve equation (22) for i = 2. Since $H_o(\omega,q,0) \neq 0$ on supp ρ_2, the solution can be obtained by means of successive approximations:

$$g_{N1}(q,\omega,h) = \sum_{k=0}^{N_1} g_{N2}^{(k)}(q,\omega,h)h^k, \tag{23}$$

where $g_{N2}^{(k)}(q,\omega,h)$ depend continuously on $h \in [0,1]$. We obtain the system of equations which enables us to find the functions $g_{N2}^{(k)}(q,\omega,h)$:

$$\sum_{|k|=0}^{N} \frac{\partial^k H}{\partial p^k}(\omega,h,q,0)\left(-ih\frac{\partial}{\partial q}\right)^k \cdot \sum_{j=0}^{N_1} g_{N2}^{(j)}(q,\omega,h)h^j = \rho_2(q,\omega) + \hat{O}(h^N). \tag{24}$$

This system can be solved recurrently, since $H(\omega,h,q,0) \neq 0$ on supp ρ_2 for sufficiently small h.

We shall seek the function $g_{N1}(q,\omega,h)$ in the form:

$$g_{N1}(q,\omega,h) = \frac{i}{h}\int_o^T \psi(q,\omega,h,\tau)d\tau, \tag{25}$$

where the function ψ satisfies the following Cauchy problem:

$$-ih\frac{\partial\psi}{\partial t} + \hat{H}\psi = \hat{O}(h^{N+2}), \quad \psi|_{t=0} = \rho_1(q,\omega) + \hat{O}(h^{N+1}). \tag{26}$$

The absorption conditions imply (see [52]) that the solution of Cauchy problem (26) exists on the segment [0,T] and that $\psi(T) = \hat{O}(h^N)$ due to (14).

Although to be precise this fact is proved in [52] only for finite initial data, nevertheless we use the initial data which are not finite. However, the proof given in [52] is suitable without essential changes if in the definition of absorption conditions one replaces the requirement of finiteness of initial conditions by the requirement that the trajectory tube projection on physical space be proper. This was made in Definition 4. Then we have:

$$\hat{H}g_{N1} = \frac{i}{h}\int_0^T \hat{H}\psi d\tau + \hat{O}(h^{N+1}) = -\int_0^T \frac{\partial\psi}{\partial\tau}\, d\tau + \hat{O}(h^{N+1}) =$$
$$= \psi(0) - \psi(T) + \hat{O}(h^{N+1}) = \rho_1(x,\omega) + \hat{O}(h^{N+1}). \tag{27}$$

Hence the solution of problem (26) is constructed and problem (18) is thus solved. Returning to initial variables, we complete the proof of Theorem 1.

2. POISSON ALGEBRAS AND NONLINEAR COMMUTATION RELATIONS

A. Poisson Algebras

Let N be a smooth manifold of dimension n, and let a smooth homomorphsim

$$\Omega : T^*N \to TN \tag{1}$$

of vector bundles over N be given. (Thus Ω maps linearly the fiber T^*_yN over the arbitrary point $y \in N$ into the fiber T_yN over the same point, see [19].) The induced homomorphism of section spaces we denote by the same letter $\mathcal{T}^*N \to \mathcal{T}N$.

Given a function $f \in C^\infty(N)$, we consider a vector field Y_f on N, given by

$$Y_f = \Omega(df), \tag{2}$$

and define in $C^\infty(N)$ the bilinear operation

$$\{f,g\} \equiv \{f,g\}_\Omega = Y_f(g), \quad f,g \in C^\infty(N). \tag{3}$$

Definition 1. The space $C^\infty(N)$, supported with bilinear operation (3), is called the Poisson algebra on N if and only if for any functions $f,g,h \in C^\infty(N)$, we have

$$\{f,g\} = -\{g,f\},$$
$$\{\{f,g\},h\} + \{\{g,h\},f\} + \{\{h,f\},g\} = 0. \tag{4}$$

We denote the Poisson algebra by $P(N) \equiv P(N,\Omega)$.

Let $(y_1,\dots,y_n)$ be local coordinates in some coordinate chart $\mathcal{U} \subset N$. Using standard coordinates, corresponding to the bases $(dy_1,\dots,dy_n)$ and $(\frac{\partial}{\partial y_1},\dots,\frac{\partial}{\partial y_n})$ in the fibers of $T^*\mathcal{U}$ and $T\mathcal{U}$, respectively, we set a matrix-valued function $\|\Omega_{ik}(y)\|_1^n$ into correspondence to the mapping Ω.

Lemma 2. The conditions (4) are equivalent to the following conditions, given in terms of local coordinates:

$$\Omega_{ik}(y) + \Omega_{ki}(y) = 0, \tag{5}$$

$$\sum_k [\Omega_{ik}(y)\frac{\partial\Omega_{js}(y)}{\partial y_k} + \Omega_{jk}(y)\frac{\partial\Omega_{si}(y)}{\partial y_k} + \Omega_{sk}(y)\frac{\partial\Omega_{ij}}{\partial y_k}] = 0. \tag{6}$$

The proof consists of simple calculation. Further, we assume that conditions (4) of Definition 1 are satisfied.

Definition 3. The function (3) is called the Poisson bracket of the functions f and g, and the vector field Y_f, given by (2), is called the Eulerian vector field correspondent to the function Y_f. Generally, the field Y will be called Eulerian if it may be locally represented in the form Y_f for some $f \in C^\infty(N)$.

Lemma 4. The Poisson bracket and the Eulerian vector field possess the following properties:

$$Y_f\{g,h\} = \{Y_f g,h\} + \{g,Y_f h\}, \tag{7}$$

$$[Y_f,Y_g] \equiv Y_f Y_g - Y_g Y_f = Y_{\{f,g\}}, \tag{8}$$

$$L_{Y_f}(\Omega) = 0. \tag{9}$$

Here L_Y is the Lie derivative along the vector field Y. In the local coordinates the Poisson bracket and Eulerian vector fields are given by the formulas

$$Y_f = \sum_{i,k=1}^{m} \Omega_{ik}(y) \frac{\partial f}{\partial y_k} \frac{\partial}{\partial y_i}, \tag{10}$$

$$\{f,g\} = \sum_{i,k=1}^{n} \Omega_{ik}(y) \frac{\partial f}{\partial y_k} \frac{\partial g}{\partial y_i}. \tag{11}$$

Proof. The equalities (10) - (11) follow directly from definitions; (7) is valid since (2) and (4) are. To prove (8) consider an arbitrary function $h \in C^\infty(M)$. Applying the commutator $[Y_f,Y_g]$ to it, we obtain

$$\begin{aligned}[Y_f,Y_g]h &= \{f,\{g,h\}\} - \{g,\{f,h\}\} = \\ &= -\{\{g,h\},f\} - \{\{h,f\},g\} = \{\{f,g\},h\} = Y_{\{f,g\}}h\end{aligned} \tag{12}$$

(here we used the Jacobi identity).

Now we prove (9). Let ϕ_t be a local one-parametric group of diffeomorphisms of N, generated by the field Y_f. We will show that $\phi_t^*\Omega = \Omega$. Really, it follows from (7) that ϕ_t preserves the Poisson bracket:

$$\phi_t^*\{g,h\} = \{\phi_t^* g,\phi_t^* h\}. \tag{13}$$

We may interpret Ω as a section of vector bundle $TN \otimes TN$; in this interpretation the formula which defines the Poisson bracket becomes

$$\{g,h\} = \langle \Omega, dg \otimes dh \rangle \tag{14}$$

(the brackets $\langle , \rangle$ denote the pairing of covariant and contravariant tensor fields). From this formula it follows that the invariance of the Poisson bracket implies the invariance of Ω. The lemma is proved.

Note 5. Contrary to the case of symplectic manifolds and Hamiltonian vector fields, we cannot consider the equality $L_Y(\Omega) = 0$ as the definition of the Eulerian vector field since even locally there may be no function $f \in C^\infty(N)$ such that $Y = Y_f$. (A trivial example: $\Omega = 0$. Then Y is arbitrary although $Y_f = 0$ for any $f \in C^\infty(N)$.)

Thus we have shown that the Eulerian vector fields on N form an algebra $Eu(N) = Eu(N,\Omega)$ and that the correspondence $f \to Y_f$ is the representation of P(N) in Eu(N). Now we prove simple assertions about the homomorphisms of Poisson algebras.

Definition 6. The homomorphism of Poisson algebras is a Lie algebra homomorphism

$$\Phi : P(N_1,\Omega_1) \to P(N_2,\Omega_2), \tag{15}$$

such that $\Phi = \phi^*$ for some smooth mapping $\phi : N_2 \to N_1$.

Lemma 7. Let $P(N)$ be a Poisson algebra and $f \in P(N)$ be an element such that the vector field Y_f generates the global group $\{\phi_t\}$ of automorphisms of N. Then $\{\phi_t^*\}$ is the group of automorphisms of Poisson algebra $P(N)$. The proof follows from (9).

Lemma 8. Let ϕ^* be a homomorphism of Poisson algebras. Then for any point $y \in N_2$ we have

$$\phi_*(\Omega_2(y)) = \Omega_1(\phi(y)). \tag{16}$$

Moreover, for any function $f \in P(N_1,\Omega_1)$, there is a relation

$$\phi_*(Y^{(2)}_{\phi^* f}(y)) = Y^{(1)}_f(\phi(y)). \tag{17}$$

Proof. Let $g,h \in P(N_1)$. Then

$$\begin{aligned} <\Omega_1, dg \otimes dh>(\phi(y)) &= \phi^*\{g,h\}(y) = \{\phi^* g, \phi^* h\}(y) = \\ &= <\Omega_2, d\phi^* g \otimes d\phi^* h>(y) = <\Omega_2, \phi^*(dg \otimes dh)>(y) \end{aligned} \tag{18}$$

and since g and h are arbitrary, we immediately obtain (16). Prove now (17). Using (16), we obtain

$$\begin{aligned} \phi_*(Y^{(2)}_{\phi^* f}(y)) &= \phi_*(\Omega_2(d\phi^* f))(y) = \\ &= \phi_*(\Omega_2(\phi^* df))(y) = \Omega_1(df)(\phi(y)) = Y^{(1)}_f(\phi(y)). \end{aligned} \tag{19}$$

The lemma is proved.

The well-known example of the Poisson algebra is the Poisson algebra of functions on a symplectic manifold M. The symplectic form ω^2 obviously defined a linear isomorphism of spaces of vector fields and differential 1-forms on M which sets into correspondence to a linear field $Y \in TM$ the differential 1-form

$$\alpha(Y) = Y \lrcorner\, \omega^2 \tag{20}$$

(the fact that α is an isomorphism follows from the independence of the form ω^2).

Denote by $\Omega : T^*M \to TM$ the inverse mapping

$$\Omega = \alpha^{-1}. \tag{21}$$

Lemma 9. The closure of the form ω^2 is equivalent to the condition (6) ((5) follows from the fact that ω^2 is an exterior form). The proof reduces to a straightforward computation.

Thus on any symplectic manifold there is a natural Poisson algebra with a nondegenerate mapping Ω. Vice versa, let the mapping $\Omega : T^*M \to TM$ defining the structure of Poisson algebras, be nondegenerate. Then M is even-dimensional, orientable, and we may define the symplectic structure on M, setting

$$\omega^2(Y,X) = \Omega^{-1}(Y)(X). \tag{22}$$

Next we study Eulerian vector fields on M. Let $f \in C^\infty(M)$. Then the field $Y = Y_f$ is defined by condition $Y = \alpha^{-1}(df)$ or

$$Y \lrcorner\, \omega^2 = df. \tag{23}$$

We have also

$$L_Y\omega^2 = d(Y \lrcorner\, \omega^2) + Y \lrcorner\, d\omega^2 = ddf = 0 \tag{24}$$

since ω^2 is closed.

On the contrary, let Y be such that $L_Y\omega^2 = 0$. Then the form $Y \lrcorner\, \omega^2$ is closed and, consequently, there exists always a function f (locally) such that (23) is satisfied and thus $Y = Y_f$.

Hence the algebra Eu(M) for the symplectic manifold M is the Lie algebra of Hamiltonian vector fields. Let $y \in M$ be an arbitrary point. By the Darboux theorem in the vicinity of y there exists a system of local coordinates, in which the form ω^2 reads

$$\omega^2 = \sum_{j=1}^n dp_j \wedge dq_j \equiv dp \wedge dq. \tag{25}$$

Lemma 10. In the coordinate system $(q_1,\ldots,q_n,p_1,\ldots,p_n)$ the Hamiltonian vector field and Poisson bracket are defined by the equalities

$$Y_f = \frac{\partial f}{\partial p}\frac{\partial}{\partial q} - \frac{\partial f}{\partial q}\frac{\partial}{\partial p} \equiv \sum_{j=1}^n \left(\frac{\partial f}{\partial p_j}\frac{\partial}{\partial q_j} - \frac{\partial f}{\partial q_j}\frac{\partial}{\partial p_j}\right), \quad \{f,g\} = Y_f(g). \tag{26}$$

Proof. Calculating $Y_f \lrcorner\, \omega^2$, where Y_f is defined by (26), we obtain

$$\begin{aligned} Y_f \lrcorner\, \omega^2 &= \sum_{j=1}^n (dq_j(Y_f) \wedge dp_j - dp_j(Y_f) \wedge dq_j) = \\ &= \sum_{j=1}^n \left(\frac{\partial f}{\partial p_j} dp_j + \frac{\partial f}{\partial q_j} dq_j\right) = df, \end{aligned} \tag{27}$$

i.e., we come to (23). The second of the equalities (26) is the immediate consequence of the first one. The lemma is proved.

B. Poisson Algebras and Commutation Relations with Small Parameter

Important examples of Poisson algebras arise in the consideration of nonlinear commutation relations with small parameter $h \to 0$. Let $\mathcal{H}$ be a Hilbert space. Assume that an n-tuple $A_1 = A_1(h),\ldots,A_n = A_n(h)$ of self-adjoint operators, depending on a small parameter $h \in [0,1]$, is given and that the commutation relations

$$[A_j,A_k] = ih\Omega_{jk}(A), \quad j,k = 1,..,n \tag{28}$$

are satisfied. Here $\Omega_{jk}(y_1,\ldots,y_n)$ are the given symbols and we use the standard notation

$$\Omega_{jk}(A) \equiv \Omega_{jk}(\overset{1}{A_1},\ldots,\overset{n}{A_n}).$$

Perform the coordinate change

$$z_1 = \phi_1(y),\ldots,z_n = \phi_n(y) \tag{29}$$

and introduce the operators

$$B_1 = \phi_1(A),\dots,B_n = \phi_n(A). \tag{30}$$

Then

$$\begin{aligned}[B_j,B_k] &= ih \sum_{r,s} \left(\frac{\partial \phi_j}{\partial y_r} \frac{\partial \phi_k}{\partial y_s} \Omega_{rs}\right)(A) + O(h^2) = \\ &= ih \sum_{r,s} \left(\frac{\partial \phi_j}{\partial y_r} \frac{\partial \phi_k}{\partial y_s} \Omega_{rs}\right)(\phi^{-1}(B)) + O(h^2)\end{aligned} \tag{31}$$

(to prove (31) it suffices to apply the formula of indexes permutation and K-formula of [52]). It follows that after the coordinate change the collection of functions $\Omega_{jk}(y)$ transforms as a contravariant tensor of rank 2. It turns out that the conditions (5) - (6) are rather natural. Namely,

$$[A_j,A_k] + [A_k,A_j] = 0,$$

so it is natural to require that the identity

$$\Omega_{jk}(y) + \Omega_{kj}(y) = 0 \tag{32}$$

hold; further, the following statement is valid.

Lemma 11. ([35]). For any symbol f there is a commutation formula

$$[A_j, f(A)] = ih \sum_{m=1}^{n} \left(\Omega_{jm} \frac{\partial f}{\partial y_m}\right)(A) + O(h). \tag{33}$$

Applying this lemma, we obtain

$$[A_s,[A_j,A_k]] = -h^2 \sum_{m=1}^{n} \left(\Omega_{sm} \frac{\partial \Omega_{jk}}{\partial y_m}\right)(A) + O(h^2). \tag{34}$$

Using the Jacobi identity for commutators, we come directly to (26).

The above considerations were not completely rigorous and played essentially the role of a hint, but we have shown the natural role of Poisson algebra in the asymptotic theory. We come now to exact discussions and formulations.

3. POISSON ALGEBRA WITH μ-STRUCTURE. LOCAL CONSIDERATIONS

In this section we give the construction of μ-structure for a Poisson algebra of functions on R^n, with the given fixed coordinate system. Item A is purely technical; we introduce some new symbol spaces which enable us to perform later the asymptotic expansions and to estimate the remainders. In item B the conditions on operators are imposed and the μ-mapping is defined. In items C and D we establish the composition formulas for the product of an element of the algebra with, respectively, another element of the algebra and a general operator, whose product lies in $\overset{0}{L}{}^{\infty}_{(\lambda_1,\dots,\lambda_n)}(R^n)$ (in particular, it may be a CRF (see Chapter 3, Section 3)). Almost all the geometric constructions were previously developed by Karasev [33,34], Karasev and Maslov [38,36] for the case of the small parameter asymptotics; however, the main idea being slightly modified, the proof technique is completely different since the methods within the cited papers to estimate the remainder do not work in the quasi-homogeneous situation.

It seems quite probable that the conditions imposed on the operators $A_1,\dots,A_n$ in item B have not taken their final form yet, and the relationships between them still need further investigation.

A. Some Auxiliary Function Spaces

Let $\lambda = (\lambda_1,\dots,\lambda_n)$ be a given n-tuple of non-negative numbers such that the set $I_+ = \{j \mid \lambda_j > 0\} \subset [n] = \{1,\dots,n\}$ is non-empty. We assume λ to be fixed throughout the subsequent exposition.

For any n-tuple $r = (r_1,\dots,r_n)$ of natural numbers consider the space $R^r \overset{def}{=} R^{r_1} \oplus R^{r_2} \oplus \dots \oplus R^{r_n}$ with coordinates $x = (x_{(1)},\dots,x_{(n)})$ divided into "blocks" or "clusters" $x_{(j)} = (x_{j1},\dots,x_{jr_j})$, $j = 1,\dots,n$, and define the action of the group R_+ on R^r by

$$\tau(x_{(1)},\dots,x_{(n)}) = (\tau^{\lambda_1}x_{(1)},\dots,\tau^{\lambda_n}x_{(n)}), \tag{1}$$

i.e., the element $\tau \in R_+$ acts as multiplication by τ^{λ_j} within the j-th "block" of variables, $j = 1,\dots,n$.

Denote by Σ_r the set of all the mappings

$$\sigma : I_+ \to N = \{1,2,\dots\}, \quad j \to \sigma_j, \tag{2}$$

such that $\sigma_j \leqslant r_j$ for all $j \in I_+$. For any $\sigma \in \Sigma_r$ we define the smooth function $\Lambda_\sigma(x) = \Lambda_\sigma(\{x_{j\sigma_j}\}_{j\in I_+})$, satisfying the conditions: (i) $\Lambda_\sigma(x) \geqslant \frac{1}{2}$ for all x; (ii) $\Lambda_\sigma(x)$ is quasi-homogeneous of degree 1 for $\Lambda_\sigma(x) \geqslant 1$, i.e.,

$$\Lambda_\sigma(\tau x) = r\Lambda_\sigma(x) \text{ for } \tau \geqslant 1,\ \Lambda_\sigma(x) \geqslant 1; \tag{3}$$

(iii) there are the two-sided estimates

$$c\Lambda_\sigma(x) \leqslant 1 + \sum_{j\in I_+} |x_{j\sigma_j}|^{1/\lambda_j} \leqslant C\Lambda_\sigma(x) \tag{4}$$

with positive constants c and C (see Chapter 3, Section 3:A for detailed construction of such functions). We also define the function $\Lambda_r(x) = \Lambda_r(\{x_{(j)}\}_{j\in I_+})$, satisfying the same conditions except that instead of (4), we have

$$c\Lambda_r(x) \leqslant 1 + \sum_{j\in I_+} \sum_{k=1}^{r_j} |x_{jk}|^{1/\lambda_j} \leqslant C\Lambda_r(x). \tag{5}$$

Set

$$\tilde{\Lambda}_r(x) = \min_{\sigma\in\Sigma_r} \Lambda_\sigma(x). \tag{6}$$

The following statements are obvious: (a) $\Lambda_r(x)$ is equivalent to $\max_{\sigma\in\Sigma_r} \Lambda_\sigma(x)$; (b) if $r_j = 1$ for all $j \in I_+$, then $\Lambda_r(x) = \tilde{\Lambda}_r(x)$.

The operator of the difference derivation $\delta_{j\ell} = \dfrac{\delta}{\delta x_{j\ell}}$, $\ell \leqslant r_j$ naturally acts in the spaces

$$\delta_{j\ell} : C^\infty(R^{(r_1,\dots,r_j,\dots,r_n)}) \to C^\infty(R^{(r_1,\dots,r_{j+1},\dots,r_n)}),$$

$$f(x_{(1)},\ldots,x_{(n)}) \to \frac{\delta}{\delta x_{j\ell}}(x_{(1)},\ldots,\tilde{x}_{(j)},\ldots,x_{(n)}) =$$

$$= (x_{j\ell} - x_{j,r+1})^{-1}(f(x_{(1)},\ldots,x_{(j)},\ldots,x_{(n)}) - f(x_{(1)},\ldots,\tilde{\tilde{x}}_{(j)},\ldots,x_{(n)})), \tag{7}$$

where $\tilde{x}_{(j)} = (x_{(j)},x_{j,r_j+1})$, $\tilde{\tilde{x}}_{(j)} = (x_{j1},\ldots,x_{j,r_j+1},\ldots,x_{jr_j})$ (x_{j,r_j+1} stands in place of $x_{j\ell}$).

Our aim is to present function subspaces in $C^\infty(R^r)$, in which the difference derivatives act in a natural way and which coincide also with $S^m_{(\lambda_1,\ldots,\lambda_n)}(R^n)$ in the case when all r_j are equal to one. Let m, m_1 be real numbers such that

$$m_1 \geqslant 0, \; m \leqslant m_1. \tag{8}$$

We denote by $\Gamma^m_{m_1}(R^r) \equiv \Gamma^m_{(\lambda_1,\ldots,\lambda_n)m_1}(R^r)$ the space of functions $f \in C^\infty(R^r)$, satisfying the estimates

$$\left|\frac{\partial^{|\alpha|}f(x)}{\partial x^\alpha}\right| \leqslant C_\alpha \Lambda_r(x)^{m_1}\tilde{\Lambda}_r(x)^{m-m_1-\langle\lambda,\alpha\rangle}, \quad |\alpha| = 0,1,2,\ldots . \tag{9}$$

In the inequality (9), $\alpha = (\alpha_{(1)},\ldots,\alpha_{(n)}) = (\alpha_{(1)},\ldots,\alpha_{(r1)},\ldots,\alpha_{n1},\ldots,\alpha_{nrn})$, and

$$\langle\lambda,\alpha\rangle = \sum_{j=1}^{n} \lambda_j|\alpha_{(j)}| \equiv \sum_{j=1}^{n}\sum_{k=1}^{r_j} \lambda_j\alpha_{jk}. \tag{10}$$

It is obvious that $\Gamma^m_{m_1}(R^r) = S^m_{(\lambda_1,\ldots,\lambda_n)}(R^n)$ if $r_1 = \ldots = r_n = 1$, no matter what m_1 is equal to.

<u>Lemma 1</u>. The operator $\delta/\delta x_{j\ell}$ acts in the spaces

$$\delta/\delta x_{j\ell} : \Gamma^m_{m_1}(R^{(r_1,\ldots,r_j,\ldots,r_n)}) \to \Gamma^{m-\lambda_j}_{m_1}(R^{(r_1,\ldots,r_j+1,\ldots,r_n)}). \tag{11}$$

<u>Proof</u>. For the sake of convenience, we denote $x_{j\ell}$ by y, x_{j,r_j+1} by z, and omit the other arguments under the function sign. We have

$$\frac{\delta f}{\delta x_{j\ell}} = \frac{f(y) - f(z)}{y - z} = \int_0^1 \frac{\partial f}{\partial y}(\tau y + (1-\tau)z)d\tau. \tag{12}$$

The equality (12) makes it evident that (11) holds if $\lambda_j = 0$ (the integrand in (12) satisfies the estimates (9) uniformly with respect to $r \in [0,1]$). Let now $\lambda_j > 0$. The space $R^{\tilde{r}}$, where $\tilde{r} = (r_1,\ldots,r_j+1,\ldots,r_n)$, may be divided into two zones $R^{\tilde{r}} = D_1 \cup D_2$, where

$$D_1 = \{\tilde{x} \in R^{\tilde{r}} \,|\, |y - z| \leqslant \varepsilon\tilde{\Lambda}_r(x)^{\lambda_j}\},$$
$$D_2 = \{\tilde{x} \in R^{\tilde{r}} \,|\, |y - z| \geqslant \varepsilon\tilde{\Lambda}_r(x)^{\lambda_j}\}, \tag{13}$$

and where $\varepsilon > 0$ is chosen small enough (see below). We claim that for ε small enough the functions $\Lambda_{\tilde{r}}(\tilde{x})$, $\Lambda_r(x)$, and $\Lambda_r(x(\tau))$, $\tau \in [0, 1]$, are equivalent in D_1 uniformly in τ,

$$\frac{\Lambda_{\tilde r}(\tilde x)}{\Lambda_r(x)} \leqslant \text{const}, \quad \frac{\Lambda_r(x)}{\Lambda_{\tilde r}(\tilde x)} \leqslant \text{const},$$
$$\frac{\Lambda_r(x)}{\Lambda_r(x(\tau))} \leqslant \text{const}, \quad \frac{\Lambda_r(x(\tau))}{\Lambda_r(x)} \leqslant \text{const}, \quad \tilde x \in D_1, \tag{14}$$

and so are the functions $\tilde\Lambda_{\tilde r}(\tilde x)$, $\tilde\Lambda_r(x)$, and $\tilde\Lambda_r(x(\tau))$, where $x(\tau)$ is obtained via replacement of $x_{j\ell}$ in x by $\tau y + (1 - \tau)z$. Indeed, it is enough to prove the equivalence of $\Lambda_\sigma(x)$ and $\Lambda_\sigma(x(\tau))$ for any $\sigma \in \Sigma_r$. We have $\Lambda_\sigma(x) = \Lambda_\sigma(x(\tau))$ if $\sigma_j \neq \ell$, and if $\sigma_j = \ell$, then by definition of D_1 and by (4)

$$|y - z| \leqslant \{\tfrac{\varepsilon}{C}(1 + |y|^{1/\lambda_j} + P)\}^{\lambda_j}, \tag{15}$$

where $P = \Sigma|x_{k\sigma_k}|^{1/\lambda_k} \geqslant 0$. (15) yields

$$|y - z| \leqslant \phi(\varepsilon)(1 + |y| + |P|^{\lambda_j}), \tag{16}$$

where ϕ is continuous, $\phi(0) = 0$. We obtain

$$1 + |\tau y + (1 - \tau)z| + |P|^{\lambda_j} \leq$$
$$\leqslant 1 + |y| + |P|^{\lambda_j} + (1 - \tau)\phi(\varepsilon)(1 + |y| + |P|^{\lambda_j}) \leqslant C(1 + |y| + |P|^{\lambda_j}); \tag{17}$$

thus

$$\Lambda_\sigma(x(\tau)) \leqslant \text{const}\cdot\Lambda_\sigma(x). \tag{18}$$

On the other hand, we have

$$1 + |y| + |P|^{\lambda_j} \leqslant 1 + |\tau y + (1 - \tau)z| + |P|^{\lambda_j} + (1 - \tau)\phi(\varepsilon)(1 + |y| + |P|^{\lambda_j}); \tag{19}$$

thus

$$\Lambda_\sigma(x(\tau)) \geqslant \text{const}\cdot\Lambda_\sigma(x), \tag{20}$$

provided that $\phi(\varepsilon) < 1$.

Thus we obtain the desired estimates in D_1 by differentiating under the integral sign in (12). To obtain the estimate in D_2 we make use of the identity

$$\frac{\partial^{|\tilde\alpha|}}{\partial x^{\tilde\alpha}}\left(\frac{\delta f}{\delta x_{j\ell}}\right) = \frac{\partial^{\beta_1 + \beta_2}}{\partial y^{\beta_1} + \partial z^{\beta_2}}\left\{\frac{f^{(\alpha)}(y) - f^{(\alpha)}(z)}{y - z}\right\}, \tag{21}$$

where α includes the derivatives with respect to all the variables except y and z. We have

$$\frac{\partial^{|\tilde\alpha|}}{\partial x^{\tilde\alpha}}\left(\frac{\delta f}{\delta x_{j\ell}}\right) = \sum_{\gamma_1=0}^{\beta_1}\sum_{\gamma_2=0}^{\beta_2} C_{\beta\gamma}(y - z)^{-1-\gamma_1-\gamma_2} \times$$
$$\times \frac{\partial^{\beta_1-\gamma_1 + \beta_2-\gamma_2}}{\partial y^{\beta_1-\gamma_1}\partial z^{\beta_2-\gamma_2}}[f^{(\alpha)}(y) - f^{(\alpha)}(z)], \tag{22}$$

where $C_{\beta\gamma}$ are constants. Only the terms with $\gamma_1 = \beta_1$ or $\gamma_2 = \beta_2$ are different from zero. Estimating the typical terms, we obtain

$$|(y - z)^{-1-\gamma_1-\beta_2}\cdot(\tfrac{\partial}{\partial y})^{\beta_1-\gamma_1}f^{(\alpha)}(y)| \leqslant \text{const}\cdot\tilde\Lambda_r(x)^{-\lambda_j(1+\gamma_1+\beta_2)}\Lambda_r(x)^{m_1} \times$$
$$\times \tilde\Lambda_r(x)^{m-m_1-<\lambda,\alpha>-(\beta_1-\gamma_1)\lambda_j} \leqslant \text{const}\cdot\Lambda_{\tilde r}(\tilde x)^{m_1}\tilde\Lambda_{\tilde r}(\tilde x)^{m-m_1-\lambda_j-<\tilde\alpha,\lambda>} \tag{23}$$

or

$$|(y-z)^{-1-\gamma_2-\beta_1}(\frac{\partial}{\partial z})^{\beta_2-\gamma_2}f^{(\alpha)}(z)| \leqslant \text{const}\cdot\tilde{\Lambda}_r(x)^{-\lambda_j(1+\beta_1+\gamma_2)}\Lambda_r(\tilde{\tilde{x}})^{m_1} \times$$
$$\times\ \tilde{\Lambda}_r(\tilde{\tilde{x}})^{m-m_1-\langle\lambda,\alpha\rangle-(\beta_2-\gamma_2)\lambda_j} \leqslant \text{const}\cdot\Lambda_{\tilde{r}}(x)^{m_1}\tilde{\Lambda}_{\tilde{r}}(x)^{m-m_1-\lambda_j-\langle\tilde{\alpha},\lambda\rangle} \tag{24}$$

(here $\tilde{\tilde{x}}$ is obtained from x by replacing y on z). Thus the required estimate holds in D_2 and the lemma is proved.

Lemma 2. The multiplication operation $(fg(x,y) = f(x)g(y))$ maps

$$\Gamma^m_{m_1}(R^r)\Gamma^{\ell}_{\ell 1}(R^{r'}) \to \Gamma^{m+\ell}_{m_1+\ell_1}(R^{r+r'}). \tag{25}$$

Lemma 3. The operation of identifying the pair of arguments within the k-th "block" $(f(x_{(1)},\ldots,x_{(k)},\ldots,x_{(n)}) \to \tilde{f}(x_{(1)};\ldots;x_{k1},\ldots,x_{k,r_k-1};\ldots;x_{(n)}) \equiv f(x_{(1)};\ldots,x_{k1},\ldots,x_{k,r_k-1}x_{k,r_k-1},\ldots;x_{(n)}))$ acts in the spaces:

$$\Gamma^m_{m_1}(R^{(r_1,\ldots,r_k,\ldots,r_n)}) \to \Gamma^m_{m_1}(R^{(r_1,\ldots,r_k-1,\ldots,r_n)}). \tag{26}$$

The proof of both lemmas is quite evident. Combining these two assertions we may obtain various statements about the products, for which the arguments in the function-factors (partially) coincide.

To end this item, we introduce the function space $\tilde{\Gamma}^m_{m_1}(R^r)$. By the definition $f(x) \in \tilde{\Gamma}^m_{m_1}(R^n)$ $(m_1 \geqslant 0,\ m \leqslant m_1)$ if and only if there exists a function $\Lambda_f(x)$, depending only on $\{x_{(j)}\}_{j\in I_+}$, satisfying the following properties: (i)

$$\Lambda_f(x) \leqslant \text{const}\cdot\Lambda_r(x),\ \Lambda_f(x) \geqslant 1,\ x \in R^r. \tag{27}$$

(ii) For some $\varepsilon > 0$ in the domain defined by inequalities

$$|x_{j\ell} - x_{jk}| \leqslant \varepsilon\Lambda_r(x)^{\lambda_j},\ j \in I_+,\ \ell,r \in \{1,\ldots,r_j\}, \tag{28}$$

the inverse inequality

$$\Lambda_r(x) \leqslant \text{const}\cdot\Lambda_f(x) \tag{29}$$

holds. (iii) The function f(x) and its derivatives satisfy the estimates

$$|f^{(\alpha)}(x)| \leqslant C_\alpha\Lambda_r(x)^{m_1}\Lambda_f(x)^{m-m_1+\Sigma_j(1-\lambda_j)\alpha_j} \tag{30}$$

for α such that

$$\Sigma(1-\lambda_j)\alpha_j + m - m_1 \leqslant 0, \tag{31}$$

and

$$|f^{(\alpha)}(x)| \leqslant C_\alpha\Lambda_r(x)^{m+\Sigma_j(1-\lambda_j)\alpha_j}, \tag{32}$$

provided that

$$\Sigma(1-\lambda_j)\alpha_j + m - m_1 > 0. \tag{33}$$

The space $\tilde{\Gamma}^m_{m_1}(R^r)$ is a generalization of $L^m_{(\lambda_1,\ldots,\lambda_n)}(R^n)$ just as $\Gamma^m_{m_1}(R^r)$ is a generalization of $S^m_{(\lambda_1,\ldots,\lambda_n)}(R^n)$. The statements of Lemmas 1-3 extend, with corresponding modifications, to the case of spaces $\tilde{\Gamma}^m_{m_1}(R^r)$.

B. Conditions on Operators and Definition of the μ-Mapping

Let a Banach (in particular, Hilbert scale) over R or Z be given (see Definition 1 of Chapter 2, Section 3:C), i.e., the collection of the Banach (resp. Hilbert) spaces $X = \{X_\delta\}$, $\delta \in R$ or Z, together with the continuous embeddings $X_\delta \subset X_\sigma$ for $\delta \geqslant \sigma$. We assume that there is a subset $D \subset \bigcap_\delta X_\delta = X_\infty$ given so that any X_δ is the completion of D in the norm $\|\cdot\|_\delta$. We slightly restrict here the general definitions of Chapter 2, Section 3:D and assume everywhere that tempered generators A in the space $X_{-\infty} = \bigcup_\delta X_\delta$ satisfy additional conditions: $D \subset D_A$, $AD \subset D$ and $\exp(iAt)D \subset D$ for any $t \in R$. Such operators A will be called merely "tempered generators (in the scale x)." We assume that the n-tuple $\lambda = (\lambda_1,\dots,\lambda_n)$ (see item A) is fixed throughout the exposition.

Let $(A_1,\dots,A_n)$ be the n-tuple of tempered generators in the scale $X = \{X_\delta\}$. We introduce some "good-behavior" conditions on this tuple. The first of these conditions was primarily used in the paper [35] in somewhat different form.

Condition 1. Let $j_1,\dots,j_m \in \{1,\dots,n\}$ be a finite sequence. For any $u \in D$ consider the D-valued function

$$u(t_1,\dots,t_m) = \exp(iAj_m t_m) \times \dots \times \exp(iAj_1 t_1)u, \quad t \in R^m. \tag{1}$$

We require that $u(t_1,\dots,t_m)$ be infinitely differentiable in any space X_δ, and the derivatives satisfy the estimates

$$\left\| \frac{\partial^\alpha u(t_1,\dots,t_m)}{\partial t^\alpha} \right\|_\delta \leqslant C_{\alpha\delta}(1+|t|)^{m_{\alpha\delta}}, \tag{2}$$

where $C_{\alpha\delta}$ depends on $u \in D$, and $m_{\alpha\delta}$ depends only on δ and $\alpha_{\tilde I_o}$, and does not depend on $\alpha_{\tilde I_+}$ (here $\tilde I_o = \{j \mid \lambda_{k_j} = 0\}$, $\tilde I_+ = [m] \setminus \tilde I_o$).

Proposition 1. Under Condition 1, the operator $f(\overset{1}{A}_{j_1},\dots,\overset{m}{A}_{j_m})$ for any $f \in L^\infty(\lambda_{j_1},\dots,\lambda_{j_m}(R^m)$ at least on D and acts in the spaces

$$f(\overset{1}{A}_{j_1},\dots,\overset{m}{A}_{j_m}) : D \to \bigcap_\delta X_\delta. \tag{3}$$

Proof. Let δ and $u \in D$ be fixed. We have, formally,

$$f(\overset{1}{A}_{j_1},\dots,\overset{m}{A}_{j_m})u = (2\pi)^{-m/2}\int \tilde f(t_1,\dots,t_m)u(t_1,\dots,t_m)dt_1,\dots,dt_m, \tag{4}$$

where $u(t_1,\dots,t_m)$ is defined by (1), and $\tilde f$ is the Fourier transform of f. What one needs to show is that the integral (4) may be regularized. So represent $f(x_1,\dots,x_m)$ in the form

$$f(x_1,\dots,x_m) = f_1(x_1,\dots,x_m)(1+x^2_{\tilde I_o})^k(1+x^2_{\tilde I_+})^\ell \tag{5}$$

with some natural k and ℓ. Since $f \in L^\infty(\lambda_{j_1},\dots,\lambda_{j_m})(R^m)$, f_1 satisfies the estimates

$$f_1^{(\gamma)}(x_1,\dots,x_m) \leqslant C(1+x^2_{\tilde I_o})^{-k}(1+x^2_{\tilde I_+})^{M-\ell+\sigma|\gamma|}, \tag{6}$$

where M and σ are some positive constants. (6) yields, by well-known properties of the Fourier transform, that the following estimate is valid:

$$|(1+|t|)^r \tilde{f}_1(t_1,\ldots,t_m)| \leqslant C, \tag{7}$$

provided that $k > \frac{1}{2}|\tilde{I}_o|$ and $M - \ell + \sigma r < -\frac{1}{2}|\tilde{I}_+|$. Using (5), we rewrite (4) as follows:

$$f(\overset{1}{A}_{j_1},\ldots,\overset{m}{A}_{j_m})u = (2\pi)^{-m/2}\int f_1(t_1,\ldots,t_m)\{(1-\frac{\partial^2}{\partial t^2_{\tilde{I}_o}})^k(1-\frac{\partial^2}{\partial t^2_{\tilde{I}_+}})^\ell \times$$
$$\times u(t_1,\ldots,t_m)\}dt_1,\ldots,dt_m = (2\pi)^{-m/2}\int \tilde{f}_1(t_1,\ldots,t_m) \times \tag{8}$$
$$\times u_{k\ell}(t_1,\ldots,t_m)dt_1,\ldots,dt_m,$$

where by (2)

$$\| u_{k\ell}(t_1,\ldots,t_m)\|_\delta \leqslant C_{k\ell}(1+|t|)^{m(k)}. \tag{9}$$

Now we choose $k > \frac{1}{2}|\tilde{I}_o|$ and then define ℓ in such a way that $M - \ell + \sigma(m(k)+m) < -\frac{1}{2}|\tilde{I}_+|$. Under this choice of k and ℓ, the integral (8) converges absolutely (in the strong sense) and gives the desired regularization. Proposition 1 is proved.

Our next condition deals with the estimates of the norm of product of semigroups, generated by $A_1,\ldots,A_n$ in pairs of spaces (X_δ, X_σ).

Condition 2. Let $j_1,\ldots,j_m \in \{1,\ldots,n\}$. Set

$$U(t_1,\ldots,t_m) \equiv U_{j_1,\ldots,j_m}(t_1,\ldots,t_m) = \exp(it_m A_{j_m}) \times \ldots \times \exp(it_1 A_{j_1}). \tag{10}$$

For any $N \geqslant 0$ there exists $N_1 \geqslant 0$ such that for any multi-index $\alpha = (\alpha_1,\ldots,\alpha_m)$ satisfying

$$\alpha_k = 0 \text{ if } \lambda_{jk} = 0,\ \sum_k \alpha_k \lambda_{jk} \leqslant N, \tag{11}$$

the following estimates are valid:

$$\left\| \frac{\partial^{|\alpha|}}{\partial t^\alpha} U(t_1,\ldots,t_m)u \right\|_{\delta-N_1} \leqslant C_{\alpha\delta}(1+|t|)^{m(N,\delta)} \|u\|_\delta \tag{12}$$

for any δ (the derivative in (12) is taken in the strong sense in $X_{\delta-N_1}$).

Next we impose the condition which makes it possible to develop the asymptotic theory.

Condition 3. (the "asymptotically diagonal spectrum" condition):

(a) Let $x \in R^r$,

$$f(x) \in \Gamma^m_{m_1}(R^r) \tag{13}$$

for some m_1. Then

$$f(\overset{\pi_{11}}{A_1},\ldots,\overset{\pi_{1r_1}}{A_1};\overset{\pi_{21}}{A_2},\ldots,\overset{\pi_{2r_2}}{A_2};\ldots;\overset{\pi_{n1}}{A_n},\ldots,\overset{\pi_{nr_n}}{A_n}) \in O^m_p(x) \tag{14}$$

(here π_{ij} are arbitrary pairwise different real numbers, and $O^m_p(x)$ denotes the set of operators $BX_{-\infty} \to X_{-\infty}$ such that for any δ, B is a continuous operator from X_δ to $X_{\delta-m}$).

(b) Given N and δ, there exists m such that for any r and m_1, the inclusion

$$f(x) \in \tilde{\Gamma}^{m}_{m_1}(R^r) \tag{15}$$

implies that

$$f(\overset{\pi_{11}}{A_1},\ldots,\overset{\pi_{nr_n}}{A_n}) : X_\delta \to X_{\delta+N} \tag{16}$$

is a continuous operator.

Definition 1. A proper tuple (of tempered generators in the scale $X = \{X_\delta\}$) is a tuple of tempered generators satisfying Conditions 1,2 and 3.

Some explanations need to be given to make the situation clearer. Conditions 1 - 3 are of complicated functional nature and it is doubtful whether they could be derived from some simple assumptions within the framework of general theory. In practice, these conditions should be verified for concrete operators to which this general theory is to be applied. Condition 3 is of crucial importance for the theory; the reason for its name is that, roughly speaking, it asserts the following: if $f(x_{11},\ldots, x_{1r_1},\ldots,x_{n1},\ldots,x_{nr_n})$ decays as $\Lambda(x) \to \infty$ in the R_+-invariant vicinity of the diagonal set $\Delta = \{x_{j1} = \ldots = x_{jr_j}$ for all $j \in I_+\}$, then the corresponding operator is a "smoothing operator" in the scale $\{X_\delta\}$.

It is not difficult to show that the 2n-tuple of operators

$$(-i\frac{\partial}{\partial x_1},\ldots,-i\frac{\partial}{\partial x_n}, x_1,\ldots,x_n)$$

in the Sobolev scale is a proper one (here $\lambda_1 = \ldots = \lambda_n = 1$, $\lambda_{n+1} = \ldots = \lambda_{2n} = 0$). It is useful to note that Conditions 1 - 3 and Definition 1 (and therefore all the subsequent arguments) depend on the choice of the tuple $(\lambda_1,\ldots,\lambda_n)$.

Let $(A_1,\ldots,A_n)$ be a proper tuple of generators in the Banach scale $X = \{X_\delta\}$, and assume that they satisfy the commutation relations

$$[A_j,A_k] = -i\omega_{jk}(\overset{1}{A_1},\ldots,\overset{n}{A_n}),\ j,k = 1,\ldots,n, \tag{17}$$

where the symbols

$$\omega_{jk}(x) \in P^{\lambda_j+\lambda_k-1}_{(\lambda_1,\ldots,\lambda_n)}(R^n) \tag{18}$$

have the asymptotic expansions

$$\omega_{jk}(x) \simeq \sum_{S=0}^{\infty} \omega_{jk}^{(S)}(x),\quad \omega_{jk}^{(S)}(x) \in O^{\lambda_j+\lambda_k-1-S}_{(\lambda_1,\ldots,\lambda_n)}(R^n), \tag{19}$$

and the functions

$$\Omega_{jk}(x) = \omega_{jk}^{(0)}(x) \in O^{\lambda_j+\lambda_k-1}_{(\lambda_1,\ldots,\lambda_n)}(R^n) \tag{20}$$

define the Poisson algebra structure

$$\{f,g\} = \sum_{j,k=1}^{n} \Omega_{jk}\frac{\partial f}{\partial x_j}\frac{\partial g}{\partial x_k} \tag{21}$$

on $C^\infty(R^n)$, i.e., the functions (20) satisfy the relations (5) and (6) of Section 2:A of the present chapter.

Definition 2. The above conditions being satisfied, we say that a μ-structure is defined over the Poisson algebra given by (21). The μ-mapping is a mapping

$$\mu : L^{\infty}_{(\lambda_1,\dots,\lambda_n)}(R^n) \to O_p(x), \qquad f(x_1,\dots,x_n) \to \mu(f) \overset{\text{def}}{=} f(\overset{1}{A_1},\dots,\overset{n}{A_n}), \tag{22}$$

where $O_p(x)$ is an algebra of operators in the scale x, defined at least on D.

In the sequel we are particularly interested in the action of the μ-mapping on a certain subspace of $L^{\infty}_{(\lambda_1,\dots,\lambda_n)}(R^n)$, namely, on the subspace $L^{\infty}_{st,(\lambda_1,\dots,\lambda_n)}$ of functions "stabilizing" at infinity. By definition,

$$L^{\infty}_{st,(\lambda_1,\dots,\lambda_n)}(R^n) = \bigcup_m L^m_{st,(\lambda_1,\dots,\lambda_n)}(R^n) \tag{23}$$

and $f(x) \in L^m_{st,(\lambda_1,\dots,\lambda_n)}(R^n)$, if and only if $f(x) \in L^m_{(\lambda_1,\dots,\lambda_n)}(R^n)$, and there exists the function $f_o(x_{I_+}) \in L^m_{(\lambda_1,\dots,\lambda_n)}(R^n)$ such that

$$|f^{(\alpha)}(x) - f_o^{(\alpha)}(x_{I_+})| \leq C_{\alpha,N}(1 + |x_{I_o}|)^{-N}\Lambda(x)^{m + \Sigma^n_{j=1}(1-\lambda_j)\alpha_j}, \qquad N, |\alpha| = 0,1,2,\dots, \tag{24}$$

i.e., f(x) stabilizes rapidly together with the derivatives as $|x_{I_o}| \to \infty$. The spaces $S^{\infty}_{st,(\lambda_1,\dots,\lambda_n)}$, $P^{\infty}_{st,(\lambda_1,\dots,\lambda_n)}$, etc. are defined in a similar way.

In what follows, we require that

$$\omega_{jk} \in P^{\lambda_j+\lambda_k-1}_{st,(\lambda_1,\dots,\lambda_n)}(R^n). \tag{25}$$

Next we come to establishing the composition formulas for operators lying in the image of the μ-mapping (22).

C. Composition Formulas for Elements of an Algebra

In this item we establish the asymptotic formulas for the composition

$$[\![f(\overset{1}{A_1},\dots,\overset{n}{A_n})]\!] \, [\![g(\overset{1}{A_1},\dots,\overset{n}{A_n})]\!] \equiv f(A) \circ g(A) \tag{1}$$

in the case when $f,g \in S^{\infty}_{(\lambda_1,\dots,\lambda_n)}(R^n)$; in particular, when f and g are asymptotically quasi-homogeneous functions.

Theorem 1. Let $f \in S^{m_1}_{(\lambda_1,\dots,\lambda_n)}(R^n)$, $g \in S^{m_2}_{(\lambda_1,\dots,\lambda_n)}(R^n)$. Then the composition (1) satisfies

$$f(A) \circ g(A) - h(A) \in O^{-\infty}_p(x) \equiv \bigcap_N O^N_p(x), \tag{2}$$

where the symbol $h(x_1,\dots,x_n) \in S^{m_1+m_2}_{(\lambda_1,\dots,\lambda_n)}(R^n)$ has the asymptotic expansion

$$h(x) \simeq F(x)g(x) + \sum_{j=1}^{\infty} B_j[f, g](x); \tag{3}$$

here $B_j[f,g] \in S^{m_1+m_2-j}_{(\lambda_1,\dots,\lambda_n)}(R^n)$ is a bilinear form

$$B_j[f,g](x) = \sum_{\substack{|\alpha| \leqslant j \\ |\beta| \leqslant j}} b_{j\alpha\beta}(x) f^{(\alpha)}(x) g^{(\beta)}(x), \tag{4}$$

where $b_{j\alpha\beta}(x) \in P^{-j+\Sigma_{k=1}^{n}\lambda_k(\alpha_k+\beta_k)}_{(\lambda_1,\dots,\lambda_n)}(R^n)$; in particular,

$$B_1[f,g] - B_1[g,f] = -i\{f,g\}, \tag{5}$$

where $\{f,g\}$ is the Poisson bracket of f and g.

Remark 1. It is easy to see that (5) is a principal term of the symbol of the operator $[f(A),g(A)]$ (the product cancels out). Thus, the theorem, in particular, asserts that the mapping $\mu : f \to \mu(f) = f(A)$ is an "almost representation" of Poisson algebra $S^{\infty}_{(\lambda_1,\dots,\lambda_n)}(R^n)$ with the Poisson bracket given by (21) of item B.

Proof. We make use of the permutation formula

$$\phi(\overset{1}{A},\overset{2}{B}) - \phi(\overset{2}{A},\overset{1}{B}) = \overset{2}{[B,A]} \frac{\delta^2\phi}{\delta x_1 \delta x_2}(\overset{1}{A},\overset{5}{A};\overset{2}{B},\overset{4}{B}). \tag{6}$$

This identity may be easily obtained using Theorems 2 and 3 of Chapter 2, Section 3:D. We have

$$\phi(\overset{1}{A},\overset{2}{B}) - \phi(\overset{2}{A},\overset{1}{B}) = \phi(\overset{1}{A},\overset{2}{B}) - \phi(\overset{3}{A},\overset{2}{B}) = (\overset{1}{A} - \overset{3}{A}) \frac{\delta\phi}{\delta x_1}(\overset{1}{A},\overset{3}{A};\overset{2}{B}) =$$

$$= \overset{3}{A}\left[\frac{\delta\phi}{\delta x_1}(\overset{1}{A},\overset{5}{A};\overset{4}{B}) - \frac{\delta\phi}{\delta x_1}(\overset{1}{A},\overset{5}{A};\overset{2}{B})\right] = \overset{3}{A}(\overset{4}{B} - \overset{2}{B}) \frac{\delta^2\phi}{\delta x_1 \delta x_2}(\overset{1}{A},\overset{5}{A};\overset{2}{B},\overset{4}{B}) = \tag{7}$$

$$= \overset{3}{[B,A]} \frac{\delta^2\phi}{\delta x_1 \delta x_2}(A_1 A_5; B_2 B_4),$$

Q.E.D.

In this formula ϕ may depend also on several other operators none of which, however, acts "between" A and B. Apply the identity (6) to the product (1), permitting successfully A_1 in f with $A_n,\dots,A_2$ in g, then coming forth to A_2 in f and so on. We obtain in this first stage:

$$f(\overset{n+1}{A_1},\dots,\overset{2n}{A_n})g(\overset{1}{A_1},\dots,\overset{n}{A_n}) = f(\overset{1}{A_1},\dots,\overset{n}{A_n})g(\overset{1}{A_1},\dots,\overset{n}{A_n}) -$$

$$- i \sum_{j<k} \omega_{jk}(\overset{k+2}{A_1},\dots,\overset{k+n+1}{A_n}) \frac{\delta f}{\delta x_j}(\overset{1}{A_1};\dots;\overset{j-1}{A_{j-1}};\overset{k+1}{A_j}\;\overset{k+n+2}{A_j}\;\overset{2n+4}{A_{j+1}},\dots,\overset{3n+3-j}{A_n}) \times \tag{8}$$

$$\times \frac{\delta g}{\delta x_k}(\overset{1}{A_1};\dots;\overset{k-1}{A_{k-1}};\overset{k}{A_k}\overset{k+n+3}{A_k};\overset{k+n+4}{A_{k+1}},\dots,\overset{2n+3}{A_n}).$$

The typical term of the sum in (7) belongs to the space $\Gamma^{m_1+m_2-1}_{S}(R^{(r_1,\dots,r_n)})$, where $r_j \in \{1,2,3,4\}$ ($S = \max\{m_1 + m_2 - 1, 0, \lambda_j + \lambda_k - 1\}$).

Next we apply the permutation procedure described above to the sum in (8). As a result, the difference derivatives transform into usual ones and the remainder appears belonging to $\Gamma^{m_1+m_2-2}_{\tilde{S}}(R^{\tilde{r}})$ for some $\tilde{r}$ and $\tilde{S}$, depending on the difference derivatives of f, g (and ω_{jk}). This procedure being repeated N times, we obtain

$$f(A) \circ g(A) = (fg)(A) + \sum_{j=1}^{N-1} B_j[f,g](A) + \hat{R}_N, \tag{9}$$

where the symbol R_N of the operator $\hat{R}_N$ belongs to the space $\Gamma^{m_1+m_2-N}_{S_N}(R^{rN})$.

Note that at any step of this process R_N is a bilinear form of the difference derivatives of f and g, and that the operators under f and g function signs always act in the proper order. This yields that when passing from R_N to R_{N+1} the order of the difference derivatives of f and g involved increases exactly by 1 and subsequently $B_j[f,g]$ depends on the derivatives of f and g of order $\leqslant j$. Summing the asymptotic series, we obtain the function h(x) satisfying (3).

Equations (3) and (9) together with Condition 3 of item B show that $f(A) \circ g(A) - h(A) \in \mathcal{O}_p^{-N}(x)$ for any N; therefore (2) holds. It remains to verify (5). We have

$$B_1[f,g] = -i \sum_{j<k} \Omega_{jk}(x) \frac{\partial f(x)}{\partial x_j} \frac{\partial g(x)}{\partial x_k} \tag{10}$$

(lower-order terms in ω_{jk} contribute to $B_j[f,g]$ with $j > 1$). Since $\Omega_{jj} \equiv 0$ and $\Omega_{jk} + \Omega_{kj} \equiv 0$, we have

$$\begin{aligned} B_1[f,g] - B_1[g,f] &= -i \sum_{j<k} \Omega_{jk}\left(\frac{\partial f}{\partial x_j}\frac{\partial g}{\partial x_k} - \frac{\partial f}{\partial x_k}\frac{\partial g}{\partial x_j}\right) = \\ &= -i \sum_{j,k} \Omega_{jk} \frac{\partial f}{\partial x_j}\frac{\partial g}{\partial x_k} = -i\{f,g\}, \text{ Q.E.D.} \end{aligned} \tag{11}$$

Theorem 1 is thereby proved.

D. Composition of Elements of an Algebra with Operators Whose Symbols Lie in $\overset{0}{L}{}^{\infty}_{(\lambda_1,\ldots,\lambda_n)}(R^n)$

This item is devoted to the solution of the problem: given the symbols

$$F(x) \in S^{\infty}_{st,(\lambda_1,\ldots,\lambda_n)}(R^n) \text{ and } \Phi(x) \in \overset{0}{L}{}^{\infty}_{st,(\lambda_1,\ldots,\lambda_n)}(R^n),$$

construct a symbol $\Phi_1(x) \in \overset{0}{L}{}^{\infty}_{st,(\lambda_1,\ldots,\lambda_n)}(R^n)$ such that

$$F(A) \circ \Phi(A) \equiv \Phi_1(A) \pmod{\mathcal{O}_p^{-\infty}(x)}. \tag{1}$$

A brief analysis shows that the argument of item C fails in the considered case and that some new techniques are necessary to provide the construction of $\Phi_1(x)$. However, the idea is quite simple (although the estimates appear to be cumbersome). We have*

$$\begin{aligned} \Phi(A) &= (2\pi)^{-n/2}\int\tilde{\Phi}(t)e^{iAt}dt = (2\pi)^{-n/2}\int F_{x\to t}[\chi(\overset{2}{x},-i\overset{1}{\frac{\partial}{\partial x}})\Phi(x)] \times \\ &\times e^{iAt}dt \pmod{\mathcal{O}_p^{-\infty}(x)} = (2\pi)^{-n/2}\int\tilde{\Phi}(t)[\chi(-i\overset{1}{\frac{\partial}{\partial t}},\overset{2}{t})e^{iAt}]dt \pmod{\mathcal{O}_p^{-\infty}(x)}, \end{aligned} \tag{2}$$

---

* Here and below we write $e^{iAt} \overset{\text{def}}{=} e^{iA_n t_n} \cdot \ldots \cdot e^{iA_1 t_1}$ for short.

where $\chi(x,p) \in \tilde{T}^0_{st,(\lambda_1,\ldots,\lambda_n)}(R^{2n})$, $\chi(x,p) = 1$ in the neighborhood of some set $K \in \mathrm{Ess}(\Phi)$.

Assume that for any $f(x,p) \in \tilde{T}^\infty_{st,(\lambda_1,\ldots,\lambda_n)}(R^{2n})$, we have managed to find a function

$$F(x,p) \overset{\text{def}}{=} L(f)(x,p) \in \tilde{T}^\infty_{st,(\lambda_1,\ldots,\lambda_n)}(R^{2n}) \tag{3}$$

such that

$$f(\overset{1}{-i\frac{\partial}{\partial t}},\overset{2}{t})e^{iAt} = F(A,t)\circ e^{iAt} + \hat{R}(t), \tag{4}$$

where $\hat{R}(t)$ is a smoothing operator in the scale X, satisfying the estimates

$$\| t^\alpha (\frac{\partial}{\partial t})^\beta \hat{R}(t) \|_{\delta \to \delta+N} \leqslant C_{\alpha\beta\delta N} \tag{5}$$

for any α, β, δ, N with $|\beta_{I_0}| = 0$. Then we obtain, modulo $O_p^{-\infty}(x)$

$$F(A)\circ\Phi(A) \equiv (2\pi)^{-n/2}\int \Phi(t)F(A)\circ \ell(\chi)(A,t)\circ e^{iAt}dt.$$

The composition $F(A)\circ\ell(\chi)(A,t)$ may be calculated using the results of item C and we obtain

$$F(A)\circ\Phi(A) \equiv (2\pi)^{-n/2}\int\tilde{\Phi}(t)H(A,t)\circ e^{iAt}dt,$$

where $H(x,p) \in \tilde{T}^\infty_{st,(\lambda_1,\ldots,\lambda_n)}(R^{2n})$. If $L^{-1}(H)$ is defined, then we obtain

$$F(A)\circ\Phi(A) \equiv (2\pi)^{-n/2}\int\tilde{\Phi}(t)\{L^{-1}(H)(\overset{1}{-i\frac{\partial}{\partial t}},\overset{2}{t})e^{iAt}\}dt;$$

that is, (1) is valid with

$$\Phi_1(x) = L^{-1}(H)(\overset{2}{x},\overset{1}{-i\frac{\partial}{\partial x}})\Phi(x). \tag{6}$$

Our argument, carrying out the above program, consists of two stages: (1') we perform some formal calculations and expansions to obtain (4); and (2') the estimates of the remainders are derived, thus validating the calculations made in the first stage.

We come to the first stage. We have

$$f(\overset{1}{-i\frac{\partial}{\partial t}},t)(e^{iAt}) = f(\overset{1}{A_1},\ldots,\overset{n}{A_n},t)e^{\overset{n}{iA_nt_n}}\times\ldots\times e^{\overset{1}{iA_1t_1}}. \tag{7}$$

We intend to find a function $F(x,t)$ such that

$$\begin{aligned} f(\overset{n}{A_1},\ldots,\overset{2n-1}{A_n},t)e^{i(\overset{n}{A_1}-\overset{n}{A_1})t_1}\times\ldots\times e^{i(\overset{2n-1}{A_n}-\overset{1}{A_n})t_n} = \\ = F(\overset{1}{A_1},\ldots,\overset{n}{A_n},t)\ (\mathrm{mod}\ O_p^{-\infty}(x)). \end{aligned} \tag{8}$$

The problem of solving (8) for $F(x,t)$ still cannot be treated by straightforward expansions like that in item C, since we are a posteriori interested in values of t of order $t_j \sim \Lambda(x)^{1-\lambda_j}$; therefore the derivative $\frac{\partial}{\partial x_j}\times$ $\times\ (e^{i(x_j-y_j)t})$ has the order $\Lambda(x)^{1-\lambda_j}$ rather than $\Lambda(x)^{-\lambda_j}$ necessary for these expansions to be applied. However, it appears that the problem may be reduced to solving a number of differential equations of the first order in t.

To perform this we introduce the system of unknown functions

$$\begin{aligned} &F_k(x_1,\dots,x_n,y_{k+1},\dots,y_n,t_1,\dots,t_n),\ k=1,\dots,n \\ &G_k(x_1,\dots,x_n,y_k,y_{k+1},\dots,y_n,t_1,\dots,t_n),\ k=2,\dots,n+1 \end{aligned} \tag{9}$$

(G_n and G_{n+1} do not depend on y) and require that the following conditions be satisfied:

$$F_1(x_1,\dots,x_n,y_2,\dots,y_n,t_1,\dots,t_n) = f(x_1,y_2,\dots,y_n,t_1,\dots,t_n); \tag{10}$$

$$G_{n+1}(x_1,\dots,x_n,t_1,\dots,t_n) = F(x_1,\dots,x_n,t_1,\dots,t_n); \tag{11}$$

$$\begin{aligned} &F_k(\overset{1}{A}_1,\dots,\overset{n}{A}_n,\overset{n+k+1}{A}_{k+1},\dots,\overset{2n}{A}_n,t_1,\dots,t_n)e^{i(\overset{2n}{A}_n-\overset{-n}{A}_n)t_n}\times\dots\times e^{i(\overset{n+k}{A}_k-\overset{-k}{A}_k)t_k} = \\ &\quad = G_{k+1}(\overset{1}{A}_1,\dots,\overset{n}{A}_n,\overset{n+k+1}{A}_{k+1},\dots,\overset{2n}{A}_n),t_1,\dots,t_n))\times \\ &\quad\times e^{i(\overset{2n}{A}_n-\overset{-n}{A}_n)t_n}\times\dots\times e^{i(\overset{n+k+1}{A}_{k+1}-\overset{-k-1}{A}_{k+1})t_{k+1}} - \hat{R}_k(t),\ k=1,\dots,n; \end{aligned} \tag{12}$$

$$\begin{aligned} &G_k(\overset{1}{A}_1,\dots,\overset{n}{A}_n,\overset{n+k}{A}_k,\dots,\overset{2n}{A}_n,t_1,\dots,t_n)e^{i(\overset{2n}{A}_n-\overset{-n}{A}_n)t_n}\times\dots\times e^{i(\overset{n+k}{A}_k-\overset{-k}{A}_k)t_k} = \\ &\quad = F_k(\overset{1}{A}_1,\dots,\overset{n}{A}_n,\overset{n+k+1}{A}_{k+1},\dots,\overset{2n}{A}_n,t_1,\dots,t_n)e^{i(\overset{2n}{A}_n-\overset{-n}{A}_n)t_n}\times \\ &\quad\times\dots\times e^{i(\overset{n+k}{A}_k-\overset{k}{A}_k)t_k} + \hat{\tilde{R}}_k(t),\ k=2,\dots,n, \end{aligned} \tag{13}$$

where $\hat{R}_k(t)$, $\hat{\tilde{R}}_k(t)$ satisfy the estimates (5).

Our scheme is as follows: we start from k = 1 and define F_1 by the equality (10). Next we solve (12) for G_2. After this successively for $k = 2,\dots,n$ we solve (13) for F_k and then (12) for G_{k+1}. Finally, F is defined by (11). The crucial point of our analysis is the solution of (12). We introduce the function $W_k(x_1,\dots,x_n,y_{k+1},y_n,t_1,\dots,t_n,t_o)$ depending on the additional parameter $t_o \in [0,t_k]$ such that

$$\begin{aligned} &\frac{d}{dt_o}\{W_k(\overset{1}{A}_1,\dots,\overset{n}{A}_n,\overset{n+k+1}{A}_{k+1},\dots,\overset{2n}{A}_n,t_1,\dots,t_n,t_o)e^{i(\overset{2n}{A}_n-\overset{-n}{A}_n)t_n}\times\dots\times \\ &\quad\times e^{i(\overset{n+k+1}{A}_{k+1}-\overset{-k-1}{A}_{k+1})t_{k+1}}e^{i(\overset{n+k}{A}_k-A_k)(t_k-t_o)}\} = \hat{R}_k(t,t_o) \end{aligned} \tag{14}$$

where $\hat{R}_k(t,t_o)$ satisfy estimates (5) uniformly in $t_o \in [0,t_k]$ and

$$W_k|_{t_o=0} = F_k; \tag{15}$$

so clearly we may set

$$G_{k+1} = W_k|_{t_o=t_k} \tag{16}$$

which yields

$$\hat{R}_k(t) = -\int_0^{t_k}\hat{R}_k(t,t_o)dt_o \in O_p^{-\infty}(x). \tag{17}$$

To solve (14) we calculate the derivative on the left-hand side of this equality and obtain

$$\frac{d}{dt_o}\{\dots\} = \Big\{ \frac{\partial W_k}{\partial t_o}(\overset{1}{A}_1,\dots,\overset{n}{A}_n,y_{k+1},\dots,y_n,t,t_o) + i(\overset{0}{A}_k - \overset{n+1}{A}_k)W_k(\overset{1}{A}_1,\dots,\overset{n}{A}_n,y_{k+1},\dots,y_n,t,t_o)\Big\} e^{i\sum_{j=k}^{n} y_j t_j}\Bigg|_{\substack{y_k=\overset{n+k}{A}_k\\ \dots\dots\\ y_n=\overset{2n}{A}_n}} \times$$
$$\times\, e^{-iA_k(t_k-t_o)}\, e^{-iA_{k+1}t_{k+1}} \times \dots \times e^{-iA_n t_n}. \tag{18}$$

We seek the formal solution for W_k vanishing the expression in the curly brackets. (Writing $y_k,\dots,y_n$ instead of $\overset{n+k}{A}_k,\dots,\overset{2n}{A}_n$ and omitting temporarily the product of exponents occurring in (18) is a convenient tool to thus shorten the notation.) We have

$$(\overset{0}{A}_k - \overset{n+1}{A}_k)W_k(\overset{1}{A}_1,\dots,\overset{n}{A}_n,y,t,t_o) =$$
$$= \sum_{j=1}^{n} [\overset{j+1}{A_j},A_k]\,\frac{\delta W_k}{\delta x_j}(\overset{1}{A}_1;\dots;\overset{j}{A}_j,\overset{j+2}{A}_j;\dots;\overset{n+2}{A}_n,y,t,t_o) = \tag{19}$$
$$= -i\sum_{j=1}^{n} \omega_{jk}(\overset{j+1}{A}_1,\dots,\overset{j+n}{A}_n)\,\frac{\delta W_k}{\delta x_j}(\overset{1}{A}_1;\dots;\overset{j}{A}_j,\overset{j+n+1}{A}_j;\dots;\overset{n}{A}_n,y,t,t_o) = \dots .$$

Continuing this process, as in item C, we obtain

$$(\overset{0}{A}_k - \overset{n+1}{A}_k)W_k(\overset{1}{A}_1,\dots,\overset{n}{A}_n,y,t,t_o) =$$
$$= \sum_{s=1}^{N-1} (L_{ks}W_k)(\overset{1}{A}_1,\dots,\overset{n}{A}_n,y,t,t_o) + \hat{Q}_{kN}(y,t,t_o) \tag{20}$$

for any natural N, where

$$L_{ks} = \sum_{|\alpha|\leqslant s} C_{ks\alpha}(x)\,\frac{\partial^{|\alpha|}}{\partial x^{\alpha}}\,,\quad C_{ks\alpha} \in P^{\lambda_k+\sum_j\alpha_j\lambda_j-s}_{(\lambda_1,\dots,\lambda_n)}(R^n). \tag{21}$$

$\hat{Q}_{kN}$ is a sum of terms, each of which has the symbol of the form

$$Q(\tilde{x},y,t,t_o) = \tilde{C}_{kN\alpha}(\tilde{x})\,\frac{\delta^{\alpha}W_k}{\delta x^{\alpha}}(\tilde{x},y,t,\tau) \tag{22}$$

with $|\alpha| \leqslant N$, $\tilde{x} \in R^{(r_1,\dots,r_n)}$,

$$\tilde{C}_{kN\alpha}(\tilde{x}) \in \Gamma^{\lambda_k+\Sigma\alpha_j\lambda_j-N}_{N(2\lambda_{max}-1)}(R^r); \tag{23}$$

here $\lambda_{max} = \max_j \lambda_j$. In particular,

$$L_{k1} = -i\sum_{j=1}^{n}\Omega_{jk}(x)\,\frac{\partial}{\partial x_k}\,. \tag{24}$$

We seek W_k in the form of the formal series

$$W_k(x,y,t,t_o) \sim \sum_{\ell=0}^{\infty} W_{k\ell}(x,y,t,t_o), \tag{25}$$

satisfying

$$\frac{\partial W}{\partial t_o} + i \sum_{j=1}^{\infty} L_{kj} W_k = 0. \tag{26}$$

In term-by-term form (26) reads

$$\frac{\partial W_{ko}}{\partial t_o} + \sum_{j=1}^{n} \Omega_{jk}(x) \frac{\partial W_{ko}}{\partial x_j} = 0, \tag{27}$$

$$\frac{\partial W_{k\ell}}{\partial t_o} + \sum_{j=1}^{n} \Omega_{jk}(x) \frac{\partial W_{k\ell}}{\partial x_j} = -i \sum_{s=0}^{\ell=1} L_{k,\ell-s+1} W_{ks}, \quad \ell = 1,2,\ldots . \tag{28}$$

Consider the system of ordinary differential equations of the first order

$$\dot{X}_j = \Omega_{kj}(X), \; j = 1,\ldots,n, \; X \in R^n. \tag{29}$$

Denote the solution of this system with the initial data

$$X_j(0) = x_j \tag{30}$$

by

$$X = X^{(k)}(x,t_o) \tag{31}$$

(here t_o is the "time variable").

Lemma 1. The solution (31) of the system (29) with initial data (30) has the property

$$X^{(k)}(\tau^{\lambda}x, \tau^{1-\lambda_k}t_o) = \tau^{\lambda}X^{(k)}(x,t_o) \tag{32}$$

for $\tau \geqslant 1$, $\Lambda(x)$ is large enough, $|t_o| \leqslant \text{const}\cdot\Lambda(x)^{1-\lambda_k}$.

Proof. Differentiating both sides of (32) with respect to t_o yields (assuming that $\Lambda(x)$ is large enough):

$$\frac{d}{dt_o}(X^{(k)}(\tau^{\lambda}x,\tau^{1-\lambda_k}t_o))_j = \tau^{1-\lambda_k}\Omega_{kj}(X^{(k)}(\tau^{\lambda}x,\tau^{1-\lambda_k}t_o)), \tag{33}$$

$$\frac{d}{dt_o}(\tau^{\lambda}X^{(k)}(x,t_o))_j = \tau^{\lambda_j}\Omega_{kj}(X^{(k)}(x,t_o)) = \tau^{1-\lambda_k}\Omega_{kj}(\tau^{\lambda}X^{(k)}(x,t_o)). \tag{34}$$

Thus both sides of (32) satisfy the same system of equations and (32) holds by the uniqueness theorem for ordinary differential equations (the fact that initial data coincide is trivial). Lemma 1 is proved.

Calculate the solution of the system of equations (27) - (28). Using the solution (30) of the system (29) and taking into account the anti-symmetry of Ω_{kj}, these equations may be written in the form

$$\frac{d}{dt_o} W_{ko}(X^{(k)}(x_o,-t_o),y,t,t_o) = 0, \tag{35}$$

$$\begin{aligned} \frac{d}{dt_o} W_k \; (X^{(k)}(x_o,-t_o),y,t,t_o) &= -i \sum_{s=0}^{\ell-1} [L_{k,\ell-s+1}W_{ks}] \times \\ &\times (X^{(k)}(x_o,-t_o),y,t,t_o), \quad \ell = 1,2,3,\ldots . \end{aligned} \tag{36}$$

From (35) we obtain

$$W_{ko}(X^{(k)}(x_o,-t_o),y,t,t_o) = W_{ko}(x_o,y,t,0) \tag{37}$$

or, resolving for x_o the system of equations $X^{(k)}(x_o,-t_o) = x$,

$$W_{ko}(x,y,t,t_o) = W_{ko}(X^{(k)}(x,t_o),y,t,0). \tag{38}$$

Also we have for $\ell = 1,2,\ldots,$

$$W_{k\ell}(X^{(k)}(x_o,-t_o),y,t,t_o) = W_{k\ell}(x_o,y,t,0) - \\ - i\int_o^{t_o} \cdot \sum_{s=0}^{\ell-1} [L_{k,\ell-s+1}W_{ks}](X^{(k)}(x_o,-t_o'),y,t,t_o')dt_o', \tag{39}$$

or

$$W_{k\ell}(x,y,t,t_o) = W_{k\ell}(X^{(k)}(x,t_o),y,t,0) - \\ - i\int_o^{t_o} \cdot \sum_{s=0}^{\ell-1} [L_{k,\ell-s+1}W_{ks}](X^{(k)}(x,t_o - t_o'),y,t,t_o')dt_o'. \tag{40}$$

However, the latter expression admits further simplification; to carry this out, we begin from $\ell = 1$, using the expression (21) for L_{ks}:

$$W_{k1}(x,y,t,t_o) = W_{k1}(X^{(k)}(x,t_o),y,t,0) - i\int_o^{t_o} [\sum_{|\alpha|\leqslant 2} C_{k2\alpha} \frac{\partial^{|\alpha|}W_{ko}}{\partial x^\alpha}] \times \\ \times (X^{(k)}(x,t_o - t_o'),y,t,t_o')dt_o' = W_{k1}(X^{(k)}(x,t_o),y,t,0) - \\ - i\int_o^{t_o} [\sum_{|\alpha|\leqslant 2} C_{k2\alpha}(X^{(k)}(x,t_o - t_o'))(({}^t \frac{\partial X^{(k)}}{\partial x}(x,t_o - t_o'))^{-1} \frac{\partial}{\partial x})^\alpha] \times \\ \times W_{ko}(X^{(k)}(x,t_o - t_o'),y,t,t_o')dt_o'. \tag{41}$$

From (38) we have

$$W_{ko}(X^{(k)}(x,t_o - t_o'),y,t,t_o') = W_{ko}(X^{(k)}(x,t_o),y,t,0) = W_{ko}(x,y,t,t_o); \tag{42}$$

thus (41) takes the form

$$W_{k1}(x,y,t,t_o) = W_{k1}(X^{(k)}(x,t_o),y,t,0) + L_{k2}(t_o)W_{ko}(x,y,t,t_o), \tag{43}$$

where $L_{k2}(t_o)$ is a differential operator of second order,

$$L_{k2}(t_o) = -i\int_o^{t_o} \sum_{|\alpha|\leqslant 2} C_{k2\alpha}(X^{(k)}(x,t_o - t_o')) \times \\ \times (({}^t \frac{\partial X^{(k)}}{\partial x}(x,t_o - t_o'))^{-1} \frac{\partial}{\partial x})^\alpha dt_o'. \tag{44}$$

Similarly we have

$$W_{k2}(x,y,t,t_o) = W_{k2}(X^{(k)}(x,t_o),y,t,0) - i\int_o^{t_o} [L_{k3}W_{ko}] \times \\ \times (X^{(k)}(x,t_o - t_o'),y,t,t_o')dt_o' - i\int_o^{t_o} [L_{k2}(W_{k1}(\tilde{x},t_o'),y,t,0) + L_{k2}(t_o') \times \\ \times W_{ko}(\tilde{x},y,t,t_o')]|_{\tilde{x}=X^{(k)}(x,t_o-t_o')} dt_o' = W_{k2}(X^{(k)}(x,t_o),y,t,0) + \\ + L_{k3}(t_o)W_{ko}(x,y,t,t_o) + L_{k2}(t_o)W_{k1}(x,y,t,t_o), \tag{45}$$

and generally,

$$W_{k\ell}(x,y,t,t_o) = W_{k\ell}(X^{(k)}(x,t_o),y,t,0) + \sum_{s=1}^{\ell=1} L_{k,\ell-s+1}(t_o)W_{ks}(x,y,t,t_o), \tag{46}$$

where $L_{ks}(t_o)$ is a differential operator of order $\leqslant s$, which may be expressed in an obvious way through $L_{ks'}$, with $s' \leqslant s$.

Next we intend to find the formal solutions for equations (13) in order to obtain the initial conditions for W_{k+1} from the end point values of $W_{k_n^1}$, $k = 1,\dots,n-1$. It is quite simple since no exponents act between $\overset{n+k}{A_1},\dots,\overset{n+k}{A_n}$ and A_k on the left-hand side of (13). We commute A_k in the k-th place, using the permutation formula from item C; we obtain the asymptotic expansion

$$W_{k+1}(x,y_{k+2},\dots,y_n,t,0) \simeq W_k(x,x_{k+1},y_{k+2},\dots,y_n,t,t_k) + \sum_{s=1}^{\infty} (L_{ks}^{(1)}W_k)(x,x_{k+1},y_{k+2},\dots,y_n,t,t_k), \tag{47}$$

where $L_{ks}^{(1)}$ are differential operators in x,y_{k+1} of the form similar to (21). Of course, (47) may be rewritten in the term-by-term form just as (26) was.

Now we are ready to obtain the formal solution for the problem (8). We begin with the principal term. Using (15), (16), (38), and (47) and taking into account that $L_{ks}^{(1)}$ in (47) contribute only to lower-order terms, we obtain successively:

$$\begin{aligned}
F_{10}(x,y_2,\dots,y_n,t) &= f(x_1,y_2,\dots,y_n,t);\\
F_{20}(x,y_3,\dots,y_n,t) &= W_{10}(x,y_2,\dots,y_n,t,t_1)|_{y_2=x_2} =\\
&= f(X_1^{(1)}(x,t_1),x_2,y_3,\dots,y_n,t);\\
F_{30}(x,y_4,\dots,y_n,t) &= W_{20}(x,y_3,\dots,y_n,t,t_2)|_{y_3=x_3} =\\
&= f(X_1^{(1)}(X^{(2)}(x,t_2),t_1),X_2^{(2)}(x,t_2),x_3,y_4,\dots,y_n,t);\\
&\dots\dots\dots\dots\dots\dots\dots\dots\dots\dots\dots\dots
\end{aligned} \tag{48}$$

and, finally, the principal term of $F(x,t)$ occurring in (8) takes the form

$$F_{(o)}(x,t) = G_{n+1,o}(x,t) = f(X(x,t),t), \tag{49}$$

where the j-th component of the mapping X is defined by

$$X_j(x,t) = X_j^{(j)}(X^{(j+1)}(\dots(X^{(n)}(x,t_n),\dots),t_{j+1}),t_j). \tag{50}$$

In a more convenient form the mapping (50) may be defined also in the following way: for any $j \in \{1,\dots,n\}$ and $t_j \in \mathbb{R}$, set

$$X^{(j,t_j)}(x) \overset{\text{def}}{=} X^{(j)}(x,t_j), \quad j = 1,\dots,n. \tag{51}$$

Define

$$\tilde{X}(x,t) = X^{(1,t_1)} \circ X^{(2,t_2)} \circ \dots \circ X^{(n,t_n)}(x). \tag{52}$$

In other words, the point $\tilde{X}(x,t)$ may be obtained if we move, beginning from x, along the trajectories of system (29) subsequently for $k = n,n-1,\dots,2,1$ during the time intervals $t_n,\dots,t_1$, respectively. $X(x,t)$ may now be defined by

$$X_j(x,t) = \tilde{X}_j(x,0,\ldots,0,t_j,t_{j+1},\ldots,t_n)\ j = 1,\ldots,n. \tag{53}$$

Thus, seeking $F(x,t)$ as the formal series

$$F(x,t) \simeq \sum_{s=0}^{\infty} F_{(s)}(x,t), \tag{54}$$

we have found the principal term of this series; it is given by (49). Now calculate the lower-order terms in (54). As for $F_{(1)}(x,t)$, we have, using (15), (16), (38), and (47):

$$\begin{aligned} F_{11}(x,y_2,\ldots,y_n,t) &= 0, \\ F_{21}(x,y_3,\ldots,y_n,t) &= W_{11}(x,y_2,\ldots,y_n,t,t_1)|_{y_2=x_2} + \\ &+ [L_{11}^{(1)} W_{10}(x,y_2,\ldots,y_n,t,t_1)]|_{y_2=x_2} = \\ &= \{[\mathcal{L}_{12}(t_1) + L_{11}^{(1)}]f(X_1^{(1)}(x,t_1),y_2,\ldots,y_n,t)\}|_{y_2=x_2}, \end{aligned} \tag{55}$$

where $\mathcal{L}_{12}(t_1)$ is a differential operator of order $\leqslant 2$ in x while $L_{11}^{(1)}$ is an operator in x and y_2. The expression in curly brackets may be rewritten as $(\tilde{\mathcal{L}}_{12}(t_1)f)(X_1^{(1)}(x,t_1),y_2,\ldots,y_n,t)$, where $\tilde{\mathcal{L}}_{12}$ is some new differential operator acting on f before the substitution $x_1 \to X_1^{(1)}(x,t_1)$. Thus we can obtain

$$F_{21}(x,y_3,\ldots,y_n,t) = (\tilde{\mathcal{L}}_{12}(t_1)f)(X_1^{(1)}(x,t_1),x_2,y_3,\ldots,y_n,t), \tag{56}$$

and further application of this technique yields

$$F_{(1)}(x,t) = (P_1 f)(X(x,t),t), \tag{57}$$

and, generally,

$$F_{(s)}(x,t) = (P_s f)(X(x,t),t), \tag{58}$$

where P_s is a differential operator in x of order $\leqslant s + 1$, and, besides,

$$F_{(s)}(\tau^{\lambda}x,\tau^{1-\lambda}t) = \tau^{m-s}F_{(s)}(x,t) \tag{59}$$

for $\tau \geqslant 1$, $\Lambda(x)$ large enough, provided that $f(\tau^{\lambda}x,\tau^{1-\lambda}t) = \tau^{m}f(x,t)$ for $\tau \geqslant 1$ and large $\Lambda(x)$ (we omit routine calculations leading to this result). If we set $P_0 = 1$, we may write the formal series solution of (8) in the form

$$F(x,t) \simeq \sum_{s=0}^{\infty} (P_s f)(X(x,t),t) = f(X(x,t),t) + \sum_{s=1}^{\infty} (P_s f)(X(x,t),t). \tag{60}$$

Our next task is to validate, under certain conditions, the expansion (60); that is, to "sum" the asymptotic series occurring throughout the calculations and to estimate the appearing remainders. Also we need the conditions under which (60) is solvable for $f(x,t)$.

Let $f(x,p) \in \tilde{T}^{m}_{st,(\lambda_1,\ldots,\lambda_n)}(\mathbb{R}^{2n})$ (recall that this means that $f(x,p) = 0$ if $|p_j| > C\Lambda(x)^{1-\lambda_j}$ for some j, where $C = C(f)$, and that

$$\left| \frac{\partial^{|\alpha|+|\beta|} f(x,p)}{\partial x^{\alpha}\partial p^{\beta}} \right| \leqslant C_{\alpha\beta}\Lambda(x)^{m-\langle\lambda,\alpha\rangle+\langle\lambda-1,\beta\rangle},\quad |\alpha|,|\beta| = 0,1,2,\ldots, \tag{61}$$

plus stabilization conditions at infinity with respect to x_{I_0}). We take for W_k, $k = 1,\ldots,n$, the partial sum of $N - 1$ terms of the series (25) where the $W_{k\ell}$'s are obtained from f recursively via (46), and substitute this expression into (14). The expression for W_k reads

$$\begin{gathered} W_k(x,y_{k+1},\ldots,y_n,t,t_o) = (L_k f)(X_1(x,t_1,\ldots,t_{k-1},t_o,0,\ldots,0), \\ X_2(x,t_1,\ldots,t_{k-1},t_o,0,\ldots,0),\ldots,X_k(x,t_1,\ldots,t_{k-1},t_o,0,\ldots,0), \\ \times\, y_{k+1},\ldots,y_n,t,t_o);\ t_o \in [0,t_k]; \end{gathered} \tag{62}$$

here L_k is some differential operator whose explicit form is of no interest to us. The remainder $\hat{R}_k(t,t_o)$ in (14) after our substitution would be a sum of a number of operators with symbols of the type (22) times the product of exponentials.

Assume that (for some given positive c_1 and c_2)

$$\begin{gathered} c_1\Lambda(x) \leqslant \Lambda(X_1(x,t_1,\ldots,t_{k-1},t_o,0,\ldots,0),\ldots,X_k(x,t_1,\ldots,t_{k-1},t_o, \\ \times\, 0,\ldots,0),x_{k+1},\ldots,x_n) \leqslant c_2\Lambda(x), \end{gathered} \tag{63}$$

provided that $(X_1(x,t_1,\ldots,t_{k-1},t_o,0,\ldots,0),\ldots,X_k(x,t_1,\ldots,t_{k-1},t_o,0,\ldots,0),x_{k+1},\ldots,x_n) \in \operatorname{supp} f$ and $t_o \in [0,t_k]$. Then, making use of Condition 3 of item B and the fact that under (63) the symbols occurring in the remainder belong to $\tilde{\Gamma}^{m-N}_{m_1}(\mathbb{R}^r)$ with various m_1 and r, we come to the following result: given δ and N_1 we always can choose N large enough, so that

$$\begin{gathered} \| t^\alpha (\tfrac{\partial}{\partial t})^\beta (\tfrac{\partial}{\partial t_o})^\gamma \hat{R}(t,t_o) \|_{\delta \to \delta+N_1} \leqslant C \text{ for } |\alpha|,|\beta|,|\gamma| \leqslant N_1, \\ |\beta_{I_o}| = 0, \gamma = 0 \text{ if } k \in I_o. \end{gathered} \tag{64}$$

Establish thus the conditions under which (63) holds. Note first that the argument of Λ in (63) is merely

$$X(x,t_1,\ldots,t_{k-1},t_o,0,\ldots,0).$$

Also it follows from Lemma 1 that the mapping $X(x,t)$ defined by (53) satisfies the identity

$$X(\tau^\lambda x, \tau^{1-\lambda} t) = \tau^\lambda X(x,t) \tag{65}$$

in any set of the form $\{|t_j| \leqslant C\Lambda(x)^{1-\lambda_j},\ j = 1,\ldots,n\}$ for $\tau \geqslant 1$, $\Lambda(x)$ large enough. In order to establish the desired conditions, perform in $X(x,t)$ the change of variables

$$t_j = \Lambda(x)^{1-\lambda_j}\theta_j,\ j = 1,\ldots,n \tag{66}$$

and

$$Z(x,\theta) = X(x,\Lambda(x)^{1-\lambda}\theta). \tag{67}$$

Clearly on any set of the form $\{|\theta_j| \leqslant C,\ j = 1,\ldots,n\}$, $Z(x,\theta)$ satisfies the condition

$$Z(\tau^\lambda x,\theta) = \tau^\lambda Z(x,\theta) \tag{68}$$

for $\tau > 1$, $\Lambda(x)$ large enough.

Lemma 2. Assume that a subset $K \subset \mathbb{R}^{2n}_{(x,\theta)}$ is given such that the following inequalities are valid for $(x,\theta) \in K$ with some positive C:

$$|\theta_j| \leqslant C, \; j = 1,\ldots,n, \tag{69}$$

$$|\partial Z_j/\partial x_k| \leqslant C\Lambda(x)^{\lambda_j-\lambda_k}, \; j,k = 1,\ldots,n, \tag{70}$$

$$\frac{DZ}{Dx} \overset{\text{def}}{=} \det \| \partial Z_j/\partial x_k \|_{j,k=1}^{n} \geqslant C^{-1}. \tag{71}$$

Then $\Lambda(x)$ and $\Lambda(Z(x,\theta))$ are equivalent on K; more precisely,

$$c_1\Lambda(x) \leqslant \Lambda(Z(x,\theta)) \leqslant c_2\Lambda(x), \tag{72}$$

where c_1 and c_2 depend only on C.

Proof. $Z(x,\theta)$ satisfies (68) for $\Lambda(x) \geqslant c_3$, where c_3 depends only on C. (72) is evident for $\Lambda(x) \leqslant c_3$, so it suffices to consider the domain where (68) holds. The Euler identity (see Proposition 1 of Chapter 3, Section 3:A) holds:

$$\lambda_j Z_j(x,\theta) = \sum_{k=1}^{n} \frac{\partial Z_j(x,\theta)}{\partial x_k} \lambda_k x_k, \; j = 1,\ldots,n. \tag{73}$$

The right inequality in (72) follows from (70) and (73) immediately. To prove the left one, we note that the matrix $\| \partial Z_j/\partial x_k \|$ is by (71) invertible and the elements of the inverse matrix B satisfy, in view of (70) and (71), the inequalities

$$|B_{jk}(x,\theta)| \leqslant c_4\Lambda(x)^{\lambda_j-\lambda_k}, \tag{74}$$

where c_4 depends only on C. We have

$$x_j = \sum_{k=1}^{n} B_{jk}(x,\theta) \frac{\lambda_k}{\lambda_j} Z_k(x,\theta), \; j \in I_+, \tag{75}$$

and therefore

$$|x_j| \leqslant c_5 \sum_{k\in I_+} \Lambda(x)^{\lambda_j-\lambda_k}|Z_k(x,\theta)|, \; j \in I_+. \tag{76}$$

It follows from (76) that for any $j \in I_+$, there exists $k = k(j) \in I_+$ such that

$$|x_j| \leqslant c_5|I_+|\Lambda(x)^{\lambda_j-\lambda_{k(j)}}|Z_{k(j)}(x,\theta)|, \; j \in I_+, \tag{77}$$

or

$$|x_j|\Lambda(x)^{\lambda_{k(j)}-\lambda_j} \leqslant c_5|I_+||Z_{k(j)}(x,\theta)|, \; j \in I_+. \tag{78}$$

Raising (78) to the power $1/\lambda_{k(j)}$ and summing over $j \in I_+$, we obtain

$$[\sum_{j\in I_+} \{|x_j|\Lambda(x)^{-\lambda_j}\}^{1/\lambda_{k(j)}}]\Lambda(x) \leqslant c_6\Lambda(Z(x,\theta)). \tag{79}$$

The expression in the square brackets in (79) is, however, greater than some positive constant on the set $\Lambda(x) \geqslant c_3$; therefore, we come to the left inequality in (72), and Lemma 2 is proved.

Assume from now on that all the systems (29) for $k = 1,\ldots,n$ satisfy the property: the solution (31) is defined for any initial data and any t_o and the following estimates are valid:

$$\left| \frac{\partial^{|\alpha|+\ell}}{\partial x^\alpha \partial t_o^\ell} (X_j^{(k)}(x,t_o) - x_j)\right| \leqslant C_{\alpha,M}\Lambda(x)^{\lambda_j-\langle\alpha,\lambda\rangle+\ell-\lambda_k} \text{ for } |t_o| \leqslant \leqslant M\Lambda(x)^{\lambda_k}, \; |\alpha|,M = 0,1,2,\ldots . \tag{80}$$

We refer to commutation relations with the principal part satisfying this condition as normal commutation relations.

Definition 1. $\mathcal{U} \subset \mathbb{R}^{2n}_{(x,p)}$ is a left normal domain if:

(a) The system of equations

$$\mathcal{X}(y,p) = x, \tag{81}$$

where $\mathcal{X}$ is given by (53), has for $(x,p) \in \mathcal{U}$ a unique differentiable solution

$$y = L(x,p) \tag{82}$$

turning into $y = x$ for $p = 0$; in particular, the Jacobian

$$\frac{D\mathcal{X}}{Dy} = \det \frac{\partial \mathcal{X}(y,p)}{\partial y}\Big|_{y=L(x,p)} \neq 0 \text{ for } (x,p) \in \mathcal{U}. \tag{83}$$

(b) For any $R > 0$ the intersection

$$\mathcal{U}_R = \mathcal{U} \cap \{(x,p) \mid |p_j| < R\Lambda(x)^{1-\lambda_j},\ j = 1,\dots,n\} \tag{84}$$

satisfies the properties: (b_1) the functions (82) satisfy the estimates

$$\left| \frac{\partial^{|\alpha|+|\beta|}(L_j(x,p)-x_j)}{\partial x^\alpha \partial p^\beta} \right| \leqslant C_{\alpha\beta R}\Lambda(x)^{\lambda_j - \langle\lambda,\alpha\rangle + \langle\lambda-1,\beta\rangle} \tag{85}$$

$$\text{for } (x,p) \in \mathcal{U}_R \text{ and } |\alpha|,|\beta| = 1,2,\dots;$$

(b_2) denote by $\tilde{\mathcal{U}}_R \subset \mathbb{R}^{2n}$ the set

$$\tilde{\mathcal{U}}_R = \{(y,p) \in \mathbb{R}^{2n} \mid \exists p^{(1)} \in \mathbb{R}^n : (x,p^{(1)}) \in \mathcal{U}_R, y = L(x,p^{(1)}),$$

$$p_1^{(1)} = p_1,\dots,p_{k-1}^{(1)} = p_{k-1}, p_k = \mu p_k^{(1)}, p_{k+1} = \dots = p_n = 0 \text{ for some} \tag{86}$$

$$k = \{1,\dots,n\} \text{ and } \mu \in [0,1]\}.$$

The conditions of Lemma 2 are satisfied for

$$(x,t) \equiv (x,\Lambda(x)^{\lambda-1}\theta) \in \tilde{\mathcal{U}}_R.$$

In what follows we usually call a left normal domain simply a "normal domain" omitting the word "left."

Lemma 3. For $C > 0$ sufficiently small, the domain

$$\mathcal{U} = \{(x,p) \in \mathbb{R}^{2n} \mid |p_j| < C\Lambda(x)^{1-\lambda_j},\ j = 1,\dots,n\} \tag{87}$$

is a normal domain.

Proof. Estimates (69) and (70) are always valid in the domain of the form (87) (the second of these possibly with another constant C). Both DZ/Dx and $D\mathcal{X}/Dy$ are equal to unity for $t = 0$ (respectively, $p = 0$); thus for C small enough these Jacobians will be greater than some positive constant. The lemma is thereby proved except for some technical details which we leave to the reader.

Let now the support of $f(x,p) \in \tilde{T}^m_{(\lambda_1,\dots,\lambda_n)}(\mathbb{R}^{2n})$ lie in a normal domain $\mathcal{U}$, $\operatorname{supp} f \subset \mathcal{U}$. It follows then that

$$(P_s f)(X(x,p),p) \in \tilde{T}^{m-s}_{(\lambda_1,\ldots,\lambda_n)}(\mathbb{R}^{2n}) \tag{88}$$

for any s and therefore we may obtain the asymptotic sum of the series (60) (we denote it also by $F(x,t) \in \tilde{T}^{m}_{(\lambda_1,\ldots,\lambda_n)}$). We claim that (4) is valid. Indeed, using the asymptotic expansions for $F(x,t)$ (and, respectively, for $W_k(x,y,t,t_o)$) up to an arbitrary order, we obtain the estimates (64) for arbitrary δ and N_1 since estimates (63) take place via Lemma 2 and condition (b_2) of Definition 1. The estimates (64), in turn, yield after simple calculations, the estimates (5) for the remainder in (4). The only thing remaining to obtain (6) is to solve the "inverse problem," i.e., to find $f(x,t)$, $F(x,t)$ being given. From (60), and (81) and (82), it is easy to see that the described solution has the form

$$f(x,p) \simeq F(L(x,p),p) + \sum_{s=1}^{\infty} (\tilde{P}_s F)(L(x,p),p), \tag{89}$$

where $\tilde{P}_s$ are differential operators of order $\leqslant s + 1$ in x and the s-th term of the series belongs to $\tilde{T}^{m-s}_{(\lambda_1,\ldots,\lambda_n)}(\mathbb{R}^{2n})$ provided that also $f(x,p) \in$ $\in \tilde{T}^{m}_{(\lambda_1,\ldots,\lambda_n)}(\mathbb{R}^{n})$. Thus we have proved the following theorem:

Theorem 1. Assume that $\Phi(x) \in \overset{0}{L}{}^{\infty}_{st,(\lambda_1,\ldots,\lambda_n)}(\mathbb{R}^{n})$ and that for some $K \in \mathrm{Ess}\,\Phi$, $K \subset U$ where $U \subset \mathbb{R}^{2n}$ is some normal domain. For any $F(x) \in$ $\in S^{\infty}_{(\lambda_1,\ldots,\lambda_n)}(\mathbb{R}^{n})$ the composition formula

$$F(A) \circ \Phi(A) = (H(\overset{2}{x},-i\overset{1}{\frac{\partial}{\partial x}})\Phi)(A) \pmod{O_p^{-\infty}(x)} \tag{90}$$

is valid, where $H(x,p) \in \tilde{T}^{\infty}_{(\lambda_1,\ldots,\lambda_n)}(\mathbb{R}^{2n})$ is any symbol such that for (x,p) in some neighborhood of K

$$H(x,p) \simeq F(L(x,p)) + \sum_{s=1}^{\infty} (\tilde{P}_s F)(L(x,p)), \tag{91}$$

where $\tilde{P}_s$ are given differential operators of order $\leqslant s + 1$, and $L(x,p)$ is a solution of the system of equations (81).

In a completely analogous way the following problem may be solved: given the symbols $F(x) \in S^{\infty}_{(\lambda_1,\ldots,\lambda_n)}(\mathbb{R}^{n})$ and $\Phi(x) \in \overset{0}{L}{}^{\infty}_{st,(\lambda_1,\ldots,\lambda_n)}(\mathbb{R}^{n})$, construct a symbol $\Phi_1(x) \in \overset{0}{L}{}^{\infty}_{(\lambda_1,\ldots,\lambda_n)}(\mathbb{R}^{n})$ such that

$$\Phi(A) \circ F(A) = \Phi_1(A) \pmod{O_p^{-\infty}(x)}. \tag{92}$$

We do not perform the calculations again but merely formulate the final result. Again, let $X^{(k)}(x,t) \equiv X^{(k,t)}(x)$ denote the solution of the system (29). We set

$$\tilde{Y}(x,t) = X^{(n,-t_n)} \circ \ldots \circ X^{(1,-t_1)}(x), \tag{93}$$

and define the mapping Y by

$$Y_j(x,t) = \tilde{Y}_j(x,t_1,\ldots,t_j,0,\ldots,0),\ j = 1,\ldots,n. \tag{94}$$

Set also

$$Z(x,\theta) = Y(x,\Lambda(x)^{1-\lambda}\theta). \tag{95}$$

Denote by

$$y = R(x,p) \tag{96}$$

the solution of the system of equations

$$Y(y,p) = x. \tag{97}$$

The domain $\mathcal{U} \subset \mathbb{R}^{2n}_{(x,p)}$ will be called right normal if it satisfies the conditions of Definition 1 with X replaced by Y, and L(x,p) by R(x,p), and $\tilde{\mathcal{U}}_R$ replaced by

$$\tilde{\mathcal{U}}_R = \{(y,p) \in \mathbb{R}^{2n} | \exists p^{(1)} : (x,p^{(1)}) \in \mathcal{U}_R, y = R(x,p^{(1)}), \tag{98}$$

$p_n^{(1)} = p_n, \ldots, p_{k+1}^{(1)} = p_{k+1}, p_k = \mu p_k^{(1)}, p_{k-1} = \ldots = p_1 = 0$ for some $k \in \{1,\ldots,n\}$ and $\mu \in [0,1]\}$. The analogue of Lemma 3 is obviously valid; the following theorem takes place:

<u>Theorem 2</u>. Assume that $\Phi(x) \in \overset{0}{L}{}^{\infty}_{st,(\lambda_1,\ldots,\lambda_n)}(\mathbb{R}^n)$ and that for some $K \in \mathrm{Ess}\,\Phi$, $K \subset \mathcal{U}$, where $\mathcal{U} \subset \mathbb{R}^{2n}$ is a right normal domain. For any $F(x) \in S^{\infty}_{(\lambda_1,\ldots,\lambda_n)}(\mathbb{R}^n)$ the composition formula

$$\Phi(A) \circ F(A) = (G(\overset{2}{x},-i\overset{1}{\frac{\partial}{\partial x}})\Phi)(A) \pmod{O_p^{-\infty}(x)} \tag{99}$$

is valid, where $G(x,p) \in \tilde{T}^{\infty}_{(\lambda_1,\ldots,\lambda_n)}(\mathbb{R}^{2n})$ is any symbol such that for (x,p) in some neighborhood of K

$$G(x,p) \simeq F(R(x,p)) + \sum_{s=1}^{\infty} (\tilde{Q}_s F)(R(x,p)), \tag{100}$$

where $\tilde{Q}_s$ are given differential operators of order $\leqslant s + 1$.

<u>Definition 2</u>. The mappings

$$F \to L(F) = H(\overset{2}{x},-i\overset{1}{\frac{\partial}{\partial x}}),\ F \in S^{\infty}_{(\lambda_1,\ldots,\lambda_n)}(\mathbb{R}^n) \tag{101}$$

and

$$F \to R(F) = G(\overset{2}{x},-i\overset{1}{\frac{\partial}{\partial x}}),\ F \in S^{\infty}_{(\lambda_1,\ldots,\lambda_n)}(\mathbb{R}^n) \tag{102}$$

constructed in Theorems 1 and 2, respectively, are called the left and right (local) regular representation for the tuple $(\overset{1}{A}_1,\ldots,\overset{n}{A}_n)$.

It should be emphasized that the operators $L(F)$ ($R(F)$) of the left (right) regular representation are therefore correctly defined (up to $L^{-\infty}_{(\lambda_1,\ldots,\lambda_n)}(\mathbb{R}^{2n})$) operators on the subspace in $\overset{0}{L}{}^{\infty}_{st,(\lambda_1,\ldots,\lambda_n)}(\mathbb{R}^n)$ consisting of functions $\Phi(x)$ such that some $K \in \mathrm{Ess}\,\Phi$ lies in the left (right) normal domain.

4. SYMPLECTIC MANIFOLD OF A POISSON ALGEBRA AND PROOF OF THE MAIN THEOREM

A. Effects of Variable Changes

Let a μ-structure on the Poisson algebra in R^n be given in the sense of Definition 2, Section 3:B, i.e., a mapping

$$\mu : f \to f(\overset{1}{A_1},\ldots,\overset{n}{A_n}),\ f \in \mathcal{L}^{\infty}_{(\lambda_1,\ldots,\lambda_n)}(R^n), \tag{1}$$

where $(A_1,\ldots,A_n)$ is a proper tuple of operators in a Banach scale X satisfying the commutation relations

$$[A_j,A_k] = -i\omega_{jk}(\overset{1}{A_1},\ldots,\overset{n}{A_n}), \tag{2}$$

where the principal part of $\omega_{jk}(x)$ equal to $\Omega_{jk}(x)/\Omega$ is the tensor defining the Poisson algebra structure.

Our final aim is to construct asymptotic (in the scale X) solutions of the equations of the type

$$f(\overset{1}{A_1},\ldots,\overset{n}{A_n})u = v, \text{ where } f \in \mathcal{P}^{\infty}_{st,(\lambda_1,\ldots,\lambda_n)}(R^{2n}); \tag{3}$$

this, in turn, requires the solution of an auxiliary Cauchy problem with subsequent integration over t (we have this construction in item B):

$$-i\frac{\partial}{\partial t} + F(\overset{1}{A_1},\ldots,\overset{n}{A_n})u = 0,\ u|_{t=0} = v \tag{4}$$

(here we assume that $F \in \mathcal{P}^1_{st,(\lambda_1,\ldots,\lambda_n)}(R^{2n})$). We seek the asymptotic resolving operator for the problem (4) in the form

$$\mathcal{U}(t) = \Phi(\overset{1}{A_1},\ldots,\overset{n}{A_n},t), \tag{5}$$

where $\Phi(x,t) \in \overset{0}{\mathcal{L}}{}^{\infty}_{st,(\lambda_1,\ldots,\lambda_n)}(R^{2n})$. Using Theorem 1 of Section 3:D, we obtain for $\Phi(x,t)$ the equation

$$-i\frac{\partial\Phi}{\partial t} + H(\overset{2}{x},-i\overset{1}{\frac{\partial}{\partial x}})\Phi \in \mathcal{L}^{-\infty}_{(\lambda_1,\ldots,\lambda_n)}(R^{2n}) \tag{6}$$

with the initial condition

$$\Phi|_{t=0} = 1, \tag{7}$$

where H(x,p) is described in the cited theorem, provided that some $K_t \in \in \mathrm{Ess}(\Phi(t))$ lies in a normal domain for the considered values of t. Now assume for a moment that f(x) is real-valued. The solution to (6) and (7) may be found with the help of the canonical operator (see Chapter 3, Section 3:B) by a method quite similar to that given in Chapter 3, Section 4:D. Thus,

$$\Phi(x,t) = [K\phi](x,t) \tag{8}$$

is a CRF associated with the proper Lagrangian manifold $L(t) \subset R^{2n}$ which is nothing other than a shift of the manifold

$$L(0) \equiv L_o = \{(x,p)\,|\,x \in R^n, p = 0\} \tag{9}$$

during the time t along the trajectories of the Hamiltonian vector field corresponding to the Hamiltonian function

$$H_1(x,p) = F_1(L(x,p)), \tag{10}$$

where $F_1 \in \mathcal{O}^1_{st,(\lambda_1,\ldots,\lambda_n)}(R^{2n})$ is the principal homogeneous part of F. For $|t|$ small enough everything is alright since L(t) lies inside the normal domain. This may not be the case for greater values of t, and the method fails once L(t) attempts to "leave" the normal domain. However, this does not necessarily mean that there is no solution for these values of t (the situation is analogous to focal points in the WKB method). This means only that the solution no longer has the form (5) with $\Phi(x,t)$ given by (8). Introducing another family of operators $(B_1,\ldots,B_n)$, functionally dependent on $A_1,\ldots,A_n$, and considering functions of these operators, gives the possibility of obtaining the solution for these values of t as well. It turns out that the development of this idea leads us directly to the consideration of the symbol of the resolving operator as a section of the canonical sheaf on a special symplectic manifold, which we call the symplectic manifold of a Poisson algebra.

We come to detailed considerations. To simplify the subsequent arguments, we assume that the following condition is satisfied for the operators:

Condition 1. If $f(x_1,\ldots,x_n) \in L^\infty_{(\lambda_1,\ldots,\lambda_n)}(R^n)$ and $f(\overset{1}{A}_1,\ldots,\overset{n}{A}_n) \in \mathcal{O}^{-\infty}_p(x)$, then $f(x_1,\ldots,x_n) \in L^{-\infty}_{(\lambda_1,\ldots,\lambda_n)}(R^n)$ $(= S^{-\infty}_{(\lambda_1,\ldots,\lambda_n)}(R^n))$. This Condition is not necessary for the validity of the subsequent theorems; however, the requirement that the μ-mapping be "asymptotically monomorphic" enables us to present much more obvious proofs and to avoid cumbersome calculations.

Lemma 1. Let $\mathcal{U} \subset R^{2n}$ be a normal domain (see Definition 1 of Section 3:D). The mapping

$$L : \mathcal{U} \to R^n, \quad (x,p) \to y = L(x,p), \tag{11}$$

where L is defined in the cited definition, preserves the Poisson brackets, i.e., for any $f,g \in C^\infty(R^n)$,

$$\{L^*f, L^*g\} = L^*\{f,g\} \text{ in } \mathcal{U} \text{ for } \Lambda(x) \text{ large enough} \tag{12}$$

(in (12) $(L^*f)(x,p) \overset{\text{def}}{=} f(L(x,p))$; $\{f,g\}$ is a Poisson bracket of f and g with respect to Ω, and $\{L^*f,L^*g\}$ is a Poisson bracket of L^*f and L^*g generated by the symplectic structure

$$\omega^2 = dp \wedge dx \tag{13}$$

in $\mathcal{U}$, i.e.,

$$\{L^*f, L^*g\} = \sum_{j=1}^n \left\{ \frac{\partial(L^*f)}{\partial p_j}\frac{\partial(L^*g)}{\partial x_j} - \frac{\partial(L^*f)}{\partial x_j}\frac{\partial(L^*g)}{\partial p_j} \right\}. \tag{14}$$

Proof. Obviously (12) is valid for any $f,g \in C^\infty(R^n)$ if and only if

$$\sum_s \left\{ \frac{\partial L_j(x,p)}{\partial p_s}\frac{\partial L_k(x,p)}{\partial x_s} - \frac{\partial L_k(x,p)}{\partial p_s}\frac{\partial L_j(x,p)}{\partial x_s} \right\} = \Omega_{jk}(L(x,p)),\ j,k = 1,\ldots,n. \tag{15}$$

To prove (15) consider any symbol $\Phi(x) \in L^{\infty}_{st,(\lambda_1,\dots,\lambda_n)}(R^{2n})$ and calculate the symbol of composition (equal to zero by commutation relations)

$$[\![(A_j \circ A_k - A_k \circ A_j + i\omega_{jk}(\overset{1}{A_1},\dots,\overset{n}{A_n})]\!] \circ \overset{1}{\Phi}(A_1,\dots,A_n) = 0 \tag{16}$$

using Theorem 1 of Section 3:D (we assume that $K \subset U$ for some $K \in \mathrm{Ess}\,\Phi$). We obtain that the symbol of operator (16) equals

$$H(\overset{2}{x}, -i\frac{\overset{1}{\partial}}{\partial x})\Phi(x), \tag{17}$$

where $H(x,p) \in \tilde{T}^{\lambda_j+\lambda_k-1}_{(\lambda_1,\dots,\lambda_n)}(R^{2n})$, and where

$$H(x,p) = H_0(x,p) + H_1(x,p), \quad H_1(x,p) \in \tilde{T}^{\lambda_j+\lambda_k-2}_{(\lambda_1,\dots,\lambda_n)}(R^{2n}), \tag{18}$$

$$\begin{aligned} H_0(x,p) = -i\sum_s \{ \frac{\partial L_j(x,p)}{\partial p_s}\frac{\partial L_k(x,p)}{\partial x_s} - \frac{\partial L_k(x,p)}{\partial p_s}\frac{\partial L_j(x,p)}{\partial x_s} \} + \\ + i\Omega_{jk}(L(x,p)) \text{ in } U \end{aligned} \tag{19}$$

(we used the composition formulas from Theorem 2 of Chapter 3, Section 3:A).

Substituting for Φ various CRF's associated with Lagrangian manifolds $L \subset U$ we deduce, using Theorem 2 of Chapter 3, Section 3:B, that $H(\tau^{\lambda}x, \tau^{1-\lambda}p)$ rapidly decays for $\tau \to \infty$. In particular, $H_0(x,p) = 0$ for $\Lambda(x)$ large enough since for these $\Lambda(x)$ we have $H_0(\tau^{\lambda}x, \tau^{1-\lambda}p) = \tau^{\lambda_j+\lambda_k-1}H_0(x,p)$, $\tau > 1$. Thus Lemma 1 is proved.

We denote by

$$\ell : U \to R^n, \tag{20}$$

the R_+-homogeneous mapping coinciding with L for large $\Lambda(x)$. In a quite analogous way we may prove the following:

Lemma 2. Let $U \subset R^{2n}$ be a right normal domain. The mapping

$$R : U \to R^n, \quad (x,p) \to y = R(x,p),$$

"anti-preserves" Poisson brackets, i.e., satisfies the property

$$\{R^*f, R^*g\} = -R^*\{f,g\} \text{ for } \Lambda(x) \text{ large enough,}$$

for any $f,g \in C^{\infty}(R^n)$. Also, if U is both right and left normal, then

$$\{R^*f, L^*g\} = 0$$

for any $f,g \in C^{\infty}(R^n)$.

Proof. We leave it to the reader as an easy exercise. As above, we denote by

$$r : U \to R^n, \tag{21}$$

where U is a right normal domain, the R_+-homogeneous mapping coincident with R for large $\Lambda(x)$.

Next we study the transformation of symbols which accompanies passage from one proper set of generators $(A_1,\dots,A_n)$ to another one, say, $(B_1,\dots,B_n)$. Let $\mu = (\mu_1,\dots,\mu_n)$ be the n-tuple of non-negative numbers satisfying the same conditions as $\lambda = (\lambda_1,\dots,\lambda_n)$. We denote $T_o = \{j|\mu_j=0|\}$, $T_+ = \{j|\mu_j > 0|\} = [n]\setminus T_o$. We assume that $(B_1,\dots,B_n)$ is a proper tuple of generators in the scale X, corresponding to the tuple $\mu = (\mu_1,\dots,\mu_n)$ and that the following conditions are satisfied:

Condition 2. (Change of variables). For $j = 1,\dots,n$

$$B_j = Y_j(\overset{1}{A_1},\dots,\overset{n}{A_n}); \tag{22}$$

the symbols $Y_j(x_1,\dots,x_n)$ satisfy the following conditions:

$$Y_j(x_1,\dots,x_n) \in P^{\mu_j}_{st,(\lambda_1,\dots,\lambda_n)}(R^n_x) \text{ for } j \in T_+, \tag{23}$$

$$Y_j(x_1,\dots,x_n) - \sum_{k\in I_o} a_{jk}x_k \in P_{st,(\lambda_1,\dots,\lambda_n)}(R^n_x) \text{ for } j \in T_o, \tag{24}$$

where a_{jk} are some given constant coefficients; symbols $Y_j(x_1,\dots,x_n)$ have the asymptotic expansions of the form

$$Y_j(x_1,\dots,x_n) \simeq \sum_{k=0}^{\infty} Y_{j,k}(x_1,\dots,x_n), \tag{25}$$

where

$$Y_{jk}(\tau^{\lambda_1}x_1,\dots,\tau^{\lambda_n}x_n) = \tau^{\mu_j-k} Y_{j,k}(x_1,\dots,x_n) \tag{26}$$

for large $\Lambda(x)$, and the difference between Y_j and the partial sum of N terms of the series on the right-hand side of (25) belongs to $P^{\mu_j-N-1}_{st,(\lambda_1,\dots,\lambda_n)} \times (R^n)$. We denote

$$y_j(x_1,\dots,x_n) = Y_{j,o}(x_1,\dots,x_n), \tag{27}$$

and require that the coordinate change $y = y(x)$ be strongly non-degenerate in the sense that

$$\left|\det \frac{\partial y}{\partial x}(x)\right| \geqslant C\Lambda(x)^{\Sigma\mu_j-\Sigma\lambda_j} \tag{28}$$

with some positive constant C.

Remark 1. Under these conditions $\Lambda(y(x))$ is equivalent to $\Lambda(x)$ (where of course $\Lambda(y)$ is constructed according to $(\mu_1,\dots,\mu_n)$). The proof is quite analogous to that of Lemma 2 of Section 3:D.

Condition 3. (Commutation relations). For $j,k = 1,\dots,n$

$$[B_j,B_k] = -i\tilde{\omega}_{jk}(\overset{1}{B_1},\dots,\overset{n}{B_n}), \tag{29}$$

where

$$\tilde{\omega}_{jk} \in P^{\mu_j+\mu_k-1}_{st,(\mu_1,\dots,\mu_n)}(R^n), \tag{30}$$

$$\tilde{\omega}_{jk} \simeq \sum_{s=0}^{\infty} \tilde{\omega}^{(s)}_{jk}, \quad \tilde{\omega}^{(s)}_{jk}(y) \in O^{\mu_j+\mu_k-s}_{st,(\mu_1,\dots,\mu_n)}(R^n). \tag{31}$$

We denote

$$\Omega_{jk}(y) = \tilde{\omega}^{(0)}_{jk}(y). \tag{32}$$

Condition 4. The 2n-tuple of operators $(A_1,\dots,A_n,B_1,\dots,B_n)$ is a proper tuple (with respect to weight tuple $(\lambda_1,\dots,\lambda_n,\mu_1,\dots,\mu_n)$); moreover, the item (b) of Condition 3 of Section 3:B remains valid for this 2n-tuple if we replace the space $\tilde{\Gamma}^m_{m_1}$ by the space $\tilde{\tilde{\Gamma}}^m_{m_1}(R)$ in the definition of which we require only (cf. Section 3:A) that

$$\Lambda(x) \leqslant C\Lambda_f(\tilde{x},\tilde{y})$$

in the R_+-invariant neighborhood of the set

$$\tilde{\Delta} = \{(\tilde{x},\tilde{y})\,|\,x_{j1} = \dots = x_{jr_j}, j \in I_+, y_{j1} = \dots = y_{jr_{n+j}}, j \in J_+,$$
$$y_j = y_j(x),\ j \in T_+\}.$$

Lemma 3. For $\Lambda(x)$ large enough, we have

$$\tilde{\Omega}_{jk}(y(x)) = \sum_{\ell,m} \Omega_{\ell m}(x)\,\frac{\partial y_j}{\partial x_\ell}\,\frac{\partial y_k}{\partial x_m}. \tag{33}$$

Proof. We calculate $[B_j,B_k]$ from (22). Modulo $O_p^{-\infty}(x)$, we have

$$[B_j,B_k] \equiv -i\{y_j,y_k\}(\overset{1}{A_1},\dots,\overset{n}{A_n}) \text{ (+ lower-order terms)}. \tag{34}$$

On the other hand, we have by (29)

$$[B_j,B_k] = -i\tilde{\omega}_{jk}(⟦\overset{1}{Y_1}(\overset{1}{A_1},\dots,\overset{n}{A_n})⟧,\dots,⟦\overset{n}{Y_n}(\overset{1}{A_1},\dots,\overset{n}{A_n})⟧). \tag{35}$$

We claim that modulo $O_p^{-\infty}(x)$ the right-hand side of (35) may be represented as the function of $(\overset{1}{A_1},\dots,\overset{n}{A_n})$:

$$\begin{aligned}\tilde{\omega}_{jk}(⟦\overset{1}{Y_1}(\overset{1}{A_1},\dots,\overset{n}{A_n})⟧,\dots,⟦\overset{n}{Y_n}(\overset{1}{A_1},\dots,\overset{n}{A_n})⟧) = \\ = \tilde{\Omega}_{jk} \circ y(\overset{1}{A_1},\dots,\overset{n}{A_n}) \text{ (+ lower-order terms)}.\end{aligned} \tag{36}$$

Indeed, (36) follows by successive application of the K-formula ([52]) and then by using Condition 4. Lemma 3 is proved.

Let the symbol $\Phi(y) \in L^{0\infty}_{st,(\mu_1,\dots,\mu_n)}(R^n)$ be given. The problem we intend to solve now is to find a symbol $\Phi_1(x) \in L^{0\infty}_{st,(\lambda_1,\dots,\lambda_n)}(R^n)$ such that

$$\Phi(\overset{1}{B_1},\dots,\overset{n}{B_n}) = \Phi_1(\overset{1}{A_1},\dots,\overset{n}{A_n}) \pmod{O_p^{-\infty}(x)} \tag{37}$$

(in particular, we obtain the tool to verify the identity (36)). To do this, we turn to the definition of $\Phi(\overset{1}{B_1},\dots,\overset{n}{B_n})$ via the Fourier transformation:

$$\Phi(\overset{1}{B_1},\dots,\overset{n}{B_n}) = (2\pi)^{-n/2}\int\tilde{\Phi}(\eta_1,\dots,\eta_n)e^{i\eta_n B_n}\dots e^{i\eta_1 B_1}d\eta_1\dots d\eta_n. \tag{38}$$

Our first step will be to find the asymptotics for the product of $e^{i\eta_n B_n}\dots e^{i\eta_1 B_1}$ of the form

$$e^{i\eta_n B_n}\dots e^{i\eta_1 B_1} = E(\eta,\overset{1}{A_1},\dots,\overset{n}{A_n}) + \hat{R}(\eta) \tag{39}$$

with the appropriate estimates for the remainder $\hat{R}(\eta)$. In some analogy with procedures performed in Section 3:D, we successively solve the definitive equations for $e^{i\eta_1 B_1},\dots,e^{i\eta_n B_n}$. More precisely, we consider the sequence of functions $E_j(\eta,t,x)$ of $(\eta,x)\in R^{2n}$, $t\in[0,1]$, $j = 1,\dots,n$, satisfying the conditions:

$$\begin{aligned} E_1(\eta,0,x) &= \chi(\eta,x),\\ E_j(\eta,0,x) &= E_{j-1}(\eta,1,x),\ j = 2,\dots,n; \end{aligned} \tag{40}$$

E_j is an asymptotic solution of the problem*

$$\begin{aligned} -i\frac{\partial}{\partial t}E_j(\eta,t,A) &= \eta_j Y_j(A)\circ E_j(\eta,t,A) + \hat{R}_j(\eta,t),\ t\in[0,1],\\ \hat{R}_j(\eta,t) &\in \mathcal{O}_p^{-\infty}(x),\ j = 1,\dots,n. \end{aligned} \tag{41}$$

In (40) $\chi(\eta,x)$ is a cut-off function,

$$\begin{aligned} \chi(\eta,x) &= 1 \text{ if } |\eta_j| \leqslant C\Lambda(x)^{1-\mu_j},\ j = 1,\dots,n;\\ \chi(\eta,x) &= 0 \text{ if } |\eta_j| \geqslant C_i\Lambda(x)^{1-\mu_j} \text{ for at least one } j; \end{aligned} \tag{42}$$

$$\left|\frac{\partial^{|\alpha|+|\beta|}\chi(\eta,x)}{\partial x^{\alpha}\partial\eta^{\beta}}\right| \leqslant C_{\alpha\beta}\Lambda(x)^{-\langle\lambda,\alpha\rangle+\langle\mu-1,\beta\rangle},\quad |\alpha|,|\beta| = 0,1,2,\dots. \tag{43}$$

For example, we may set $\chi(\eta,x) = \zeta(\eta\Lambda(x)^{\mu-1})$ where ζ is an appropriate finite function.

Assuming that $E_j(\eta,t,x)\in \overset{0}{L}{}^{\infty}_{(\lambda_1,\dots,\lambda_n)}(R^n)$ for any η and t under consideration and that some $K_j\in \mathrm{Ess}(E_j]$ lies in the normal subset $\mathcal{U}$, we obtain, via Theorem 1 of Section 3:D, the following equation for $E_j(\eta,t,x)$:

$$-i\frac{\partial}{\partial t}E_j(\eta,t,x) = \eta_j H_j(\overset{2}{x},-i\overset{1}{\frac{\partial}{\partial x}})E_j(\eta,t,x) \pmod{L^{-\infty}_{(\lambda_1,\dots,\lambda_n)}(R^n)}, \tag{44}$$

where the principal symbol of $H_j(\overset{2}{x},-i\overset{1}{\frac{\partial}{\partial x}})$ in the neighborhood of K_j equals $y_j(L(x,p))$. Rigorously speaking, Theorem 1 of Section 3:D was established only for symbols belonging to the space $S^{\infty}_{(\lambda_1,\dots,\lambda_n)}(R^n)$ and for $j\in T_0$; $y_j(x)$ does not belong to this space. However, using the representation (24) it is easy to show that this theorem remains valid for $Y_j(x)$, $j\in T_0$.

To produce the asymptotic solution to (44) we make use of the theory of canonical operators on the proper Lagrangian manifold developed in Chapter 3, Section 3:B. However, here we need Lagrangian manifolds depending on the parameter η, also involved in the R_+-action, so we indicate the main modifications which have to be made in definitions and estimates of Chapter 3, Section 3:B.

---

* Here we again use the shortened notation $f(A) = f(\overset{1}{A_1},\dots,\overset{n}{A_n})$.

First of all, Definition 1 of that item reads now: $L \subset R^{2n}_{(x,p)} \times R^n_\eta$ is a proper R_+-invariant family of Lagrangian manifolds if:

(a) $L(\eta_o) = L \cap \{\eta = \eta_o\}$ is a Lagrangian manifold in $R^{2n}_{(x,p)}$ for any $\eta_o \in R^n$;

(b) L is R_+-invariant (the action of R_+ is defined by)

$$\tau(x,p,\eta) = (\tau^\lambda x, \tau^{1-\lambda} p, \tau^{1-\mu}\eta), \tag{45}$$

in obvious abbreviations;

(c) the inequalities

$$|p_j| \leqslant C\Lambda(x)^{1-\lambda_j}, \quad |\eta_j| \leqslant C\Lambda(x)^{1-\mu_j} \tag{46}$$

hold on L with some constant C.

Next, the formula of Definition 2

$$S_I = \lambda_I x_I p_I + (\lambda_{\bar I} - 1)p_{\bar I} x_{\bar I} \tag{47}$$

for the action in the canonical chart with the coordinates $(x_I, p_{\bar I})$, is now valid only for $\eta = 0$; generally S_I must be defined as the unique homogeneous solution of the Pfaff equation

$$dS_I = p_I dx_{\bar I} - x_{\bar I} dp_{\bar I}. \tag{48}$$

As for estimates, the conditions on ψ to belong to $\overset{0}{L}{}^m_{(\lambda_1,\ldots,\lambda_n)}(R^n)$ now read

$$\left| \frac{\partial^{|\alpha|+|\beta|}\psi}{\partial x^\alpha \partial \eta^\beta} \right| \leqslant C_{\alpha\beta}\Lambda(x)^{m+\langle 1-\lambda,\alpha\rangle + \langle \mu,\beta\rangle} \text{ for } |\alpha|,|\beta| = 0,1,2,\ldots,$$
$$|\alpha_{T_o}| = 0. \tag{49}$$

We do not indicate here numerous minor modifications still necessary; all the theorems of Chapter 3, Section 3:B remain valid in the new context.

Let us solve (44) successively for $j = 1,2,\ldots,n$. Consider the family of Lagrangian manifolds L_o, given by the relations

$$p = 0, \quad |\eta_j| < C_1\Lambda(x)^{1-\mu_j}, \quad j = 1,\ldots,n, \tag{50}$$

and let L_t, $t \in [0,1]$ be the shift of L_o along the trajectories of the Hamiltonian system

$$\dot x = -\eta_1 \frac{\partial}{\partial p}(y_1(L(x,p))), \quad \dot p = +\eta_1 \frac{\partial}{\partial x}(y_1(L(x,p))), \tag{51}$$

corresponding to the Hamiltonian function $-\eta_1 y_1(L(x,p))$. Note that for C_1 small enough the trajectories of (51) are necessarily defined for $t \in [0,1]$, and L_t is a proper family of Lagrangian manifolds lying completely in $\mathcal{U}$. Note also that since L_o obviously satisfies the quantization conditions, so does L_t for any t. Thus we may find $E_1(\eta,t,x)$ in the form (cf. Chapter 3, Section 4:D)

$$E_1(\eta,t,x) = [K_t\phi](\eta,t,x), \tag{52}$$

where K is a canonical operator on L_t, and ϕ satisfies the transport equation and the initial data

$$\phi|_{t=0} = \chi(\eta,x) \tag{53}$$

(we assume that the measure chosen on L_o has the form $dx_1 \wedge \ldots \wedge dx_n$). Then $E_1(\eta,t,x)$ thus constructed satisfies the equation

$$-i\frac{\partial}{\partial t}E_1(\eta,t,x) = \eta_1 H_1(\overset{2}{x},-i\overset{1}{\frac{\partial}{\partial x}})E_1(\eta,t,x) + R_1(\eta,t,x), \tag{54}$$

where R_1 satisfies the estimates (49) uniformly in $t \in [0,1]$.

We proceed by induction on j. Once (44) is solved for E_{j-1}, we impose for E_j the initial condition (40) and solve (44) for E_j, using the Hamiltonian function $-\eta_j Y_j(L(x,p))$. We obtain

$$-i\frac{\partial}{\partial t}E_j(\eta,t,x) = \eta_j H_j(\overset{2}{x},-i\overset{1}{\frac{\partial}{\partial x}})E_j(\eta,t,x) + R_j(\eta,t,x), \tag{55}$$

where R_j satisfies the estimates (49) uniformly in $t \in [0,1]$. From (55) and Theorem 1 of Chapter 4, Section 3:D it follows easily that

$$-i\frac{\partial}{\partial t}E_j(\eta,t,A) = \eta_j Y_j(A) \circ E_j(\eta,t,A) + \hat{R}_j(\eta,t), \tag{56}$$

where uniformly in

$$\left\| \eta^\gamma \frac{\partial^\alpha \hat{R}_j}{\partial \eta^\alpha} \right\|_{\delta \to \delta+N} \leqslant C_{\alpha\delta\gamma N} \text{ for any } \delta, N \text{ and} \qquad |\alpha|,|\gamma| = 0,1,2,\ldots,\ |\alpha_{J_o}| = 0. \tag{57}$$

Taking into account (22), we obtain

$$-i\frac{\partial}{\partial t}E_j(\eta,t,A) = \eta_j B_j \circ E_j(\eta,t,A) + \hat{Q}_j(\eta,t), \tag{58}$$

where $\hat{Q}_j(\eta,t)$ satisfies the same estimates (57). Next we obtain, using the fact that B_j is a tempered generator,

$$E_j(\eta,1,A) - e^{iB_j\eta_j}E_j(\eta,0,A) = ie^{i\eta_j B_j}\int_0^1 e^{-i\eta_j B_j \tau}\hat{Q}_j(\eta,t)dt; \tag{59}$$

thus

$$E_j(\eta,1,A) = e^{iB_j\eta_j} \circ E_j(\eta,0,A) + \hat{Q}_j(\eta), \tag{60}$$

where $\hat{Q}_j(\eta)$ again satisfies (57). Combining these estimates, we obtain, setting

$$E(\eta,x) = E_n(\eta,1,x), \tag{61}$$

that

$$e^{i\eta_n B_n} \times \ldots \times e^{i\eta_1 B_1} = E(\eta,A) + \hat{R}(\eta) + e^{i\eta_n B_n} \times \ldots \times e^{i\eta_1 B_1}(1 - \chi(\eta,A)), \tag{62}$$

where $\hat{R}(\eta)$ satisfies the estimates (57).

Make further transformations on the identity (62). Then we have

$$\chi(\eta,A) = \chi_1(\eta,B) + \hat{\chi}_2(\eta), \tag{63}$$

where $\hat{\chi}_2(\eta)$ satisfies the estimates (57),

$$\chi_1(\eta,y) \in \tilde{T}^0_{(\mu_1,\dots,\mu_n)}(\mathbb{R}^{2n}), \tag{64}$$

and may be expanded into the asymptotic series

$$\chi_1(\eta,y) \simeq \sum_{j=0}^{\infty} \chi_{1j}(\eta,y), \quad \chi_{1j}(\eta,y) \in \tilde{T}^{-j}_{(\mu_1,\dots,\mu_n)}(\mathbb{R}^{2n}), \tag{65}$$

where

$$\chi_{1o}(\eta,y(x)) = \chi(\eta,x), \quad \chi_{1j}(\eta,y(x)) = (\pi_j\chi)(\eta,x), \quad j = 1,2,\dots, \tag{66}$$

where π_j are differential operators.

To prove (63) it suffices to use the K-formula (cf. proof of Lemma 3) recursively. Thus

$$e^{i\eta_n B_n} \times \dots \times e^{i\eta_1 B_1} = E(\eta,A) + e^{i\eta_n B_n} \times \dots \times e^{i\eta_1 B_1} \circ (1-\chi(\eta,B)) + \hat{R}(\eta), \tag{67}$$

where $\hat{R}(\eta)$ satisfies (57) (but is different from that in (62)). Substituting (67) into (38) we obtain, using Theorem 2 of Section 3:D,

$$\begin{aligned}\Phi(B) = (2\pi)^{-n/2}\int\tilde{\Phi}(\eta)E(\eta,A)d\eta + (2\pi)^{-n/2}\int[(1-\chi_2(\overset{2}{y},-i\overset{1}{\frac{\partial}{\partial y}}))\Phi](\eta) \times \\ \times e^{i\eta_n B_n} \times \dots \times e^{i\eta_1 B_1} d\eta + (2\pi)^{-n/2}\int\hat{R}(\eta)\tilde{\Phi}(\eta)d\eta,\end{aligned} \tag{68}$$

where $\chi_2(y,\eta) \in \tilde{T}^0_{(\mu_1,\dots,\mu_n)}(\mathbb{R}^{2n})$ has the asymptotic expansion

$$\chi_2(y,\eta) = \chi_1(\eta,R_B(y,\eta)) + (Q_1\chi_1)(\eta,R_B(y,\eta)) + \dots, \tag{69}$$

where $Q_1,\dots,$ are differential operators and $R_B(y,\eta)$ is the mapping constructed in Theorem 2 of Section 3:D (according to the tuple of generators $(\overset{1}{B}_1,\dots,\overset{n}{B}_n)$).

The third term in the sum (68) belongs to the space $O_p^{-\infty}(x)$. We come to the following conclusion:

$$\Phi(B) = \Phi_1(A) \pmod{O_p^{-\infty}(x)}, \tag{70}$$

where

$$\Phi_1(x) = (2\pi)^{-n/2}\int\tilde{\Phi}(\eta)E(\eta,x)d\eta, \tag{71}$$

provided that there is some subset $K \in \mathrm{Ess}(\Phi)$ such that $K \cap \mathrm{supp}(1-\chi_2 \times (y,\eta)) = \emptyset$. Evidently, this is always the fact if the set $\{|\eta_j| \leqslant C\Lambda(y)^{1-\mu_j}\}$ belongs to $\mathrm{Ess}(\Phi)$, where C is small enough.

Study now more thoroughly the transformation (71). The kernel $E(\eta,x)$ has locally the form

$$E(\eta,x) = \left(\frac{i}{2\pi}\right)^{|\bar{I}|/2}\int e^{i[S_I(\eta,x_I,p_{\bar{I}})+x_{\bar{I}}p_{\bar{I}}]} a(\eta,x_I,p_{\bar{I}})dp_{\bar{I}}, \tag{72}$$

where S_I is quasi-homogeneous of degree 1, and a is asymptotically quasi-homogeneous. The function S_I is equal to

$$S_I = S - x_{\bar{I}}p_{\bar{I}}, \tag{73}$$

where all the functions are considered as functions on $L(\eta)$, and

$$dS = pdx \tag{74}$$

on $L(\eta)$ (η being fixed!). If we consider S as a function on L, η no longer being fixed, we have obviously

$$dS = pdx + Ad\eta, \tag{75}$$

where $A = (A_1,\ldots,A_n)$ is a certain set of functions on L. Consider the immersion $j = L \to \mathbb{R}^{2n}_{(x,p)} \times \mathbb{R}^{2n}_{(y,\eta)}$, given by (recall that $L \subset \mathbb{R}^{2n}_{(x,p)} \times \mathbb{R}^{2n}_{\eta}$):

$$j : (x,p,\eta) \to (x,p,A,\eta). \tag{76}$$

In view of (75) this immersion defines a Lagrangian manifold in $\mathbb{R}^{2n}_{(x,p)} \times \mathbb{R}^{2n}_{(y,\eta)}$ (with respect to the form $\Omega^2 = dp \wedge dx - dq \wedge dy$), which we again denote by L. Consider the subset of L given by the inequalities

$$|\eta_j| < \varepsilon\Lambda(x)^{1-\mu_j},\ j = 1,\ldots,n. \tag{77}$$

We claim that for $\varepsilon > 0$ small enough this subset is the graph of the proper $\mathbb{R}_+$-invariant canonical transformation $g : (y,\eta) \to (x,p)$. Indeed, it is not hard to calculate $y_j = A_j$ in (75) at the point $\{\eta = 0\}$.

To perform this, fix $\eta_1 = \ldots = \eta_{j-1} = \eta_{j+1} = \ldots = \eta_n = 0$, and let η_j vary in the vicinity of zero. Then, from (50), p varies in the vicinity of zero and the coordinate system (η,x) may be chosen on L in this neighborhood. The Hamilton-Jacobi equation for $S(0,\ldots,\eta_j,\ldots,0,x)$ reads

$$\frac{\partial S}{\partial \eta_j} = y_j(L_A(x,p)), \tag{78}$$

i.e.,

$$A_j = y_j(L_A(x,p)); \tag{79}$$

in particular,

$$A_j|_{\eta=0} = y_j(L_A(x,0)) = y_j(x). \tag{80}$$

Thus

$$\frac{\partial^2 S}{\partial\eta\partial x}\Big|_{\eta=0} = \frac{\partial y}{\partial x}, \tag{81}$$

and thus $\det \frac{\partial^2 S}{\partial\eta\partial x} \neq 0$ for $\eta = 0$. It follows that L is the graph of canonical transformation in some neighborhood (77) and, obviously, this canonical transformation is a proper one (see (28) and Remark 1).

Considering all that is presented above, we come to the following theorem.

Theorem 1. Assume that a μ-structure on a Poisson algebra in $\mathbb{R}^n$ is given and that the variable change and operators $B_1,\ldots, B_n$ satisfying the Conditions 2 - 4 are given. There exists for a $\mathbb{R}_+$-invariant two-sided normal domain $\mathcal{U} \subset \mathbb{R}^{2n}_{(y,\eta)}$ necessarily containing the set $\{\eta_j < \varepsilon\Lambda(y)^{1-\mu_j}\}$ for $\varepsilon > 0$ small enough, a proper $\mathbb{R}_+$-invariant canonical transformation

$$g : \mathcal{U} \to \mathbb{R}^{2n}_{(x,p)}, \tag{82}$$

and an asymptotically (modulo $L^{-\infty}_{(\mu_1,\ldots,\mu_n)}(\mathbb{R}^{2n})$) invertible pseudo-differential operator $\hat{K} \in \hat{T}^{\infty}_{(\mu_1,\ldots,\mu_n)}$ such that for any function $\Phi(y) \in L^{\infty}_{st,(\mu_1,\ldots,\mu_n)}$ such that K $\in \mathcal{U}$ for some K Ess$\in (\Phi)$, the formula is valid

$$\Phi(\overset{1}{B_1},\ldots,\overset{n}{B_n}) = [T(g)\hat{K}\Phi](\overset{1}{A_1},\ldots,\overset{n}{A_n}) \pmod{O_p^{-\infty}(x)}, \tag{83}$$

where the operator T(g) is defined in Section 3:C.

The canonical transformation g possesses the properties:

$$g(y,0) = (x(y),0), \tag{84}$$

where $x = x(y)$ is the solution of the equation $y = y(x)$, and

$$\begin{aligned} x(\ell_B(y,\eta)) &= \ell_A(g(y,\eta)), \\ x(r_B(y,\eta)) &= r_A(g(y,\eta)), \quad (y,\eta) \in \mathcal{U}. \end{aligned} \tag{85}$$

Only the proof of identities (85) needs a hint. Consider any symbol $\Phi(y)$ satisfying the condition of the theorem and also any symbol $H(y) \in \in S^{\infty}_{st,(\mu_1,\ldots,\mu_n)}(\mathbb{R}^n)$. The composition $H(B) \circ \Phi(B)$ may be calculated in two different ways, using Theorem 1 of Section 3:D and the present theorem. Substituting various CRF's for $\Phi(y)$, we easily obtain that the principal homogeneous part of $H(L_B(y,\eta))$ and $H(y(L_A(g(y,\eta))))$ are equal, from where the first of identities (85) immediately follows. The second of the identities is proved in an analogous way.

B. Globalization and Proof of the Main Theorem

Let a manifold N, diffeomorphic to $\mathbb{R}^n$, be given with the smooth action of the group $\mathbb{R}_+$ and the Poisson algebra structure homogeneous of degree -1 (i.e., the Poisson bracket $\{f,g\}$ is homogeneous of degree $m + \ell - 1$ if f and g are homogeneous of degrees m and ℓ respectively). Let also a smooth function

$$\Lambda \equiv \Lambda_N : N \to [1/2,\infty) \tag{1}$$

be given, homogeneous of degree 1 for $\Lambda \geqslant 1$.

We assume that the following conditions are satisfied:

(i) A finite set of coordinate systems

$$\begin{aligned} x^{(1)} &= (x_1^{(1)},\ldots,x_n^{(1)}), \\ &\ldots\ldots\ldots\ldots\ldots\ldots \\ x^{(k)} &= (x_1^{(k)},\ldots,x_n^{(k)}), \\ &\ldots\ldots\ldots\ldots\ldots\ldots \\ x^{(N)} &= (x_1^{(N)},\ldots,x_n^{(N)}) \end{aligned} \tag{2}$$

is given on N. Each of these coordinate systems covers N, and the action of the group $\mathbb{R}_+$ is given by

$$\tau x^{(k)} = (\tau^{\lambda_1^{(k)}} x_1^{(k)},\ldots,\tau^{\lambda_n^{(k)}} x_n^{(k)}) \equiv \tau^{\lambda^{(k)}} x^{(k)}, \tag{3}$$

where

$$\lambda^{(k)} = (\lambda_1^{(k)},\ldots,\lambda_n^{(k)}) \tag{4}$$

is a tuple of non-negative numbers such that

$$I_+^{(k)} = \{j \in [n] | \lambda_j^{(k)} > 0\} \tag{5}$$

is a non-empty set (we denote also

$$I_o^{(k)} = [n] - I_+^{(k)} = \{j \in [n] | \lambda_j^{(k)} = 0\}). \tag{6}$$

The inequalities

$$c\Lambda(x^{(k)}) \leqslant \Lambda \leqslant C\Lambda(x^{(k)}) \tag{7}$$

hold with positive constants c and C (here $\Lambda(x^{(k)})$ is defined according to the tuple $\lambda^{(k)}$).

(ii) To any coordinate system $x^{(k)}$, $k = 1,\ldots,N$, the proper tuple of generators

$$A^{(k)} = (A_1^{(k)},\ldots,A_n^{(k)}) \tag{8}$$

is taken into correspondence, so that the local μ-structure

$$\mu_k : f \to f(\overset{1}{A}{}_1^{(k)},\ldots,\overset{n}{A}{}_n^{(k)}), \; f \in L^\infty_{(\lambda_1^{(k)},\ldots,\lambda_n^{(k)})}(\mathbb{R}^n) \tag{9}$$

in the sense of Section 3:B is defined, where the operators $A_1^{(k)},\ldots,A_n^{(k)}$ satisfy the commutation relations

$$[A_j^{(k)}, A_\ell^{(k)}] = -i\omega_{j\ell}^{(k)}(\overset{1}{A}{}_1^{(k)},\ldots,\overset{n}{A}{}_n^{(k)}), \tag{10}$$

and the principal part $\Omega_{j\ell}^{(k)}$ of $\omega_{j\ell}^{(k)}(x^{(k)})$ equals the corresponding component of the tensor, defining the Poisson algebra structure in the coordinates $x^{(k)}$. We assume also that Condition 1 of the item A is valid for any of the tuples $A^{(k)}$.

(iii) For any pair $x^{(k)}, x^{(\ell)}$ of coordinate systems, the corresponding operator tuples are related as described in Conditions 2 - 4 of item A. In more detail, the notations will be

$$A_j^{(\ell)} = \chi_j^{(\ell k)}(\overset{1}{A}{}_1^{(k)},\ldots,\overset{n}{A}{}_n^{(k)}), \; j = 1,\ldots,n, \tag{11}$$

where

$$\chi_j^{(\ell k)} \in P^{\lambda_j^{(\ell)}}_{st,(\lambda_1^{(k)},\ldots,\lambda_n^{(k)})}(\mathbb{R}^n), \; j \in I_+^{(\ell)}, \tag{12}$$

$$\chi_j^{(\ell k)} - \sum_{s \in I_o^{(k)}} a_{js}^{(\ell k)} x_s^{(k)} \in P^0_{st,(\lambda_1^{(k)},\ldots,\lambda_n^{(k)})}(\mathbb{R}^n), \; j \in I_o^{(\ell)} \tag{13}$$

with given constants $a_{js}^{(\ell k)}$; symbols $\chi_j^{(\ell k)}$ have asymptotic expansions in the homogeneous part for large $\Lambda(x^{(k)})$ functions with integer step; the principal homogeneous part of $\chi_j^{(\ell k)}$ equals $x_j^{(\ell)} \circ (x^{(k)})^{-1}$ (note that this, together with conditions (12) - (13), implies serious restrictions on the possible coordinate changes).

Let for any $k \in \{1,\ldots,N\}$, a two-sided $\mathbb{R}_+$-invariant normal domain $V_k \subset \mathbb{R}^{2n}_{(x^{(k)},p^{(k)})}$ be fixed. By Theorem 1 of item A, for any pair

$k,\ell \in \{1,\dots,N\}$ the proper $\mathbb{R}_+$-invariant canonical transformation

$$g_{\ell k} : v_{k\ell} \to v_{\ell k}, \tag{14}$$

where $v_{k\ell} \subset v_k$, $v_{\ell k} \subset v_\ell$, is defined so that

$$\Phi(\overset{1}{A}{}_1^{(k)},\dots,\overset{n}{A}{}_n^{(k)}) = (T_{\ell k}\Phi)(\overset{1}{A}{}_1^{(\ell)},\dots,\overset{n}{A}{}_n^{(\ell)}) \pmod{O_p^{-\infty}(x)}, \tag{15}$$

if $K \subset v_{k\ell}$ for some $K \in \mathrm{Ess}\,\Phi$;

$$T_{\ell k} = T(g_{\ell k}) \circ \hat{K}_{\ell k}, \tag{16}$$

where $\hat{K}_{\ell k} \in \hat{T}^{\infty}_{(\lambda_1^{(k)},\dots,\lambda_n^{(k)})}$ is a given invertible pseudo-differential operator. Standard argument treating Condition 1 of item A shows that

$$(T_{\ell k} \circ T_{ks})\Phi = T_{\ell s}\Phi \pmod{L^{-\infty}_{(\lambda_1^{(\ell)},\dots,\lambda_n^{(\ell)})}(\mathbb{R}^n)}, \tag{17}$$

once $K \subset v_{sk} \cap g_{ks}^{-1}(v_{k\ell})$ for some $K \in \mathrm{Ess}(\Phi)$ and, in particular,

$$g_{\ell k} \circ g_{ks} = g_{\ell s} \tag{18}$$

on the common domain of definition of the right- and left-hand sides of (18).

The equality (18) and equalities (84) - (85) of item A lead to the following conclusion. There is a symplectic $\mathbb{R}_+$-manifold of dimension 2n and the mappings

$$\ell : \mu \to N, \tag{19}$$

$$r : M \to N, \tag{20}$$

$$j : N \to M, \tag{21}$$

such that: (a) these mappings are R_+-invariant and smooth except, possibly, the set $\{0\} \subset M$ of fixed points of R_+-action; (b) ℓ preserves the Poisson bracket while r anti-preserves it, and ℓ and r are in involution, i.e.,

$$\{\ell^* f, \ell^* g\} = \ell^*\{f,g\}, \tag{22}$$

$$\{r^* f, r^* g\} = r^*\{g,f\}, \tag{23}$$

$$\{r^* f, \ell^* g\} = 0 \tag{24}$$

for $f,g \in C^\infty(N)$ (here on the left we have the Poisson bracket on the symplectic manifold, and on the right, on the manifold N).

The mapping j is given in local coordinates by

$$j(x^{(k)}) = (x^{(k)},0), \tag{25}$$

and satisfies

$$\ell \circ j = r \circ j = id_N \tag{26}$$

(the identity mapping of N).

The manifold M is obtained from v_k - s via transition functions $g_{\ell k}$. Thus, there is a covering

$$\bigcup_{\ell=1}^{N} \mathcal{U}_\ell = M \tag{27}$$

and coordinate canonical $\mathbb{R}_+$-mappings

$$\nu_\ell : \mathcal{U}_\ell \to v_\ell, \tag{28}$$

such that

$$g_{\ell k} = \nu_\ell \circ \nu_k^{-1}, \tag{29}$$

and the mappings (19) - (20) in local coordinates are defined via Theorems 1 and 2 of Section 3:D. We may define the function

$$\Lambda_M : M \to [1/2,\infty), \tag{30}$$

setting

$$\Lambda_M = \ell^* \Lambda_N. \tag{31}$$

Then we require that

(iv) $v_{k\ell} \subset v_k$ are chosen "properly," i.e., in such a way that M is a separable manifold (roughly speaking, this means that $v_{k\ell}$ are maximal possible ones).

(v) $M \setminus \{0\}$ is a proper symplectic $\mathbb{R}_+$-manifold (in particular, the atlas $\{\mathcal{U}_\ell\}$ may be inscribed into appropriate atlas $\{\tilde{\mathcal{U}}_\ell\}$, so that $(\mathcal{U}_\ell, \tilde{\mathcal{U}}_\ell)$ will be a proper pair for any ℓ).

(vi) For any function $f \in C^\infty(N)$ and any trajectory $\gamma : [0,T] \to N$ of the Eulerian vector field corresponding to f, there exists the trajectory $\tilde{\gamma}$: $: [0,T] \to M$ of the Hamiltonian vector field of $\ell^* f$, such that

$$\tilde{\gamma}(0) = j(\gamma(0)) \tag{32}$$

(and, consequently, $\ell(\tilde{\gamma}(t)) \equiv \gamma(t)$, $t \in [0,T]$).

Definition 1. Let the Conditions (i) - (vi) be satisfied. We say then that a global μ-structure on the Poisson algebra on N is given, and that M is a symplectic manifold of this Poisson algebra.

Remark 1. Condition (vi) is the crucial one, since all other conditions are always satisfied in the local version (one should merely set N = 1). Denote

$$A_j = A_j^{(1)}, \quad \lambda_j = \lambda_j^{(1)}, \quad j = 1,2,\ldots,n, \tag{33}$$

and let $f \in P^m_{st,(\lambda_1,\ldots,\lambda_n)}(\mathbb{R}^n)$,

$$\hat{f} = f(\overset{1}{A}_1,\ldots,\overset{n}{A}_n). \tag{34}$$

Definition 2. The function

$$H = \ell^*(\Lambda(x)^{-m+1} f)_1 \tag{35}$$

on M, where the subscript "1" denotes the principal homogeneous part of the function in parentheses, is called a left Hamiltonian function, or simply a left Hamiltonian, of the operator $\hat{f}$.

Now we are able to formulate and prove the main theorem of this book, namely, the quasi-invertibility theorem.

Theorem 1 (Main Theorem). Let the Hamiltonian function H of the operator $\hat{f} = f(\overset{1}{A_1},\dots,\overset{n}{A_n})$ satisfy the absorption conditions of Chapter 3, Section 4 for some $\varepsilon > 0$, $T > 0$ and the initial manifold given by the immersion

$$j : N_o \to \mu,$$

where $N_o \subset N$ is an $\mathbb{R}_+$-invariant neighborhood of the set of zeros of the principal part of f having the form

$$N_o = \{x \in N \mid |x_j - x_{oj}| < \varepsilon,\ j \in I_o,\ |x_j - x_{oj}| \leqslant \varepsilon\Lambda(x)^{\lambda_j},\ j \in I_+, f(x_o) = 0\}.$$

Then the operator $\hat{f}$ has a right quasi-inverse operator $\hat{G}$, i.e., an operator such that

$$\hat{f} \circ \hat{G} = 1 + \hat{R}, \tag{36}$$

where $\hat{R} \in \mathcal{O}_p^{-\infty}(x)$.

Remark 2. The theorem on the left quasi-inverse operator is formulated and proved in a quite analogous way.

Proof. The most essential work was performed already in Chapters 3 and 4, and it remains to gather all the results and to complete the argument.

The operators $T_{\ell k}$ given by (16) satisfy the cocyclicity condition modulo $L^{-\infty}$ and therefore define a sheaf, $\mathcal{F}_+$, over M (this sheaf slightly differs from that considered in Chapter 3; the operator $\hat{K}_{\ell k}$ is present in the definition of transition operators, but since $\hat{K}_{\ell k}$ is almost an invertible pseudo-differential operator, all the results hold in this case automatically).

Using the condition (15), we may define the mapping

$$\mu^{\#} : \mathcal{F}_{st,+}(M) \to \mathcal{O}_p^{\infty}(x)/\mathcal{O}_p^{-\infty}(x), \tag{37}$$

where $\mathcal{F}_{st,+}(M)$ is the set of sections with representatives "stabilizing at infinity," by means of the formula

$$\{\Phi_\ell\}_{\ell=1}^N \to \sum_{\ell=1}^N \Phi_\ell(\overset{1}{A_1^{(\ell)}},\dots,\overset{n}{A_n^{(\ell)}}); \tag{38}$$

the condition (15) guarantees that the mapping (38) admits factorization. If

$$F(x) \in \mathcal{P}^m_{(\lambda_1,\dots,\lambda_n)}(\mathbb{R}^n) \tag{39}$$

and

$$\Phi = [\{\Phi_\ell\}_{\ell=1}^N] \in \mathcal{F}_{st,+}(M), \tag{40}$$

then the composition $F(\overset{1}{A_1},\dots,\overset{n}{A_n}) \circ \mu^{\#}(\Phi)$ may be modulo $\mathcal{O}_p^{-\infty}(x)$ calculated in the following way. We have (all calculations modulo $\mathcal{O}_p^{-\infty}(x)$):

$$
\begin{aligned}
&F(\overset{1}{A}_1,\ldots,\overset{n}{A}_n) \circ \sum_{\ell=1}^{N} \Phi_\ell(\overset{1}{A}{}_1^{(\ell)},\ldots,\overset{n}{A}{}_n^{(\ell)}) = \\
&= \sum_{\ell=1}^{N} [\![F_\ell(\overset{1}{A}{}_1^{(\ell)},\ldots,\overset{n}{A}{}_n^{(\ell)})]\!] [\![\Phi_\ell(\overset{1}{A}{}_1^{(\ell)},\ldots,\overset{n}{A}{}_n^{(\ell)})]\!] = \\
&= \sum_{\ell=1}^{N} [\![(H_\ell(\overset{2}{x}{}^{(\ell)}, -i \overset{1}{\frac{\partial}{\partial x^{(\ell)}}})\Phi_\ell)(\overset{1}{A}{}_1^{(\ell)},\ldots,\overset{n}{A}{}_n^{(\ell)})]\!] ,
\end{aligned}
\tag{41}
$$

where the principal symbol of $H_\ell(\overset{2}{x}{}^{(\ell)}, -i \overset{1}{\frac{\partial}{\partial x^{(\ell)}}})$ equals

$$[(\ell^* f) \circ \nu_\ell^{-1}](x^{(\ell)}, p^{(\ell)}), \tag{42}$$

where f is the principal homogeneous part of F. Thus,

$$F(A) \circ \mu^{\#}(\Phi) = \mu^{\#}(\hat{H}(\Phi)), \tag{43}$$

where $\hat{H}$ is a pseudo-differential operator in $\mathcal{F}_{st,+}(M)$ with the principal symbol $\ell^* f$.

Turn now to the quasi-inversion problem posed in Theorem 1. Multiplying f(A) by $\Lambda(A)^{-m+1}$ from the right, one reduces the problem to the case m = 1. We seek $\hat{G}$ in the form

$$\hat{G} = \phi(A) - i\int_0^T \hat{\psi}(t)dt, \tag{44}$$

where $\phi(A)$ satisfies modulo $\mathcal{O}_p^{-\infty}(x)$ the problem

$$f(A) \circ \phi(A) = 1 - \rho(A), \tag{45}$$

and modulo $\mathcal{O}_p^{-\infty}(x)$

$$-i\,\frac{\partial\hat{\psi}(t)}{\partial t} + f(A) \circ \hat{\psi}(t) = 0, \tag{46}$$

$$\hat{\psi}(0) = \rho(A). \tag{47}$$

Here the symbol $\rho(x)$ is chosen in such a way that

$$\rho(x) = 1 \tag{48}$$

in the $\mathbb{R}_+$-invariant neighborhood of the set of zeros of the principal part of the function f and supp $\rho(x) \subset N_0$.

The equation (45) is easily solved by means of the composition theorems presented in Section 3:B. We have

$$\phi(x) = (1 - \rho(x))/f_1(x)\ (+\text{ lower-order term}). \tag{49}$$

where $f_1(x)$ is the principal homogeneous part of f.

Further, we seek the solution of (46) - (47) in the form

$$\hat{\psi}(t) = \mu^{\#}\psi(t), \tag{50}$$

or, more precisely, as some representative of $\mu^{\#}(\psi(t))$, and obtain for $\psi(t) \in \mathcal{F}_{st,+}(M)$ the Cauchy problem

$$-i\,\frac{\partial\psi}{\partial t} + \hat{H}\psi = 0, \tag{51}$$

$$\psi(0) = K_o\rho, \tag{52}$$

where K_o is the canonical operator on the Lagrangian manifold $j : N_o \to M$ and $\hat{H}$ is a pseudo-differential operator in $F_{st,+}(M)$ with the principal symbol $\ell^* f_1$.

Theorem 1 of Section 3:D guarantees that the problem (51) - (52) has a solution and that $\psi(T) = 0$. Therefore, substituting the operator (44) into (36), we obtain

$$\begin{aligned} \hat{f} \circ \hat{G} &= 1 - \rho(A) - i\int_o^T \hat{f} \circ \hat{\psi}(t)dt = 1 - \rho(A) - \int_o^T \frac{d}{dt}\hat{\psi}(t)dt = \\ &= 1 - \rho(A) + \rho(A) - \hat{\psi}(T) = 1 \pmod{O_p^{-\infty}(x)}. \end{aligned} \tag{53}$$

The theorem is therefore proved.

References

1. R. F. V. Anderson, "The Weyl functional calculus," J. Funct. Anal., 4, 240-267 (1969).
2. V. I. Arnold, Mathematical Methods in Classical Mechanics, Springer-Verlag (1978).
3. V. I. Arnold, "On a characteristic class occurring in the condition of quantization," Funkts. Anal. Prilozhen., 1, No. 1, 1-14 (1967).
4. L. Auslander and B. Kostant, "Quantization and unitary representations of solvable Lie groups," Bull. Am. Math. Soc., 73, 692-695 (1967).
5. M. V. Babich, "Asymptotics with respect to large parameter for the solution of Klein-Gordon-Fock equation with discontinuous initial data," J. Sov. Math., 28, No. 5 (1985).
6. F. A. Berezin, The Method of Second Quantization, Academic Press (1966).
7. A. Borel and F. Hirzebruch, "Characteristic classes and homogeneous spaces II," Am. J. Math., 81, No. 2, 315-382 (1959).
8. L. Boutet de Monvel and V. Guillemin, "The spectral theory of Toeplitz operators," Ann. Math. Stud., 99 (1981).
9. C. H. Cook and H. R. Fisher, "Uniform convergence structures," Math. Ann., No. 173, 290-306 (1967).
10. J. Ĉzyz, "On geometric quantization and its connections with the Maslov theory," Reports in Math. Phys., 15, No. 1, 57-97 (1979).
11. J. J. Duistermaat, "Oscillatory integrals, Lagrangian immersions, and unfolding of singularities," Commun. Pure Appl. Math., 27, No. 2 (1974).
12. J. J. Duistermaat and L. Hörmander, "Fourier integral operators II," Acta Math., 128, 183-269 (1972).
13. N. Dunford and J. T. Schwartz, Linear Operators, Vol. II, Interscience (1963).
14. V. I. Feigin, "Two algebras of PDO in R^n and some applications," Trans. Moscow Math. Soc., 8(36) (1979).
15. R. P. Feynman, "An operator calculus having applications in quantum electrodynamics," Phys. Rev., 84, No. 2, 108-128 (1951).
16. G. B. Folland, "Subelliptic estimates and function spaces on nilpotent Lie groups," Ark. Math., 13, No. 2, 161-207 (1975).
17. G. B. Folland and E. M. Stein, "Estimates for ∂_b complex and analysis on the Heisenberg group," Commun. Pure Appl. Math., 27, No. 4, 429 (1974).
18. B. Gaveau, "Systemes dynamiques associes a certains opérateurs hypoelliptiques," Bull. Sci. Math., 102, No. 3, 203-229 (1978).

19. C. Godbillon, Geometrie Differentialle et Mecanique Analitique, Hermann, Paris (1969).
20. R. Godement, Topologie Algebrique et Theorie des Faisceaux, Actualités Sci. Ind., No. 1252, Hermann, Paris (1958).
21. A. Grigis, "Propogation des singularités le long de courbes microbicharacteristiques pour des opérateurs pseudo-differentiels a characteristiques doubles I," Ann. Sci. Norm. Sup., 15, No. 1, 147-159 (1982).
22. V. V. Grushin, "On a class of hypoelliptic operators," Mat. Sb., 83 (125), No. 3, 456-473 (1970) [English transl. in Math. USSR Sb., 12 (1970)].
23. V. Guillemin and S. Sternberg, "Homogeneous quantization and multiplicities of group representations," J. Funct. Anal., 47, No. 3, 344-380 (1982).
24. V. Guillemin and S. Sternberg, "Some problems in integral geometry and some related problems in microlocal analysis," Am. J. Math., 101, 915-955 (1979).
25. V. Guillemin and G. Uhlmann, "Oscillatory integrals with singular symbols," Duke Math. J., 48, No. 1, 251-267 (1981).
26. B. Helfer and J. Nourrigat, "Hypoellipticite pour des opérateurs sur des groups de Lie nilpotents," C. R. Acad. Sci., AB287, No. 6, A395-A398 (1978).
27. H. Hess, "On a geometric quantization scheme generalizing those of Kostant-Souriau and Ĉzyz," Lect. Notes Phys., 139, 1-35 (1981).
28. E. Hille and R. S. Phillips, Functional Analysis and Semigroups, N.Y. (1948).
29. L. Hörmander, "Fourier integral operators I," Acta Math., 127, 79-183 (1971).
30. L. Hörmander, "Pseudo-differential operators," Commun. Pure Appl. Math. 18, 501-517 (1965).
31. V. Ya. Ivrij, "Wave fronts of the solutions of certain pseudo-differential equations," Funkts. Anal. Prilozhen., 10, No. 2, 71-72 (1976).
32. M. V. Karasev, "Path integral and quasi-classical asymptotics on Lie groups," Teor. Mat. Fiz., 30, No. 1, 41-47 (1977).
33. M. V. Karasev, "Asymptotic spectrum and front of oscillations for operators with nonlinear commutation relations," Dokl. Akad. Nauk SSSR, 243, No. 1, 15-18 (1978).
34. M. V. Karasev, "Maslov quantization conditions in higher cohomologies and analogues of Lie theory objects for canonical fibrations of symplectic manifolds," MIEM, 64 pp., deposited VINITI, Dep No. 1092-82, 1093-82 (1981).
35. M. V. Karasev and V. P. Maslov," Algebras with general commutation relations and their applications II," J. Sov. Math., 15, No. 3 (1981).
36. M. V. Karasev and V. P. Maslov, "Global asymptotic operators of regular representation," Dokl. Akad. Nauk SSSR, 257, No. 1, 33-38 (1981).
37. M. V. Karasev and V. P. Maslov, "Quantization of symplectic manifolds with conic points," Teor. Mat. Fiz., No. 2, 263-269 (1982).
38. M. V. Karasev and V. P. Maslov, "Pseudo-differential operators in general symplectic manifolds," Izv. Akad. Nauk SSSR, Ser. Mat., 47, No. 5 (1983).
39. M. V. Karasev and V. E. Nazaikinskii, "On quantization of rapidly oscillating symbols," Mat. Sb., 106, No. 2, 183-213 (1978) [English transl., Math. USSR Sb., 34, No. 6 (1978)].
40. T. Kato, Perturbation Theory for Linear Operators, Springer-Verlag (1966).
41. A. A. Kirillov, "Unitary representations of nilpotent Lie groups, Usp. Mat. Nauk, 17, No. 4, 57-101 (1962).
42. A. A. Kirillov, "Characters of unitary representations of Lie groups," Funkts. Anal. Prilozhen., 2, No. 2, 40-45 (1968).

43. A. A. Kirillov, "Construction of irreducible unitary representations of Lie groups," Vestn. Mosk. Gos. Univ., Mat. Mekh., 2 (1970).
44. A. A. Kirillov, Elements of Representation Theory, Springer-Verlag (1976).
45. J. J. Kohn and L. Nirenberg, "An algebra of pseudo-differential operators," Commun. Pure Appl. Math., 18, 269-305 (1965).
46. B. Kostant, "Quantization and unitary representations," Lect. Notes Math., 170, 87-208, Springer Verlag (1970).
47. S. G. Krein and A. M. Shichvatov, "Linear differential equations on Lie groups," Funkts. Anal. Prilozhen., 4, No. 1, 61-62 (1970).
48. J. Leray, "Analyse lagrangienne et mecanique quantique," Seminaire College de France, Paris (1976-1977).
49. B. M. Levitan, Theory of Generalized Shift Operators [in Russian], Nauka, Moscow (1973).
50. V. P. Maslov, Theory of Perturbations and Asymptotic Methods [in Russian], Moscow State Univ. (1965) [French translation, Theorie des perturbations et methodes asymptotiques, Dunod, Paris (1972)].
51. V. P. Maslov, "Theory of bicharacteristics for difference schemes," Usp. Mat. Nauk, 22, No. 4 (1968).
52. V. P. Maslov, Operator Methods [in Russian], Nauka, Moscow (1973); English transl., Operational Methods, Mir, Moscow (1976); Spanish transl., Metodes Operatorios, Mir, Moscow (1981).
53. V. P. Maslov, "Application of ordered operators method for obtaining exact solutions," Teor. Mat. Fiz., 33, No. 2, 185-209 (1977).
54. V. P. Maslov and V. E. Nazaikinskii, "Algebras with general commutation relations and their applications I," J. Sov. Math., 15, No. 3 (1981).
55. V. P. Maslov and V. E. Nazaikinskii, "Asymptotics for equations with singularities in characteristics," Izv. Akad. Nauk SSSR, Ser. Math., 45, No. 5, 1049-1087 (1981) [English transl., Bull. Acad. Sci. USSR, Math. Ser., 19, No. 2, 315-347 (1982)].
56. V. P. Maslov and M. V. Fedoryuk, Quasi-Classical Approximation for the Equations of Quantum Mechanics [in Russian], Nauka, Moscow (1976) [English transl., Semi-Classical Approximation in Quantum Mechanics, Reidel Publ. Corp. (1981)].
57. A. Melin and J. Sjostrand, "Fourier integral operators with complex-valued phase functions," Springer Lecture Notes, 459, 120-223 (1976).
58. R. Meirose and G. Uhlmann, "Lagrangian intersection and the Cauchy problem," Commun. Pure Appl. Math., 32, 483-513 (1979).
59. A. S. Miscenko, B. Yu. Sternin, and V. E. Shatalov, Canonical Maslov Operator. Complex Theory [in Russian], MIEM, Moscow (1974).
60. A. S. Miscenko, B. Yu. Sternin, and V. E. Shatalov, Lagrangian Manifolds and the Canonical Operator Method [in Russian], Nauka, Moscow (1978).
61. V. E. Nazaikinskii, V. G. Oshmyan, B. Yu. Sternin, and V. E. Shatalov, "Integral Fourier operators and canonical operator," Usp. Mat. Nauk, 36, No. 2, 81-140 (1981) [English transl., Russian Math. Surveys, 36 (1981)].
62. E. Nelson, "Analytic vectors," Ann. Math., 70, 572-615 (1959).
63. J. G. Nosmas, "Parametrix du probléme de Cauchy pour une classe d'opérateurs à characteristiques de multiplicité variable," C. R. Acad. Sci., A285-B285, No. 16, A1065-A1068 (1977).
64. O. A. Oleynik, "Linear equations of second order with non-negative characteristic form," Mat. Sb., 69, No. 1, 111-140 (1966) [English transl., Am. Math. Soc., Transl., 65 (1967)].
65. O. A. Oleynik and E. V. Radkevic, "Equations of second order with non-negative characteristic form," Mathematical Analysis, VINITI AN SSSR (1969) and (1971) [English transl., Plenum Press (1977)].

66. V. M. Petkov and V. Ya. Ivrij, "Necessary conditions of Cauchy problem correctness for non-strictly hyperbolic equations," Usp. Mat. Nauk, 29, No. 5, 3-40 (1974) [English transl., Russian Math. Surveys, 29, (1974)].
67. E. V. Radkevic, "Hyperbolic operators with multiple characteristics," Mat. Sb., 79, 193-216 (1969) [English transl., Math. USSR Sb., 8 (1969)].
68. C. Rockland, "Hypoellipticity on the Heisenberg group-representation-theoretic criteria," Trans. Am. Math. Soc., 240, 1-52 (1978).
69. L. P. Rotschild, "A criterion for hypoellipticity of operators constructed from vector fields," Commun. Part. Diff. Equations, 4, No. 6, 645-699 (1979).
70. L. P. Rotschild and E. M. Stein, "Hypoelliptic differential operators and nilpotent groups," Acta Math., 137, 247-320 (1976).
71. W. Rudin, Functional Analysis, McGraw-Hill (1973).
72. M. A. Shubin, Pseudo-Differential Operators and Spectral Theory [in Russian], Nauka, Moscow (1978).
73. J.-M. Souriau, "Quantification geometrique," Commun. Math. Phys., 1, 374-398 (1966).
74. J.-M. Souriau, Structure des systems dynamiques, Paris, Dunod (1970).
75. J.-M. Souriau, Geometric sympletique et physique mathematique, Paris (1975).
76. F. Treves, Introduction to Pseudo-Differential and Fourier Integral Operators, Vols. I-II, Plenum Press, N.Y. (1980).
77. H. F. Trotter, "Approximation of semigroups of operators," Pac. J. Math., 8, No. 4, 887-920 (1958).
78. H. F. Trotter, "On the product of semigroups of operators," Proc. Am. Math. Soc., I, No. 4, 545-551 (1959).
79. K. Yosida, Functional Analysis, Springer-Verlag (1965).